AF820772

Urban Operations

Plus

Sniper Training

Department of Defense

Pentagon
Publishing

Field Manual
No. 3-06

Headquarters
Department of the Army
Washington, DC, 26 October 2006

Urban Operations

Contents

Distribution Restriction: Approved for public release; distribution is unlimited.

***This publication supersedes FM 3-06, 1 June 2003.**

Figures

Historical Vignettes

Preface

Doctrine provides a military organization with a common philosophy, a language, a purpose, and unity of effort. Rather than establishing a set of hard and fast rules, the objective of doctrine is to foster initiative and creative thinking. To this end, FM 3-06 discusses major Army operations in an urban environment. This environment, consisting of complex terrain, a concentrated population, and an infrastructure of systems, is an operational environment in which Army forces will operate. In the future, it may be the predominant operational environment. Each urban operation is unique and will differ because of the multitude of combinations presented by the threat, the urban area itself, the major operation of which it may be part (or the focus), and the fluidity of societal and geopolitical considerations. Therefore, there will always exist an innate tension between Army doctrine, the actual context of the urban operation, and future realities. Commanders must strike the proper balance between maintaining the capability to respond to current threats and preparing for future challenges.

PURPOSE

This manual provides the analytical tools for evaluating an urban operation to determine if the operation is necessary for overall mission success. It also provides the means to understanding and determining the impacts of the urban environment on military operations and provides information on managing, taking advantage of, and mitigating the effects of those impacts as appropriate. As such, this manual demonstrates how to apply the doctrinal principles in FM 3-0 to this unique environment.

SCOPE

Chapter 1 introduces theoretical and historical perspectives of urban operations that serve as the underlying basis for the rest of the manual. While this manual has incorporated lessons learned from recent and ongoing operations, it has deliberately taken a broad and varied historical perspective in order to remain relevant to future threats and circumstance differing from those that the Army currently faces. Chapter 2 discusses the characteristics of urban centers and populations as well as their impact on operations. It is unlikely that Army forces will ever operate in a benign urban environment; therefore, Chapter 3 discusses the varied nature of potential urban threats. Chapter 4 describes the effects of the urban environment on warfighting functions and tactics. An understanding of the complexities of the urban environment, the nature of the enemy as an adaptive, learning opponent, and the effects of the environment on warfighting capabilities and skills is essential to sound decision making. Chapters 5 and 6 discuss the potential costs of urban operations and risk reduction measures that the commander and his staff must consider early in their planning. These chapters also outline an urban operational framework and specific urban considerations that create the foundations necessary for successfully applying operational doctrine to an urban environment.

The second part of the manual (Chapters 7 – 10) discusses how urban operations are conducted and resourced. Urban operations include major offensive and defensive operations in urban environments as well as stability or civil support operations ranging from peace operations and counterterrorism to disaster relief and humanitarian assistance. For the different types of operations—offense, defense, and stability or civil support—the purpose, characteristics, organization, and considerations are discussed. However, commanders consider that most urban operations will normally involve the simultaneous execution of offense and defense with stability or civil support (although proportional emphasis will shift over time) and plan accordingly.

APPLICABILITY

This manual is intended for Army commanders and their staffs at the brigade level and above. It addresses full spectrum operations that Army units will execute in urban settings. However, users should also consult JP 3-06 for specific joint information. Additionally, users should be familiar with FM 3-06.1, FM 3-06.11, TC 90-1, and urban operations chapters, appendices, or sections found in other infantry, armor, combined arms, and proponent field manuals for the tactics, techniques, and procedures (TTP) and appropriate proponent information necessary to conduct tactical urban operations at the brigade level and below. This publication applies to the Active Army, the Army National Guard (ARNG)/Army National Guard of the United States (ARNGUS), and United States Army Reserve unless otherwise stated.

ADMINISTRATIVE INSTRUCTIONS

When this FM 3-06 was approved, doctrine was incomplete for incorporating stability operations and civil support operations in place of stability operations and support operations. This manual incorporates stability and reconstruction and civil support operations doctrine found in FM 1 and FMI 5-0.1; it will be revised to reflect the future updates of FM 3-0 and FM 3-07 and, when published, FM 3-28. (Note: Stability and reconstruction operations has been redesignated stability operations to comply with Department of Defense Directive 3000.05.)

Chapter 2 defines "city" according to a population size. However, in historical vignettes and accounts, the term "city" may be applied in its common usage without specific regard to size to maintain conformity with most other historical reports.

In this manual, the term "threat" is applied broadly to include an enemy force (conventional or unconventional), an armed belligerent in a peace operation, antagonistic or unfriendly elements of the civilian population, or some other hazardous condition in the urban environment that negatively influences mission accomplishment.

The term military operations on urban terrain (MOUT) is replaced by urban operations (UO). MOUT is an acronym from FM 90-10, which is superseded by this manual.

Otherwise, the glossary lists most terms used in FM 3-06 that have joint or Army definitions. Where Army and joint definitions are different, (Army) follows the term. Definitions for which FM 3-06 is the proponent manual (the authority) are marked with an asterisk (*). The proponent or amplifying manual for other terms is listed in parentheses after the definition.

The manual attempts to incorporate a broad range of historical vignettes into each chapter where the account supports the doctrinal line of reasoning. Two historical vignettes, however, were included as appendices (A and C) because of their longer lengths.

Unless this publication states otherwise, masculine nouns or pronouns do not refer exclusively to men.

This publication contains copyrighted material.

The proponent for this publication is the United States Army Training and Doctrine Command. Send written comments and recommended changes on DA Form 2028 (Recommended Changes to Publications and Blank Forms) directly to Commander, U.S. Army Combined Arms Center and Fort Leavenworth, Combined Arms Doctrine Directorate, ATTN: ATZL-CD, (FM 3-06), 201 Reynolds Avenue, Fort Leavenworth, KS 66027-2337. Send comments and recommendations by e-mail to webb-cadd@leavenworth.army.mil. Follow the DA Form 2028 format or submit an electronic DA Form 2028.

This page intentionally left blank.

Chapter 1
The Urban Outlook

Today's security environment demands more from Army leaders than ever before. Army leaders must not only be able to lead Soldiers but also influence other people. They must be able to work with members of other Services and governmental agencies. They must win the willing cooperation of multinational partners, both military and civilian. But ultimately, the Army demands self-aware and adaptive leaders who can compel enemies to surrender in war and master the circumstances facing them in peace. Victory and success depend on the effectiveness of these leaders' organizations. Developing effective organizations requires hard, realistic, and relevant training.

FM 1

Given the prevalence of large cities throughout the world, Army forces will likely be required to conduct operations in, around, and over large urban areas. These operations will normally be in support of a joint force commander (JFC) conducting military operations pursuant to U.S. national security policy. This manual is designed to facilitate the planning and conduct of the full range and spectrum of land operations in a complex urban environment. Each urban environment and urban operation is unique; prescribing a specific doctrinal "solution" for each situation is impossible. Instead, this manual provides a framework to commanders and their staffs for understanding the urban environment, for analyzing and deciding whether or not to initiate urban operations (UO), and for applying operational doctrine to this complex environment. It also provides a broad base of historical vignettes and examples to help develop a refined analytical perspective and stimulate thought. The manual also includes some planning points and tactics and techniques to assist in preparing for and conducting UO. This information provides a foundation for approaching major UO. Combined with other joint and Army doctrine, this information also will help commanders and their staffs learn to adapt and succeed in this challenging environment.

THE PROSPECT OF URBAN OPERATIONS

1-1. The world is undergoing massive urbanization. Although exceptions exist, an overall trend of migration from rural to urban areas is occurring throughout the globe. (Australia, one of the world's most urbanized countries, is actually becoming less urbanized.) This trend is especially evident in developing nations. Combined with the exponential growth of the global population in the last quarter century, this migration has created massive urban areas that hold the centers of population, government, and economics in their respective regions. In many cases, rapid urbanization has overburdened already weak infrastructures, scarce resources, and a fragile economic base. As urbanization has changed the demographic landscape, potential enemies recognize the inherent danger and complexity of this environment to the attacker, and may view it as their best chance to negate the technological and firepower advantages of modernized opponents. Given the global population trends and the likely strategies and tactics of future threats, Army forces will likely conduct operations in, around, and over urban areas—not as a matter of fate, but as a deliberate choice linked to national security objectives and strategy, and at a time, place, and method of the commander's choosing.

Army Urban Operations

Army forces conduct UO either as one component of a larger operation or as a single operation focused totally on a specific urban environment. Major Army UO are often part of a joint and multinational effort requiring rigorous interagency and civil-military coordination that typically includes the full spectrum of military operations. Commanders of Army major operations must determine if UO are essential to mission accomplishment. If so, commanders must carefully integrate the operations into campaign planning to support the operational objectives of the JFC.

Army leaders conducting UO must—

- Understand the urban environment to determine decisive points.
- Shape the operation to set the conditions for success.
- Precisely mass the effects of combat power to thoroughly engage the decisive points that lead to centers of gravity.
- Continually consolidate gains essential to the retention of the initiative.
- Transition the urban area to the control of another force or agency or, ultimately, back to legitimate and functioning civilian control.

AN URBAN PERSPECTIVE

1-2. As a subset of all Army operations, UO are operations focused on an urban environment. UO include full spectrum operations—offensive, defensive, and stability or civil support—that may be executed, either sequentially or (more likely) simultaneously, during the conduct of a single urban operation (see figure 1-1). UO may be the sole mission of the commander or one of several tasks nested in a larger operation. Regardless of the types of operations conducted or whether the urban area is the single focus of the operation or only one component of a larger operation or campaign, the complex urban environment significantly affects the overall conduct of the mission.

1-3. When conceptualizing urban operations, commanders must understand two important terms: urban area and urban environment. The first is a subset of the second. An ***urban area*** **is a topographical complex where man-made construction or high population density is the dominant feature**. Focusing on urban areas means concentrating on the physical aspects of the area and their effects on weapons, equipment, line-of-sight, and tactics, techniques, and procedures. The ***urban environment*** **includes the physical aspects of the urban area as well as the complex and dynamic interaction and relationships between its key components—the terrain (natural and man-made), the society, and the supporting infrastructure—as an overlapping and interdependent system of systems.**

1-4. Importantly, commanders must also understand and consider that critical elements of the infrastructure may lie far beyond the area's physical confines. For example, the generating source providing power to the urban energy system is part of that system but may be located well outside of the urban area. Similarly, effects of the interaction between components of the infrastructure, located both inside and outside the urban area, extend well into smaller, neighboring urban areas and surrounding rural areas and often form their political, economic, and cultural focus. Understanding the total urban environment is essential to planning and conducting full spectrum urban operations.

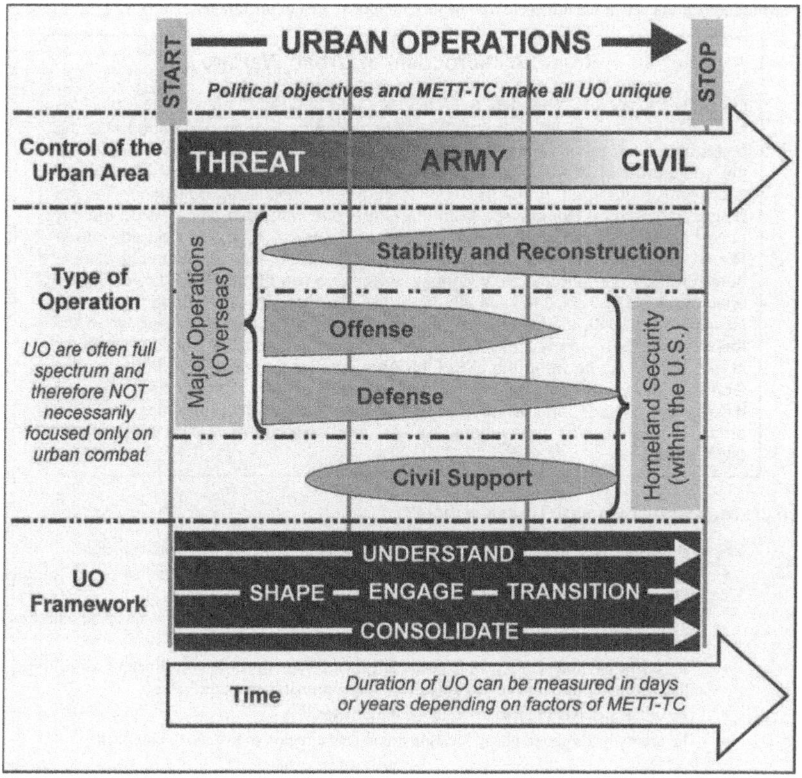

Figure 1-1. Full spectrum urban operations

HISTORICAL SIGNIFICANCE OF URBAN AREAS IN WARFARE

1-5. Urban areas always have been central to, or have significantly influenced, military operations. One of the first urban-centered battles was the siege of Troy at the beginning of Greek history. Moreover, much of the history of early Greece revolved around wars between its city-states or with Persia and centered on the conquest, siege, or blockade of cities. Five hundred years later, the Roman Empire replaced Greece as the dominant world power; although, urban areas remained central to the Roman method of warfare. Even Rome's history can be viewed as a microcosm of urban warfare over the past two thousand years. Though military operations within the physical confines of many of these historic urban areas were not the norm, the focus of these operations was their conquest or control.

1-6. Although Rome last saw combat in 1944, urban areas have been no less prominent in warfare since that time. Seoul in Korea, Beirut in Lebanon, Panama City in Panama, Grozny in Chechnya, Sarajevo in Bosnia-Herzegovina, Kabul in Afghanistan, and Baghdad in Iraq have been centers of conflict in the last 50 years. Urban areas, now more pervasive than ever before, will continue to be essential to successful operational and strategic warfighting. Today, armies cannot expect to execute major military operations without the influence of the urban environments within their area of operations.

> **Rome: A Microcosm of Urban Warfare**
>
> During two millennia, Rome has been the center of at least 12 battles. The Gauls lay siege to Rome first in 387 BC. That first siege lasted six months and ended after the barbarians burnt much of the city. The surviving patrician families paid a ransom for the withdrawal of Brennus' army. From 408 to 410 AD, the Goth leader, Alaric, successfully besieged Rome no less than three times. The Byzantine General Belisarius captured Rome twice from the Goths and withstood siege inside the city once between 536 and 549. Five hundred years later in 1084, Norman adventurer Robert Guiscard captured medieval Rome and sacked the city during a dispute between the Pope and the Holy Roman Empire. Forces of the Holy Roman Empire again stormed and captured the city to punish the Pope in 1527. During the Italian Revolution in 1849, a French army supporting the Pope captured the city from the Italian revolutionary army under Garibaldi. In 1944, the last military action took place in and around Rome when the U.S. Fifth Army captured the city from the retreating German army. Rome's turbulent history—fought over ethnic and religious differences, prestige, and military necessity—demonstrates the importance of urban areas in warfare and the various causes and combatants within this complex environment.

STRATEGIC IMPORTANCE OF URBAN AREAS

1-7. Several reasons have attracted (and continue to attract) armies to combat in urban areas:

- A military force chooses to position itself in an urban area to capitalize on the perceived defensive advantages offered by the environment. In contrast, an opposing force, by analyzing the factors of the situation, determines that it must enter the urban area to attack and destroy its enemy (or devote essential combat power to their isolation).
- The urban environment's people (their allegiance and support), infrastructure, capabilities, or other resources have or can be of significant operational or strategic value.
- The urban area has significant symbolic importance.
- The urban area's geographical location dominates a region or avenue of approach.

1-8. Russia's 1994 experience in Chechnya illustrates an increasingly important motivation for conducting urban operations. The Chechen rebels, after failing to engage Russian forces outside the city, chose to turn Grozny into the main battlefield. Leaders of the defeated Chechen conventional forces recognized that fighting in the urban area provided them their best chance for success. The complexities of urban combat and the perceived advantages of defending an urban area mitigated their numerical and technological inferiority. The urban area provided the Chechens protection from fires, resources, interior lines, and covered and concealed positions and movement. Given such advantages offered by the environment, smaller or less-sophisticated military forces have similarly chosen to fight in urban areas either as a deliberate strategy or to escape certain destruction in open terrain.

1-9. Such advantages of operating in an urban environment also prompt forces to conduct an urban operation to facilitate a larger campaign plan and decisive battle in another location. The urban operation can focus the enemy on the urban area and allow other forces to conduct operations elsewhere. From a defensive perspective, an urban defense may gain time and space to reorganize forces in new defensive positions, to divert enemy forces from other critical tasks, or to prepare to conduct offensive operations. To some extent, these reasons motivated Soviet forces defending Leningrad and Stalingrad from the Germans in World War II. The stubborn defense permitted the Soviets to reorganize for later offensive operations. From an offensive perspective, an attack on an urban area may be a shaping operation used to divert enemy resources from the decisive operation that will follow.

1-10. Armies also fight in an urban area to obtain some critical feature or resource in the area, such as a port facility. The desire to control an important seaport and access to the Persian Gulf largely motivated the Iranian and Iraqi struggle for Basra in the 1980s. Earlier, in 1944, British forces fought German units in Arnhem for control of the Rhine River Bridge. Other key infrastructure of the urban environment may have operational or strategic significance and can compel military forces to attack or defend the area. As urban areas account for an increasing share of a country's national income, often generating over 50 percent of gross national product, the strategic implications for their control or influence become even greater.

1-11. Urban areas are often located on terrain that dominates a region or an avenue of approach. In these cases, offensive armies capture these areas to proceed with security to another objective. Conversely, defensive forces commonly defend the area to deny the area of operations. To illustrate, Cassino, Italy during World War II stood astride the critical highway approach up the Liri valley to Rome. The allies had to attack and capture the monastery to facilitate the allied offensive north. Cassino's location made bypassing virtually impossible. Likewise, in the early 1980's, Israeli army urban operations in Beirut were a result of its strategic location near the Israeli security zone; various Arab insurgent and terrorist groups used Beirut as a base for attacks against Israel. Beirut evolved as the major base of the Palestine Liberation Organization, a major opponent of Israel. Beirut's location made it a security threat to Israel and thus compelled several major Israeli operations in the urban area (see Appendix A).

1-12. Another reason for engaging in urban operations is the symbolic—historical, cultural, political, and even economic—importance of many urban areas. Often, capital cities—such as Rome, Paris, Seoul, Berlin, and Baghdad—are identified as the strategic centers of gravity of their respective nations. Possessing or threatening these urban areas may impact directly on the outcome of a conflict. The objective of Germany's wars with France in 1870 and 1914 was ultimately Paris. Napoleon's 1812 campaign had as its objective Moscow, as did Hitler's 1941 offensive into Russia. The objective of the Soviet 1945 offensive was Berlin, and the North Vietnamese 1975 offensive had as its objective the South's capital of Saigon. Still, history also reminds us that commanders must assess the sustainability and decisiveness of operations directed toward these "prestige" objectives. For example, in 1812, Napoleon captured Moscow but had to evacuate it within 30 days. He lacked supplies and shelter, failed to destroy the Russian Army, and failed to defeat the political will of the Czar and the people. Similarly, the North Korean occupation of Seoul during the Korean War was equally indecisive.

U.S. Army's Experience in Urban Operations

1-13. The U.S. Army has a varied history of conducting urban operations. The American Revolution saw the Army conduct several urban operations. These operations included the unsuccessful defense of New York, the successful attack on Trenton, and the decisive siege and attack on British forces at Yorktown. The Mexican War also had a successful assault on the fortified city of Monterey and the decisive siege of Mexico City. During the American Civil War, the armies, in the tradition of Napoleonic maneuver warfare, avoided urban areas and fought in the open. However, the opposing armies frequently made urban areas their objective because of their importance as railheads. Success in the siege of several key urban areas—Vicksburg, Atlanta, and Petersburg—contributed to the Northern victory.

1-14. Following the Civil War, the U.S. Army faced no large-scale urban combat for several generations. The Indian Wars, the Spanish-American War, the Philippine Insurrection, and even World War I did not require the Army to fight in large urban areas. Between the Civil War and World War II, the U.S. Army fought in several urban areas worldwide supporting U.S. commitments. These limited urban combat operations were small but essential parts of what are currently called urban stability operations. From 1900 to 1901, the Army provided public security for a sector of Peking, China of around 50,000 inhabitants. The Army conducted UO and, in the course of the operation, the 9th U.S. Infantry suffered 20-percent casualties while fighting in Tientsin. Punitive expeditions to places such as Siberia, Cuba, Philippines, Central America, and Mexico put the Army in various urban situations that required using military power, notably, the occupation and security of Vera Cruz, Mexico in 1914. In the context of these smaller-scale contingencies, UO became a staple of U.S. Army employment.

1-15. World War II forced the Army to grapple with the issues of large-scale urban combat almost immediately. In his 1941 defense of the Philippines, General MacArthur examined how to defend Manila.

Manila represented a large, modern, friendly urban area, which was the capital city of a close U.S. ally. Defending the urban area posed numerous challenges. Ultimately General MacArthur determined that he could best conduct its defense outside the city by defeating the enemy forces in combat on the invasion beaches or shortly after they landed. When Japanese forces defeated MacArthur's Philippine Army in a series of engagements, MacArthur had to decide how best to protect the friendly populace of Manila. He had two choices: abandoning the city or waging a costly defense that would likely result in the city's destruction, thousands of noncombatant casualties, and no operational advantage. He had little choice but to declare Manila an open city and move his forces to Bataan to wage an operational defense in the vain hope that a counteroffensive could relieve his isolated force. On 2 January 1942, Japanese forces entered Manila unopposed.

1-16. Had General MacArthur decided to defend Manila, his forces would have found scant doctrine in the Army regarding how to fight in an urban area. Doctrine for urban operations did not appear until early 1944, when faced with the possibility of fighting through the larger urban areas of Western Europe. At this time the U.S. Army published FM 31-50. This manual had the first formal discussion of how the Army viewed urban combat. It was based on the Army's limited experiences in the Mediterranean theater and the study of German and Soviet experiences on the Eastern front.

1-17. FM 31-50 emphasized a deliberate pace, individual and small unit initiative, the liberal use of direct and indirect firepower, and decentralized command and execution. It focused on the urban area (as opposed to the environment); however, it did include policies towards the noncombatants. The manual was also focused at the regimental combat team level. Complementing the doctrine of FM 31-50 was the 1944 operations manual, FM 100-5. This latter manual rightly emphasized the importance of combined arms actions and the need for extensive reconnaissance of prepared and defended cities. The Army successfully implemented this doctrine in several major instances of urban combat, most notably the capture of the first German city, Aachen, and hundreds of small-scale urban assaults on cities, towns, and villages across France, the Benelux, and Germany. Army forces also successfully employed this urban combat doctrine during the liberation of Manila in 1945.

1-18. The legacy of this era of Army operations was an effective tactical solution to urban offensive combat: isolate the urban area, seize a foothold, and expand the foothold block by block until occupying the entire urban area and destroying the enemy. The doctrine's emphasis on firepower kept friendly casualties to a minimum. Unfortunately, when enemy forces stoutly defended the urban area, the emphasis on firepower resulted in the area's virtual destruction and high casualties among noncombatants.

1-19. The doctrinal approach honed in World War II remained the accepted Army approach to urban combat to the century's end. The last successful implementation occurred during the Korean War and the final liberation of Seoul. The Vietnam conflict did not offer the Army opportunities or the requirement to practice urban combat or test and refine doctrine on a large scale. The largest urban battle, Hue, was a chaotic tactical battle that validated most of the historical lessons of urban combat without generating any new doctrinal insights for large-scale urban warfare.

1-20. From the mid-1950s through the 1990s, the Army conducted UO in the U.S. in support of civil authorities during civil unrest and anti-Vietnam protests. Some operations involved numerous active and Reserve Component forces engaged in restoring public order. The Detroit riots of 1967 and the Los Angeles riots of 1992 required the commitments of active and National Guard units. In 1968, the Army deployed over 35,000 troops to Washington D.C., Chicago, and Baltimore following the death of Dr. Martin Luther King, Jr.

1-21. In the 1970s and 1980s, Army doctrine predominantly focused on urban areas and successfully fighting a conventional ground war against Soviet and Warsaw Pact forces in Central Europe. FM 90-10 (1979) described how to conduct urban operations against Soviet forces in Germany. Its concepts were never tested other than in simulation, and its approach to urban combat was not substantially different from that practiced by the Army since World War II. Despite previous doctrine's admonition to avoid cities, the Army has had to fight in them in diverse circumstances.

1-22. In the 1990s and early 21st century, commanders and planners began to more carefully consider the impact of urban environments on the overall conduct of campaigns and major operations. Maneuver warfare characterized Operation Desert Storm in 1991; however, urban environments became prominent during subsequent operations in Somalia, the Balkans, and again in Iraq. (During Operation Iraqi Freedom in 2003, for example, Baghdad was considered as a strategic center of gravity for planning.) This was a result of adversaries recognizing the asymmetric advantages to be gained by fighting superior U.S. forces in an urban environment as well as a general worldwide trend of increasing urbanization. The evolving nature of Army operations in the urban environment was recognized in the Army doctrine of full spectrum operations. No longer could combat operations be considered exclusively as a separate phase of the major operation. Instead, full spectrum operations—combat and stability or civil support—would be conducted simultaneously. It also became clear to Army commanders that while one type of operation can often dominate for a period, simultaneous, full-spectrum urban operations were now the norm and an in-depth understanding of complex urban societies essential to overall success.

MODERN ARMY URBAN OPERATIONS

1-23. Modern urban operations will be full spectrum operations executed jointly and will often contain a multinational and interagency component. They will span the entire range of possible applications of military power. At higher echelons, these separations are often viewed as levels or scales of intensity. For the tactical units conducting urban operations, these divisions appear indistinct, as the intensity is often high despite where the operation falls within the range of military operations.

JOINT, INTERAGENCY, AND MULTINATIONAL

1-24. The urban environment is too multifaceted for a single-service, single-agency, or single-dimensional solution. Generating desired effects and avoiding unintended negative consequences in this complex environment requires careful integration of joint (and often multinational) forces and interagency capabilities throughout all phases of the operation. Effective interagency collaboration will help plan effects, supporting actions, and measures of effectiveness to ensure that military actions complement diplomatic, economic, and informational activities.

1-25. Joint urban operations (JUOs) in which Army forces are a major component will be primarily land operations. These operations may take place within the context of a joint campaign conducted by a joint force land component commander or a joint task force (JTF) commander. Or they may be an Army operation under an ARFOR commander who himself operates for a JFC, depending on the organization of the theater's joint command structure. In the later case, the JFC will manage joint issues in the urban area.

1-26. The JFC conducting JUOs will focus on effectively organizing his forces for UO and tasking them in accordance with their service capabilities. His guide for the conduct of the JUO will be the joint operational tasks described in JP 3-0. JP 3 06 will provide the JTF commander specific guidance regarding the conduct of joint operational tasks in the urban environment. Army commanders will execute tasks assigned by the JFC and advise him on using Army forces and capabilities. Army commanders will also ensure that Army UO are nested within the JFC's concept of operations. Also, the ARFOR commander will request support through the JFC from other service and functional commanders who have urban capabilities critical to the success of Army UO. See Appendix D for more information on joint capabilities in an urban environment.

1-27. The military conflict will not be an end to itself. It is inevitably a means of transition from a perceived unsatisfactory state of affairs to an improved end. Therefore, urban operations will require the careful orchestration of military and civilian capabilities to achieve success. An early identification of potential requirements will allow for the proper allocation of responsibilities between military and civilian agencies and a clear understanding of what is possible for each to achieve. Such understanding will preclude the establishment of false perceptions from all concerned, but particularly from the urban population and the public at large.

1-28. Interagency cooperation will include the Army as part of the Department of Defense, elements of the Department of State, and other various governmental and nongovernmental organizations. Urban operations demand the expertise and abilities of various organizations and a synergistic unity of effort if strategic goals aimed at bringing about a more satisfactory political, social, diplomatic, economic, and military situation are to be achieved. However, there will often be a gap in time before bringing many governmental capabilities to bear that the Army and other military forces—to include coalition partners—will have to fill. Routine interagency training, planning, and coordination will help to decrease that gap in time to its minimum. As it is likely that civilian organization leaders will be in charge during some phases of the overall operation, they should lead planning at those times. A careful survey of urban requirements and rapid communication to participating civilian organizations will also help speed civilian capabilities to where they are needed most.

Winning the Peace in Iraq:
The Requirement for Full Spectrum Operations

We found that if we concentrated solely on establishing a large security force and targeted counterinsurgent combat operations—and only after that was accomplished, worked toward establishing a sustainable infrastructure supported by a strong government developing a free-market system—we would have waited too long. The outcome of a sequential plan allowed insurgent leaders to gain a competitive advantage through solidifying the psychological and structural support of the populace.

Further, those who viewed the attainment of security solely as a function of military action alone were mistaken. A gun on every street corner, although visually appealing, provides only a short-term solution and does not equate to long-term security grounded in democratic process. Our observation was born not from idealism, but because it creates the essence of true security, protecting not only our soldiers, but Iraq, the region, and, consequently, our homeland.

Major General Peter Chiarelli
Commander, 1st Cavalry Division

FULL SPECTRUM OPERATIONS

1-29. Army forces will conduct full spectrum operations within urban areas. Army commanders conduct full spectrum operations abroad by executing offensive, defensive, and stability urban operations as part of an the joint, interagency, and multinational effort described above. Army forces within the United States and its territories conduct full spectrum operations by combining offensive, defensive, and civil support operations to support homeland security. The situation will mandate that one type of operation—offense, defense, stability, or civil support—dominates the urban operation. Depending on whether the operation is overseas or within the United States, however, commanders will often find themselves executing offensive and defensive operations and stability or civil support operations simultaneously. In fact, waiting until all combat operations are concluded before beginning stability or civil support operations often results in lost, sometimes irretrievable, opportunities. Unsurprisingly, the mission determines the dominant type of operation, with the other types of Army operations conducted to shape the AO for mission success. The dominant type of operation will vary between different urban areas even in the same campaign.

Offense

1-30. Against a large conventional enemy in a major urban area with a large civil population present, offensive operations will likely require a greater commitment of Army resources than in other environments. Urban offensive operations will also incur the greatest risks to Army forces and noncombatants. Within defensive and stability or civil support operations, forces may conduct tactical offensive UO, such as counterattacks to maintain the initiative or raids to eliminate elements disrupting the stability or civil support operation.

Defense

1-31. Defensive UO are generally conducted as a shaping operation within a larger major operation. These temporary operations often set conditions for successful offensive operations, stability operations, or civil support operations. Commanders often conduct defensive UO within other types of operations to protect essential facilities in the urban area, protect flanks against counterattack, prevent the breakout of isolated enemies, or protect valuable supply bases or vulnerable convoy routes. Army forces conducting defensive UO must creatively use the environment to enhance their combat power.

1-32. In UO, essential facilities will likely include urban ports and airfields required by sister services to support ground operations. Therefore, Army commanders will need to carefully plan integrated air and base defense operations with air and naval component commanders. Protection requirements increase dramatically as air and naval assets are most vulnerable when aircraft are on the ground and ships are docked at port. Without a carefully synchronized and integrated base defense plan, Army commanders risk sporadic or total loss of support by air and naval forces—forces essential to successful UO.

Stability

1-33. Stability operations in an urban environment require offensive and defensive operations, combined with other tasks unique to each stability operation. Army forces conduct urban stability operations for various reasons, including noncombatant evacuation operations, peace operations, or support to insurgencies (see Chapter 9). Urban stability operations will require an offensive capability to destroy any military capability that overtly threatens its objectives before that military threat can adversely affect the operation. Army forces employ defensive capabilities to safeguard themselves as well as secure critical places, populations, or infrastructure in the urban area. Commanders may also employ defensive capabilities to separate and protect one faction from another. Various stability operations will also require the distribution of food or aid and the protection or assistance of agencies conducting economic or humanitarian activities.

Civil Support

1-34. Army civil support operations in an urban environment aid other U.S. agencies and organizations in mitigating the consequences of natural and man-made disasters. In response to the disaster, civil support operations require the equipment, personnel, or organizational abilities of Army forces rather than the Army's combat capabilities. In a civil support mission, these capabilities often involve Army transportation, medical, quartermaster, or engineer forces. Unless conducted in conjunction with homeland defense, urban civil support operations will seldom require combat; however, commanders must determine if force protection threats exist that could hamper Army civil support operations. During homeland security operations, defensive and offensive capabilities will be required to defeat hostile armed forces.

PREPARING FOR FUTURE URBAN OPERATIONS

1-35. To operate successfully in a complex urban environment requires a thorough understanding of the urban environment and rigorous, realistic UO training. Training should be conducted by the complete combined arms team and cover full spectrum operations to include appropriate tactics, techniques, and procedures (see FM 3-06.11 and TC 90-1). It should also replicate—

- The psychological impact of intense, close combat against a well-trained, relentless, and adaptive enemy.
- The effects of noncombatants—including governmental and nongovernmental organizations and agencies—in close proximity to Army forces. This necessitates—
  - An in-depth understanding of culture and its effects on perceptions.
  - An understanding of civil administration and governance.
  - The ability to mediate and negotiate with civilians including the ability to effectively communicate through an interpreter.
  - The development and use of flexible, effective, and understandable rules of engagement.
- A complex intelligence environment requiring lower-echelon units to collect and forward essential information to higher echelons for rapid synthesis into timely and useable intelligence for all levels of command. Understanding the multifaceted urban environment necessitates a bottom-fed approach to developing intelligence (instead of a top-fed approach more common and efficient for open terrain and conventional threats). It also emphasizes the need for intelligence reach and a truly collaborative approach to the development and sharing of intelligence.
- The communications challenges imposed by the environment as well as the need to transmit large volumes of information and data.
- The medical and logistic problems associated with operations in an urban area including constant threat interdiction against lines of communications and sustainment bases.

Developing Joint, Interagency, and Multinational Operability

1-36. UO training extends from the individual Soldier to the joint level including the integration of conventional and special operations forces. Preparedness also includes enhancing interoperability in regards to urban multinational and interagency operations. Training, as well as campaign and contingency plan creation, should include significant and sustained participation by civilian agencies. Fruitful multinational and interagency relationships must be cultivated before the onset of operations, that is, before Soldiers and their coalition counterparts are making decisions—many with strategic implications and often when they are under fire. Joint, interagency, and multinational collaboration will help design effects, supporting actions, and measures of effectiveness necessary to ensure that military actions in urban environments complement the diplomatic, informational, and economic activities necessary to achieve strategic objectives.

Conducting Live, Virtual, and Constructive Training

1-37. Force preparedness mandates integrating the actual use of urban terrain, exercises at urban training sites, simulations, or any combination into tactical- and operational-level intra- and interservice training. This type of multi-faceted training will help commanders develop a better understanding of the complexity of the urban environment and enable them to execute missions across full spectrum operations. Careful use of these facilities will also allow Army forces to fully integrate urban operations within self-development, institutional, and operational training.

Ensuring Every Soldier is an Urban Warrior

1-38. In a complex urban environment, every Soldier—regardless of branch or military occupational specialty—must be committed and prepared to close with and kill or capture threat forces in an urban environment. Every Soldier must also be prepared to effectively interact with the urban area's noncombatant population and assist in his unit's intelligence collection efforts. Each urban operation will be unique and commanders—given an opportunity to surge and conduct mission-specific post-mobilization and in-theater training—will need to conduct an analysis of the tasks requiring emphasis or modification to fit the conditions of the operational environment. In UO, every soldier will likely be required to—

- Perform advanced rifle marksmanship to include advanced firing positions, short-range marksmanship, and night firing techniques (unassisted and with the use of optics).
- Operate their unit's crew-served weapons.
- Conduct urban reconnaissance and combat patrolling.
- Enter and clear buildings and rooms as part of an urban attack or cordon and search operation.
- Defend an urban area.
- Act as a member of a combat convoy (including specific drivers training).
- Control civil disturbances.
- Navigate in an urban area.

1-39. While not all-inclusive and necessarily urban-specific, other critical individual and collective UO tasks (often modified for the urban environment) might include—

- Conduct troop-leading procedures.
- React to contact, ambush, snipers, indirect fire, and improvised explosive devices.
- Establish an observation point, personnel or vehicle checkpoint, or roadblock.
- Secure a disabled vehicle or downed aircraft.
- Call for indirect fire and close air support.
- Create and employ explosive charges.
- Handle detainees and enemy prisoners of war.
- Treat and evacuate casualties.
- Accurately report information.
- Understand the society and culture specific to the area of operations.
- Use basic commands and phrases in the region's dominant language.
- Conduct tactical questioning.
- Interact with the media.
- Conduct thorough after-action reviews.

1-40. Commanders must understand that all Soldiers will require urban-specific equipment to conduct many urban-specific tasks. Critically, commanders cannot expect supporting forces to fight alongside and support ground maneuver units in an urban environment without the appropriate equipment and training.

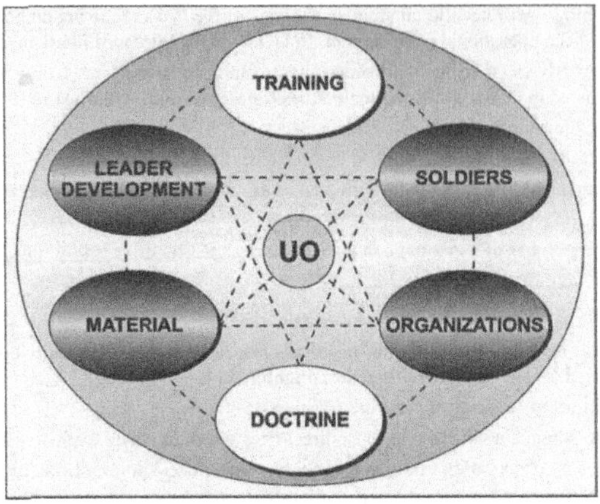

Figure 1-2. UO and DOTMLPF

Continual Adaptation and Innovation

1-41. Realistic UO training (as well as the conduct of real world operations) has the added benefit of identifying operational requirements and resultant changes necessary in our doctrine, organizations, training, materiel, leadership and education, personnel, and facilities (DOTMLPF) (see figure 1-2 and FM 100-11). While technology (materiel) and organizational changes are important, confident Soldiers (personnel) remain the decisive means for success. Technology and organizational changes will be a critical enabler to attain better understanding of the urban environment, enhance command and control on the noncontiguous battlefield, and achieve the agile, simultaneous, and precise effects required in UO. In the future, technology may lead to a radically new operational concept and approach to UO. Still, competent leaders and well-trained, disciplined, and adaptive Soldiers will remain the decisive means for the Army to succeed in this complex and multidimensional urban environment.

Chapter 2

Understanding the Urban Environment

From a planning perspective, commanders view cities not just as a topographic feature but as dynamic entities that include hostile forces, local population, and infrastructure. Planning for urban operations requires careful IPB, with particular emphasis on the three-dimensional nature of the topography and the intricate social structure of the population.

FM 3-0

Of all the environments in which to conduct operations, the urban environment confronts Army commanders with a combination of difficulties rarely found elsewhere. Its distinct characteristics result from an intricate topography and high population density. The topography's complexity stems from the man-made features and supporting infrastructure superimposed on the natural terrain. Hundreds, thousands, or millions of civilians may be near or intermingled with soldiers—friendly and enemy. This second factor, and the human dimension it represents, is potentially the most important and perplexing for commanders and their staffs to understand and evaluate. To this end, this chapter provides information essential to understanding the urban environment and conducting an effective intelligence preparation of the battlefield (see FM 2-01.3, FMI 2-91.4, and Appendix B).

Although urban areas possess general similarities, each environment is distinct and will react to and affect the presence and operations of Army forces differently. A tactical technique effective in one area may not be effective in another area due to physical differences, such as street patterns or the type of building construction. An Army policy popular with one urban group may cause resentment and hostility in another due to diverse cultural differences. All difficulties potentially exist, and they increase the complexity for Army forces operating in urban areas. These difficulties range from conventional military forces to disease and starvation (see Chapter 3) to a pervasive media—often acutely present in intricate combinations. Thus, commanders at all levels must make extraordinary efforts to assess and understand their particular urban environment to plan, prepare for, and execute effective urban operations (UO).

A COMPLEX ENVIRONMENT

2-1. Urban areas vary depending on their history, the cultures of their inhabitants, their economic development, the local climate, available building materials, and many other factors. This variety exists not only among urban areas but also within any particular area. The ever-changing mix of natural and man-made features in urban areas present commanders with some of the most difficult terrain in which to conduct military operations.

2-2. Although urban areas possess similar characteristics, no two are identical. The sprawl of Los Angeles, for example, bears little physical resemblance to New Delhi. Societal characteristics most significantly affect each area's uniqueness and complexity. While complex, information about the terrain, its potential effects on operations, and how it changes over time may be determined with some degree of certainty. However, the human dimension is much more difficult to understand and assess, particularly its effects on military operations. Like any environment, the side that can best understand and exploit the positive and mitigate the negative effects of the urban environment has the best chance of success.

2-3. Whether a large metropolis or a small village, each urban environment has identifiable components that constantly change and interact. This "system of systems" consists of the terrain, the society, and the infrastructure that links the two (see figure 2-1).

2-4. These systems are not separate and distinct categories, but rather overlapping and interdependent systems, acting dynamically with each other. Thoroughly analyzing these elements, along with the other factors of mission, enemy, weather, troops and support available, and time—

- Contributes to the accuracy of the commanders' situational understanding.
- Potentially lessens the number and cost of close combat engagements.
- Allows commanders to develop courses of action that apply appropriate resources against decisive points.

2-5. In stability operations and civil support operations, this understanding allows commanders to engage and dominate the decisive points critical to maintaining peace or restoring normalcy to the urban environment. Although each system is categorized into subordinate components or subsystems, commanders must often "step back" and visualize each system, the complex urban environment, and their area of operations (AO) as a unified whole. This "systems thinking" aids commanders in uncovering key relationships and intersections that can help reveal centers of gravity (COGs) and decisive points.

2-6. To comprehend the urban environment and its components to the fullest extent possible, commanders must carefully integrate and employ tactical reconnaissance forces, special operations forces (SOF)—to include psychological operations (PSYOP) and civil affairs units—and a myriad of other human intelligence (HUMINT) assets and regional, language, and cultural experts. The societal aspects and integrating infrastructure will challenge commanders' assessment and understanding. These aspects will also require greater dependence on nonmilitary and nongovernmental organizations (NGOs) and host-nation agencies for their information, knowledge, and expertise. This last consideration requires commanders to develop effective techniques and procedures for coordinating, interacting, and, to the greatest extent possible, synchronizing activities with these agencies.

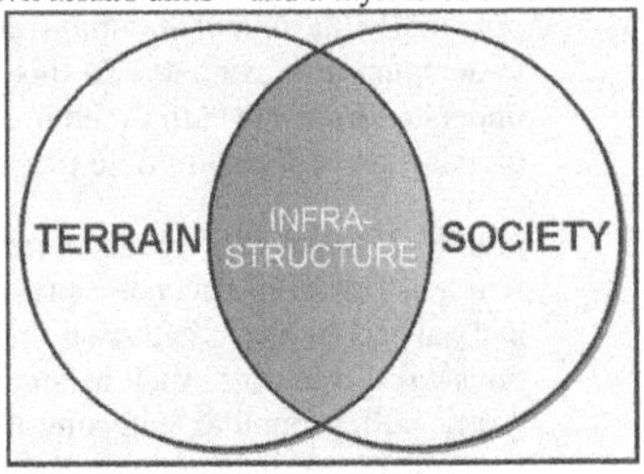

Figure 2-1. Keys to understanding the urban environment

URBAN TERRAIN

2-7. Although complex and difficult to penetrate with many intelligence, surveillance, and reconnaissance (ISR) assets, the terrain is the most recognizable aspect of an urban area. Truly understanding it, however, requires comprehending its multidimensional nature. The terrain consists of natural and man-made features, with man-made features dominating; an analysis considers both. Buildings, streets, and other infrastructure have varied patterns, forms, and sizes. The infinite ways in which these factors can intertwine make it difficult to describe a "typical" urban area. However, these various factors provide a framework for understanding the complex terrain in an urban area. Furthermore, man-made features significantly affect military systems and Soldiers, and thus tactics and operations. General effects on urban operations are discussed in this chapter. Specific effects on warfighting functions (see Chapters 4 and 10) and the spectrum of operations (see Chapters 7, 8, and 9) are interwoven throughout the manual.

MULTIDIMENSIONAL BATTLEFIELD

2-8. Urban areas present an extraordinary blend of horizontal, vertical, interior, exterior, and subterranean forms superimposed on the natural relief, drainage, and vegetation. An urban area may appear dwarfed on a map by the surrounding countryside. In fact, the size and extent of the urban area of operations is many times that of a similarly sized portion of undeveloped natural terrain. A multi-storied building may take up the same surface area as a small field, but each story or floor contains approximately an equal area as the ground upon which it sits. In effect, a ten-story building can have eleven times more defensible area than "bare" ground—ten floors and the roof. It is the sheer volume and density created by this urban geometry that makes UO resource intensive in time, manpower, and materiel.

2-9. Like natural disasters, UO can radically alter the physical characteristics of the urban terrain in ways not experienced in other environments. They may cause (either intentionally or not) uncontrollable fires or the loss of electricity. A power outage can cause flooding (especially in subsurface areas) by shutting down pumping stations. Entire buildings may be destroyed, eliminating reference points, leaving large piles of rubble, altering fields of fire, and making movement and transportation extremely difficult. Additionally, buildings and other urban structures, damaged but not destroyed, can become (or remain) effective obstacles and possible booby traps. Even without enemy exploitation, their weakened construction and unstable structure increase the risk of injury to Soldiers and civilians moving within them. (Engineer expertise will often be needed to determine whether the buildings can support occupation by Army forces or civilians.) Yet, even the total collapse of a building may not eliminate its defenders. Of additional concern, the likely presence of toxic industrial material (TIM) can create additional obstacles and health hazards.

2-10. Commanders in other environments normally address the depth, breadth, and height of their AO in terms of two areas: airspace and surface. In an urban environment, they broaden their scope to include supersurface and subsurface areas (see figure 2-2) that voluminously extend the commanders area of operations. Although spatially separated, each area may be used as an avenue of approach or mobility corridor, line of communications (LOC), or engagement area.

2-11. Supersurface and subsurface areas magnify the complexity of the urban physical environment. Commanders must consider activities that occur outside buildings and subterranean areas (the external space) as well as the activities happening unseen in buildings and subterranean systems (the internal space). This internal volume further challenges command, control, and intelligence collection activities and increases the combat power required to conduct UO. Commanders must develop methods and techniques to help themselves, their staffs, and their subordinate commanders and staffs to represent, visualize, and reference these multiple dimensions. Increasing the difficulty, such dimensions can change rapidly simply due to continued urban growth or, as described earlier, the effects of nature and UO themselves.

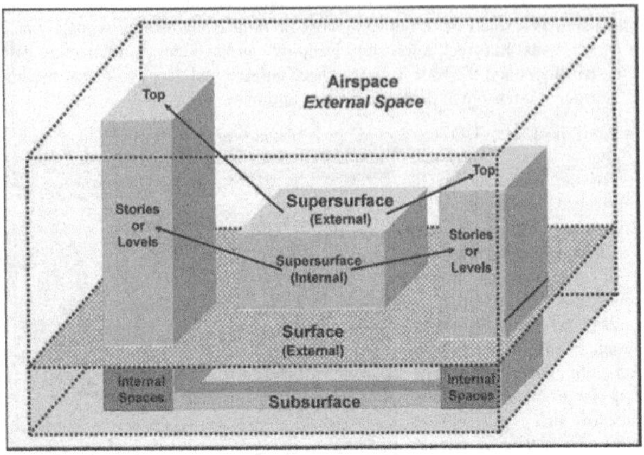

Figure 2-2. The multidimensional urban battlefield

Airspace

2-12. As in all other environments, aircraft and aerial munitions use the airspace as rapid avenues of approach in urbanized areas. Forces can use aviation assets for observation and reconnaissance, aerial attack, or high-speed insertion and extraction of Soldiers, supplies, and equipment. Some surface obstacles in an urban area, such as rubble, do not affect flight (though they may prevent the take-off and landing of aircraft). Buildings of varying height and the increased density of towers, signs, power lines, and other urban constructions, however, create obstacles to flight and the trajectory of many munitions (masking). Similarly, these obstacles can restrict a pilot's line of sight as well as physically limit low-altitude maneuverability in the urban airspace. Excellent cover and concealment afforded enemy gunners in an urban area increases aviation vulnerability to small arms and man-portable air defense systems (MANPADS), particularly when supporting ground forces. The potential for a high volume of air traffic (military and civilian) over and within urban airspace (including fixed-wing, rotary-wing, and unmanned aircraft systems) may become another significant hazard and necessitate increased airspace command and control measures.

Surface

2-13. Surface areas apply to exterior ground-level areas, such as parking lots, airfields, highways, streets, sidewalks, fields, and parks. They often provide primary avenues of approach and the means for rapid advance. However, buildings and other structures often canalize forces moving along them. As such, obstacles on urban surface areas usually have more effect than those in open terrain since bypass often requires entering and transiting buildings or radical changes to selected routes. Where urban areas border the ocean or sea, large lakes, and major rivers, the surface of these bodies of water may provide key friendly and threat avenues of approach or essential LOCs—a significant consideration for Army commanders. As such, amphibious, river-crossing, and river operations may be integral parts of the overall urban operation (see FM 3-05.212 and FM 3-90.12).

2-14. Larger open areas—such as stadiums, sports fields, school playgrounds, and parking lots—are often critical areas during urban operations. They can provide locations for displaced civilians, interrogation centers, and prisoner of war holding facilities. These areas also can afford suitable aircraft landing and pickup zones and artillery firing locations. They can provide logistic support areas and aerial resupply possibilities because they are often centrally located. Finally, large open areas (and immense or unusually shaped structures) within urban areas are often easier to see—especially from the air—and can serve as excellent target reference points from which to shift or control fires.

Supersurface

2-15. These areas include the internal floors or levels (intrasurface areas) and external roofs or tops of buildings, stadiums, towers, or other vertical structures. They can provide cover and concealment; limit or enhance observation and fields of fire; and restrict, canalize, or block movement. However, forces can move within and between supersurface areas creating additional, though normally secondary, avenues of approach. Rooftops may offer ideal locations for landing helicopters for small-scale air assaults and aerial resupply. First, however, engineers must analyze buildings for their structural integrity and obstacles. Such obstacles include electrical wires, antennas, and enemy-emplaced mines (although personnel may be inserted by jumping, rappelling, or fast roping from a hovering helicopter and extracted by hoist mechanisms). Some rooftops are designed as helipads. Roofs and other supersurface areas may also provide excellent locations for snipers; lightweight, handheld antitank weapons; MANPADS; and communications retransmission sites. They enable top-down attacks against the weakest points of armored vehicles and unsuspecting aircraft. Overall, elevated firing positions reduce the value of any cover in surrounding open areas and permit engagement at close range with less risk of immediate close assault. This area (and the subsurface area) requires commanders to think, plan, and execute ground operations vertically as well as horizontally. In this latter regard, UO share strong similarities with mountain operations (see FM 3-97.6).

Subsurface

2-16. Subsurface areas are below the surface level. They may serve as secondary and, in fewer instances, primary avenues of approach at lower tactical levels. When thoroughly reconnoitered and controlled, they offer excellent covered and concealed LOCs for moving supplies and evacuating casualties. They may also provide sites for caching and stockpiling supplies. Subsurface areas include subterranean areas such as subways, mines, tunnels, sewers, drainage systems, cellars, civil defense shelters, and other various underground utility systems. In older cities, they may include ancient hand-dug tunnels and catacombs. Both attacker and defender can use subsurface areas to gain surprise and maneuver against the rear and flanks of a threat and to conduct ambushes. However, these areas are often the most restrictive and easiest to defend or block. Their effectiveness depends on superior knowledge of their existence and overall design. Army commanders may also need to consider potential avenues of approach afforded by the subsurface areas of rivers and major bodies of water that border urban areas. This particularly applies when operating as part of a joint task force (JTF) task organized with SOF or when opposing a threat with similar capabilities.

Major Urban Patterns

2-17. Four major urban patterns (satellite, network, linear, and segment) can influence UO (see figure 2-3). Central to two of the patterns (satellite and network) is the hub or dominant urban area or pattern around which outlying urban areas or patterns radiate. (A segmented urban area, because it tends to be a larger urban area, can often be a hub.) In offensive and defensive operations, the hub serves as a pivot or strong point; as such, it can become a major obstacle to an attacker. If the attacker chooses to bypass the urban area (hub) located along his axis of advance without first isolating the area, he may expose his flank or LOC to attack from the hub as well as dependent urban areas or subordinate satellite patterns. Because the focus of stability and civil support operations is normally on people, commanders should understand the value and influence of the hub to the economic, political, or cultural well being of the surrounding area. Generally the larger the hub, the greater influence it has on satellite urban areas and surrounding rural areas. Commanders must remember that urban areas are not islands; all are connected to the surrounding rural (and other urban) areas through fluid and permeable boundaries and LOCs.

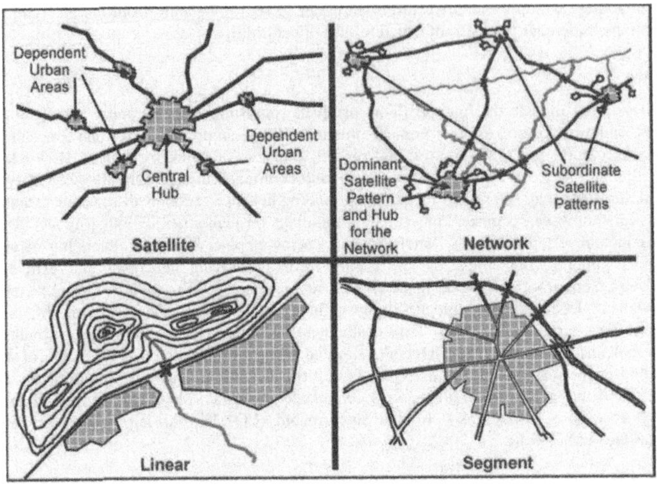

Figure 2-3. Major urban patterns

Satellite Pattern

2-18. This common pattern consists of a central hub surrounded by smaller, dependent urban areas. LOCs tend to converge on the hub. Outlying areas often support the principal urban area at the hub with means of reinforcement, resupply, and evacuation. In some instances, they may serve as mutually supporting battle positions. Commanders should consider the effects of the outlying urban areas on operations within the hub, and, conversely, the effects of operations within the hub on outlying urban areas. Information operations (IO), for example, targeted primarily at key leaders and other civilians located within the hub of a satellite pattern may subsequently influence civilians in outlying urban areas and achieve necessary effects without having to commit specific resources to these outlying areas.

Network Pattern

2-19. The network pattern represents the interlocking of the primary hubs of subordinate satellite patterns. Its elements are more self-sufficient and less supportive of each other, although a dominant hub may exist. Major LOCs in a network extend more than in a satellite pattern and take more of a rectangular rather than a convergent form. Its natural terrain may vary more than in a single satellite array. Operations in one area may or may not easily influence, or be influenced by, other urban areas in the pattern.

Linear Pattern

2-20. Potentially a subelement of the previous two patterns, the linear pattern may form one ray of the satellite pattern or be found along connecting links between the hubs of a network. Most frequently, this pattern results from the stringing of minor urban areas along a confined natural terrain corridor, such as an elongated valley, a body of water, or a man-made communications route. In offensive and defensive operations, this latter form of the linear pattern facilitates developing a series of strong defensive positions in depth, effectively blocking or delaying an attacking force moving along the canalized terrain.

Segment Pattern

2-21. When dominant natural terrain, such as a river or man-made features (canals, major highways, or railways), divides an urban area, it creates a segmented pattern. This pattern often makes it easier for commanders to assign areas of operations to subordinate commanders. However, this pattern may fragment operations and increase risk to an operation requiring mutual support between subordinate units. Still, the segmented urban areas may allow commanders to isolate threats more easily in these areas and focus operations within segments that contain their decisive points. Although an integral part of the whole (the urban area), each segment may develop distinct social, economic, cultural, and political characteristics. This social segmenting may benefit commanders faced with limited assets to influence or control the urban populace. After thoroughly analyzing the society, they may be able to focus IO and populace and resources control measures against only specific segments that affect decisive operations. Commanders may need only to isolate other segments or may need to just monitor for any significant changes in the attitudes, beliefs, or actions of the civilians located there.

LESSER STREET PATTERNS

2-22. Lesser patterns in the urban area result from the layout of the streets, roads, highways, and other thoroughfares. They evolve from influences of natural terrain, the original designer's personal prejudices, and the changing needs of the inhabitants. Street patterns (and widths) influence all warfighting functions; however, they greatly affect movement and maneuver, command and control, and sustainment. (In some portions of older Middle Eastern urban areas, the labyrinths of streets were designed only to allow two loaded donkeys to pass each other; tanks are too wide.) Urban areas can display any of three basic patterns and their combinations: radial, grid, and irregular (see figure 2-4).

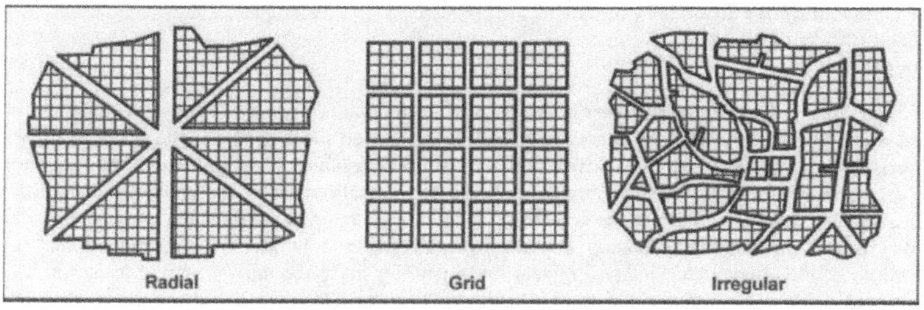

Figure 2-4. Basic internal street patterns

Radial

2-23. Societies of highly concentrated religious or secular power often construct urban areas with a radial design: all primary thoroughfares radiating out from the center of power. Urban areas with this design may signal an important historical aspect in the overall analysis of the urban society. Terrain permitting, these streets may extend outward in a complete circle or may form a semicircle or arc when a focal point abuts a natural barrier, such as a coastline or mountain. To increase mobility and traffic flow, societies often add concentric loops or rings to larger radial patterns. Unless commanders carefully plan boundaries, routes, and axes of advance, their subordinate units' movement or maneuver may be inadvertently funneled toward the center of urban areas with this pattern resulting in congestion, loss of momentum, and an increased potential for ambush or fratricide.

Grid

2-24. The most adaptable and universal form for urban areas is the grid or rectangular pattern: lines of streets at right angles to one another forming blocks similar to the pattern of a chessboard. A grid pattern can fill in and eventually take over an original radial pattern. Grid patterns often appear to ease the assignment of boundaries for subordinates units. However, commanders also consider how the natural terrain influences operations and the establishment of graphic control measures. They also consider the influence of the buildings and other structures lining these streets, such as their height and construction, before assigning boundaries and developing other control measures. Commanders should also consider the following when developing urban graphic control measures:

- Describing boundaries, phase lines, checkpoints, and other graphic control measures by easily recognizable features is as important in urban areas as elsewhere. While easily identifiable urban structures such as unusually tall or oddly-shaped buildings, cemeteries, stadiums, or prominent rail or highway interchanges can be useful references, available natural features are a better descriptor than man-made features that may be altered or unrecognizable. As an aid to air-to-ground coordination, commanders should select features that can be identified by both ground and air forces. Those that help in controlling ground forces may not be easily visible from the air and vice versa.
- Commanders should also consider whether a boundary along an easily-recognizable terrain feature, such as a river, will also be easy to identify by the threat who may seek to "find the seam" and exploit the likely control and coordination difficulties associated with boundaries, especially between higher-level units. This often requires commanders to carefully position a control measure away from the key feature to provide a designated subordinate force with the terrain and space necessary to control the feature. On the other hand, commanders working closely with local authorities during stability and civil support operations may not need to thoroughly understand the physical effect of street patterns on the assignment of boundaries as they might for combat UO. Instead, commanders may choose to assign boundaries overlaid on existing geopolitical boundaries used by local agencies to increase interoperability and aid in unity of effort.

Irregular

2-25. In most urban areas, regardless of the original intent, plan, or vision, existing street patterns emerge from successive plans overlaid one on another. Some are well planned to fit with previous plans while others a haphazard response to explosive urban growth. The result may mix patterns. Urban engineers and planners may specifically design irregular patterns for aesthetic reasons (as in many suburban housing developments) or to conform to marked terrain relief. Irregular street patterns may alert commanders and analysts that the underlying natural terrain may exert greater influence over operations than in other portions of the urban area. Finally, irregular street patterns make the movement and maneuver of forces less predictable. However, a labyrinth of irregular or "twisting" street patterns may increase the possibility of fratricide particularly for units that are trained or accustomed only to grid patterns.

An Urban Model

2-26. Throughout the world, urban areas have similar form and function. In form, urban areas contain like characteristics, readily divisible into distinct sections or areas. Functionally, they tend to be the centers of population, finance, politics, transportation, industry, and culture. While urban areas may be modeled by several different means, figure 2-5 illustrates the general forms and internal functions. Some forms and functions may overlap. For example, high-rise buildings are located in core areas as well as in outlying areas and may be used for residential purposes. With the rapid urbanization associated with developing nations, the areas displayed in this urban model often manifest themselves less clearly there than in developed nations.

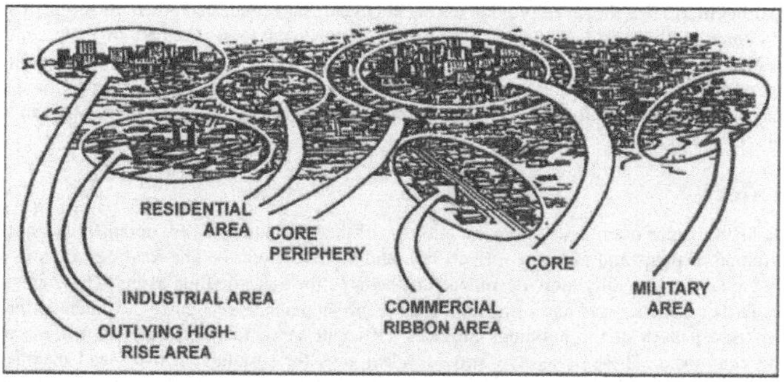

Figure 2-5. Urban functional zones

2-27. This analysis helps to determine, in general terms, potential advantages and disadvantages each portion of the urban area may have toward accomplishing the urban operation. However, construction materials and methods can vary drastically. Commanders and their staff will often need to identify specific building types and construction and understand weapons effects on them. If a commander desires precise effects, the chosen munitions or weapons system must be sufficiently accurate, capable of penetrating the target structure (without exiting the other side), and achieve effects within. Often noncombatants, critical infrastructure, or protected targets are in the vicinity. Commanders may need to determine if the surrounding walls or structures will sufficiently absorb or negate the blast or thermal effects of the weapon. Regardless, understanding the structure and composition of buildings and other structures in the urban AO may be necessary to allow commanders to determine the best means to accomplish the mission.

Core

2-28. The core is the heart of the urban area, the downtown or central business district. Relatively small and compact, it contains a large percentage of the urban area's shops, offices, and public institutions. Often, it houses the headquarters for commercial and financial activities and contains important cultural, historical, and governmental buildings. These activities prefer the core because of its accessibility. As the focal point of the transportation network, residents find the core the easiest part of the urban area to reach. It normally has the densest concentration of multistory buildings and subterranean features (underground parking garages, underground shopping centers, and basements).

2-29. High-rise buildings, varying greatly in height (possibly 50 stories above ground and four stories below ground), make up the cores of many of today's urban areas. Buildings routinely abut one another, with little or no setback from the sidewalks. Building height and density (except in outlying high-rise areas) often decreases from the core to the edge of the residential areas, while the amount of open areas frequently increases. Modern urban planning allows for more open spaces between buildings than found in the cores of older urban areas. Most core areas have undergone constant redevelopment resulting in various types of construction. Commonly, brick buildings abound in the oldest part of the core; framed, heavy-clad structures in the next oldest part; and a concentration of framed, light-clad buildings in the newest part. The outer edge of the core, the core periphery, has ordinarily undergone less change than the core resulting in buildings of uniform height (commonly two to three stories in towns and five to ten stories in larger urban areas).

2-30. Generally, offensive operations focused in core areas (even when effectively isolated) will require greater resources—particularly manpower, time, and information—than in many other parts of the urban area. Mounted maneuver often proves more difficult in core areas because of fewer open areas, buildings closer to the streets, and an increased density of civilian vehicles. Razed buildings in central core areas (especially high-rise buildings) become greater obstacles to mobility as they can collapse on and easily block thoroughfares. Rubble piles provide excellent covered and concealed positions for dismounted threat forces. Consequently, commanders often use more dismounted forces as part of their combined arms operations. Conversely, the core may be critical to urban defensive operations, particularly older areas of heavier construction that afford greater protection. Despite potential difficulties, the core area may be key to accomplishing many stability or civil support missions since it houses much of the human activity that occurs in the urban area.

Industrial Area

2-31. Industrial areas often develop on the outskirts of the urban areas where commercial transportation is easiest (along airfields and major sea, river, rail, and highway routes). The road networks in and around industrial areas are generally more developed and suitable for transportation assets. These areas will likely displace farther from the core and residential areas as urban planners recognize the potential threat of TIM. The dispersed pattern of the buildings provides sufficient space for large cargoes, trucks, and materiel handling equipment. These areas may provide ideal sites for sustainment bases and maintenance sites. While older heavier-clad structures may be found, new construction generally consists of low, large, flat-roofed factory and warehouse buildings with large parking areas and work yards. These structures generally have steel frame and lightweight exterior walls. Multistory structures usually have reinforced concrete floors and ceilings.

Toxic Industrial Chemical	Industrial/Commercial Uses
Ammonia	Commercial Refrigerant, Fertilizer and Food Production, Petroleum, Explosives, Other Chemicals
Arsine	Semiconductor Industry
Boron Trichloride	Organic Catalyst, Soldering Magnesium
Boron Trifluoride	Chemical Catalyst, Aluminum Refining
Carbon Disulfide	Industrial Solvent, Dry Cleaning, Agriculture, Petroleum, Electroplating
Chlorine	Potable Water, Disinfectants, Metal Treatment, Plastics & Rubber
Diborane	Plastics and Rubber
Ethylene Oxide	Industrial Alcohols, Fumigant, Industrial Sterilant
Fluorine	Uranium Processing, Rocket Fuel
Formaldehyde	Plastics, Fertilizers, Preservative/Corrosion Inhibitor, Fungicide and Germicide, Pesticide, Pharmaceuticals
Fuming Nitric Acid	Fertilizers, Explosives, Metal Processing, Pesticides, Rocket Fuel
Hydrogen Bromide	Chemical Industry, Pharmaceuticals
Hydrogen Chloride	Fabrics, Semiconductors
Hydrogen Cyanide	Pesticides, Other Chemicals, Pharmaceuticals, Electroplating
Hydrogen Fluoride	Glass Production, Chemical Catalyst
Hydrogen Sulfide	Metallurgy, Agricultural Disinfectant
Phosgene	Dyes, Pharmaceuticals, Herbicides & Insecticides
Phosphorus Trichloride	Metallurgy, Pesticides and Germicides, Gasoline Additive
Sulfur Dioxide	Paper, Food Processing, Ice Production, Disinfectant, Leather Processing
Sulfuric Acid	Fertilizers, Petroleum, Iron and Steel Production, Battery Electrolyte
Tungsten	Electronics, Other Chemicals

Figure 2-6. Some toxic industrial chemicals and their industrial or commercial uses

2-32. Toxic industrial chemicals and other TIM may be transported through an urban area (by rail, barge, truck, or pipeline) or found stored throughout. However, larger concentrations will exist in industrial areas, and their presence should concern Army forces operating near them. Some TIM are heavier than air and tend to settle in low-lying and subsurface areas.

2-33. Each year, over 70,000 different chemicals are produced, processed, or consumed globally. An estimated 25,000 commercial facilities around the world produce, process, or store chemicals that have a legitimate industrial use yet are also classified as chemical warfare agents. Many other chemicals (not classified as weapons) may still be sufficiently hazardous to pose a considerable threat to Army forces and civilians in urban areas as choking agents or asphyxiates, flammables or incendiaries, water contaminants, low-grade blister or nerve agents, or debilitating irritants. These chemicals can be released either accidentally or deliberately. For example, on 2 December 1984, nearly 40 tons of methylisocyanate used to produce pesticides leaked from a storage tank at Bhopal, India, killing thousands and injuring hundreds of thousands. Figure 2-6 contains a small sampling of other toxic industrial chemicals along with their industrial or commercial usage that commanders may encounter in an urban area. The most common chemicals that may pose an immediate risk to Army forces are highly toxic irritant gases such as ammonia, chlorine, hydrogen chloride, and sulfur dioxide.

2-34. Standard chemical defense equipment may not protect against (and chemical detection devices may fail to detect) many toxic industrial chemicals. Therefore, the risk to Soldiers operating near the chemicals may increase. Commanders must vigilantly identify these potential hazards, carefully consider them as part of their overall vulnerability analysis, factor the analysis into their risk assessment, and execute necessary contamination avoidance measures. (Local urban fire fighters may be a critical source of information for determining the likely locations of TIM.) Any assessment includes the chance that toxic industrial chemicals may be deliberately released by a threat to gain advantage or accidentally released by friendly actions (see FM 3-11.9, FM 3-11.14, FM 3-11.21, and the U.S. Department of Transportation's current version of the Emergency Response Guidebook).

Outlying High-Rise Area

2-35. High-rise areas consist of multistoried apartments, commercial offices, and businesses separated by large open areas, such as parking lots, parks, and individual one-story buildings. High-rise buildings are generally of framed, light-clad construction with thin walls of brick, lightweight concrete, or glass. The automobile, mass transit systems, and improved road networks encourage these areas to grow and function further from the urban core.

2-36. Similar to the urban core, units given the mission to clear these areas, or even portions therein, will need more resources—most notably personnel and time—to accomplish their mission. Commanders should consider courses of action that isolate these entire areas, multiple sections within these areas, or even individual buildings before assigning tasks. Without careful consideration and analysis, some tasks in these areas could unintentionally—but rapidly—drain a unit's resources or unhinge other portions of the major operation. When defending, commanders who can integrate these areas in the defense will present the attacker with similar resource problems and may be appropriate in a defense to delay. However, defending commanders must ensure that the defense is arranged so that this portion cannot be easily isolated and bypassed. Defensive positions in structures may require extensive reinforcement due to light-clad construction.

Residential Area

2-37. Residential areas can be found dispersed throughout the urban area; however, large suburban areas (or sprawl) normally form on the outskirts. Residential areas often consist of row houses or single-family dwellings set in a grid or ringed pattern in a planned development project. Yards, gardens, trees, and fences may separate the buildings in a residential area. In some areas of the world, residential areas may be located in high walled compounds with houses built right up to the edge of the street. Modern residential construction is more often of light-clad, framed wood construction, or brick, however; residential homes formed by poured or pre-cast concrete can be found throughout many parts of the world. The combined population of surrounding suburban areas often far outnumbers that of the urban area proper. Specific suburbs tend toward homogeneity based on ethnicity, religion, economics, or some other social aspect. Commanders must locate and analyze these areas to determine their impact on operations—often the most critical importance is the people located there (see the subsequent discussion in this chapter on the urban society).

2-38. In offensive and defensive operations, commanders should determine whether operations pose an unacceptable physical risk to civilians. If so, they may have to relocate civilians to a safer area, perhaps another residential area. If not, commanders may implement a "stay-put" policy for that area and attempt to isolate the effects of the operation from them. During civil support and stability operations, residential locations may be the initial focal point for operations since most of the permanent population is located there.

2-39. This area also contains an urban phenomenon known as shantytowns. These areas are commonly on unoccupied, low-value land in and around many urban areas in underdeveloped countries. Shantytowns may contain over 50-percent of the total urban population. They usually lack streets and public utilities. The lean-to structures tend to be irregularly laid out, connected by walking paths, and made of any scrap material available: lumber, brick, sheet metal, cardboard, cloth, or vegetation. The random arrangement of structures, the absence of formal street naming and numbering, and often the lack of easily identifiable buildings and terrain create challenges. These challenges include navigation, coordination, and the transmission of accurate information and intelligence. Depending on the operation, the temporary nature of the structures can also mean that mobility can be either more or less restricted than other sections of the urban area. A military force may easily knock down and traverse structures without affecting mobility at all. However, their destruction may cause unacceptable civilian casualties, in which case mobility becomes more restrictive as the narrow paths often do not accommodate vehicular traffic. Similarly, the makeshift materials inhibit weapons effects less than many other parts of the urban area built more solidly. A tank round, for example, may go much farther and injure many more noncombatants than in an area where the primary building material is stone. Regardless, commanders should consider the effects of their operations in this area, to include vehicles and weapons, as the weak structures increase the risk of fratricide, civilian casualties, and large, rapidly spreading fires.

Commercial Ribbon Area

2-40. Commercial ribbon areas are rows of stores, shops, and restaurants built along both sides of major streets that run through and between urban areas. These same types of areas often develop along the roads that connect one urban area to another (strip areas). The buildings uniformly stand two to three stories tall (about one story taller than the dwellings on the streets behind them).

Military Area

2-41. Fortifications and military installations may be found in or near urban areas throughout the world. Historically, they may have been the "seed" responsible for initiating the growth of the present-day urban area. Many countries possess long coastlines and borders with potentially hostile neighbors. To meet their defensive needs, they developed coastal and border defense works that include extensive subsurface facilities, many contiguous to urban areas. North Korea, for example, has built numerous hardened artillery, missile, and command and control facilities along both its coasts and the demilitarized zone which separates them from South Korea. While some fortifications may be inactive (particularly ancient fortifications), they can be rapidly activated and modified by threats (or Army forces) to accomplish their original role. Even if not activated, they may still serve as choke points and major obstacles to movement and maneuver.

2-42. Permanent-type fortifications can be made of earth, wood, rock, brick, concrete, steel-reinforced concrete, or any combination of the above. Some contemporary constructions have been built subsurface and employ heavy armor, major caliber weapons, internal communications, service facilities, and chemical, biological, radiological, and nuclear overpressure systems. Because they have been built specifically for military purposes, commanders and planners should carefully consider the effects of these military constructions on the conduct of UO.

URBAN SOCIETY

2-43. Although intricate, understanding the urban terrain is relatively straightforward in comparison to comprehending the multifaceted nature of urban society. UO often require Army forces to operate in close proximity to a high density of civilians. Even evacuated areas can have a stay-behind population in the tens of thousands. This population's presence, attitudes, actions, communications with the media, and needs may affect the conduct of operations. Homogeneity decreases drastically as the size of the urban area increases. Commanders must take into account the characteristics of a population whose beliefs and interests vary based on factors (see figure 2-7). Analysis and understanding of these societal factors is critical to a successful IO campaign and, thus, the entire operation.

2-44. Civilian populations continually influence, to varying degrees, operations conducted in an urban area. Thoroughly understanding these societal aspects and avoiding "mirror-imaging"—overlaying one's own values and thought processes on top of the person or group one is trying to understand—will help to accurately anticipate civilian actions and response. For example, while bribery and nepotism are not accepted norms in U.S. society, they may be customary practices in the foreign urban environment in which Army forces are operating. If commanders attempt to understand incidences of bribery and nepotism using their own societal norms and values as a basis, they are likely to misinterpret the significance of the events and implement an inappropriate course of action. Critically, however, the development of an in-depth understanding of the urban society does not mean that commanders and their Soldiers should adopt the culture of the foreign urban society. While understanding and respect of the urban area's culture is crucial to successful UO, Soldiers function best in dealing with the urban society when they act in accordance with American culture and values.

Figure 2-7. Key aspects of the urban society

POTENTIAL CENTER OF GRAVITY

2-45. A COG during an urban operation, particularly in stability and civil support operations, may be the civilian inhabitants themselves—specifically their behavior. However, supportive behavior is generally an advantage in any type of operation. Correspondingly, neutral behavior toward friendly forces is an advantage over hostile behavior. (However, a neutral population normally will not readily provide information to Army forces—often critical to understanding the urban environment and threat. In this light, neutrality can be seen as a disadvantage.) To influence or control their behavior, commanders must first understand the society's complex nature and character. Second, they must understand and accept that every military action (or inaction) may influence the relationship between the urban population and Army forces, and, by extension, mission success or failure. Lastly, they must understand that Army forces may play only a supporting (but essential) role as part of an integrated and synchronized multiagency effort focusing all aspects of national power—diplomatic, informational, military, and economic. With this awareness, commanders can take one or more actions:

- Coordinate and plan operations.
- Implement effective civil-military programs.

- Take the immediate action necessary to maintain support of a friendly populace, neutralize or gain the support of hostile or neutral factions, or do any combination of these activities to achieve precise effects and accomplish the mission.

Without this understanding, commanders increase the risk that their actions, particularly concerning the urban population, may not have the intended and necessary effects.

2-46. Although the factor of civil considerations takes on added significance in UO, it is just one that commanders evaluate. Sometimes, as in most urban counterinsurgency operations, it may be the most important factor to consider as a COG. At other times it may be the least important as to be almost negligible. Its importance is not constant; it changes over time (like all factors). At the beginning of the operation, civil considerations may not be essential to mission accomplishment, but as the operation progresses this factor's importance to success may increase. In other circumstances, the opposite may be true. Overall, commanders should consider three objectives regarding the civilians of the urban area:

- Minimize their interference with urban operations. In offensive and defensive operations this means moving them away from combat operations or establishing measures to shield them from its effects. In all operations, it often requires centralizing them in one or more locations—which may mean keeping them where they are.
- Maximize their support of Army, joint, and multinational forces and government agencies.
- Observe the necessary legal, moral, and humanitarian obligations.

General Population Size

2-47. Urban areas are commonly classified according to the general size of their population instead of landmass. Figure 2-8 lists one method of defining and categorizing urban areas by various population sizes.

2-48. These categories are useful to establish commonality and standardize terms that shape ideas, discussion, and concepts. Smaller populations usually suggest homogeneity among the inhabitants. Homogeneity can make consensus or compromise easier to achieve because fewer opposing viewpoints exist. Given this homogeneity, effects of change are more certain and often easier to determine. However, homogenous does not mean identical. If major social divisions do exist (either physical or ideological), commanders can more easily determine those divisions and their fundamental causes with smaller populations. Treating an urban population as a completely homogenous entity can lead to false assumptions, cultural misunderstandings, and poor situational understanding.

Category	Population
Village	3,000 or less.
Town	Over 3,000 to 100,000.
City	Over 100,000 to 1 million.
Metropolis	Over 1 million to 10 million.
Megalopolis	Over 10 million.

Figure 2-8. Urban areas by population size

2-49. As urban areas expand, the urban patterns begin to blur and the social complexity increases. For example, as satellite patterns continue to grow, the LOCs between a central hub and outlying urban areas may develop and begin to assume a linear urban pattern. Simultaneously, a hub and outlying urban areas may continue to expand until they merge into a single, large metropolis. On a larger scale, a network pattern can grow and unite as a single megalopolis. This growth physically unites smaller urban areas but cannot force conformity of needs and beliefs. It also increases the physical and social complexity of an urban area.

Group Size, Location, and Composition

2-50. Understanding how specific elements of the urban society affect operations (and vice versa), begins with analyzing their size, location, and composition (see figure 2-9). Because commanders must minimize civilian casualties, size and location (without regard to composition) are important initial demographic considerations. After determining the presence and numbers of civilians relative to decisive points, commanders can then decide whether civilian proximity and density represent a significant risk to their mission—refugees clogging LOCs, for example. If civilians are the primary focus of the operation, as in many stability and civil support operations, this same analysis may help to determine decisive points. In this analysis, commanders should consider that urban areas, on many levels, are in constant motion. The densities of circulating people and other traffic often vary according to cultural events or the time of day, such as religious holidays or sporting events and rush hours or market times. Therefore in planning urban operations, commanders must consider the timing, rhythms, or patterns of the population and their vehicular movements in the urban area. Identifying and understanding trends and patterns of activity (and disruptions to them) may provide critical information to commanders.

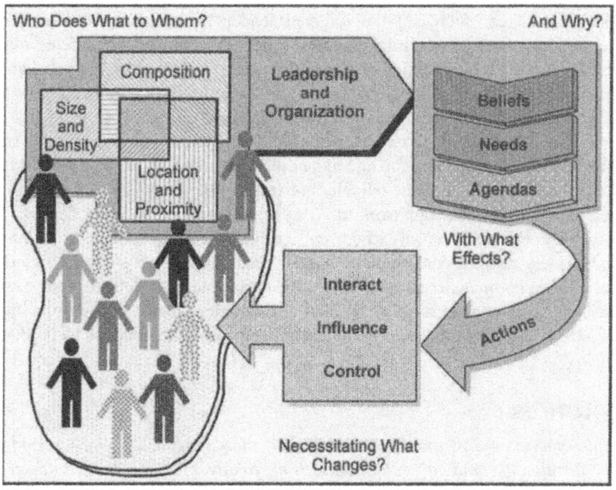

Figure 2-9. Simplified analysis of urban society

2-51. Commanders normally determine the composition of, or the identifiable groups or organizations within, the civilian urban population. Groups may be categorized by race, religion, national origin, tribe, clan, economic or social class, party affiliation, education level, union memberships, age, gender, occupation, or any other significant social demographic. Physical and ideological overlaps (and divisions) often exist between groups. Overlaps may provide early focus for analysis and suggest ways to affect more than one group simultaneously. In some cases, groups may have radically different ideologies but are (or can be) united by a single characteristic. Commanders must understand the intricacies of "who does what to whom." Such understanding furthers identifying the urban society's sources of power, influence (both formal and informal), and decisive points that hold the keys to controlling or protecting this potential COG. (See also the discussion of competing power structures in Chapter 3.) Commanders should have expert, detailed, and current knowledge and information to avoid developing simple formulas of social interaction that may actively mislead and add to a flawed course of action.

Leadership and Organization

2-52. Commanders must also understand how authority and responsibility is held or shared within and between each of the identified groups; they must understand leadership and the social hierarchy. For groups to exert meaningful influence, leadership provides vision, direction, and organized coherence. This leadership can be a function of personality as well as organization. Some groups depend on a charismatic leader to provide cohesion; although in some cultures, the spokesman is not the leader. Others de-emphasize individual leadership and provide redundancy and replacement in decision making. Others combine elements of both these types of leadership and organization. Based solely on personality, a leader may centralize power or, while still being in ultimate control, decentralize decision making and execution to subordinates. In contrast, a single person may head a group while a ruling council actually makes and executes policy. Groups centered on one leader (which may or may not be the officially designated leader) can often produce decisions and initiate actions rapidly but are vulnerable to disruptions if key personalities are removed or co-opted. Groups with shared or redundant leadership take longer to make decisions yet are more resistant to change and outside influence.

2-53. In UO, particularly stability operations, Army commanders and leaders at all levels will devote considerable effort to identifying and cultivating relationships with civilian leaders in their AO. This civilian leadership will include political, religious, tribal or clan, ethnic, and economic leaders. Commanders should consider that their attention toward and discussion with identified leaders may increase (or, some instance, decrease) the targeted leaders' prestige and power. While this may be intentional, commanders must often ensure that the leaders that they chose to deal with are the legitimate and accepted in the eyes of the urban population. Otherwise, they may further imbalance an already weak power structure and exacerbate the situation. In unique circumstances, commanders may need to identify and interact with the leadership of criminal organizations. These civilian leaders will be the means to affect change and allow the urban populace to understand and accept the purpose behind Army operations. As importantly, they will be the conduit for understanding the urban society's sentiments, perceptions, and reactions to Army operations. As such, reliable and trustworthy linguists that can not only interpret the language but can serve as cultural advisors to Army forces will be in great demand. Keeping the local population objectively informed not only of current Army operations but also of the intent and desired end state will often be a key task for commanders. A poor communications effort and a lack of understanding of intent may alienate a commander and cause significant problems in future negotiations. It is imperative that communications be clear and effective and that all Army leaders are aware of its implications. See FM 4-05.401 and FM 3-07 for the effective use of interpreters and the conduct of productive meetings and negotiations.

Interests and Actions

> *Me and Somalia against the world, me and my clan against Somalia, me and my family against the clan, me and my brother against my family, me against my brother.*
>
> Somali Proverb

2-54. Identifying and analyzing groups also helps commanders focus on specific segments of the urban society to determine their beliefs, needs, and agendas. It also helps commanders determine how those interests motivate groups to future action (or inaction)—previous patterns of activity are critical in this regard. This analysis seeks to determine why groups (and their leaders) act as they do. Commanders should consider political, economic, cultural, and religious factors in this analysis. These factors affect all groups to some extent and often provide the basis for their beliefs, needs (actual or perceived), and subsequent behavior. Size and location considerations also apply to each group to help determine to what extent its beliefs or ideologies, needs, and actions may impact the urban operation. However, size and proximity may not accurately indicate actual or potential capabilities. Individuals, small groups, and groups located some distance from the actual conduct of the urban operation may be able to influence large portions of the population. These individuals or groups may have a capability disproportionate to their size and proximity—especially against objectives that are not terrain oriented (as in the case of many stability operations).

Interaction, Influence, or Control

2-55. Commanders must cultivate an understanding of a group's—

- Size, location (and proximity to operations), and composition (to include leadership and organization).
- Interests.
- Capabilities.
- Potential actions (intent) and their effects—if any—on operations.

2-56. Then they can develop or modify courses of action as appropriate. Certain courses of action may be needed to improve the interaction between Army forces and civilians (and between other agencies) to accomplish common goals. Others may be needed to influence favorable support, stabilize neutral groups, or neutralize hostile groups. Still others may require more forceful means to control and protect—never to punish—civilians. Courses of action may include—

- Establishing buffer zones and restricted areas.
- Setting up checkpoints and roadblocks with other travel restrictions on people and goods.
- Screening civilians.
- Conducting negotiations (directly or as a mediator).
- Providing or protecting rations, water, and other critical resources.
- Restoring or improving specific, key infrastructure.
- Enforcing curfews.
- Inspecting facilities.
- Directing amnesty programs.
- Conducting internment and resettlement operations.
- Maintaining a "stay-put" policy.

2-57. Commanders must remember that many measures will require significant resources that may initially be beyond the capabilities of the Army force to impose and enforce. (Where possible, commanders should attempt to use host-nation security forces and local urban law enforcement to control activities.) Other elements of the environment, terrain and infrastructure, may fragment efforts and make it difficult to impose controls throughout the area. A careful assessment and understanding of the urban society's interests (beliefs, needs, and agendas) is essential before implementing any populace and resources control measures. Otherwise inappropriate controls, particularly if they are improperly perceived as punishment, may only aggravate the situation. Finally, an appropriate course of action may require no specific action towards the urban society. In most cases, training and discipline, grounded in cultural understanding and sensitivity, will help mitigate many potential adverse effects resulting from military-civilian interaction. (Commanders should review FM 3-19.15 and FM 3-19.40 for additional civilian control measures and considerations.)

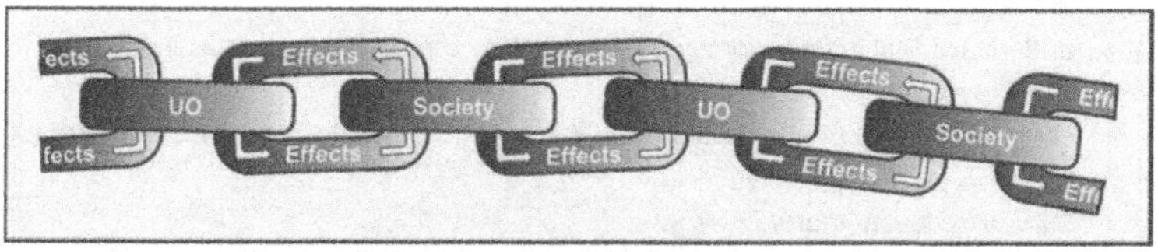

Figure 2-10. UO society chain of effects

A CHAIN OF EFFECTS

2-58. Since urban society is so dynamic and the relationship between various elements so complex, commanders must continually assess how their operations will affect the society and vice versa (see figure 2-10). Specifically, they assess how effectively their measures improve interaction with, influence of, and control over civilians. There is always a difference between intended and actual effects of a specific course of action. Nowhere is this more prominent than dealing with the urban society. This chain of effects frustrates understanding during UO. Therefore, commanders must continuously monitor these effects to make decisions and modifications while planning, preparing, executing, sustaining, and transitioning UO. (Commanders may conduct informal and formal polling activities to assess and understand changing civilian perceptions.) Initially certain aspects of the society, such as religion, may not affect the operation. However, if the threat successfully shapes the perceptions of the urban populace that Army forces are biased against their religion or religious beliefs (or at least critical segments are affected by propaganda), this element may become extremely important. In this instance, the urban commander may need to adjust his IO (to include PSYOP), public affairs (PA) activities, and CMO to counter this propaganda while diverting other combat power to control the populace. Overall, commanders must understand and account for second, third, and higher order effects of their actions and decisions.

Understanding the Effects of Unit and Soldier Actions in Iraq

During OPERATION IRAQI FREEDOM, an armor task force commander described his methodology as that of plotting and measuring everything. After a positive or negative event, he would have his staff evaluate all actions they had conducted before, during, and after the event. This would allow him to correlate activities with outcomes and develop tactics, techniques, and procedures for future success. While not all actions easily correlated to events, this methodological approach helped establish a base line of comparative success for his task force.

Some units assessed the Iraqi culture and developed practical guidelines or rules of interaction for their soldiers to follow as they conducted their missions. These guidelines encompassed a list of actions that their Soldiers should and should not do when dealing with the Iraqi populace. The guiding principle behind these actions was not only to "win the hearts and minds" of the Iraqi people but, more accurately, to establish legitimacy and credibility through appropriate Soldier actions that reflected a combined understanding of the mission and the Iraqi culture.

DO:

- Separate the men from the women and children during cordon and searches.
- Talk to the oldest or most senior male when civilian cooperation is needed.
- Bring something (for example, water, food, or replacement locks) to give residents to compensate for damages when the search ends up being a "dry hole."
- Learn Arabic phrases; say hello and greet people on the street.
- Use female Soldiers to search and talk to Iraqi females.
- Cordon and knock whenever possible instead of kicking in doors needlessly.
- Understand that in Iraq, women do not have the same social status as men.

DO NOT:

- Force a detainee's head to the ground or use your feet to hold someone to the ground.
- Stare at or touch Iraqi women.
- Show obvious disrespect to suspects in front of family members.
- Manhandle the Koran or other religious objects.
- Commit to anything; instead of agreeing to (or denying) an Iraqi's request, it is good practice to say "In'sha Allah" (If God wills it).

URBAN INFRASTRUCTURE

2-59. Urban infrastructures are those systems that support urban inhabitants and their economy. They form the essential link between the physical terrain and the urban society. During urban stability and civil support operations, restoration or repair of urban infrastructure will often be decisive to mission accomplishment. During urban combat operations, destroying, controlling, or protecting vital parts of the infrastructure may be a necessary shaping operation that can isolate a threat from potential sources of support. A threat force operating in an urban area may rely on the area's water, electricity, and sources of bulk fuel to support his forces. This is true particularly when his bases or facilities are physically located in or near the area. Isolating this threat from these sources may require him to generate his own electricity and transport his own water and fuel from outside the urban area. To transport supplies, the threat may rely on roads, airfields, sea or river lanes, and rail lines. Controlling these critical transportation nodes may prevent the threat from resupplying his forces. The control of key radio, television, and newspaper facilities may isolate him from the urban populace (another potential source of support).

INTERDEPENDENCE

2-60. Commanders must understand that destroying or disrupting any portion of the urban infrastructure can have a cascading effect (either intentional or unintentional) on the other elements of the infrastructure. Yet, they may be able to gain an operational advantage while minimizing unwanted and unintended effects. Commanders can control, seize, or secure an essential facility or structure by using precision munitions, electronic disruption of communications, or SOF and conventional ground forces. To gain this advantage, commanders will rely more on the expertise of Army engineer and civil affairs units; local urban engineers, planners, and public works employees; and others with infrastructure-specific expertise. After understanding the technical aspects of the area's systems and subsystems, commanders can then develop the best course of action.

A SYSTEM OF SYSTEMS

2-61. Hundreds of systems may exist. Each system has a critical role in the smooth functioning of the urban area. Simple or complex, all systems fit into six broad categories (see figure 2-11). Commanders should analyze key facilities in each category and determine their role and importance throughout all phases of the urban operation. As there is much overlap between infrastructure systems, this analysis considers each system individually and in relation to others to determine an appropriate course of action toward it.

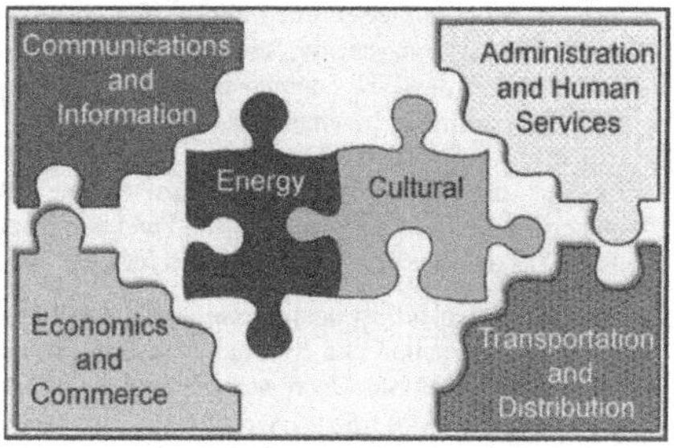

Figure 2-11. Urban infrastructure

A COMBINATION OF STRUCTURES AND PEOPLE

2-62. As depicted earlier in figure 2-1, each element of the infrastructure consists of both a physical (terrain) and human component. For example, the physical component of the electrical segment of the energy infrastructure consists of power stations, substations, a distribution network of lines and wires, and necessary vehicles and repair supplies and equipment. The human component of this same segment consists of the supervisors, engineers, linemen, electricians, and others who operate the system. Commanders must understand and recognize the physical and human components in their assessments.

POTENTIAL IMPACT ON FUTURE OPERATIONS

2-63. Destroying or incapacitating of any of these elements may impact future operations and inhabitants of the urban area. Destroying urban infrastructure during initial phases of an operation may require commanders to assume responsibility for repair, maintenance and clean up, and operation of those same facilities later. Although exceptions will exist, commanders cannot destroy or significantly damage the infrastructure of a foreign urban center during operations and expect the population to remain friendly to U.S. or allied forces. On the other hand, early repair or restoration of critical or essential infrastructure may improve civil-military relations, speed transition back to competent civilian authorities, and, overall, aid in successful mission accomplishment. Still, support from the urban society (albeit of increased importance in UO) is only one factor that commanders weigh while developing appropriate courses of action.

Resource Intensive

2-64. Requirements to protect, restore, or maintain critical infrastructure may divert substantial amounts of resources and manpower needed elsewhere and place additional constraints on subordinate commanders. Civilian infrastructure is often more difficult to secure and defend than military infrastructure. The potentially large and sprawling nature of many systems (such as water, power, transportation, communications, and government), make their protection a challenge. Yet, the infrastructure of an urban area may provide commanders with essential logistics and support. Therefore, the initial expenditure of time and other resources may be necessary to support concurrent or future operations. Legal considerations, however, may affect using the infrastructure and acquiring the urban area's goods and services. Commanders, their staffs, and subordinates (often down to the individual soldier) must know their limits concerning Army authority to commandeer civilian supplies or equipment to facilitate mission accomplishment (see the legal support discussion in Chapter 9). In stability and civil support operations, the safeguard or restoration of critical urban infrastructure for military or civilian use may be a decisive point in the overall operation.

2-65. Keys to understanding the magnitude of the resources and manpower required to restore the infrastructure are an initial infrastructure assessment and, as soon as practical afterward, a detailed infrastructure survey. An initial assessment provides the commander immediate feedback concerning the status of basic services needed to meet the urgent needs of the urban population. The systems assessed are based on mission, enemy, terrain and weather, troops and support available, time available, civil considerations, and the commander's vision of the overall end state. The infrastructure assessment, while typically performed by engineers, may be accomplished by, or in conjunction with, others with sufficient expertise to provide the type and quality of information required. These others may include civil affairs, medical, and chemical personnel. Those tasked with this assessment should routinely consult other Army and coalition forces and governmental and nongovernmental agencies currently operating in the urban area as well as the urban civilian leadership for their informed input.

2-66. While an infrastructure assessment functions to support the resolution of immediate challenges to urban reconstruction and restoration, it also provides the initial basis for determining the conditions for successful transition. However, commanders and planners must continually expand and refine their understanding. As a necessary follow-on, commanders initiate a detailed infrastructure survey. This survey is normally conducted by U.S. Army Corps of Engineers personnel assigned to forward engineer support teams. As with the assessment, the commander should incorporate other technical specialty personnel in the survey team to enhance the quality and accuracy of the product (see FM 3-34.250).

Communications and Information

2-67. This system is comprised of the facilities and the formal and informal means to transmit information and data from place to place. Understanding communication and information infrastructure of an urban area is important because it ultimately controls the flow of information to the population and the enemy. It includes—

- Telecommunications, such as telephone (to include wireless), telegraph, radio, television, and computer systems.
- Police, fire, and rescue communications systems.
- Public address, loudspeaker, and emergency alert systems.
- The postal system.
- Newspapers, magazines, billboards and posters, banners, graffiti, and other forms of print media.
- The informal human interaction that conveys information such as messengers, open-air speeches and protests, and everyday conversations.
- Other inventive informal means such as burning tires and honking horns.

2-68. Perhaps more than any other element of the infrastructure, communications and information link all the other elements in an interdependent "system of systems." It is a critical enabler that helps coordinate, organize, and manage urban activities and influence and control the urban society. Army commanders are acutely aware of the impact that a loss or degradation in communications has on their own operations. The urban environment experiences similar impacts to communication failures; however, urban governments and administrations are generally less prepared to deal with a collapsed communications and information infrastructure than are trained Army forces.

2-69. Militarily, a functioning urban communications and information system can serve as an alternate for both friendly and threat forces and can be easily secured with civilian, off-the-shelf technologies. Threats may make use of commercial systems intertwined with legitimate civilian users, making it unpalatable to prevent use of these assets. Forces can also use these systems to influence public opinion, gain intelligence information, support deception efforts, or otherwise support IO.

Increasing Impact of Computers

2-70. In many urban areas, computers link other elements of the urban infrastructure. They link functions and systems in the urban area and connect the area to other parts of the world. This latter aspect creates important implications for commanders of a major operation. Operations involving this cybernetic function may produce undesirable effects on a greater scale than initially intended. For example, commanders may be able to close or obstruct an urban area's banking system; however, this system may impact the international monetary exchange with unwanted or even unknown effects. The authority to conduct these types of IO will often be retained at the strategic level.

Pervasive Media

2-71. The media is central to the communications and information infrastructure and a critical operational concern. Compared to other operational environments (jungles, deserts, mountains, and cold weather areas), the media has more access to urban operations. This is due largely to airports, sea and river ports, and major road networks; ready access to power sources and telecommunications facilities; as well as access to existing local media structures. Hence, media presence may be pervasive and IO even more critical to success in UO than operations in many other environments.

A Complex Relationship

2-72. A complex relationship exists among information, the public, and policy formulation. Although the degree and manner in which public opinion shapes government policy are difficult to accurately determine, negative visual images of military operations presented by the media can change political objectives and, subsequently, military objectives. As important, media reporting can influence civilian activity in an urban AO to either the advantage or disadvantage of the commander.

Induce Cooperation Through Credibility

2-73. Commanders do not control the media; however, they monitor the flow of information that the news media receives and subsequently reports. Consequently, commanders should plan and execute PA operations that will induce cooperation between the media and Army forces. Successful relations between urban Army forces and the news media evolve from regular interaction based on credibility and trust. More information is usually better than less, except when the release of such information may jeopardize security and the success of the operations and threaten the safety of Soldiers. However, commanders cannot simply withhold information to protect the command from embarrassment. They consider media interests as part of the normal planning process and work to ensure that information presented to the news media is accurate, timely, and consistent with operations security. Since the media will likely arrive in the urban area before the conduct of operations, early deployment of PA assets may be critical. Commanders should synchronize PA activities with CMO and PSYOP. Such action eliminates duplicated effort and ensures a unity of purpose consistent with the IO concept of support (see Chapter 4).

2-74. Failure to provide sufficient information can hamper a commander's ability to conduct the mission. Commanders cannot refuse to deal with particular news media because they consistently report a negative image of Army forces and operations. Poor relationships with any media can result in inaccurate and even biased reporting. Such reporting can cause a public reaction that influences the ability to achieve operational objectives. During the Russian 1994-95 battle against Chechen separatists in Grozny, for example, the Russian military refused to communicate with reporters. The media reported primarily from the perspective of the Chechen rebels. This encouraged both local and international support for the rebels. It also allowed the Chechens, who lacked sophisticated information systems, to use the media to broadcast operational guidance to their forces. (During their second Chechnya campaign of 1999-2000, Russia learned this lesson well and the Russian view of the war dominated domestic public opinion.) On the other hand, successfully engaging the media can serve as a force multiplier. The Army's open and responsive interaction with the media during peacekeeping operations in Bosnian urban areas helped to explain the challenges and successes of Army forces in the Balkans to the public. This helped maintain domestic, international, and local political support for NATO operations and, with a successful command information program, helped maintain Soldiers' morale.

TRANSPORTATION AND DISTRIBUTION

2-75. This element of the infrastructure consists of—

- Networked highways and railways to include bridges, subways and tunnels, underpasses and overpasses, ferries, and fords.
- Ports, harbors, and inland waterways.
- Airports, seaplane stations, and heliports.
- Mass transit.
- Cableways and tramways.
- Transport companies and delivery services that facilitate the movement of supplies, equipment, and people.

Similar to communications and information, this facet provides the physical link to all other elements of the infrastructure.

2-76. Army forces deploying into a theater of operations depend on ports and airfields; seizure and protection of these critical transportation nodes may impact the projection of combat power. Once in theater, transportation and distribution systems in the urban area can contribute greatly to the movement of forces, maneuver, and logistic operations throughout the entire AO. Control of decisive points in this infrastructure may be important to the military operation and to the normal functioning of the urban area (and surrounding rural areas). Supplies traveling through the transportation and distribution system may be military-specific supplies (such as ammunition and repair parts) and supplies for both the military and urban population (such as food, medicine, oil, and gas). The system may also support the movement of military forces and the urban area's population (for which it was designed). Therefore, commanders of a major operation may have to develop innovative methods that limit the transit of threat supplies and reinforcements while facilitating the movement of their own resources and those of civilians. This last consideration attempts to minimize hardship and promote normalcy in the urban area and will increase in significance as the need for legitimacy increases.

2-77. Most urban areas (particularly in developing countries) have two forms of transportation and distribution systems that exist simultaneously: a formal system and an informal or paratransit system. Large organizations, bureaucracy, imported technology, scheduled services, and fixed fares or rates characterize formal systems. Low barriers to entry; family and individual entrepreneur organizations; adapted technology; flexible routes, destinations, and times of service; and negotiated prices characterize the informal system. The informal system is more decentralized and covers a much greater portion of the urban area than the formal system. The informal transportation and distribution system often includes a waterborne element, is more likely to function through turbulence and conflict, and can extend hundreds of kilometers beyond the urban area. Accordingly, commanders should understand both systems to establish effective movement control.

ENERGY

2-78. The energy system provides the power to run the urban area. It consists of the industries and facilities that produce, store, and distribute electricity, coal, oil, wood, and natural gas. This area also encompasses alternate energy sources, such as nuclear, solar, hydroelectric, and geothermal power. Energy is needed for industrial production and is therefore vital to economics and commerce. Among many other things, this system also provides the fuels to heat, cool, and light homes and hospitals, cook and preserve food, power communications, and run the transportation necessary to move people and their supplies throughout the urban area. Loss of an important energy source such as electricity or gasoline, especially for those accustomed to having it, will become an immense area of discontent that the commander of a major urban operation will need to quickly address. Therefore some threats, particularly terrorists and insurgents, may actively target this element of the urban area's infrastructure to erode support for civilian authorities and Army forces.

2-79. Sources of energy may be tens or hundreds of miles away from the urban area itself. Therefore, commanders may exert control without applying combat power directly to the urban area itself by controlling or destroying the source (power generation or refinement plant) or the method of distribution (pipelines or power lines). With electrical energy that cannot be stored in any sizable amount, the latter may be the best means as most major urban areas receive this energy from more than one source in a network of power grids. However, control may be as simple as securing a power station or plant and turning off switches or removing a vital component that could later be restored. On the other hand, lengthy pipelines and power lines may compound security and protection of this element of the infrastructure.

2-80. The number of nations that have invested in nuclear power and nuclear research is increasing. With this increase, the potential for Army forces to operate in urban areas that include (or are near) these facilities also increases. Damage to one of these facilities and potential radiation hazards will present special challenges to commanders of a major operation. To safeguard friendly forces and civilians, commanders will need to employ a blend of peacetime and tactical nuclear contamination avoidance principles (see FM 3-11.14).

ECONOMICS AND COMMERCE

2-81. This system encompasses—

- Business and financial centers to include stores, shops, restaurants, hotels, marketplaces, banks, trading centers, and business offices.
- Recreational facilities such as amusement parks, golf courses, and stadiums.
- Outlying industrial, mineral, and agricultural features to include strip malls, farms, food processing and storage centers, manufacturing plants, mines, and mills.

2-82. An essential aspect of this area during operations may be the political sensitivity of U.S. or allied industries investing and operating in a foreign country, particularly during stability operations. An enemy or a disgruntled civilian population may attack or disrupt commercial activities as a political statement against the United States or our allies. Food production facilities also may assist commanders in Army food services and may be an essential concern during relief operations. During long-term stability operations, visible, material, and tangible economic progress consisting of the creation or restoration (and protection) of businesses, agriculture, and overall jobs will often be critical to—

- Generating or maintaining the urban population's support to Army forces and operations.
- Reducing support to threat forces and operations to include eliminating civilians as a potential manpower pool for insurgent or terrorist organizations and activities.
- Lowering other hostile civilian activities such as protests and riots.
- Transitioning the urban area back to legitimate civilian responsibility and control.

2-83. This element of the infrastructure also consists of the production and storage of toxic industrial chemicals used in agriculture (insecticides, herbicides, and fertilizers), manufacturing, cleaning, and research (to include biological agents). Fertilizer plants may be of specific concern as they contribute to providing a key material in terrorist and insurgent bomb-making activities. A thorough analysis of this element of the infrastructure may also be essential to understanding how urban insurgencies are funded and supported. This helps commanders to understand the true organization of the insurgency as well as to suggest methods to isolate insurgents from their economic or financial support. In their overall assessment of this area of the infrastructure, commanders should also consider the activities and influence of criminal organizations or elements.

Administration and Human Services

2-84. This wide-ranging system covers urban administrative organizations and service functions concerned with an urban area's public governance, health, safety, and welfare. Together, it encompasses—

- Governmental services that include embassies and diplomatic organizations.
- Activities that manage vital records, such as birth certificates and deeds.
- The judicial system.
- Hospitals and other medical services and facilities.
- Public housing and shelter.
- Water supply systems.
- Waste and hazardous material storage and processing facilities.
- Emergency and first-responder services such as police, fire, and rescue.
- Prisons.
- Welfare and social service systems.

Cultural

2-85. This system encompasses many organizations and structures that provide the urban populace with its social identity and reflect its culture. (This infrastructure system overlaps with many recreational facilities included under the economics and commerce infrastructure. For example, an urban society may radically follow soccer matches and teams. Hence, soccer stadiums relate to the society's cultural infrastructure.) Some of these facilities, particularly religious structures, will be protected targets and others may require security and law enforcement protection from looting and pilferage. However, commanders will need to quickly educate, inform, and continually remind the urban populace (and the media) that cultural infrastructure may lose its protected status when used by threats for military purposes. Cultural infrastructure may include—

- Religious organizations, places of worship, and shrines.
- Schools and universities.
- Museums and archeological sites.
- Historic monuments.
- Libraries.
- Theaters.

RESTORING AND PROTECTING ESSENTIAL SERVICES

2-86. Losing the support of essential elements of the infrastructure will have an immediate, destabilizing, and life-threatening impact on the inhabitants of the urban area. In stability and civil support operations, numerous parts of the administrative and human services and energy infrastructure often rise to critical importance before all other elements. Again, however, complete restoration of these essential services is often a lengthy, resource-intensive civil-military operation. (Following the end of major combat operations of OPERATION IRAQI FREEDOM, most units developed one or more of their logical lines of operations oriented along the restoration or improvement of urban infrastructure and essential services. An acronym used by many units to focus and track critical activities within this line of operation was sewer, water, electricity, and trash (SWET). Later, other units modified this acronym to SWEAT and then SWEAT-MS to include concerns for restoring academics [or schools], revitalizing medical facilities, and establishing security [police and host-nation security forces].) Of critical importance will be a simultaneous IO campaign that includes efforts to help ensure that the urban population develops realistic expectations about Army abilities to restore their essential urban infrastructure.

Understanding the Urban Environment: Paris – 1944

The summer of 1944 confronted German General Dietrich von Choltitz with a dilemma. As military commander of greater Paris, he was to eliminate French Resistance internal to the city while defending against approaching Allied units, missions for which he had insufficient forces. (Notably, General Eisenhower, the commander of the Allied forces, wanted to bypass Paris to sustain the offensive. Seizing the French capital would task his forces with the support of tens of thousands of civilians. Eisenhower was nevertheless ordered to capture the city. A political decision resulted in civilian assistance taking precedence over combat operations.)

Choltitz's situation was further complicated by Hitler's demand that he destroy the city, an action the general saw as needlessly destructive (and infeasible given his scant resources). Choltitz's seniors directed the preparation, and later the destruction, of Paris's 45 Seine River bridges. They were the only remaining crossing points over that waterway given Allied bombing of others outside the French capital. Premature destruction would trap German forces defending to their north, a second-order effect that Choltitz used to justify his disobedience of orders demanding the bridges' demolition.

The German general also recognized that some mission-critical elements were part of Paris's social rather than physical infrastructure: the leadership of the various resistance groups and the relationships between them. Choltitz understood that he lacked resources to defeat the many separate factions; he therefore chose the unorthodox (asymmetric) approach of accepting an intermediary's offer of a truce with these groups. Such an agreement provided some measure of the stability needed while Choltitz awaited promised reinforcements. Further, he realized that the resistance factions were by no means united in their goals. Communist elements sought a much different end than those looking toward a de Gaulle-led postwar government. A truce thus set the French Communists (who sought an uprising so as to legitimize their claims to power) against others trying to buy time until Allied forces arrived, forces that included Free French units supportive of de Gaulle.

Although his defense of the capital failed, Choltitz succeeded in harboring his available resources, reducing the effectiveness of the resistance organizations fighting his soldiers, and maintaining withdrawal routes for units north of the Seine. The German commander's analysis in support of these efforts was effective in part because of his insightful (1) identification of critical points that included elements of terrain, citizenry, and infrastructure; (2) understanding of the relationships between these parts; and (3) use of an asymmetric approach to address his lack of sufficient force to otherwise handle the urban densities that challenged him.

This page intentionally left blank.

Chapter 3
Understanding the Urban Threat

... [T]he United States could be forced to intervene in unexpected crises against opponents with a wide range of capabilities. Moreover, these interventions may take place in distant regions where urban environments, other complex terrain, and varied climatic conditions present major operational challenges.

Quadrennial Defense Review Report, 30 September 2001

As the strategic environment has become less stable, more uncertain, and more dangerous, Army forces must be trained and ready to address persistent and evolving urban threats. These threats range from regional conventional military forces, paramilitary forces, guerrillas, and insurgents to terrorists, criminal groups, and angry crowds. These threats can hide in plain sight and become indistinguishable from the noncombatant urban population that may help shield, protect, and sustain them. Although uncertain about events, Army forces can be clear about trends. Increasingly, the Army will face threats that severely differ in doctrine, organization, and equipment, are skilled at developing and adapting techniques to counter Army tactics, techniques, and procedures (TTP), and can fully interact with the three other components of the urban battlefield—terrain, society, and infrastructure. In urban operations, commanders must broaden their concept of the threat to include natural disasters, hunger and starvation, and rampant disease. Further, commanders must plan to contend with many passive urban threats, such as psychological illnesses and toxic industrial materials (TIM). These threats may be found in isolation, but most likely commanders will encounter them in various combinations. Moreover, each new threat will pose a different combination and likely have new capabilities that previous opponents lacked.

ASYMMETRICAL AND ADAPTIVE

3-1. An emphasis on asymmetry to offset U.S. military capabilities has emerged as a significant trend among potential threats and become an integral part of threat objectives and tactics (discussed below). Though, this is not to imply that all future urban threats will fight and operate asymmetrically. Some may be asymmetrical; others may not. Future threats may be a symmetric conventional military; others may be paramilitary, insurgent, or even a nonmilitary threat such as a criminal organization. Asymmetry results when one opponent has dissimilar capabilities—values, organization, training, or equipment—that the other cannot immediately counter. It is not a new concept. It naturally evolves from a sound mission, enemy, terrain and weather, troops and support available, time available, civil considerations (METT TC) analysis by an intelligent, freethinking, and adaptive threat. These asymmetric approaches will include the most advanced, commercially-available technology innovatively applied and mixed with crude, simple, and unsophisticated weapons and TTP. The success of these approaches often depends upon the predictability of the forces on which they are employed. Thus, Army commanders must consider whether their operations and activities lend themselves to pattern analysis by threat forces and, if so, they may need to vary their battle rhythm and take other steps necessary to become less predictable.

WEAPONS OF MASS DESTRUCTION

3-2. A chief asymmetric means of engaging the national power of the United States is to employ weapons of mass destruction (WMD) against the United States or its allies. These weapons can be used against military forces and include high-yield explosives as well as nuclear, biological, and chemical weapons. Operations in urban areas may require concentrating forces and may create a lucrative target for a threat that possesses fewer numbers and less equipment.

3-3. A threat's WMD use will adversely affect the Army's abilities to conduct urban operations (UO) to various degrees. For example, the intervening structures and the effects of urban microclimates complicate the ability to detect and identify radiological, chemical, or biological attacks from a standoff distance. Also, the individual Soldier's ability to recognize his leaders, understand oral and visual commands, and operate increasingly sophisticated equipment is difficult when wearing protective clothing and equipment—particularly if his training proficiency is low. Despite the increased challenges and complexity, Army forces have the best (or better) training and equipment necessary to respond to such an attack compared with most armies around the world or with any civilian organization.

3-4. Although initial casualties could be high, the public can accept military casualties before those of civilians. Therefore, threats may gain an initial tactical advantage but would achieve less asymmetric benefit by directly attacking Army forces. Instead, they may attempt to achieve a strategic advantage by employing WMD against U.S. or allied civilian populations. In doing so, threats hope to use political sensitivity to high civilian casualties to reduce popular support for the United States or its allies. The chance of these attacks occurring in an urban area increases because—

- The area facilitates weapon effects and camouflages delivery means.
- The dense civilian population ensures a high casualty rate.
- The attack (or even the threat of attack) often will receive more publicity and public attention.
- The urban infrastructure is especially vulnerable to WMD, (particularly the systems of the economics and commerce infrastructure located in large urban areas) and may have far-reaching national and global effects.

THREAT OBJECTIVES

3-5. The threat will seek to achieve several key objectives when opposing Army forces operating in an urban environment (see figure 3-1). These objectives focus more on how a threat might fight in an urban area rather than specifically who the threat might be or in what region of the world the conflict might occur. These objectives are more easily achieved in an urban environment due to—

- The high costs in time, material, and manpower involved in UO.
- The limiting effects of urban areas on many technological advantages.
- The proximity of airfields and ports to urban areas.
- The potential moral dilemmas created by exposing numerous civilians to harm or injury.
- The ability of the media to observe and report the threat's version of events.

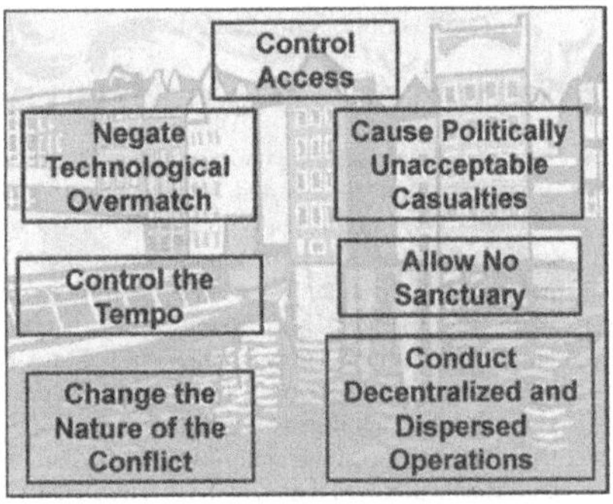

Figure 3-1. Threat objectives

3-6. These objectives complement and overlap each other; however, at their core is the threat need to defeat an enemy of superior numbers, technology, or both.

CONTROL ACCESS

> *... Iraq made no direct effort to impede the buildup in the theater. Planners should consider what might have happened if Iraq had attempted a strategy based on denying access to the region. Planners might also wonder what the outcome would have been if Iraq had attacked US forces in Kuwait before they were ready for the running start. The point is that the conditions in [US Central Command] in 2003 are unlikely to be replicated elsewhere.*
>
> *On Point: The United States Army in Operation IRAQI FREEDOM*

3-7. The Army may not be located where future conflicts are fought. Thus, the Army maintains the ability to rapidly project and sustain combat power over long distances and time spans. This capability demands that Army forces quickly gain and maintain control of seaports or aerial ports of embarkation or debarkation, particularly where the density of U.S. basing and en route infrastructure is low. Commanders gain control of these ports by unopposed (assisted or unassisted) or forcible entry operations. In either case, these phased-entry operations may present potential vulnerabilities, particularly—

- Unsuitable composition of initial or early entry forces lacking necessary combat power for immediate decisive operations.
- Initial command and control difficulties and an immature situational understanding.
- Weak initial port or base defenses without the necessary integration of naval and air component forces.
- Lack of developed logistic support.

3-8. Consequently, threats may attack during initial force projection operations to oppose, delay, disrupt, or otherwise manipulate the build-up of essential combat power into a theater of operations. These attacks may occur anywhere deploying Army forces are located, at overseas bases, at home stations, and even in military communities. Increasingly, deployment facilities such as airfields and ports exist as integral components of urban areas. Threats will invariably use the complex and concealing nature of these urban areas, coupled with the vulnerabilities, to create favorable conditions for their attacks.

NEGATE TECHNOLOGICAL OVERMATCH

3-9. Threats will always strive to force engagements at a time and place most advantageous to them. They may locate military forces and vital military capabilities in urban areas to achieve sanctuary from the effects of Army capabilities and make Army forces and systems more vulnerable to less-sophisticated weapons.

3-10. The clutter of the physical structures, electromagnetic radiation, and population diminishes Army capabilities. This clutter makes it difficult for Army forces to acquire and effectively engage targets at long ranges. In urban areas, the terrain often allows a threat to operate in closer proximity to friendly forces. Therefore, the threat may "hug" friendly forces to avoid the effects of high-firepower standoff weapon systems and degrade their ability to gain or maintain a thorough common operational picture. Additionally, this threat tactic attempts to inhibit friendly commanders from employing some weapon systems and munitions for fear of fratricide. Threats will also seek to hide and operate in culturally sensitive areas and sites, such as places of worship and schools, to establish temporary safe havens and refuge from Army forces operating within the urban environment.

CONTROL THE TEMPO

3-11. Threats will try to achieve a decisive advantage by setting and controlling the tempo. To prevent the Army's entry into theater, threats may try to create a high operational tempo to take advantage of the inherent weaknesses in power projection operations outlined earlier. As other efforts deny entry, threats may seize the initiative, achieve surprise, and exploit the tempo differential by attacking with heavy conventional forces potentially possessing greater firepower and more rapid ground mobility than the Army's initial-entry forces.

3-12. If they cannot deny entry or end the conflict quickly, threats may use any preparations made in the initial high-tempo period to prolong the event, aiming to degrade U.S. or allied commitment. (Some threats may conduct operationally planning in terms of decades rather than years.) The complex nature of the urban environment slows operations conducted in, around, and over these areas. Threats may maximize this characteristic by fighting (conventionally or unconventionally) for key urban complexes and infrastructure, forcing friendly forces to operate within these areas. If Army operations focus on one or more urban areas, the overall campaign slows. However, even when UO make up only one component of a much larger campaign, they may consume valuable resources needed for other operations and delay the entire campaign. (See also the discussion of Tempo in Chapter 7.)

Tempo

The battle for Aachen, Germany, in the fall of 1944, developed during the U.S. First Army's offensive to breach the Westwall fortifications. Aachen, the ancient capital of Charlemagne, had symbolic political and psychological significance for the Germans and Americans. Furthermore, it was the first city on German soil to face an assault by the Allies. Consequently, the symbolic importance of this first major battle in Germany ensured bitter resistance against American attackers. The Germans surrendered only after the city was destroyed. Expected to take a few days, instead, the battle took weeks. Although the Army had achieved a clear tactical victory, the German defense of Aachen cost the First Army valuable time and resources, and delayed the planned attack to the Rhine River.

Change the Nature of the Conflict

3-13. Threats may attempt to change the fundamental nature of the urban conflict to exploit ambiguous or tenuous political-military objectives. Many nations must gain and maintain domestic popular support to use their armies for political objectives. The threat may attempt to change the nature of the conflict by modifying its strategy and tactics, the environment, or any combination, ultimately hoping to reduce friendly popular support. For example, introducing an urban terrorist threat to U.S. civilians or Soldiers not directly engaged in operations changes the nature of the conflict. This type of threat may not have been an initial consideration, and this change may reverse public support for the operation. Another example, growing U.S. coalition combat power may cause the threat to switch from open maneuver warfare to urban insurgency operations to avoid decisive combat with superior forces and achieve a stalemate. Originally expecting a quick solution or victory, the political leadership may now envision a longer deployment with less chance of lasting success.

Cause Politically Unacceptable Casualties

3-14. Threat forces may gain an advantage against superior friendly forces by capitalizing on a perceived weakness of many Western nations: the inability to endure continuous losses or casualties for other than vital national interests or losses for which they are psychologically unprepared. A secondary U.S. interest may equate to national survival on the part of a threat. Therefore, the threat (particularly with fanatical leadership) may willingly sacrifice excessive amounts of money, equipment, and people (soldiers and civilians) to achieve victory. Threats may attempt to weaken U.S. resolve and national will to sustain the deployment or conflict by inflicting highly visible, embarrassing, and if possible, large losses on Army forces, even at the cost of similar losses to themselves. Many threat forces will use UO to inflict mass casualties and destroy premier Army weapon and information systems. The physical characteristics of the urban environment support these ambush techniques. Light infantry or insurgents with readily obtainable, hand-held antiarmor weapons can effectively attack armored vehicles and helicopters, no matter how sophisticated, in an urban area.

ALLOW NO SANCTUARY

3-15. Threats will attempt to deny Army forces safe haven anytime and anywhere. Terrorism and insurgency may be some of the tactics used to deny sanctuary to Army forces. They will attack Army forces anywhere, particularly while operating in urban areas where the fear from being attacked from any quarter is often greater. Threats may be or employ state-sponsored or independent terrorists, well equipped and motivated to accomplish their assigned missions.

3-16. Military buildings, facilities, and installations in urban areas are particularly vulnerable to high-yield explosive munitions as well as other clever means to create large explosions. The close-in nature of urban areas, large populations, and high volume of vehicle traffic provide a good environment for target reconnaissance, explosives positioning (conventional or high-yield), and cover for an attack. These attacks will likely be preceded by extensive, careful reconnaissance, necessitating a solid friendly counterterrorism, counterinsurgency, and counterintelligence effort.

CONDUCT DECENTRALIZED AND DISPERSED OPERATIONS

3-17. To a certain extent, decentralized and dispersed operations are an integral part of all threat objectives. However, this concept warrants separate emphasis as an objective since threat forces will likely place great significance on it on future urban battlefields. Both dispersed and decentralized approaches seek to reduce threat vulnerabilities to air power and precision-guided munitions (PGM) while increasing their agility, flexibility, and overall maneuverability in an urban environment.

3-18. Urban terrain tends to fragment and separate forces that operate in it. Threat forces recognize this characteristic, accept it, and make it work to their advantage. They conduct operations from dispersed urban locations to reduce their vulnerability to friendly decisive operations and massed firepower. Although separated, threat forces will attempt to retain the ability to assemble and mass quickly so to strike as opportunities present themselves. Once threat forces complete the operation, they will return to separate locations to avoid potential counterattack. The fluidity and seemingly disjointed appearance of these UO threats will challenge friendly efforts to conduct templating and pattern analysis. Ambushes (air and ground) will be used to deny friendly ground and air reconnaissance of their dispersed locations.

3-19. Dispersed operations normally depend on good command and control to achieve synchronization and massed effects. Threat forces also understand the debilitating effects of the urban terrain on communications and the execution of operations. When they cannot mass their forces or effects, they will depend on decentralized operations to achieve their objectives. They will operate autonomously, guided only by a higher authority's purpose and intent. These operations make them even less vulnerable to massed attacks and PGM as smaller threat forces do not present an objective or target that will allow friendly decisive operations. Again, pattern analysis and templating, although essential, will be extremely difficult. This objective may prolongs the conflict but is central to achieving the other threat objectives.

3-20. When the threat employs decentralized operations, is effectively isolated, or both, Army and coalition forces are much more likely to be able to share lessons learned and emergent TTP than a dispersed threat. Threat isolation, either imposed by Army forces or as a result of the threat's desire to remain concealed from Army intelligence, surveillance, and reconnaissance (ISR) capabilities, reduces the threat's ability to freely and easily exchange information and quickly adapt. As learning organizations, this allows Army forces to develop dominance over a threat's weapons and tactics for longer, albeit temporary, periods of time. This ability to rapidly train and learn is an advantage the Army and coalition forces must maintain while denying the same to future threats.

URBAN THREAT TACTICS

3-21. Urban areas provide a potentially casualty-producing and stress-inducing environment ideally suited for using specific urban tactics. Moreover, urban areas can provide threats with an unmatched degree of cover and concealment from friendly ISR and firepower systems. Understanding how these tactics will be used often requires a concurrent understanding of the threat's culture, customs, language, and philosophical ideals. While active urban threats may vary widely, many techniques will be common to all.

Figure 3-2 outlines a set of tactics available to potential threats opposing mission accomplishment in urban areas. Army forces may use many of these tactics, except those that violate the law, ethics, and morals, to defeat urban threats. Moreover, using asymmetric means is not the sole domain of the threat. Army commanders can also leverage capabilities, create conditions, and plan operations to develop their own asymmetric advantages to accomplish the mission.

USE THE POPULATION TO ADVANTAGE

3-22. Many urban areas may be too large to evacuate completely (if at all). Even if desirable, a military force may have no place to safeguard and secure the inhabitants. Therefore, future UO may see large segments of the populace remain. Offensive and defensive operations may be constrained not only by the terrain, but also by the presence of many civilians. Army forces involved in urban stability and civil support operations will certainly conduct missions in and amongst the residents. These residents may restrict operations and, when gathered in large numbers, may (even without initial hostile intent) present a critical force protection issue for the commander.

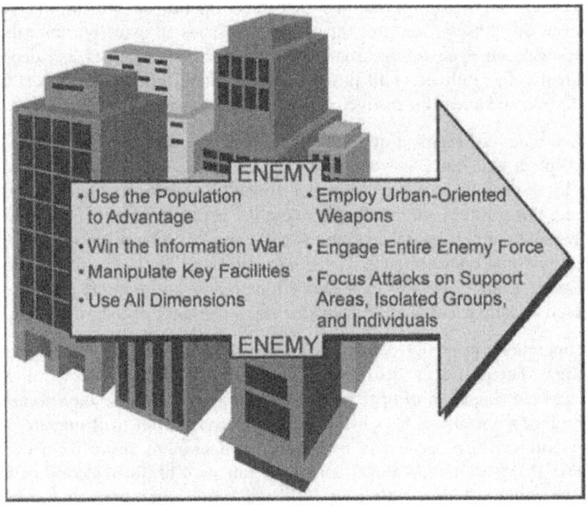

Figure 3-2. Urban threat tactics

Use as Key Terrain and Concealment

3-23. From the threat standpoint, the populace is similar to key terrain: the side that manages it best has an advantage. Threat forces may gain this advantage by using civilians as camouflage, concealment, and a means of deception. Guerrilla and terrorist elements may look no different from any other member of the community. Many foreign conventional and paramilitary troops—in addition to terrorists and insurgents—often have a "civilian" look. Western military forces originally adopted the clean-shaven and close-cut hair standards to combat disease and infection, but future opponents may not adhere to those standards. Instead, they may adopt grooming standards, civilian-looking clothing, and other "nonmilitary" characteristics to make themselves indistinguishable from the civilians. To avoid defeat or capture, threats will often throw down their weapons and attempt to blend into the urban population. Therefore, commanders must adopt effective populace and resource control measures to limit, identify, and track threat movements through the urban area of operation.

Identifying Threats from Noncombatants

During Russia's 1994-95 conflict with Chechnya, Russian forces had difficulty identifying Chechen guerrilla forces from Grozny's noncombatant population. Because their dress and ethnic appearance was identical to that of the urban populace, Chechen soldiers could freely walk around the city, suddenly disappear, and then abruptly reappear firing their weapons from basements, windows, or dark alleyways. To distinguish fighters from peaceful city dwellers, Russian forces began looking at men's shoulders to see if they were bruised (from firing weapons) and their forearms to see if there was burned hair or flesh (from the extraction of cartridges). They closely examined their clothing and smelled for gunpowder residue. To identify a Chechen artilleryman, Russian soldiers checked for glossy spots left by artillery and mortar rounds on the bends and cuffs of sleeves. They also turned pockets inside out to check for a shiny, silvery-leaden hue indicating the former presence of small arms ammunition. Russian forces also recognized a grenade launcher operator or mortar man from fibers and crumpled pieces of gun cotton on their clothing. U.S. Army commanders may need to develop similar, imaginative techniques and rapidly integrate technological means to identify and track threat forces such as—

- Military working dogs, ion spray, and electronic vapor tracers to locate explosives (or their former presence).
- Biometrics—measurable biological characteristics—such as fingerprints, voiceprints, facial scans, and retina scans.
- Electronic and chemical tagging devices.
- Monitoring hospitals for injuries that indicate participation in combat activities. For example, improper use of rocket-propelled grenade launchers may result in significant burns to the lower part of the body due to the back blast.

Gain Cover, Protection, and Increased Mobility

3-24. Threat forces may attempt to gain cover by using the urban inhabitants as human shields. With this increase in protection, they simultaneously increase their mobility. They recognize the Army's focus on developing and applying rules of engagement (ROE). They will take advantage of the restraining effects of international law and the Army ethical values to enhance their mobility in proximity to friendly positions. Knowing the Army's reluctance to cause noncombatant casualties and collateral damage, threats may operate in areas containing civilians and essential facilities to restrict the Army's use of massed or nonprecision firepower. For example, they may use school buses and ambulances to transport fighters and equipment about the urban battlefield. They may also employ "rent-a-crowds"—civilians paid or incited to demonstrate against military forces—armed only with sticks, stones, and Molotov cocktails (a potential asymmetric challenge). Threats can also hide military equipment within sensitive infrastructure—for example, caching weapons in places of worship, schools, or medical facilities. As Army and coalition forces potentially respond by targeting and firing upon these locations, threats will make use of these reactions for their possible propaganda value.

Make Moral Responsibilities a Weakness

3-25. Depending on their successes, threats may use these tactics and skillful information operations that attack national will and coalition sensitivities in an attempt to force the Army to establish more restrictive ROE. For example, a teenaged fighter coerced or enticed to fire a rifle or rocket-propelled grenade launcher can be quickly stripped of his weapon once dead and displayed at the local hospital as a child victim of coalition forces. Threat forces may also take advantage of the Army's moral responsibilities. By herding refugees into friendly controlled areas, threat forces try to make the civilians a burden on the Army's logistic and security resources. Similarly, some threats may attack to kill NGOs and force the remainder of them out of the area to achieve the same burden effect. At the same time, these threats hope to cause the population to lose confidence in the Army's (and potentially the government's) ability to protect and provide them essential services.

3-26. Threat forces, on the other hand, may not abide by international agreements, such as the Geneva conventions. Threats may not take prisoners unless they can be ransomed or made part of a local prisoner exchange. They may even execute friendly prisoners in front of the media to show their "strength" and, more importantly, to cause friendly forces to overreact and lose their legitimacy. Threat forces can then use such an overreaction to unite others with their cause. Pretending to be civilian noncombatants, they may also feign injury or pretend to need assistance (such as a disabled taxi) with the primary motive of luring Army forces into a deliberate ambush. While the threat may gain a short-term advantage by using ethical responsibilities as a weapon, the Army commander's strict adherence to moral principles will, in the long run, contribute greatly to threat defeat—particularly during long-term urban stability operations.

Acquire Intelligence and Logistic Support

3-27. Indigenous threat forces can normally use the local population for intelligence and logistic support far more effectively than can an alien army. Threat forces may manipulate local hires serving among U.S. Soldiers, such as those contracted by the Army for base operation purposes or translator duties. In addition, refugees moving through friendly controlled sectors may provide the threat with information on friendly dispositions, readiness, and intent. (Women and children may be specifically used for this purpose since U.S. Soldiers' cultural bias can create an incorrect perception that these noncombatants pose no threat to security.) Even friendly residents may become unwitting or unwilling informants, providing an enemy or a hostile with vital information on friendly activities, dispositions, and capabilities. However, a threat employing particularly cruel, abusive, or repressive measures may easily turn certain groups in the urban area against them, even when they share a common history, culture, and ethnicity with the civilians. This is more likely in those areas with higher population densities.

3-28. Threat forces may also seek to use some nongovernmental organizations (NGOs). They may try to obtain relief supplies either through the organizations' legitimate relief operations or as a target for theft. Some organizations may even be fronts for weapons, food, ammunition, money, and fighters. For example, during Russia's second conflict in Chechnya (1999-2000), documents purportedly found in Grozny by the Russians listed nations such as Sudan, Nigeria, Niger, and Ivory Coast as sending fighters to Chechnya under the guise of the International Islamic Relief Organization. (Chechen fighters also disguised themselves as Red Cross workers.) This deception increases the need for strict security and force protection measures, close coordination with NGOs operating in urban areas, and closer monitoring of suspect organizations' activities by civil affairs personnel.

Win the Information War

3-29. Threat forces will try to win the information war as much as they will directly oppose Army UO. Threat urban campaigns need not be tactical military successes. They need only to weaken legitimacy and make the opposition's campaign appear unpalatable to domestic, U.S., and world support. As a critical part of a threat's overall information operations, they will use the ever-present media to tell their story. Portable video cameras, commercial radios, and cellular telephones, available and easily concealed, will be as important to many threat actors as weapons and ammunition. Internet access, already firmly established in many urban areas, provides the means to easily disseminate threat propaganda, misinformation, and disinformation through web sites and electronic mail. The internet also allows threats to conduct their own distance learning and education. Threats will make good use of the instantaneity and global reach of television and radio broadcasts, particularly local and international media, to obtain information of immediate tactical value and broadcast their own operational and tactical guidance. Hackers, covered and concealed in the interior spaces of the urban area, may gain access to U.S. sites to manipulate and obtain information to the threat's advantage.

Information and the Media

The media coverage of the urban battle for Hue, South Vietnam, although only one of hundreds of different attacks of the Tet Offensive, affected the will of both the American people and their political leadership. On January 31, 1968, two North Vietnamese Army (NVA)/Vietcong (VC) regiments and two sapper battalions, moving rapidly and with the element of surprise, attacked and seized part of the walled city (Citadel) of Hue. It was the third largest city in South Vietnam, the former capital of a united Vietnam, the capital of Thua Thien province, and a spiritual and cultural center. Initially intending to hold the city for seven days, the NVA/VC retained portions of the city for approximately three weeks against determined U.S. and South Vietnamese attempts to retake it.

Hue marked a revolution in the coverage of war by modern media. It was the first time Americans could sit at home and watch an ongoing battle on the evening news. One of the most intense and savage battles of the Vietnam conflict, it was televised every evening for almost a month. Although the battle for Hue was a tactical victory for the United States, the North Vietnamese clearly achieved strategic success by searing the American consciousness with the high costs of urban warfare. Had U.S. leaders made winning the information war a central part of the overall campaign plan—for example, exposing the American people to the NVA's brutality by publicizing the civilian executions in Hue—civilian support for the war may have been bolstered and a different outcome achieved. See Chapter 7 for a more detailed account of the battle for Hue.

Manipulate Key Facilities

3-30. Threat forces will attempt to identify and quickly seize control of critical components of the urban area to help shape the operational environment. Urban telephone exchanges can provide threats with simple and reliable communications that can be easily secured with off-the-shelf technologies. Existing public address and loudspeaker systems can be used to incite riots or openly command and control dispersed threat forces. (Local language may provide threats with a measure of communications security.) Sewage treatment plants and flood control machinery can be used to implement WMD strategies or to make sections of the urban area uninhabitable. Media stations significantly improve the information operations abilities of the controlling force. Power generation and transmission sites provide means to control significant aspects of civilian society over a large area. Insurgents will attempt to destroy key facilities and critical infrastructure in order to shape the populace's perception that the existing government is incapable of providing for their safety and meeting their basic needs.

Use All Dimensions

3-31. Threats will think and operate throughout the depth, breadth, and height (including supersurface and subsurface areas) of the urban environment. However, threats will often choose to fight from within interior spaces rather than out in the streets and other open areas. Conventional lateral boundaries will often not apply as threat forces control some stories of the same building while friendly forces control others. The use of all three dimensions of an urban area makes identification, reporting, and targeting of enemy locations more difficult for forces accustomed and trained to acquire and engage targets in primarily two dimensions. Because of these target identification and acquisition difficulties, threats will likely use decoys in the urban area to cause erroneous assessments of their combat capabilities, strength, and disposition. They will also attempt to use these decoys to absorb expensive and limited precision-guided munitions as well as cause misallocation of other critical Army resources.

3-32. Roofs and other supersurface areas provide urban threats with excellent observation points and battle positions above the maximum elevation of many weapons. Shots from upper floors strike armored vehicles in vulnerable points. Basements and other subsurface areas also provide firing points below many weapons' minimum depressions and strike at another weakness in most armor. Sewers and subways may provide covered and concealed access throughout the area of operations.

EMPLOY URBAN-ORIENTED WEAPONS

3-33. Whether purpose-built or adapted, many weapons are more useful in an urban environment while others may have significant disadvantages. Urban threat weapons are much like the nature of urbanization and the urban environment: inventive and varied. Many threats will integrate widely available off-the-shelf technologies into their weapon systems and armed forces. However, sniper rifles and small, man-portable, fire-and-forget weapons and demolitions and other improvised explosive devices (IEDs), to include suicide and car bombs, will likely dominate the urban environment. Figure 3-3 lists examples of threat weapons favored in UO.

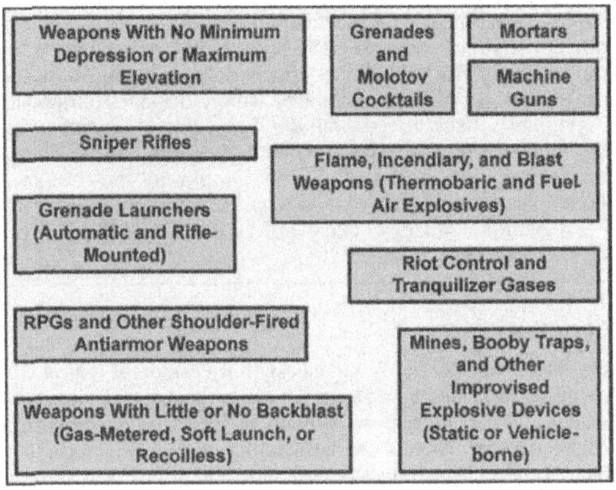

Figure 3-3. Favored threat weapons

ENGAGE ENTIRE ENEMY FORCE

3-34. Threats may attempt to keep all or significant portions of Army forces engaged in continuous operations to increase their susceptibility to stress induced illnesses. UO, by their nature, produce an inordinate number of combat-stress casualties. Continuous operations exacerbate this problem. Threat forces that employ this tactic will often maintain a large reserve or achieve respite by hiding among the civilian population to minimize the psychological impacts on their own forces.

3-35. To accomplish this key tactic, threat UO will likely involve decentralized maneuver, precision fires, and simultaneous operations involving unconventional and special purpose forces and guerilla, insurgent, and terrorist tactics. Threats may also attempt to bait Army forces by ambush or other means with a primary focus of attacking response or reaction forces. Overall, threat forces will take advantage of any exposed weakness in weapons, equipment, and TTP and they will cultivate the flexibility to engage in battles as favorable opportunities present themselves.

FOCUS ATTACKS ON SUPPORT AREAS, ISOLATED GROUPS, AND INDIVIDUALS

3-36. To supplement the previous tactic, threat forces will seek to target support areas, small groups, leaders and their headquarters, and unaware Soldiers. Their focus on resupply and convoy operations, casualty evacuation, and other sustainment activities, coupled with the compartmented terrain, navigational challenges, and multiple three-dimensional avenues of approach often makes these locations and Soldiers more susceptible to surprise raids and ambushes. Attacks on these areas and groups are conducted to erode the Army's ability to sustain UO, to inflict maximum casualties, and to induce psychological stress. These attacks can be mitigated by careful, regular evaluation of choke points and other restrictive terrain, regular awareness training for units and individuals operating in or transiting through potential incident-prone areas, well-planned and protected convoys, and thorough after-action analysis of incidents.

NEGATIVE EFFECTS OF URBANIZATION

3-37. Many urban areas are the engines for increased industrialism and economic growth as an expanding population provides the labor for manufacturing and service needs. However, rapid and inadequately planned growth can result in undesirable consequences. Uncontrolled urbanization may result in an infrastructure and economic base unable to support the growing population. A large transient, ill-housed, and idle population in a close geographic space may produce strife. Classes, cultures, ethnic groups, and races that might otherwise peacefully coexist can clash under the stress of survival. Uncontrolled urban growth has resulted in the negative effects listed in figure 3-4. In many urban stability and civil support operations, these may be the primary "threats" to mission accomplishment.

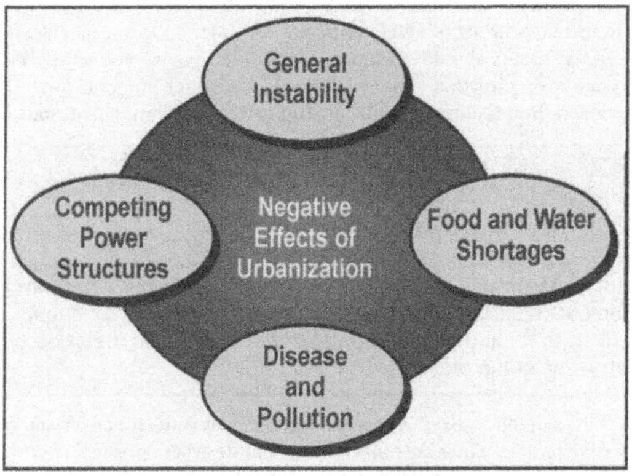

Figure 3-4. Negative effects of urbanization

3-38. Not all urban areas prevail as inherently unstable or hotbeds for unrest. Urban growth due to migration may remove sources of conflict, or it may provide the catalyst for violence. Commanders should recognize the possible effects of uncontrolled urbanization. During their intelligence preparation of the battlefield (IPB), they should determine if these conditions exist. Throughout mission analysis and the development of courses of action, commanders should consider the impact (if any) on their operations. At the same time, they must recognize that UO may create similar problems that may affect the current operation as well as the overall campaign.

General Instability

3-39. Urbanization can enhance stability by generating industrialization and economic growth resulting in more jobs, a higher overall standard of living, and an educated, relatively satisfied populace. However, the population dynamics associated with urbanization can also have an opposite, destabilizing effect. Radical population growth may create overcrowding and generate or aggravate resource and quality of life issues. Intense and destructive competition for employment, housing, and social status may develop in this climate of economic deprivation. The inability of some governments to handle these problems—

- Makes their urban areas potential sources of unrest.
- Increases the likelihood of the Army's involvement in stability operations.
- Complicates all operations conducted in such an urban environment.

Weak civil administrations—perhaps weakened by the effects of previous UO—have difficulty controlling their society, safeguarding their military armaments, and preventing their urban areas from serving as sanctuaries to terrorists, insurgents, and criminal organizations.

3-40. Urbanization in developing countries warrants more concern. Intense migration and growth, coupled with the forced closeness of people once separated by the rural countryside, may stress already struggling institutions, hasten conflict, and lead to overall instability. Commanders must understand that UO, depending on the operation, may either cause massive population movement out of or into urban areas.

3-41. Urban areas with a large youth population may also help to generate conditions for instability. Rural-to-urban migrants tend to be relatively young. In 1999, Cairo, for example, had more than 40 percent of its population younger than 15 years. Young urban populations generate enormous demands for social resources, primarily education and jobs. Even a strong urban economy may fold under the economic expectations of a tremendous influx of young migrants. Disorder and violence may result as hostiles (many nonstate actors) easily mobilize and manipulate the idle young to act politically and criminally. Urbanization and population growth are more dangerous when they combine to produce a cohort of young urban dwellers separated from traditional social controls, such as village elders and clan leaders.

Cultural and Religious Instability

The 1992 bombing of the Babri Masid Mosque in Ayodya, India, enflamed an already intense cultural and religious rivalry between Hindus and Muslims and led to rioting throughout many Indian urban areas. Of the 1,500 who died in conflicts and riots, almost 95 percent died in urban areas. The violence struck Ahmedabab and Bombay most seriously, with related acts of murder, gang rapes, and arson occurring months after the destruction of the mosque.

3-42. Ethnic, religious, and other social issues may become the vents for anger and frustration produced by the high tension of urban life. Major acts of violence and destruction, such as occurred in 1992 in India, can directly threaten a nation's security. Army forces may have to conduct large-scale, stability operations to promote peace and protect national interests. In these cases, all levels of command will be particularly concerned with maintaining impartiality and perceived legitimacy.

Food and Water Shortages

3-43. Rapid urbanization, primarily in developing nations, may lead to severe food shortages that could influence Army forces (or lead to their use). Such shortages may cause instability, massive migration, revolts, or increased support of armed opposition groups. Armed factions may target NGOs that supply aid as a means of furthering dissatisfaction among the populace. In effect, food may become a weapon. Deployed troops may need to provide or support humanitarian food aid networks to keep the humanitarian situation from escalating.

3-44. However, planners must be wary of the unintended consequences of well-intentioned humanitarian assistance operations. For example, providing free, safe food may alleviate starvation, but could also undercut the local agricultural system by reducing demand in the market. If the food is distributed through urban centers, urbanization could increase, further reducing the food supply and adding to the existing strains on the infrastructure. Unless they are trying to use them as a means to separate combatants from noncombatants, commanders should normally use centralized feeding centers as a last resort. Instead, Army forces should bring the food closest to the population to encourage civilians to stay in their homes. If safe areas or camps are created, they should be designed for use over as short a time as is feasible. The general rule should be to return the urban population to their homes as soon as possible. Army forces conducting domestic support or foreign humanitarian assistance operations that cannot maintain the safe food supplies may find the frustrations and hostility of the local population focused on them.

3-45. Water shortages (and quality) are becoming a serious problem in many regions. Sewage, industrial waste, other forms of pollution and deliberate contamination pose threats to existing water supplies. Commanders operating in an urban environment need to know the water supply origins and its treatment, purification, distribution, and vulnerabilities. Before beginning operations, commanders must know if they are providing water for the noncombatants as well as their own forces. Across the spectrum of operations, controlling and protecting a limited water supply is, or may become, an essential operational consideration during UO.

Food and Water Shortages

Countries as varied as Indonesia and Algeria exported their food surpluses only two generations ago but now import up to two-thirds of their basic staples. This cycle has resulted in many countries, which once exported agricultural products, facing the growing cost of imports to feed their urban populations. Estimates predict that in the 2010 timeframe, at least 65 countries (including 30 of Africa's 51 countries) may depend completely on food imports. For some countries, the outlook is even worse. The Democratic Republic of Congo (Zaire), once a net food exporter, faces severe malnutrition and utter starvation for a large portion of its population.

Other estimates predict that by 2025, 2.7 to 3.5 billion people will live in water-deficient countries. Even now in urban areas such as Jakarta, Indonesia, for example, urban authorities cannot provide the necessary water to support its growing population. Less than half of Jakarta's population is supplied by the urban area's water infrastructure; the remainder of the area's residents rely on water from wells and other groundwater supplies that lie only a few feet below the surface. These alternate sources need to be sufficiently boiled or residents risk increased disease. In the northern part of Jakarta, land is actually sinking as the urban area's groundwater supply is overused by its residents. Houses on low-lying areas have to be rebuilt every few years to keep them above sea level. Meanwhile, saltwater from the Java Sea is seeping into the land polluting the remaining groundwater and further exacerbating the situation.

DISEASE AND POLLUTION

3-46. Urban areas frequently spawn epidemics; therefore, widespread disease may pose a significant threat to Army forces that operate there. In many developing nations, rapid urbanization has occurred without a corresponding upgrade, expansion, or even development of adequate sewage and water systems. Some urban areas have only one toilet for every 750 people or more. In these areas, hundreds of thousands live much as they would in poor villages, yet so confined as to ensure high transmission rates for airborne, waterborne, sexually transmitted, and contact-transmitted microbes.

3-47. In urban areas lacking adequate trash and waste management infrastructure, insect-spread diseases proliferate. Mosquitoes that breed in polluted water, open water tanks, and irrigated urban gardens carry malaria and dengue fever—the leading causes of sickness and death from infectious disease in Latin America and Africa. The problem compounds with growing numbers of bacteria resistant to various antibiotics, a shortage of trained medical personnel, inadequate or insufficient medical facilities and supplies, and unclean agricultural and food-processing practices.

3-48. Pollution also creates critical health problems in developing areas and a potential health risk for intervening Army forces. For example, urban areas in China have recorded five to ten times the levels of sulfur dioxide found in the air of urban areas in the developed world. In parts of Poland, toxic waste has so polluted the land and water that ten percent of the babies have birth defects. Pollution may cause immediate health problems but more often, the insidious effects appear months or years after exposure. As discussed earlier, UO may contribute, either intentionally or unintentionally, to an increase in pollution. Destruction of industrial complexes that use, produce, and store hazardous material may produce toxic gas and smoke pollutants and create significant health concerns to exposed Soldiers and civilians. Soldiers forced to fight within the confines of subsurface areas such as sewers and drainage systems further increases their exposure to contaminants.

3-49. Commanders should initiate force health protection planning early, including analysis of the medical threat and other critical medical information requirements during the IPB process. A medical surveillance system should monitor the daily status of Army personnel throughout the operation. In preparation, all personnel must receive a predeployment medical examination. This exam establishes an accurate baseline health status of the force and ensures that Army forces do not introduce new diseases to an urban area, possibly exacerbating the situation. Conversely, Soldiers not immune to native viruses or possessing a weakened immune system due to continuous operations and the stress associated with UO may put Army forces at a significant disadvantage. An outbreak of plague during an operation would have an effect similar to a chemical or biological attack. The closer that Army forces operate to civilians (the humanitarian assistance operations conducted in Port-au-Prince, Haiti, and Mogadishu, Somalia, for example), the more probable that these situations may occur. (See Chapter 9.)

COMPETING POWER STRUCTURES

3-50. Many groups can exist that become strong enough to rival the power of the governing officials and eventually turn the urban area into a system of divergent and competing power structures. These groups can consist of insurgent forces, a merchant class or an economic elite, criminal organizations, or some other significant source of power such as religious organizations (see Chapter 10), clans, or tribes (figure 3-5). In the absence of a legitimate authority, armed factions headed by "warlords" may vie to fill the power void. Sometimes these groups or organizations, normally at odds with each other, may form alliances to achieve specific goals. Commanders must recognize, identify, and understand these alternate urban power bases and, if necessary, develop engagement strategies to neutralize or harness them to accomplish the Army mission. Four of these groups, not previously covered or covered in a subsequent chapter, are discussed in greater detail below.

Figure 3-5. Competing power structures

Urban Insurgencies

3-51. As urban migration increases in the developing world, rural guerrillas appear to follow. This transition of insurgencies from rural to urban areas occurs because urban areas offer a rich field of targets for insurgent attacks. People immediately notice any disruption of urban infrastructure, thus having great propaganda value. A concentrated urban population whose allegiance and support are critical to the success of an insurgency is often more susceptible to propaganda, political organization, and terrorism. Insurgents can easily arrange mass demonstrations using available communications facilities, both overt and covert. Travel is effortless and large urban populations provide cover and concealment. On the whole, urban areas, particularly those suffering the effects of general instability outlined earlier, may provide a fertile environment for guerrillas to apply formerly rural insurgent strategies. However, even with a primarily rural-based insurgency, operations in urban areas will often be essential to the insurgents' strategy as they offer them distinct opportunities to disrupt, discredit, and demoralize the government (see FM 3-05.20 and FMI 3-07.22).

Merchant Class or Economic Elite

3-52. Urban areas normally possess a merchant class or economic elite as part of their social structure. In some urban areas, they may carry more power than the local or central state government. They may isolate themselves physically and socially from the sprawling poor yet wield enormous power over the country's political and economic activities. The degree of economic separation between the merchant class and the poor may be small but still socially or politically significant.

3-53. In a vastly impoverished area where the economy of the urban area is severely disrupted, the merchant class will often continue to operate and function and, as a result, achieve a measure of influence. To continue to operate under acute economic turmoil, they may form alliances in criminal organizations and secure loyalties within the government. Outside resources introduced into a crisis area (such as food, water, fuel, and pharmaceuticals) take on increased value, may replace currency as the medium for exchange, and often become the means to amass and hold wealth. One primary way to obtain wealth is to steal it.

3-54. In some turbulent situations that lead to the need for stability operations, commanders may harness the power of the merchant class as a force for peace and stability instead of one that uses crime to achieve economic goals. For example, in a relief situation, instead of competing with the merchant class by distributing food directly to the needy and possibly creating an environment of looting and black marketeering, it may be possible to monetize food. Food assistance from donor governments could be sold to merchants at an attractive price so they have a reliable source of supply. This could, in turn, create a healthy economic system and separate merchants from criminals and gangs.

Criminal Organizations

3-55. Organized criminal groups have grown common in urban areas; have also become an important part of the urban social structure (gangs for example); and can exert considerable influence on governments, people, and military forces conducting UO. These organizations can threaten the successful completion of urban operations both directly and indirectly. Criminals and criminal organizations may directly target Army forces, stealing supplies and extorting money or contracts. Criminals may also violently confront Army forces during mission execution. During many UO, particularly during or following combat, civil disturbances, or large natural disasters, looting (organized or unorganized) may become of critical concern.

For a portion of the population, crime may be the only reliable source of income and resources to support their families. Therefore, UO will often require a combined law enforcement and military response.

3-56. Some large criminal organizations relying on international connections often have better resources and equipment than their insurgent counterparts. (A prime example is drug traffickers.) Their large financial resources, long-reaching connections, and ruthlessness provide them the means to corrupt or intimidate local officials and government institutions. Their tactics parallel those of insurgents. They have developed an intuitive cultural understanding of slum neighborhoods and the ability to lure civilians into criminal activities. They have also mastered the management of mobs. They recruit teenagers and young adults in their efforts against rivals and authorities, just as insurgents muster armies from the youth of rural villages. In many developing nations, there exists an alliance between insurgents and organized criminal groups. In these alliances, the insurgents defend the criminals and the criminals fund the insurgents.

3-57. Excessive criminal activity can undermine the commander's efforts to establish a sense of security among the remainder of the urban populace. Commanders will rely on the ability of their military police (MP) units to assess and understand the potential influence that crime and criminal organizations may have on UO. MPs—in coordination with local police—will need to quickly identify law enforcement gaps to include dysfunctional police organizations and stations and inadequate prison systems. Urban police records may be an important intelligence database to assist in this assessment and may be particularly useful in counterterrorism and counterinsurgency operations. Success in ensuring that law and order is maintained or reestablishing law and order will directly affect the population's perceptions and support of Army operations. Also, MP intelligence operations will contribute immeasurably to the overall IPB effort and aid in understanding of the complexities of the urban society (see FM 3-19.50).

Crime and Criminal Organizations

Crime and poverty plague urban areas such as Rio de Janeiro, Brazil's second largest urban area, and would affect military operations conducted in their limits. Rio has some of the nation's highest negative urban indicators: the largest number of slum dwellers (1 million), the highest murder rate (1 of 700 residents per year), and the highest kidnapping rate (4 per week). In 1989, the homicide rate of the urban area was three times higher than New York City's, and the rate of urban violence has continued to rise. Therefore, law enforcement management may be a critical issue for Army forces operating in urban environments similar to that of Rio de Janeiro.

However, criminal elements or organizations may not always work against Army commanders. They can be co-opted or influenced to serve friendly objectives. For example, during World War II the U.S. Navy worked covertly with the Mafia in New York City to secure the New York harbor from German U-boats believed to be torpedoing ships there. The Mafia controlled most dock activities in New York harbor and was perfectly positioned to monitor other subversive waterfront activity. This capability provided needed information to the Navy for its counterintelligence and security tasks. New York civil authorities, therefore, agreed to permit a Navy-Mafia alliance to operate at the port for the greater good of the country. Although the Mafia was not the preferred ally of the Navy, it had the capability to protect U.S. ships and the interest (patriotism) to help in the war effort. In those circumstances, the temporary alliance worked (see also the civilian threat discussion in Appendix B).

Armed Factions

3-58. A characteristic of some stability operations has been the deterioration or complete collapse of political authority in the country or urban area in crisis. In some cases, "warlords" or the leaders of various armed factions have attempted to fill the power vacuum (see Appendix C for an example). These individuals often have no particular claim to legitimacy. Their power issues from their weapons, not necessarily from their political skills, human services provided, or popular consent (although they have some popular support to remain in their relative position of authority). Others, however, may already have a substantial base of support (for example, a religious base) and then seek to achieve their goals through violent means. In dealing with the leaders of these armed factions during stability operations, it may appear that there are three general courses of action. The first is to aggressively locate and target the leaders and their forces militarily. The second is to ignore them completely, and the third is to meet and work with them diplomatically. Depending on METT-TC, commanders may reduce some of the risks involved in these first three courses of action by combining the last two to create a fourth approach. Nevertheless, the technique chosen must clearly support political and military objectives and lead to the desired end state.

3-59. In some cases, the leaders of armed factions pose a direct threat to Army forces, the urban populace, and the overall accomplishment of the mission. Therefore, it may become necessary to target them and their forces for capture or destruction. Yet, commanders should generally avoid framing the success of a major operation on the death or capture of one or more individuals. Even with increased ISR capabilities, commanders may find an individual manhunt difficult or impossible to achieve. Instead, it may be better to seek to isolate warlords and their forces militarily, while simultaneously seeking to marginalize their influence through a combination of political, economic, and informational means. This does not necessarily mean that commanders must abandon the search if the death or capture the faction leader will contribute to achieving the mission. However, commanders must maintain the capability to strike with speed and precision if the threat leader's location is discovered in the clutter of the complex urban environment.

3-60. In some situations, commanders may attempt to marginalize warlords by ignoring them. However, refusal to acknowledge warlords may increase the threat to Army forces and NGOs. Their militias may attack Army forces to achieve recognition or simply due to misunderstanding or inherent friction between armed forces. On the other hand, dealing with them may provide legitimacy to the exclusion of other elements of the urban population such as professional groups (for example, doctors or teachers), religious leaders, and traditional clan or tribal chiefs—which may have a greater claim to legitimacy and better form the foundation for a reconstituted urban society.

3-61. A compromise between the last two options may offer the best chance for success. In this fourth course of action, commanders can generally recognize these warlords and work with them diplomatically to decrease the risk of armed conflict and civilian casualties; however, this recognition can be kept at staff levels to avoid bestowing any perceived legitimacy on them. Instead, commanders themselves may visibly meet the other elements of society that have a more legitimate claim to political, social, or economic leadership. Inevitably, commanders may need to meet with warlords. In those circumstances, clan or tribal elders, and others who represent traditional authority, should attend the meetings. Moreover, commanders should ask for and give deliberate consideration to their opinions above those of the leaders of the armed factions.

This page intentionally left blank.

Chapter 4

Understanding the Urban Environment's Effects on Warfighting Functions and Tactics

War is, above all things, an art, employing science in all its branches as its servant, but depending first and chiefly upon the skill of the artisan. It has its own rules, but not one of them is rigid and invariable. As new implements are devised, new methods result in its mechanical execution; but over and above all its mechanical appliances, it rests upon the complex factors of human nature, which cannot be reduced to formulas and rules.

Captain Francis V. Greene, 1883

Commanders of major operations should understand the potential effects that the urban environment may have on warfighting functions. They should also understand the possible effects that the urban environment may have on lower-level tactics to properly plan, prepare, and execute major operations that may include UO. Otherwise, commanders may ask their subordinates to achieve effects, accomplish objectives, or adhere to a timetable that is unsupportable due to the constraints imposed by the urban environment. Commanders and their staffs must do more than simply understand the impossible; rather, they must apply the art and science of warfighting to the urban environment and determine what it will take to make it possible.

WARFIGHTING FUNCTIONS

4-1. Understanding the potential effects of the urban environment on warfighting functions permits the urban commander to better visualize his operational environment. With this appreciation, he can conduct a more thorough assessment and thereby determine the most efficient and effective means of employing Army forces. The staff should be intimately familiar with effects in their area of expertise and use that knowledge to understand the problem and develop creative and innovative solutions to achieve their commander's intent.

WARFIGHTING FUNCTIONS

- Intelligence
- Movement and maneuver
- Fire support
- Protection
- Sustainment
- Command and control

INTELLIGENCE

4-2. The intelligence function facilitates understanding of the threat and the environment. The urban environment affects this critical function in many ways. Impacts of the environment on the intelligence function include degraded reconnaissance capability; more difficult IPB process; and an increased importance of credible HUMINT (including the contribution of local civilian liaisons), and an established intelligence reach capability. The Army forces' response to these effects can result in timely, accurate, and actionable intelligence that permits the effective application of other warfighting functions to the mission within the urban environment (see FM 2-0).

Degraded Reconnaissance and Surveillance Capability

4-3. The physical environment creates a major challenge to the intelligence function. The man-made construction in the urban areas provides nearly complete cover and concealment for threats. Although improving many sensor capabilities cannot penetrate the subsurface facilities and much of the space within supersurface areas. The mass of buildings can also defuse electronic signatures. Tall buildings shield movement within urban canyons from aerial observation except from directly overhead. Urban threats may be less technology dependent and may thwart some signals intelligence efforts simply by turning off their radios and using messengers. Threat forces will likely use elements of the civilian telecommunications infrastructure for C2. These systems may include traditional landline phones, cellular telephones, and computer-to-computer or Internet data communications. Most urban telecommunications systems use buried fiber or cables or employ modern digital signaling technology. Such systems are difficult to intercept and exploit at the tactical level.

4-4. From the above, it is evident that these characteristics make it more difficult for the intelligence function to use electronic means to determine threat dispositions and, in offensive and defensive UO, identify decisive points leading to centers of gravity. While the environment limits some typical collection methods, all enemy electronic and human activity creates some form of observable signature and exposes the enemy to potential collection. Seeking ways to take advantage of these vulnerabilities will provide the commander an information advantage over his opponent.

Challenging IPB Process

4-5. The sheer complexity of the environment also challenges the intelligence function. The intelligence function applies the IPB process to the urban environment in accordance with Army doctrine (see Appendix B). With more data points for the IPB process to identify, evaluate, and monitor, this application becomes more demanding. The human and societal aspects of the environment and the physical complexity primarily cause this difference. Relationships between aspects of the environment, built on an immense infrastructure of formal and informal systems connecting the population to the urban area, are usually less familiar to analysts. Thus, the urban environment often requires more specifically-focused intelligence resources to plan, prepare for, execute, and assess operations than in other environments.

4-6. Compounding the challenges is the relative incongruity of all urban environments. No two urban areas are alike physically, in population, or in infrastructure. Thus, experience in one urban area with a particular population and pattern of infrastructure does not readily transfer to another urban area. Any experience in UO is valuable and normally serves as a starting point for analysis, but the intelligence function cannot assume (and treat as fact) that patterns of behavior and the relationships in one urban area mirror another urban area. The opposite is as likely to hold true. The intelligence function will have to study each urban area individually to determine how it works and understand its complex relationships.

4-7. Each characteristic of the urban environment—terrain, society, and infrastructure—is dynamic and can change radically in response to UO or external influences. Civilian populations pose a special challenge to commanders conducting UO. Civilians react to, interact with, and influence to varying degrees Army forces. Commanders must know and account for the potential influence these populations may have on their operations. Intelligence analysts must revisit or continuously monitor the critical points looking for changes, relationships, and patterns.

4-8. The actions of Army forces will affect, positively or negatively, their relationship with the urban population and, hence, mission success. NGOs may deliberately or inadvertently influence civilians. The intelligence function can monitor and predict the reactions of the civil population. However, accurate predictive analysis of a large population requires specific training and extensive cultural and regional expertise.

Increased Importance of Human Intelligence

> ***HUMINT*** is the collection of foreign information—by a trained HUMINT collector—from people and multimedia to identify elements, intentions, composition, strength, dispositions, tactics, equipment, personnel, and capabilities. It uses human sources as a tool, and a variety of collection methods, both passively and actively, to collect information. FM 2-0

4-9. The intelligence function adjusts to the degradation of its technical intelligence gathering systems by increasing emphasis on HUMINT in UO. HUMINT operations may be the primary and most productive intelligence source in UO. In urban offensive and defensive operations, HUMINT gathers information from refugees, immigrants and former citizens (especially previous civil administrators), civilian contractors, and military personnel who have operated in the area. Credible intelligence of this type can help meet requirements, provide more detail, and alleviate some of the need to physically penetrate the urban area with reconnaissance forces. In many urban operations where HUMINT is the primary source of intelligence, acting on single-source reporting is a constant pitfall. Yet, situations may arise where commanders must weigh the consequences of inaction against any potential negative consequences resulting from acting on uncorroborated, single-source information. (See also the Human Capabilities discussion under the Urban ISR portion of Chapter 5.)

4-10. In urban stability operations, HUMINT identifies threats and monitors the intentions and attitudes of the population. A chief source of information contributing to the development of accurate HUMINT, particularly at the tactical level, is reconnaissance forces—especially small-unit dismounted patrols. Urban reconnaissance forces and patrols should be thoroughly and routinely debriefed by unit intelligence personnel to obtain information that aids in developing a clearer picture of the threat and the urban environment. Reliable and trustworthy HUMINT is particularly important in foreign internal defense, counterterrorism, and support to counterdrug operations. Leaders must organize intelligence resources appropriately, and learn and apply valuable techniques, such as pattern and link analysis (see FM 34-3). Additionally, Soldiers, as part of reconnaissance and patrolling training, should be taught to handle captured documents, weapons, material, and equipment as legal evidence much like military and civilian police. Proper "evidence" handling is often a critical intelligence concern in counterterrorism and counterinsurgency operations.

Developing Local Liaisons

4-11. Whenever Soldiers encounter the urban populace, the resulting interaction may become an important source of information the commander can use to answer questions about the threat and the urban environment. While military intelligence units are the primary collectors and processors of HUMINT, commanders are not likely to have enough trained HUMINT Soldiers to satisfy their requirements—particularly in a larger urban environment and during longer-term stability operations. Therefore, commanders may need to cultivate and establish local civilian associations to provide relevant information for decision making and to support the overall HUMINT effort.

4-12. Urban liaisons can be developed through positive civil-military interaction with the urban populace. Critical information may be acquired through interface with the urban leadership (both formal and informal), administration officials, business owners, host-nation support workers, inhabitants along a unit's patrol route, pedestrians at a checkpoint, civilian detainees, or any other human source willing to volunteer information to Army forces or who respond positively to tactical questioning. (Noncombatants are never coerced to provide information.) Commanders may also direct unit leaders to conduct liaisons with specific local leaders and key members of the community to obtain command directed information. Critical information may also come from other U.S. and coalition forces and intelligence organizations operating near or within the commander's AO. To this end, commanders should ensure that collectors operating in an urban area coordinate and deconflict activities and, if possible, outbrief subordinate, geographically-responsible commanders with any relevant information that may affect their current operations. Any relevant information obtained incident to civilian liaison activities should be routinely provided to intelligence staffs not only to gain assistance in verifying the credibility of the information but to share the information with all affected echelons and units.

4-13. NGOs operating in urban areas can also be especially beneficial resources for credible and relevant information about the urban environment. (However, they are generally not a good source for information about the threat since providing such information can violate their neutrality thereby making it difficult for them to achieve their humanitarian aid objectives.) During the 1999 fighting in Kosovo, for example, the Red Cross provided the most accurate figures regarding the number of Kosovar refugees, helping U.S. and other coalition forces to estimate the appropriate level of support required to handle their needs. In addition to a developed understanding of the current needs of the local urban populace, NGOs may also have—

- A network of influential associations.
- Historical archives.
- Extensive understanding of the urban infrastructure.
- Key knowledge of political and economic influences.
- A keen awareness of significant changes in the urban environment.
- Insight into the current security situation.
- Up-to-date web sites and maps.

4-14. While productive civilian associations may become long term, they should not be confused with HUMINT source operations. Only trained HUMINT personnel can recruit and task sources to seek out threat information. Information obtained from these societal connections is normally incidental to other civil-military relationships. For example, as part of infrastructure repair in an urban stability operation, a commander may be instrumental in obtaining a generator for a local hospital. Within the context of this relationship, the commander may develop a rapport with one or more of the hospital's administrators or health practitioners. These civilians may be inclined to provide valuable information about the threat and the urban environment—often on a continuing basis. In any civil-military relationship, however, commanders ensure that the information provided is not tied to promises of assistance or that such assistance is in any way perceived as a means to purchase civilian loyalty.

4-15. Commanders also understand that repeated interaction with any one individual may put that individual and his family in danger from threat forces. Before this potential danger becomes a reality, they should refer their civilian connections to trained HUMINT personnel for protection and continued exploitation. In addition to civilian protection considerations, commanders may also deem it necessary to turn their civilian associations over to trained HUMINT collectors anytime during the relationship if they consider the information that the contact is providing (or may provide) is credible, relevant, and—

- Provides essential threat information on a repetitive basis.
- Helps answer higher-level CCIR.
- Affects operations in another AO.
- Requires interrogation or monetary compensation to obtain.

However, turning a liaison over to trained HUMINT teams does not necessarily preclude maintaining a continued, albeit a more guarded, relationship with the individual.

4-16. In developing these civilian liaisons essential to understanding the urban environment, commanders must avoid the distinct possibility of conducting unofficial source operations by non-HUMINT Soldiers. While prohibited by regulatory guidance (see Defense Intelligence Agency Manual 58-11 for requirements and restrictions for source operations), such actions also run the additional risks of—

- Obtaining unevaluated information that cannot be crosschecked with and verified by other sources of information.
- Creating inequities that result from illegally rewarding contracts which can undermine HUMINT Soldiers who are constrained by intelligence contingency fund regulations.
- Disrupting ongoing HUMINT operations when different sources are seen to be treated differently by non-HUMINT Soldiers vice HUMINT Soldiers.
- Providing non-HUMINT and HUMINT Soldiers with the same information potentially leading to a false confirmation of information.

- Increasing the likelihood that untrained Soldiers may fall victim to a threat deception and misinformation.

Established Intelligence Reach

4-17. Understanding the complex urban environment, particularly the infrastructure and the society, will require more sources of information beyond a unit's organic intelligence capabilities. Therefore, commanders will have to make extensive use of intelligence reach to access information and conduct collaboration and information sharing with other units, organizations, and individual subject matter experts. Before deployment (and throughout the operation), units should establish a comprehensive directory of intelligence reach resources. These resources may include national, joint, Army, foreign, commercial, and university research programs. (Prior to deployment for OPERATION IRAQI FREEDOM, some units established contacts within the local community outside their bases such as police, fire department, and government officials that expanded their reach once in theater—particularly for information regarding civilian infrastructure and urban administration.) Once deployed, intelligence reach includes effective information sharing and collaboration among adjacent units, sister services, coalition partners, and other governmental and nongovernmental agencies operating in the area. Effective information sharing and collaboration requires common network analysis software and databases to be used among all Army forces and, if possible, other governmental agencies.

MOVEMENT AND MANEUVER

4-18. Army maneuver forces—infantry, armor, cavalry, and attack aviation—move to achieve a position of advantage. Entire urban areas or specific threat forces located within may be isolated from affecting other operations and then bypassed. However, when the situation requires entering the urban area to accomplish the mission—such as when bypassed urban threat forces interdict critical lines of communications (LOCs), the environment will significantly affect the Army's ability to move and maneuver (see figure 4-1). These negative effects include canalization, compartmentalization, and increased vulnerability. However, tactics and techniques equip Army forces to overcome these challenges and maneuver successfully. One tactic, effective combined arms task organization, includes an increased dismounted maneuver capability combined with armor and combat engineers, continuous operations, and technological enhancements.

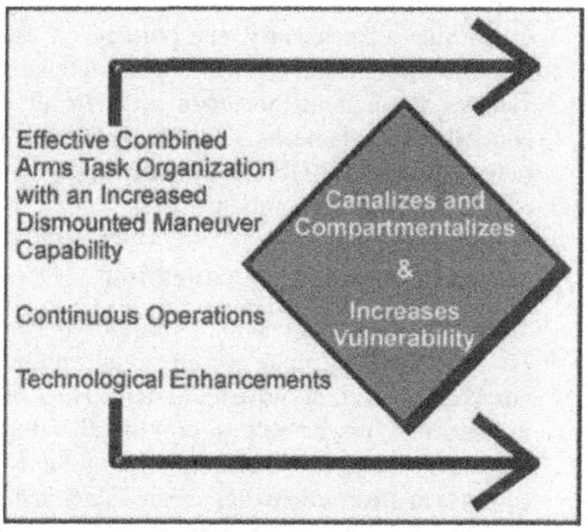

Figure 4-1. Urban maneuver challenges and means to overcome them

4-19. In all UO, mobility operations may allow civilian traffic and commerce to resume, letting the urban area return to some semblance of normalcy (often a critical objective). In stability operations, mobility often focuses on keeping lines of communications open and reducing the threat of explosive hazards to Soldiers and civilians. In civil support operations, mobility may focus on removing storm debris or reducing obstacles caused by destroyed property.

Canalization and Compartmentalization

4-20. The urban terrain will often canalize and compartment forces—and their fires—moving and maneuvering through it. Buildings pose obstacles to both mounted and dismounted movement, forcing units to be canalized along streets. The buildings also block movement between streets, thus compartmenting units. Fires are canalized into open and unmasked areas where vision is unobstructed, producing concentrated fire zones and areas, especially at road intersections and in front of defended positions. Hence, changing directions, repositioning committed forces, reinforcing forces in contact, bypassing threats, and maneuvering to the threat flank become extremely difficult. Units often breach obstacles to help solve this problem. Using helicopters to quickly move forces, both forward into contact and to rear areas as part of repositioning, also permits Army forces to overcome some terrain constraints.

4-21. The canalized and compartmented effects can slow movement and maneuver considerably. However, slowed movement also results from the physically demanding soldier tasks required in an urban environment. Soldiers operate dismounted across rubble and hard surfaces. Operating in three dimensions, they constantly move up the supersurface areas of building interiors and down into basements, cellars, and other subsurface areas. They breach many obstacles and use upper-body strength, ropes, and ladders to scale heights. The inability to see into the next room, floor, or building magnifies stress. The resulting fatigue slows the overall rate of Army force movement and maneuver.

Increased Vulnerability

4-22. The urban environment increases the vulnerability of Army forces executing movement and maneuver in offensive, defensive, and stability or civil support operations. Both the physical terrain and the urban population provide threat cover and concealment. Air movement and maneuver is vulnerable for many of the same reasons. In offensive or defensive operations, enemy forces can remain undetected in buildings and in position to ambush Army forces. Consistent with METT-TC, Army forces should clear buildings along maneuver routes prior to mounted movement along those axes. Failure to clear routes (and effectively mark cleared portions) may expose mounted movement to ambush at close range. Movement back across streets and obstacles may be difficult particularly if the element of surprise was essential in the initial crossing or breach. The same buildings also provide cover and concealment to enemy air defense capabilities—particularly man-portable air defense systems that can be fired from multiple positions hidden amongst the clutter of fires, lights, smoke and dust, and easily concealed and transported in civilian vehicles throughout the urban area. In all operations, but especially stability operations, civilians can conceal threat elements. The threat can then initiate offensive operations against Army forces from close range and where ROE will hamper applying combat power. Thus, maneuver through a dense population can be a high-risk operation.

Combined Arms Task Organization

4-23. Effective combined arms task organization ensures that forces are task organized with infantry—the essential building block for all organizations conducting UO. Infantry protects mounted elements as the combined arms unit moves and maneuvers through the urban area. (In some urban situations, mechanized infantry may not be able to provide dismounted support beyond support to its own vehicles—tanks may require the support of additional light infantry). The infantry destroys the enemy in buildings, bunkers, and subsurface areas where they cannot be defeated by mounted forces and prevents infiltration of threat forces back into hard-won urban terrain. Field artillery aids in dismounted and mounted (to include air) maneuver by suppressing known and suspected enemy positions with precision fires. Attack aviation make best use of their standoff capabilities (see later fire support discussion) and aircraft speed to conduct running and diving fires; in UO, hovering fire is generally avoided. Armored elements protect Soldiers from small arms fire and destroy or suppress enemy positions with precise, direct fire. Carefully protected artillery may also be used in this direct fire role. Armored forces and attack helicopters also can facilitate maneuver through shock action that can have a psychological effect, particularly against less well-trained threats and, in discrete instances, hostile crowds. (Although, commanders consider that the "intimidation" value of any method erodes quickly with its repetitive use.)

4-24. Combined arms also ensure that combat engineers support dismounted maneuver by assisting in covered and concealed maneuver through buildings and off exposed streets. In addition to combat engineers, explosive ordnance disposal teams, military police, chemical personnel, and other with essential expertise to conduct mobility missions significantly reduces mobility and maneuver challenges (see FM 3-34.2). Urban buildings are often obstacles to movement and mobility. Combat engineers, trained and equipped for UO, can turn these obstacles into an advantage by breaching them with "mouse holes" made by explosives, sledgehammers, bulldozers or armored vehicles, or high-strength (diamond or carbide-tipped) cutting devices. These breaches permit dismounted movement through buildings under both cover and concealment.

4-25. Combat engineers must also be trained and equipped to facilitate mounted mobility in the urban environment. Generally, buildings restrict mounted movement to the compartmented and canalized streets. Threats can block streets with roadblocks ranging from sophisticated log and concrete cribs reinforced with antitank and antipersonnel mines to the expedient use of cars, buses, and trucks to create obstacles. Combat engineers must be capable of breaching these obstacles to maintain the coherence of the combined arms team (mounted and dismounted). Combat engineers should be forward, often task organized down to platoon level, and have the expertise and equipment to rapidly reduce point obstacles. It even may be necessary that every armored vehicle (or section of two vehicles) be task organized with an associated engineer squad and vehicle. Because of the increased density and hardness of many urban building and construction materials, heavy engineer equipment (such as the D9 bulldozer) will be in great demand to accomplish mobility (and countermobility, and survivability) functions in an urban environment. However, commanders will need to consider increased protection requirements and the availability of equipment transport to move these slower-moving engineer assets around the urban battlefield.

4-26. A major difference of UO combined arms is in proportion and organization. Although based an accurate METT-TC assessment, UO will often require an increased proportion of dismounted infantry and engineer capabilities; armor may not be required in the same high numbers. As significant, genuine combined arms urban operations are required at lower tactical levels where small, well trained and led units will dominate. Commonly, company level will require true combined arms capability and may include combat engineers, military intelligence, reconnaissance, and artillery. Combined arms teams can then form at platoon and squad levels. Because of this, larger units will need more CA, military intelligence, and combat engineers than habitually attached for combat in more open or less restrictive terrain.

4-27. In determining the appropriate task organization, commanders consider a suitable span of control for subordinate commanders. They also consider the potential of dissipating a unit's combat power, capabilities, and synergy by breaking a unit up into smaller units in an attempt to ensure subordinate maneuver units have a complete combined arms capability. For example, an additional engineer battalion may be task organized to a brigade combat team. In turn, the brigade combat team may task organize this battalion into engineer companies under the control of their subordinate maneuver battalions. If this type of organization continues, maneuver companies may end up with an engineer platoon and, in the end; maneuver platoons each having an engineer squad. Ultimately, a combined arms capability may have been established at lower tactical levels but the parent maneuver unit (in this example, the brigade combat team) may have lost the ability to conduct larger engineer operations without having to re-task organize and potentially disrupt current operations and established relationships. As a guide, urban commanders may consider task organizing to create combined arms organizations at lower tactical levels when operations are predominately offensive or defensive, and bringing those assets back under their own control when the operation transitions to predominately stability or civil support operations.

Continuous Operations and Technology Enhancements

4-28. Two other means to improve Army forces' ability to move and maneuver in urban terrain is through continuous operations and the leveraging of technology, such as the Army's night operations capability. Historically, urban battles have been fought primarily during daylight because of technological limitations and fatigue. By utilizing night vision technologies, accurate situational understanding, a common operational picture (COP), training, and rotated units, Army forces can defeat threats who use the same soldiers in day and night operations and who are less well-equipped and adept at night operations. Night operations are also a means of mitigating the air defense threat against air maneuver. Continuous operations through night maneuver with fresh forces are challenging, but it can overcome many advantages that a stationary force has against maneuver in the urban environment. However, commanders should also consider that streetlights, fires, background illumination (as well as dark building interiors without ambient light), the increased heat absorption of many urban structures, and the skillful use of searchlights by threat forces may limit the effectiveness of night vision devices and make thermal imagery identification difficult.

Countermobility

4-29. Countermobility capabilities in urban terrain are also an essential consideration in all UO. In defensive operations, commanders use countermobility capability to control where the enemy moves in the urban area. Repositioning defensive forces in the urban area can be difficult and obstacles are essential to limiting the enemy's maneuver options. During offensive operations, countermobility protects exposed flanks and air assaulting forces from counterattack. In stability and civil support operations, countermobility operations may take the form of constructing barriers to assist in populace and resources control at critical urban locations.

FIRE SUPPORT

4-30. The fire support function includes the collective and coordinated use of several means to attack targets in the urban area (see Appendix D for joint capabilities). These means include target acquisition data, indirect fire weapons, rotary- and fixed-wing aircraft, offensive information operations (IO), and other lethal and nonlethal means. The urban environment affects these components of the fires function and their employment.

Target Acquisition

4-31. Target acquisition in an urban environment faces several challenges. First, forces have difficulty penetrating the urban environment's increased cover and concealment using sensors and reconnaissance. Acquiring targeting information and tracking targets throughout the depth of the urban area may prove challenging. Moving personnel or vehicular targets are normally easiest to acquire. However, the cover and concealment provided by urban terrain gives moving targets short exposure times requiring firing systems to act rapidly on targeting data. Targeting of opposing indirect fire units by acquisition radar may work more effectively in urban terrain because of the necessary high angles of indirect fire. The urban environment presents similar difficulties for battle damage assessment.

4-32. Targeting challenges are met by innovatively integrating reconnaissance capabilities. These capabilities include SOF, cavalry, unmanned aircraft systems, and aerial observers as well as the standard reconnaissance assets. More artillery systems may need to be used to ensure the responsiveness (rather than the weight) of fires. Positioning numerous artillery systems reduces the dead space (as discussed below) and permits units to establish more direct sensor-to-shooter links.

The Targeting Process

4-33. Heightened concerns for collateral damage will require that commanders pay particular attention to their targeting process. This process ensures that all available combat power, both lethal and nonlethal (including offensive IO), is effectively integrated and synchronized to accomplish the mission. Commanders ensure that techniques and procedures are in place, rehearsed, and understood by all members of their staffs. Additionally, the C2 system must be responsive and agile; otherwise, an elusive and adaptable threat will likely disappear before units can employ the appropriate weapon systems. In an urban area, even 10-digit grid coordinates may not be sufficient to accurately identify targets as buildings may be connected to each other—often throughout the entire block. Target locations, in addition to grid coordinates, may need to routinely include the street address, number of stories, shape, color, or any other distinguishing characteristic essential for ground and air forces to achieve targeting precision. A common urban reference system with graphics, reference points, and other control measures adequate for both ground and air forces may also help facilitate identification of targets and facilitate the rapid clearance of fires.

4-34. Greater concerns exist for the safety and health (environmental matters) of the urban populace and the protection of critical infrastructure and cultural structures. Hence, CA and staff judge advocates (see Chapters 5 and 10 respectively) will play a greater role for the expert advice they can provide regarding these elements of the urban environment. Nonetheless, all members of the staff ensure that operations minimize collateral damage. That responsibility does not end with identifying potential collateral damage; the goal, as always, is successful mission accomplishment. Again, staffs are guided by the commander's intent and work to develop courses of action that incorporate collateral damage concerns (short- and long-term) yet accomplish the mission. This requires a keen understanding of the legal issues and both friendly and enemy weapon systems' effects in an urban environment.

Urban Effects on Fire Support Systems

4-35. Both the physical and human components of the urban area affect how units use fire support weapon systems. The physical aspects of the urban environment, such as the heights and concentration of buildings, may cause significant masking and dead space (see figure 4-2). Intervening buildings that stand three or more stories tall hinder close indirect fire support. Tall buildings can potentially mask several blocks of area along the gun-target line of artillery. (For low-angle artillery fire, dead space is about five times the height of the building behind which the target sits.) The potential for collateral damage to adjacent buildings may also prevent engagement with artillery. Such damage might cause noncombatant and friendly troop casualties and unintentional (and unwanted) rubbling. Commanders can offset these effects by carefully placing artillery positions, repositioning artillery as targets change, and using mortars. Mortars have a steep angle of fall and short minimum ranges as a high-angle alternative to field artillery fire. (In comparison to artillery, dead space for mortar fire is only about one-half the height of the building.) For fixed-winged aircraft, precision munitions and weapons with low explosive yields and near-vertical impact angles resulting in bomb burial can significantly reduce collateral damage. Collateral damage concerns may also cause commanders to—

- Maintain approval authority for some sensitive or protected targets (churches or mosques, for example) at higher echelons of command.
- Restrict attacks to certain times of day.
- Give warning prior to an attack so that noncombatants can evacuate the area.
- Incorporate indigenous forces into the operation.
- Abort an attack unless the required level of precision effects can be achieved.
- Prepare specific branches and sequels to the IO plan to inform the populace why the collateral damage was justified. These plans may include filming or otherwise documenting the operation to thwart threat propaganda and claims of excessive collateral damage.
- Develop and rehearse detailed staff battle drills that address clearance of fires in an urban environment.

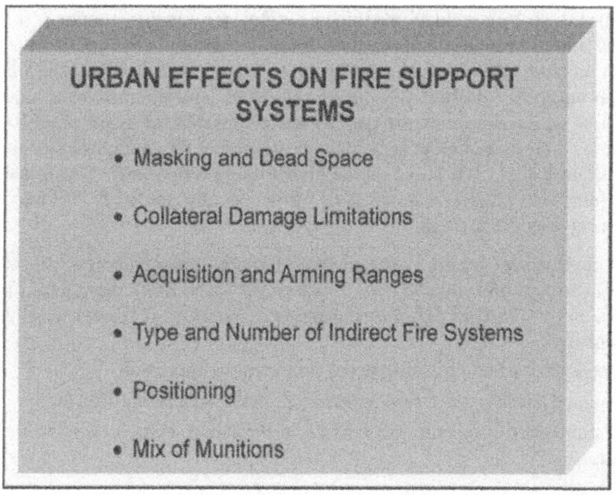

Figure 4-2. Urban effects on fire support systems

4-36. Vertical structures interrupt line of sight (LOS) and create corridors of visibility along street axes. The result is thereby shortened acquisition and arming ranges for supporting fires from attack helicopters and subsequently affected engagement techniques and delivery options. Pilots maintain LOS long enough to acquire targets, achieve weapons delivery solutions, and fly to those parameters. For example, tube-launched, optically tracked, wire-guided heavy antitank missile systems require 65 meters to arm. Similarly, the Hellfire missile requires at least 500 meters to reliably arm and stabilize on the intended target. Thus, attack helicopters firing from longer ranges actually improve the probability of a hit. Poor weather and heavy smoke and dust rising from urban fires and explosions may hinder target identification, laser designation, and guidance for rotary- and fixed-winged aircraft. Poor air-to-ground communications may also hinder effective use of airpower. The close proximity of friendly units and noncombatants requires units to agree on, thoroughly disseminate, and rehearse clear techniques and procedures for marking target and friendly locations. The ability for ground units to "talk-on" aircraft using a common reference system described earlier helps expedite aerial target acquisition (and helps mitigate potential fratricide). FM 3-06.1 details other aviation TTP in an urban environment.

4-37. The urban environment also affects the type and number of indirect fire weapon systems employed. Commanders may prefer high-angle fire because of its ability to fire in close proximity to friendly occupied buildings. Tactically, commanders may consider reinforcing units in UO with mortar platoons from reserve units. This will increase the number of systems available to support maneuver units. Unguided Multiple Launch Rocket Systems (MLRSs) may be of limited use in urban areas due to their exceptional destructive capabilities and the potential for collateral damage. However, commanders may use unguided MLRSs to effectively isolate the urban area from outside influence. Commanders may also employ field artillery systems as independent sections, particularly self-propelled systems, in the direct-fire role; decreasing volume and increasing precision of artillery fire helps minimize collateral damage. While discretely applying the effects of high-explosive and concrete-piercing munitions, these self-propelled systems take advantage of the mobility and limited protection of their armored vehicles.

4-38. The urban area may also affect the positioning of artillery. Sufficient space may not exist to place battery or platoon positions with the proper unmasked gun line. This may mandate moving and positioning artillery in sections while still massing fires on specific targets. Commanders must protect artillery systems, particularly when organized into small sections. Threats to artillery include raids and snipers. Therefore, maneuver and firing units will have to place increased emphasis on securing their positions and other appropriate force protection measures.

4-39. The mix of munitions used by indirect fire systems will change somewhat in urban areas. Units will likely request more precision-guided munitions (PGM) for artillery systems to target small enemy positions, such as snipers or machine guns, while limiting collateral damage. Currently, only conventional tube artillery, not mortars, has this capability. However, large expanses of polished, flat reflective surfaces common in urban areas may degrade laser designation for these munitions (as well as attack helicopter PGM). The vertical nature amplifies the geometrical constraints of many precision munitions. Remote designators need to be close enough to accurately designate but far enough away not to be acquired by the PGM during its flight path. PGMs based on the global positioning system (for instance, guided MLRS or the Air Force's joint direct attack munitions) or other optically guided PGMs may be more effective if urban terrain hinders laser designation.

4-40. The urban environment greatly affects the use of nonprecision munitions. Building height may cause variable time fuses to arm prematurely. Tall buildings may also mask the effects of illumination rounds. Units may choose not to use dual-purpose conventional munitions if (similar considerations apply to Air Force cluster bombs)—

- The enemy has several building floors for overhead protection.
- Dismounted friendly units need rapid access to the area being fired on.
- Large numbers of civilians will operate in the target areas soon after combat operations have ceased.

4-41. Depending on the building construction, commanders may prohibit or limit illumination, smoke, and other munitions because of fire hazards. In particular instances, they may specifically use them for that effect. Structure fires in an urban area are difficult to control and may affect friendly units. Conventional high-explosive munitions may work best against concrete, steel, stone, and other reinforced structures. When not used in the direct-fire role, a greater mass of indirect fire is often required to achieve desired effects. Commanders balance firepower and collateral damage since the rubbling caused by massive indirect fires may adversely affect a unit's ability to maneuver and provide a threat with additional cover and concealment.

4-42. Nonlethal weapons, munitions, and devices can help commanders maintain the desired balance of force protection, mission accomplishment, and safety of noncombatants by expanding the number of options available when deadly force may be problematic. As additional nonlethal capabilities are developed, they are routinely considered for their applicability to UO. In determining their use and employment, commanders, in addition to any previous experience at using these weapons, munitions, and devices, consider—

- **Risk.** The use of nonlethal weapons in situations where lethal force is more appropriate may drastically increase the risk to Army forces.
- **Threat Perspective.** A threat may interpret the use of nonlethal weapons as a reluctance to use force and embolden him to adopt courses of action that he would not otherwise use.
- **Legal Concerns.** Laws or international agreements may restrict or prohibit their use (see Chapter 10).
- **Environmental Concerns.** Environmental interests may also limit their use.
- **Public Opinion.** The apparent suffering caused by nonlethal weapons, especially when there are no combat casualties with which to contrast it, may arouse adverse public opinion.

PROTECTION

4-43. The protection function includes those tasks and systems that preserve the force so that commanders can apply maximum combat power. Preserving the force includes enhancing survivability and properly planned and executed air and missile defense as well as defensive IO (see IO discussion in Chapter 5) and chemical, biological, radiological, nuclear, and high-yield explosive (CBRNE) counterproliferation and consequence management activities (see CBRNE discussion in Chapter 9).

Survivability

4-44. Survivability in the urban environment is a significant force multiplier. Properly positioned Army forces can take advantage of the increased survivability afforded by the physical terrain. Even a limited engineer effort can significantly enhance the combat power of small Army forces. In stability operations, properly planned and constructed survivability positions can enable small groups of Soldiers to withstand the assaults of large mobs, sniping, and indirect fire. Well-protected support bases are often critically essential to minimizing casualties during long-term stability operations and can become a key engineer task.

4-45. While executing major combat operations or campaigns, in particular defensive operations, well planned and resourced engineer efforts can enhance the survivability characteristics of the urban area. These efforts, though still requiring significant time and materiel, can establish defensive strong points more quickly and with greater protection than can be done in more open terrain. Skillfully integrating the strong point into the urban defense greatly increases the overall effectiveness of the defense disproportionately to the number of forces actually occupying the strong point (see Chapter 8).

4-46. Commanders increase survivability by ensuring that all Soldiers have necessary protective equipment and are trained and disciplined in their use. In addition to standard equipment such as helmets, gloves, boots, and chemical protective overgarments, commanders should ensure, as necessary, availability of other protective equipment and material such as—

- Body armor.
- Goggles or ballistic eye protection.
- Knee and elbow protectors.
- Riot control equipment such as batons, face masks, and shields.

- Barrier material such as preformed concrete barriers, wire, sandbags, and fencing material.
- Up-armored or hardened vehicles.
- Fire extinguishers and other fire-fighting equipment.
- Immunizations.

4-47. The Army's urban survivability operations can become complex if the Army is tasked to support survivability operations for civilians. Such operations can range from constructing civil defense shelters or evacuating the population to assisting the population in preparing for or reacting to the use of weapons of mass destruction. However, Army forces are not organized or equipped to support a major urban area's requirements as well as its own mission needs. Normally, Army forces can render this type of support only as a focused mission using a unique, specially equipped task organization.

Air and Missile Defense

4-48. The air and missile defense protects the force from air surveillance and air and missile attack. This system uses—

- The careful massing of air and missile defense combat power at points critical to the urban operation.
- The proper mix of air defense weapon and sensor systems.
- Matched (or greater) mobility to the supported force.
- The integration of the air defense plan into the overall urban operation.
- The integration of Army systems with those of joint and multinational forces.

4-49. Properly planned and executed air and missile defense prevents air threats from interdicting friendly forces and frees the commander to synchronize maneuver and other elements of firepower. Even in a major combat operation or campaign, the enemy will likely have limited air and missile capabilities and so seek to achieve the greatest payoff for the use of these systems. Attacking Army forces and facilities promises the greatest likelihood of achieving results, making urban areas the most likely targets for air and missile attack.

Rotary- and Fixed-Winged Aircraft

4-50. Enemy rotary-wing aircraft can be used in various roles to include air assault, fire support, and combat service support. Some threats may use unmanned aircraft systems to obtain intelligence and target acquisition data on friendly forces. Increased air mobility limitations and targeting difficulties may cause enemy fixed-wing aircraft to target key logistics, C2 nodes, and troop concentrations outside the urban area, simultaneously attacking key infrastructure both in and out of the urban area.

Increased Missile Threat

4-51. The intermediate range missile capability of potential threats has increased to be the most likely air threat to an urban area. Urban areas, particularly friendly or allied, make the most attractive targets because of the sometimes-limited accuracy of these systems. By firing missiles at an urban area, a threat seeks three possible objectives:

- Inflict casualties and materiel damage on military forces.
- Inflict casualties and materiel damage on the urban population.
- Undermine the confidence or trust of the civil population (particularly if allied) in the ability of Army forces to protect them.

4-52. If facing a missile threat, commanders conducting UO work closely with civil authorities (as well as joint and multinational forces) to integrate the Army warning system with civil defense mechanisms. Similarly, Army forces may support urban agencies reacting to a missile attack with medical and medical evacuation support, survivor recovery and assistance in damaged areas, and crowd control augmentation of local police forces. Before such an attack, Army engineers might assist and advise urban officials on how to construct shelters.

Increased Security of Assets

4-53. When defending against an air or missile threat in a neutral or hostile urban environment, air defense assets are concerned with security. Separating air defense locations from high population and traffic centers, as well as augmenting these positions with defending forces, can prevent or defeat threat efforts to neutralize them. Additionally, increased density of UO means increased concentration of all friendly and enemy systems engaged in air and counter-air operations. This density may increase friend and foe identification challenges, air space management challenges, and the overall risk in the conduct of air operations. Finally, limited air defense assets, difficulties in providing mutual support between systems, potential mobility limitations, and other effects of the urban environment increase the need for (and effectiveness of) a combined arms approach to air defense (see FM 44-8).

SUSTAINMENT

4-54. The sustainment function incorporates support activities and technical service specialties, to include maximizing available urban infrastructure and contracted logistics support. It provides the physical means with which forces operate. Properly conducted, the sustainment function ensures freedom of action, extends operational reach, and prolongs endurance. Commanders conducting sustainment to support full spectrum operations must understand the diverse logistic requirements of units conducting UO. They must also understand how the environment (to include the population) can impact sustainment support. These requirements range from minimal to extensive, requiring Army forces to potentially provide or coordinate all life support essentials to a large urban population.

4-55. Commanders and staffs consider and plan for Army sustainment operations that are based in a major urban area. These operations are located in major urban areas to exploit air- and seaports, maintenance and storage facilities, transportation networks, host-nation contracting opportunities, and labor support. These operations are also UO. The commander gains additional factors to consider from basing the sustainment operation in an urban environment. See Chapter 10 for a detailed discussion of urban sustainment considerations.

COMMAND AND CONTROL

> *Fighting in a city is much more involved than fighting in the field. Here the "big chiefs" have practically no influence on the officers and squad leaders commanding the units and subunits.*
>
> Soviet General Vasili Chuikov
> during the 1942-43 Battle for Stalingrad

4-56. The command and control function is the related tasks and systems that support the commander in exercising authority and direction. The urban environment influences both the commander and his C2 system (which includes INFOSYS). The leader's ability to physically see the battlefield, his interaction with the human component of the environment, and his intellectual flexibility in the face of change all impact the mission. The C2 system faces difficulties placed on the tactical Internet and system hardware by the urban environment, by the increased volume of information, and by requirements to support the dynamic decision making necessary to execute successful UO.

Unity of Command

4-57. Although severely challenged, the principle of unity of command remains essential to UO. However, the number of tasks and the size of the urban area often require that Army forces operate noncontiguously. Noncontiguous operations stress the C2 system and challenge the commander's ability to unify the actions of his subordinates, apply the full force of his combat power, and achieve success. To apply this crucial principle in an urban environment requires centralized planning, mission orders, and highly decentralized execution. The method of C2 that best supports UO is mission command (see FM 6-0). Mission command permits subordinates to be innovative and operate independently according to clear orders and intent as well as clearly articulated ROE. These orders and ROE guide subordinates to make the right decision when facing—

- A determined, resolute, and adaptive threat.
- A complex, multidimensional battlefield.
- Intermittent or complete loss of communications.
- Numerous potentially hostile civilians close to military operations.
- The constant critique of the media and military pundits.

4-58. Decentralized execution allows commanders to focus on the overall situation—a situation that requires constant assessment and coordination with other forces and agencies—instead of the numerous details of lower-level tactical situations. Fundamentally, this concept of C2 requires commanders who can accept risk and trust in the initiative, judgment, and tactical and technical competence of their subordinate leaders. Many times, it requires commanders to exercise a degree of patience as subordinate commanders and leaders apply mental agility to novel situations.

Political and Media Impact

4-59. Commanders of a major operation consider how the need to maintain a heightened awareness of the political situation may affect their exercise of C2. A magnified political awareness and media sensitivity may create a desire to micromanage and rely solely on detailed command. Reliance on this method may create tactical leaders afraid to act decisively and with speed and determination—waiting instead for expected guidance from a higher-level commander. Threats may capitalize on this hesitation by conducting operations faster than Army forces can react. Mission orders that express the overarching political objectives and the impact of inappropriate actions, combined with training and trust, will decrease the need for detailed command. Leaders must reduce a complex political concept to its simplest form, particularly at the small-unit level. Even a basic understanding will help curtail potentially damaging political actions and enable subordinates to make the often instantaneous decisions required in UO—decisions that support military and political objectives.

Commander's Visualization

I heard small-arms fire and RPG explosions and felt shrapnel hit the vehicle.... Land navigation at this time was impossible; every time I tried to look out, I was thrown in a different direction.... At this time, I was totally disoriented and had not realized we were on our own.

Captain Mark Hollis
"Platoon Under Fire"

4-60. Leaders at all levels need to see the battlefield to lead Soldiers, make effective decisions, and give direction. Sensors and other surveillance and reconnaissance assets alone cannot provide all the information regarding the urban environment that commanders will need. The focus of lead elements narrows rapidly once in contact with a hostile force limiting their assessment to the local area. Therefore, tactical commanders will not be able to observe operations from long, stand-off ranges. Their personal observation remains as critical in urban areas as elsewhere and helps to preclude commanders from demanding their subordinates accomplish a task or advance at a rate inconsistent with the immediate situation. In urban offensive and defensive operations, seeing the battlefield requires that commanders move themselves and their command posts forward to positions that may be more exposed to risk. Thus, commanders modify their C2 system capabilities to make them smaller, reduce their signature, and increase their mobility. Because of the greater threat to C2, security efforts may be more intense.

4-61. In stability operations, commanders often intervene personally to reassure the urban population and community and faction leaders about the intentions of Army forces. In these type operations, threats may attack leaders to gain the greatest payoff with the least expenditure of resources. Commanders carefully evaluate risk and potential benefits of such exposure. These risks however, cannot stop them from seeing the battlefield, personally intervening in situations as appropriate, and leading their Soldiers.

4-62. Commander's visualization also requires having an accurate understanding of friendly and enemy locations, detailed maps, other appropriate intelligence products, and INFOSYS that accurately depict the urban environment and help establish a COP. The reliability of these items is as important to planning major operations as it is to tactical-level operations. The commander of the major operation ensure that subordinate tactical-level commanders have the necessary products to achieve accurate situational understanding and dominate the urban environment as subordinate commands often lack the personnel or assets to develop these products. Frequently, satellite or aerial imagery is requested to compensate for the drastic changes that can occur due to UO, natural disasters, and outdated or imprecise maps. (Even maps developed and maintained by urban area's administrative activities may not be up-to-date. Extensive and continually expanding shantytowns, for example, may not be mapped at all. Maps may have even been purposefully distorted or critical detail intentionally omitted. The systems used to transliterate some languages such as Arabic and Chinese to Anglicized alphabets often result in the same location being spelled several—and frequently considerably different—ways. Maps may also assign names to features that are completely different than those used by locals to refer to them.)

4-63. Other critical intelligence products needed in the COP include overlays or gridded reference graphics. (Whenever possible, gridded reference graphics should conform to standard military grid reference system formats to reduce the probability of error when entering target coordinates into targeting systems that use global positioning systems.) These products should be developed and distributed to all participants prior to the UO. Overall, their focus should be on ease of reference and usefulness for all forces—ground and air (see Appendix B). Overlays and graphics can also portray important societal information or urban infrastructure, such as—

- Religious, ethnic, racial, or other significant and identifiable social divisions.
- Locations of police, fire, and emergency medical services and their boundaries or zones of coverage.
- Protected structures such as places of worship, hospitals, or other historical and culturally significant buildings or locations.
- Underground subway, tunnel, sewer, or water systems.
- Bridges, elevated roadways, and rail lines.
- Electrical generation (to include nuclear) and gas storage and production facilities and their distribution lines.
- Water and sewage treatment facilities.
- Telephone exchanges and television and radio stations.
- Toxic industrial material locations.

Mental Flexibility

4-64. Commanders conducting UO must remain mentally flexible. Situations can change rapidly because of the complexity of the human dimension. Typical of the change is a stability operation that suddenly requires the use of force. Commanders must be capable of quickly adjusting their mental focus from a noncombat to combat situation. Equally important is dealing with populations during combat operations. Consequently, commanders must also be capable of rapidly adjusting plans and orders for sudden stability or civil support tasks that emerge during or soon after a combat mission. In developing their vision, commanders must consider the second- and third-order effects of UO.

Information Systems

4-65. The urban environment also challenges INFOSYS that support the commander, especially communications. Urban structures, materials, densities, and configurations (such as urban canyons) and power constraints associated with man-portable radios significantly degrade frequency modulation (FM) communications. This causes problems at brigade-level and below where commanders rely heavily on constant FM radio contact with subordinates. Tactical communication problems might also cause an inability to maintain a COP, to give orders and guidance, to request support, or to coordinate and synchronize elements of the combined arms team. Communication problems in urban areas can prevent the achievement of information superiority and contribute directly to mission failure. In UO, allocating critical or high-value communication assets will be significant and essential to the main effort.

4-66. In an urban environment, units and staffs properly prepare for and mitigate the communication problems in urban areas (see figure 4-3). Adequate communications, in most cases, are ensured by—

- Training in and use of retransmission and relay sites and equipment, which may include unmanned aircraft systems (UAS).
- Airborne command posts, satellite communications, high-frequency radios, and other redundant communication platforms and systems.
- Careful positioning of commanders, command posts, and antennas to take advantage of urban terrain characteristics.
- Detailed communications analysis for movement from one AO to another due to the likely density of units operating in the urban environment.

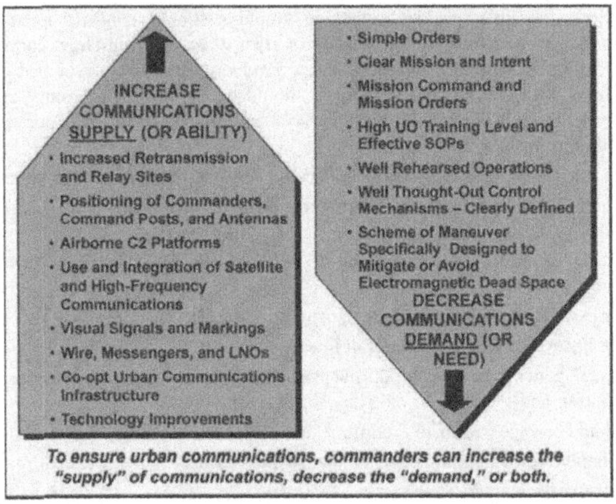

Figure 4-3. Methods to overcome urban communications challenges

4-67. Standing operating procedures (SOPs) for visual markings (both day and night) may assist in command and control. These SOPs indicate unit locations and other essential information. They coordinate with units across common boundaries. Given adequate consideration to limitations on multinational capabilities, these SOPs may assist in command and control and preclude fratricide incidents resulting from loss of FM communications. However, visual signals, including pyrotechnics, are less effective in buildings and enclosed spaces.

4-68. In defensive, stability, or civil support operations, positions often do not change as frequently as in offensive operations. Urban commanders then rely more on military wire (properly camouflaged amongst the civilian communications infrastructure), commercial communications, and messengers. Even in combat, some if not all of the urban area's organic communications structure remains intact for Army use. For example, every building may have one or more telephone distribution boxes that can control hundreds of individual telephone lines. Setting up wire communications using these points is relatively simple but, like all wire communications, is susceptible to wire-tapping. Cellular telephones can usually work well in urban areas; however, locating and destroying the repeater stations or other land-based elements of the cellular telephone system (or the effects of natural disasters) can easily disable them. (Cell phones may be a critical and only means to rapidly communicate with key civilian organizations and important community leaders.) Consequently, the C2 system may use these alternatives to FM communications but with proper operations and physical security procedures in place.

Example of Simple Communications Innovation: Israel's Six-Day War – 1967

In the 1967 battle for Jerusalem, the Israeli Defense Force placed flags on top of cleared buildings so that aircraft providing close air support could monitor the Israeli forward line of troops. They also used a spotlight during the night to mark specific buildings as close air support targets.

4-69. Units will likely use multiple means to communicate throughout the urban area. Hence, commanders emphasize proper operations security procedures (OPSEC) despite the level of security provided by the communications system. This emphasis helps to lessen the probability that Soldiers will inadvertently compromise essential information as they switch from one mode of communications to another (for example, from secure FM radio to unsecured cellular telephones or from classified to unclassified Internet domains).

4-70. Command posts above brigade-level ensure that they can communicate in an urban area without significant disruption. In stability and civil support operations, immediate and reliable communications between tactical and strategic levels may be necessary. Higher commanders anticipate that although the urban area does not significantly challenge their INFOSYS, the area may severely challenge systems at the lower tactical levels. For this reason, information flow from lower to higher may take longer. If the situation is not acceptable, the higher headquarters takes steps to mitigate it, such as increasing the number of liaison officers operating with units engaged in decisive operations. In some instances, the scheme of maneuver may be specifically designed to account for communications interference, propagation characteristics, and electromagnetic dead space. (However, this will require more time, resources, and a detailed urban communications IPB.)

4-71. Finally, urban areas can overload the INFOSYS with information. UO across the spectrum of operations can generate large volumes of information when crises threaten. This sheer volume can easily overwhelm UO commanders and command posts, and the information conduit connecting the two. Training prepares command posts to handle this volume of information and to filter the critical from the merely informative. Staffs must create products (visual or textual) that help their commanders understand the urban environment, not just present them information to know.

KEY TACTICAL CONSIDERATIONS

4-72. Commanders and planners of major UO must thoroughly understand the tactical urban battle. Especially, they must understand the effects of the environment on men, equipment, and systems. The complexity of urban environment changes and often compresses many tactical factors typically considered in the planning process. Figure 4-4 shows some of these compressed factors, however, commanders and their staffs should carefully review FM 3-06.11 for doctrine to support tactical urban operations.

Figure 4-4. Compressed tactical factors

TIME

4-73. The time available to think and act is compressed in urban combat operations. The tactical engagements that comprise battles and major UO are often quick and decisive; therefore, higher-level commanders require the ability to conduct battle command on the move so that their decision making remains correspondingly fast. The impact of decisions (or lack of) and the outcome of battle can occur in mere minutes. Often the amount of information and the number of decisions can overwhelm the overall ability of INFOSYS to respond. Commanders have little time to influence tactical actions with resources kept in reserve. Reserves and fire support assets are close to the point of decision so that they can respond in time to make a difference. The terrain causes C2 challenges that further inhibit commanders from responding quickly to changes in the situation. Small unit leaders receive training that emphasizes understanding the commander's intent so that they can recognize tactical opportunities and can act quickly to take advantage of them.

DISTANCES AND DENSITY

4-74. Distances in UO are compressed to correspond to the density of threat forces and noncombatants. In open terrain, squads, platoons, and companies may be able to control or influence thousands of meters of space. In UO, large buildings can absorb the efforts of several companies or battalions. Crowds of thousands can assemble in areas of a few hundred meters requiring correspondingly large forces for control. Maximum engagement ranges, as influenced by the urban terrain, are usually closer. Units may require field artillery for direct fire at targets ranging fewer than a hundred meters. Commanders and staffs understand the telescoping nature of the battlefield, the density of threat forces, and the density of noncombatants. In addition to the actual conduct of urban tactical operations, these factors will directly affect training, planning, force deployment, and strength.

4-75. Time-distance considerations are especially important throughout planning cycles. Though distances may be short, the physical nature of the environment can drastically change the planning factors for unit movements. The advance of a battalion may be measured in hundreds of meters per day. Thus, all time and distance calculations that relate to sequencing of forces, synchronizing combat power and other capacities, and making decisions require reevaluation based on the urban conditions.

4-76. Although it may appear counterintuitive, airspace above the urban area may be also be severely compressed as multiple fixed-wing, rotary-wing, and indirect fires (including multinational assets) compete for the same space. Increasing this density is the use of airspace by civilian aircraft and the proliferation of tactical UAS. Due to the potential for a high volume of air traffic, commanders and planers pay close attention to the integration and deconfliction of airspace over urban areas. Commanders should consider specific techniques and procedures for UAS which may include—

- Defining airspace control areas.
- Regulating flight times.
- Creating altitude restrictions.
- Including UAS on the air tasking order.

COMBAT POWER

4-77. The urban terrain can also compress combat power. This terrain increases the utility and effects of some weapons and systems, increasing overall combat power. One system that dramatically demonstrates this effect is the sniper. In open terrain, snipers slightly influence operations. In UO, snipers—well concealed, positioned, and protected—can take on significance disproportionate to their combat capability in other situations (see FM 23-10).

4-78. The density of ground combat power in a given size area is also increased because of the effect of the terrain on ranges. The complex terrain precludes standoff engagement from extended ranges by dispersed forces. Commanders often position weapon systems closer together and at shorter ranges to mass effects on the same target. Thus, commanders may position armored vehicles, which typically position themselves Hundreds of meters from friendly troops and other vehicles, within a few meters of each other to provide mutual support. Targets, which in open terrain are engaged at thousands of meters, are engaged in tens of meters on the urban battlefield.

4-79. The dense clutter of the urban environment also affects target acquisition. Systems, such as radar optimized for open terrain, will not be able to acquire targets as effectively. Decreased acquisition capability equates to diminished combat power. It may also require increasing the density of acquisition systems to compensate for reduced capability.

4-80. Finally, the density of combat power may also increase the vulnerability of Army forces. Many Army systems are protected from enemy systems at longer ranges. The number of enemy systems that can threaten Army forces at a short range increases dramatically. Lack of dispersal will make it more likely that multiple Army systems can be targeted by a single enemy threat.

LEVELS OF WAR

4-81. The levels of war are also compressed in the urban area. The tactical actions of individuals and small units can directly influence operational and even national and strategic objectives. Conversely, the decisions of the President can directly affect the conduct of tactical operations. UO have short cause and effect links between the tactical, operational, and strategic levels of operations. Because of the close media scrutiny of UO, the President can sometimes observe the actions of platoons in real time. For example, the media may film a platoon applying nonlethal force for crowd control. The President can view that film on the nightly news before the platoon even disengages from the action, much less reports formally through the various levels of command. If appropriate, the President can decide and direct the strategic and operational commanders to adjust ROE before the platoon has reported. Therefore, commanders at all levels must understand the urban environment's potential compressive effects on the levels of war. A major impact of these effects can be a lower tolerance for tactical errors and a greater need for detailed planning and precision in execution and weapons' effects (lethal and nonlethal).

DECISION MAKING

4-82. The nature of the urban environment compresses the time available to make decisions and increases the number of decisions to make. This is particularly true at the lower tactical levels. Units observing an urban AO face more potential unknowns than in other situations. A large structure presents many more potential firing positions that are observed than simpler terrain. Movement in one of those windows forces the soldier or unit to quickly make a decision regarding the nature of the target—deciding whether it is a threat or a noncombatant. Incorporating combatant and noncombatant discriminatory considerations into all live-fire training will improve Soldiers' ability to make these critical judgment decisions. Overall, lack of understanding regarding the urban environment requires commanders to rely more on analytic decision making, while greater understanding facilitates adaptability and allows commanders to make rapid and intuitive decisions while still accounting for second, third, and higher-order effects (see Figure 4-5 and FM 6-0). Realistically, however, all decisions made during urban operations will likely require a combination of analytic and intuitive decision-making abilities.

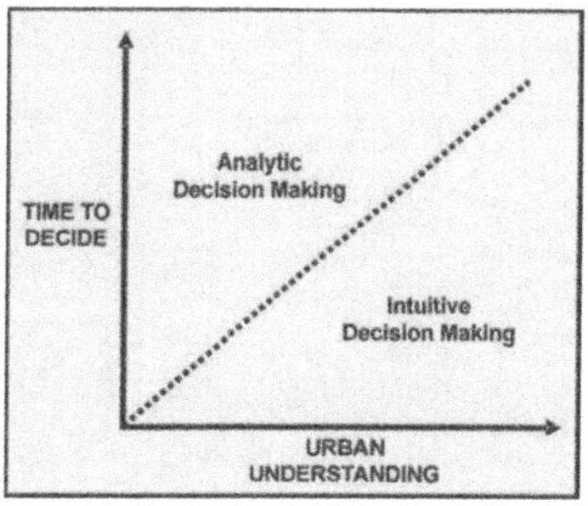

Figure 4-5. Urban understanding and decision making

This page intentionally left blank.

Chapter 5
Contemplating Urban Operations

We based all our further calculations on the most unfavorable assumptions: the inevitability of heavy and prolonged fighting in the streets of Berlin, the possibility of German counter-attacks from outside the ring of encirclement from the west and south-west, restoration of the enemy's defence to the west of Berlin and the consequent need to continue the offensive.

General of the Army, S. M. Shtemenko
describing operational level planning for taking Berlin
The Soviet General Staff at War

In any potential situation and in any area, Army commanders will likely need to assess and understand the relevance and impact of one or more urban areas on their operations. They will also need to determine whether full spectrum urban operations (UO) will be essential to mission accomplishment. UO may be the commander's sole focus or only one of several tasks nested in an even larger operation. Although UO potentially can be conducted as a single battle, engagement, or strike, they will more often be conducted as a major operation requiring joint resources. Such actions result from the increasing sizes of urban areas. Army commanders of a major urban operation must then ensure that UO clearly support the operational objectives of the joint force commander (JFC), requesting and appropriately integrating critical joint resources. Whether the urban operation is the major operation itself or one of many tasks in a larger operation, Army commanders must understand and thoroughly shape the conditions so subordinate tactical commanders can engage and dominate in this complex environment.

A ***major operation*** is a series of tactical actions (battles, engagements, strikes) conducted by various combat forces of a single or several services, coordinated in time and place, to accomplish operational, and sometimes strategic objectives in an operational area.

DETERMINING THE NECESSITY OF URBAN OPERATIONS

5-1. Early in planning, commanders of a major operation must address the necessity and feasibility of conducting operations in urban areas located throughout their areas of operations (AOs). Chapter 1 discussed strategic and operational considerations that compel forces to operate in urban areas. These reasons include the location and intent of the threat force; critical infrastructure or capabilities that are operationally or strategically valuable; the geographic location of an urban area; and the area's political, economic, or cultural significance. Additionally, humanitarian concerns may require the control of an urban area or necessitate operations within it. Several considerations exist, however, that may make UO unnecessary, unwarranted, or prohibitively inefficient. When determining whether to operate in an urban environment, commanders must consider the operational (and accidental) risks and balance them with mission benefits. The factors shown in Figure 5-1 highlight some measures to evaluate the risks associated with UO.

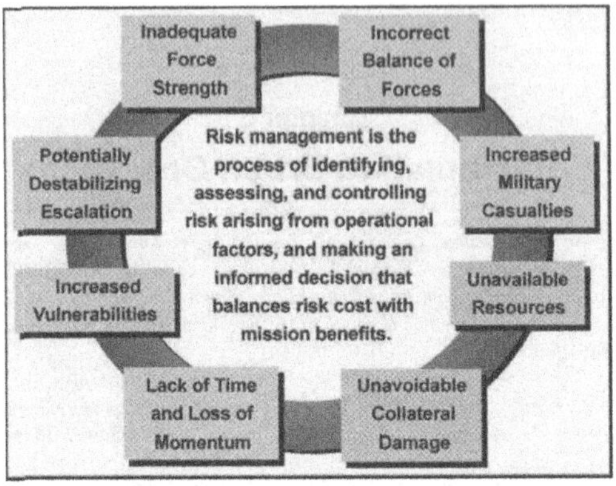

Figure 5-1. Risk management and the associated risks with urban operations

INADEQUATE FORCE STRENGTH

5-2. When facing prospective UO, commanders must consider if they have the necessary troops available to conduct the operation properly and within acceptable risk. Under normal circumstances, large urban areas require many forces merely to establish control. The New York City police department has over forty thousand officers and hundreds of support staff simply to conduct peacetime law enforcement. Major UO, particularly those that are opposed, will often require a significant number of forces. Offensive missions, for example, can require three to five times greater troop density than for similar missions in open terrain. If commanders lack sufficient force to conduct effective operations, they may postpone or consider not initiating those operations until they have the necessary strength. Commanders should also add to their analysis the requirements for troop strength elsewhere in the AO. Additionally, commanders must consider the number (and type) of forces required to make the transition from urban offensive and defensive to stability or civil support operations when determining overall force requirements—not just the number of forces required to realize objectives for major combat operations.

INCORRECT BALANCE OF FORCES

5-3. Along with force strength, commanders must consider the type and balance of forces available. This consideration includes an assessment of their level of training in urban operations. Generally, UO put a premium on well-trained, dismounted infantry units. The operational or tactical necessity to clear threat forces from the dense urban environment, hold hard-won terrain, and interact with the urban population greatly increase dismounted requirements. Therefore, Army forces conducting UO are often force tailored to include a larger infantry component. In addition, special operations forces (SOF) are invaluable in UO. SOF include psychological operations (PSYOP) and civil affairs (CA) forces. They should always be considered as part of the task organization.

5-4. UO must include combined arms to ensure tactical success in combat. Although masses of heavy forces may not be required, successful UO require the complementary effects of all Army forces. Even if an urban operation is unlikely to involve offensive and defensive operations, armor, combat engineers, and field artillery may be essential to mobility and force protection. In urban stability and civil support operations, successful mission accomplishment requires more robust CA organizations. They are also valuable in urban offensive and defensive operations. While commanders may have sufficient combat and combat support forces, they may lack enough sustainment forces to provide the support necessary to maintain the tempo. Again, commanders without balanced types of forces, to include their proficiency in operating in urban environments, should consider alternatives to UO or delaying UO until proper force types are trained and available in sufficient numbers.

INCREASED MILITARY CASUALTIES

5-5. Casualties in UO are more likely than in operations in other environments. In urban offense and defense, friendly and threat forces often engage at close range with little space to maneuver. The urban terrain provides numerous advantages to the urban defender; higher casualties occur among troops on the offensive, where frontal assaults may be the only tactical option. Conversely, defenders with limited ability to withdraw can also suffer high casualties when isolated and attacked. Casualties can be more difficult to prevent in urban stability operations because of the dense complex terrain, the close proximity of the urban population, and the possible difficulty in distinguishing friend from foe. The potential for high casualties and the subsequent need for casualty evacuation under difficult circumstances make the positioning and availability of adequate medical resources another important consideration. Additionally, high-intensity urban combat and the potential for increased stress casualties may require additional units to allow for adequate unit rotations so that Soldiers receive the rest they require.

5-6. Though casualties occur in all operations, commanders should recognize the likelihood of more casualties during large-scale or high-intensity UO. During the battle for Hue in 1968, for example, many company-size units suffered more than 60 percent casualties in only a few days of offensive operations. Commanders conducting urban stability operations must know the casualty risk and how it relates to national and strategic objectives. While a lower risk normally exists in stability operations than in offensive and defensive operations, just one casualty may adversely impact the success of the stability mission. A realistic understanding of the risk and the nature of casualties resulting from UO critically affect the decision-making process. If commanders assess the casualty risk as high, they must ensure that their higher headquarters understands their assessment and that the objectives sought within the urban area are commensurate with the anticipated risk.

UNAVAILABLE RESOURCES

5-7. Offensive and defensive operations in an urban environment put a premium on certain types of munitions and equipment. Forces may want to use vast amounts of precision munitions in the urban environment. At the tactical level, they will likely use more munitions than during operations in other environments. These munitions include—

- Grenades (fragmentation, concussion, stun, riot control, and smoke).
- Mortar ammunition (due to its rate of fire, responsiveness, and high-angle fire characteristic).
- Explosives.
- Small arms.

5-8. Soldiers need access to special equipment necessary to execute small-unit tactics effectively. In urban stability and civil support operations, this equipment may include antiriot gear, such as batons, protective clothing, and other nonlethal crowd-control devices. In urban offensive and defensive operations, special equipment can include sniper rifles, scaling ladders, knee and elbow pads, and breaching equipment such as door busters, bolt cutters, and sledgehammers. Soldiers can conduct UO with standard clothing and military equipment. However, failure to equip them with the right types and quantities of munitions and special equipment will make mission success more difficult and costly. When commanders consider whether to conduct UO, they must evaluate the ability of logistics to provide the necessary resources (see Chapter 10). Considerations must include the ability to supply all Soldiers regardless of branch or military occupational specialty with urban-specific equipment.

UNAVOIDABLE COLLATERAL DAMAGE

5-9. UO require an expanded view of risk assessment. When considering risk to Army, joint, and multinational forces, commanders must also analyze the risk to the area's population and infrastructure. This comprehensive analysis includes the second- and third-order effects of significant civil casualties and infrastructure damage. Collateral damage can influence world and domestic opinion of military operations and thus directly affect ongoing operations. It also influences the postconflict physical environment and attitudes of the population. Negative impressions of the civilian population caused by collateral damage can take generations to overcome. Destroying an urban area to save it is not a viable course of action for Army commanders. The density of civilian populations in urban areas and the multidimensional nature of the environment make it more likely that even accurate attacks with precision weapons will injure noncombatants. Unavoidable collateral damage of sufficient magnitude may justify avoiding UO, which, though it may be tactically successful, may run counter to national and strategic objectives.

Lack of Time and Loss of Momentum

5-10. Commanders conducting major operations must analyze the time required to conduct UO successfully. UO can be time consuming and can require larger quantities of resources. The density of the environment, the need for additional time to conduct a thorough reconnaissance, the additional stress and physical exertion imposed on Army forces operating in urban areas, and the potential requirements to care for the needs of the urban population consume time and slow momentum. Commanders cannot permit UO conducted as a shaping operation to divert resources from the decisive operation. Nor can they allow UO to interrupt critical time lines, unnecessarily slow tempo, or delay the overall operation. Threat forces may conduct UO with the primary purpose of causing these effects. Commanders must recognize that time generally works against political and military objectives and, hence, they must develop plans and operations to avoid or minimize UO that might unacceptably delay or disrupt the larger, decisive operation. Once commanders achieve major combat objectives, however, they will likely need to shift resources and focus on the urban areas that they previously isolated and bypassed.

Increased Vulnerabilities

5-11. Commanders must weigh the potential for increased vulnerabilities when executing UO. The density of the environment makes protection (safety, field discipline, force protection, and especially fratricide avoidance) much more difficult. Forces operating in a large urban area increase their risk of isolation and defeat in detail. Joint capabilities, such as air power, may work less effectively to support a close urban battle than in some other environments. Thus, responding to unexpected situations or augmenting disadvantageous force ratios when applying joint capabilities is significantly more difficult. Although organized, trained, and equipped for success in any environment, the Army vulnerability to weapons of mass destruction (WMD) increases when forces concentrate to conduct UO. Commanders may consider not committing forces or limiting the size of a force committed to an urban area because of increased vulnerability to (and likelihood of) attack by WMD.

5-12. Fratricide avoidance is a matter of concern for commanders in all operations. The complex urban terrain and density of participating forces coupled with typical battlefield effects—smoke, dust, burning fires—and weather effects—fog, snow, rain, and clouds—immensely increase the potential for urban fratricide. Additionally, safety requirements make it difficult to fully replicate training conditions that allow leaders to become more aware of the conditions that contribute to fratricide. The effects of fratricide can be devastating to UO and spread deeply within the Army force. Critical effects include—

- Needless loss of combat power.
- Decreased confidence in leadership, weapons, and equipment. These lead to a loss in initiative and aggressiveness, failure to use supporting combat systems, and hesitation to conduct limited visibility operations.
- Disrupted operations and decreased tempo.
- General degradation of cohesion and morale.

5-13. Therefore, commanders should increase fratricide awareness and emphasis on prevention measures during UO. Causes can be procedural, technical, or a combination of the two and include—

- Combat identification failures due to poor situational understanding, lack of communication, failure to effectively coordinate, and short engagement ranges coupled with the need for quick reaction.

- Location errors involving either the target or enemy forces due to poor situational understanding.
- Inappropriate command and control and fire support coordinating measures; a failure to receive, understand, or adhere to these measures.
- Imprecise weapons and munitions effects such as, an antitank round that penetrates several walls before exploding near friendly forces.

Exacerbating these difficulties will be the likelihood of Army forces conducting operations with (or within proximity of) SOF, coalition forces, and indigenous security forces including local police. (During OPERATION IRAQI FREEDOM, the head of the Australian Defence Forces SOF in Iraq made it a point to carefully coordinate with U.S. Army forces—down to the company level—whenever his personnel would be in an American AO. He felt the danger of fratricide was a much greater risk than the loss of operational security.)

POTENTIALLY DESTABILIZING ESCALATION

5-14. In the urban environment, Army forces cannot avoid close contact with enemy forces and civilians that may potentially become hostiles. In urban stability and civil support operations, commanders should consider the chance of this contact escalating into confrontation and violence, which may become destabilizing. This consideration may delay, limit, or altogether preclude UO using Army forces.

CONSIDER ALTERNATIVES AND RISK REDUCTION MEASURES

5-15. Since UO are often high risk, commanders should consider courses of action that provide alternatives. When the objective of an urban operation is a facility, commanders should consider replicating that facility outside of the urban area. For example, a critical requirement for an airfield to sustain operations may lead commanders to consider UO to seize or secure one located in an urban area. However, if adequate resources exist (especially time and adequate general engineering support), Army forces may build an airfield outside of the urban area and eliminate the need to conduct the urban operation. Similarly, logistics over-the-shore operations may be an alternative to seizing a port facility. In some situations, the objective of UO may be to protect a political organization such as a government. Relocating the government, its institutions, and its personnel to a safer area may be possible. Commanders can also design an operation to avoid an urban area. For example, if an urban area dominates a particular avenue of approach, use a different avenue of approach. Using a different avenue of approach differs from isolating and bypassing because the entire operation specifically makes the urban area irrelevant.

5-16. If commanders execute UO, they must assess potential hazards, and then they develop controls to either eliminate or reduce the risks to Army forces. The first means to offset risk is always to ensure a thorough understanding of the urban environment and its effects on operations by all members of the force. Other measures to mitigate risk may include—

- Detailed planning to include thorough intelligence preparation of the battlefield and the development of appropriate branches and sequels.
- Integrated, accurate, and timely intelligence, surveillance, and reconnaissance (ISR).
- Clear missions and intent, which includes a well-articulated end state that looks beyond the cessation of combat operations.
- Sufficient reserves and rotation of forces.
- Vigilant physical security precautions to include increased use of barriers and other defenses, particularly when urban areas are used as support areas.
- Operative communications and other information systems (INFOSYS).
- Effective populace and resources control measures.
- Comprehensive and flexible rules of engagement (ROE) continuously reviewed to ensure they remain adequate for the situation.
- Sufficient control measures (which often include a common urban reference system) and standard marking and identification techniques that adequately consider limited visibility concerns for both air and ground forces. Measures should allow commanders to satisfactorily

control UO and minimize fratricide without unreasonably restricting subordinate commanders' ability to accomplish assigned missions. Commanders must ensure that all subordinate units thoroughly disseminate any approved nonstandard reference systems.

- Proper targeting procedures (including effective fire support coordinating measures and a streamlined legal review of targets), positive identification of targets, and controlled clearance of fires. The goal is achievement of precise (yet rapid) effects with both lethal and nonlethal means. In close air support, positive air-to-ground communications are essential to coordinate and authenticate markings.
- Well-synchronized information operations (IO) that begin before introducing Army forces into the urban environment and well through transition. Commanders should emphasize vigilant operations security (OPSEC) particularly when operating closely with the media, nongovernmental organizations (NGOs), and elements of the civilian population.
- Active and effective integrating, synchronizing, and coordinating among all forces, agencies, and organizations involved in the operation. Commanders should allow adequate planning and rehearsal time for subordinates.
- Responsive, sustainable, and flexible urban sustainment.
- Forces well trained in joint, interagency, multinational, and combined arms UO.
- The creation of adaptable, learning organizations. This requires thorough after-action analyses conducted during actual operations as well as after training exercises. In addition to official Army sites (such as the Center for Army Lessons Learned and the Battle Command Knowledge System), commanders must create a unit-level system to allow hard-won, lessons learned and tactics developed to be immediately passed on to other units and soldiers—even in the midst of an operation. This system may be technology-based, procedural, or both. (During OPERATION IRAQI FREEDOM, the 1st Cavalry Division developed an effective web-based knowledge network—called CAVNET—that allowed them to actively capture and share lessons learned among subordinate units.)

INTEGRATION INTO LAND OPERATIONS

5-17. The commander of the major operation, after determining that urban operations are required, will then integrate the urban operation into his overall operation. He does this by articulating his intent and concept for the urban operation to his subordinates. The commander of the major operation must also set the conditions for successful tactical urban operations by his subordinates. He should define ROE, focus ISR efforts, task organize his capabilities, ensure information superiority, design the operational framework, and initiate and sustain effective coordination with other agencies and organizations (see FM 6-0).

CONCEPT OF THE OPERATION

5-18. The commander's concept of the operation should address all operationally important urban areas in his AO. It should also articulate his vision of the urban operation through directions to his staff and subordinates. Subordinate commanders should address urban areas that the higher commander does not specifically address. The commander's concept should discuss each urban area in terms of task and purpose (see FM 5-0). The commander should also describes his vision of the situation's end state in terms of—

- The threat.
- The urban environment (terrain, society, and infrastructure).
- Friendly forces.
- The conditions necessary to transition control of urban areas within his AO to another agency or back to legitimate civilian control.

RULES OF ENGAGEMENT

5-19. National- or joint-level command authorities may develop urban-specific ROE. If not, Army commanders, as part of their assessment, must determine if urban-specific ROE are required for their situation and provide supplemental ROE. However, commanders must forward any conflicts or incongruities to their higher headquarters for immediate resolution.

5-20. Developing effective ROE relies on thoroughly understanding the national and strategic environment and objectives. It also relies on understanding how to conduct urban operations at the tactical level including weapons effects. For example, broad ROE may result in significant collateral damage and civilian casualties. Even in a major operation or campaign, significant collateral damage caused during UO can make postcombat operations difficult. Such damage may even change national and international public opinion or threaten the achievement of national and strategic objectives. In contrast, restrictive ROE can hamper tactical operations causing mission failure, higher friendly casualties, or both. ROE are often part of essential elements of friendly information (EEFI), protected to reduce the potential for threat exploitation. Even in a limited urban operation, ROE will frequently need to change as circumstances warrant. Therefore, commanders should plan ROE "branches" for anticipated changes in the operational environment.

5-21. In urban operations, ROE are flexible, detailed, and understandable. They should preclude the indiscriminate use of deadly force while allowing soldiers latitude to finish the mission and defend themselves. ROE should recognize that the urban area is not homogenous and may vary according to the key elements of the threat and environment: terrain, society, and infrastructure. To be effective, ROE are consistent throughout the force (an increased challenge in multinational urban operations), and soldiers are thoroughly trained and familiar with them.

Enemy Considerations

5-22. The nature of an urban enemy affects ROE as well. Commanders must consider the type of enemy weapon systems, the degree of defensive preparation, the ability to target enemy vulnerabilities with precision systems, and the ability to distinguish combatant from noncombatant.

Terrain Considerations

5-23. ROE may vary according to the terrain or physical attributes of an urban area. Physical factors may drive the ROE to preclude certain types of munitions. For example, if the construction of a portion of the area is sensitive to fire, then ROE may preclude using incendiary munitions in that area. The ROE may lift this prohibition when units move into areas of masonry construction. Toxic industrial chemicals or radiological contaminants in an industrial area may also affect ROE.

Societal Considerations

5-24. The societal or human dimension of the urban environment will often affect ROE the most. Commanders must base the ROE development on a thorough understanding of the civilian population and threat. They evaluate the loyalty of the population, its dynamic involvement in activities that affects the operation, and its size and physical location. A population that is present and supports Army forces will likely elicit more restrictive ROE than a hostile population actively supporting forces opposing the Army forces. A neutral population, not actively involved in behavior affecting Army forces, supports consideration of more restrictive ROE. In all cases, ROE conforms to the law of war. However, ROE may be much more restrictive than the law of war requires.

5-25. The location of the population also affects ROE. The evacuation or consolidation of noncombatants into controlled, safe areas may result in less restrictive ROE. A U.S. or allied population that remains in the urban area conducting routine business in and amongst Army forces during noncombat UO will normally require the most stringent ROE.

Infrastructure Considerations

5-26. Commanders must consider the urban infrastructure when developing ROE. Urban infrastructure vital to current or future Army operations may dictate that commanders adjust ROE to ensure that critical elements of the infrastructure remain intact during the conduct of operations. If Army forces conduct an urban operation to capture port facilities, for example, the ROE should address damage to the key facilities that are the objective of the operation.

RESOURCE ALLOCATION

5-27. Commanders of a major operation must ensure that subordinate tactical commanders have the resources necessary to conduct UO effectively. They assign appropriate forces to subordinate commanders tasked to conduct UO; support them with Army forces at the operational level; and request and coordinate their support by joint resources.

Task Organization

5-28. Task organizing subordinate units for urban operations will depend largely on the nature of the operation. Some units, however, are always part of the task organization to ensure the success of UO. Infantry, CA, aviation, military police, PSYOP, military intelligence, and engineers are units required for all urban operations across full spectrum operations. Other type forces—such as armor, artillery, and chemical—have essential roles in specific types of urban operations but may be less applicable across the spectrum of operations. Commanders and staffs of a major operation must understand their mission, the particular urban environment in which they operate, and the general effects of the environment across the warfighting functions to allocate the appropriate forces to their tactical commanders. See Chapter 4 for details.

Operational-Level Support

5-29. Commanders of a major operation should also support the tactical commander with forces remaining under their direct control. These forces can include Army SOF, such as CA, PSYOP, and Special Forces, ground and air cavalry, aviation, logistics, engineers, and communications support. These forces may not be under operational control of the supported command, but their efforts are synchronized and coordinated.

Coordinating and Requesting Joint Support

5-30. Commanders of a major operation will often provide forces to the JFC as well as receive assets. They will also coordinate for and integrate joint assets to support the tactical battle. These assets will usually include air support, such as close air support, tactical airlift, and aerial reconnaissance and surveillance. Intelligence support comes in the form of reach to strategic and national intelligence capabilities and to space-based systems. This reach to space assets provides reliable, robust long-range communications, environmental monitoring, ISR, positioning and navigation, and warning of enemy missile launch. Special operations capabilities can assist the tactical mission with special reconnaissance and direct action (including special operations aviation) against high-payoff targets. Joint resources also provide the Army forces augmentation by Marine ground forces. In coastal areas, Navy forces and Coast Guard elements assist Army forces with security, sealift, and fire support. Additionally, the potential agility, reach, and power projection provided by Navy forces and platforms, combined with their self-sustaining capability, can play a vital role in supporting UO. This is especially true when UO are conducted in the littorals, require forcible entry operations, and when host nation infrastructure is severely taxed or destroyed. Commanders of a major operation will coordinate with the JFC regarding available joint resources and their allocation. They then ensure that their efforts coordinate with and complement those of tactical Army forces in the urban area. Appendix D discusses the potential contribution of joint capabilities to Army UO.

URBAN ISR

5-31. Commanders at all levels require accurate and timely information to conduct assessments for successful urban operations. This is critical to planning and execution. Senior commanders have a large role in coordinating the urban ISR effort. National strategic sources (as well as open sources) provide most of the information that commanders and staffs require on the characteristics of the human dimension, the physical properties of the terrain, and the infrastructure. The general characteristics of these aspects of the urban environment do not change drastically over time, with one exception. Military operations or natural disasters can change physical characteristics radically. Analysts can obtain crucial information through diligent research of intelligence databases and open sources. However, the disposition and composition of the urban threat is time sensitive and not likely to be discovered through this type of investigation. Due to the effects of the urban environment, deceptive efforts may influence the threat more easily. The urban population is dynamic and critical aspects of this dimension must be updated or confirmed as a prelude to urban operations. Surveillance and reconnaissance (including geospatial assets) provide relevant information regarding threat dispositions, composition and the state of the population, and the specifics of the urban terrain. Successful urban operations depend on the successful conduct of urban reconnaissance. (See also the earlier discussion of effects on the intelligence and command and control warfighting functions in Chapter 4. While counterintelligence (CI) is an essential ISR component, it will be covered later in this chapter under IO.)

Challenges

5-32. The most significant challenge to urban ISR is physical. The physical organization and complexity of the urban terrain, both man-made and natural, challenges national strategic, operational, and tactical ISR capabilities. Commanders should understand the challenges when planning and allocating time and resources to their ISR efforts. They acknowledge that subordinate commanders will face similar challenges. Therefore, commanders consider subordinate capabilities, limitations, and needs when planning, requesting, allocating, and prioritizing ISR assets and capabilities.

Imagery Capabilities

5-33. A significant national, strategic, and operational ISR capability is imagery. Even at the tactical level, units employ unmanned aircraft systems (UAS) and vehicle and dismounted electro-optics, commercial cameras (still and video), and co-opt existing civilian surveillance systems to gain situational understanding. However, the structures of the urban area significantly degrade the information that imagery acquires and may make it susceptible to physical deception measures.

5-34. Imagery, however, is an excellent source regarding the arrangement and nature of many other physical aspects. It can be used to create updated mapping products and provide significant detail of major portions of the infrastructure. Imagery can also reveal what may be happening in structures through detailed study of patterns and other exterior indicators. Yet, the bulk of a skillful threat's forces, well positioned and concealed inside or underneath structures in the urban area, are largely immune from rapid detection by overhead imaging systems. The volume of civilian movement in an urban area will itself provide a degree of camouflage and increase the difficulty of employing pattern analysis. The success in 1999 of the Yugoslavian army concealing heavy forces when confronting NATO indicates the limits of these assets to penetrate an urban area. This leaves a requirement to supplement overhead imagery in the urban environment with a reliable vehicle-mounted or handheld imaging system. For example, thermal sights on a vehicle patrolling an urban street during limited visibility may note the hot engine of a vehicle on the side of the road, possibly indicating suspicious activity. Simply providing patrols with digital cameras or video cameras can also help provide needed information and fill in information gaps.

5-35. The requirement for high altitude and low oblique imagery in more open terrain has been supplemented by a growing need for detailed, close-up, and nearly vertical overhead shots of specific portions of the urban area. This forces collection managers and system operators to move expensive and scarce imagery collection systems closer to the target or to employ flight profiles that place them at greater risk from threat air defense systems. While gun-camera images from aircraft can provide a stand-off reconnaissance platform, UAS imagery may be one of the fastest, least risky methods by which commanders can conduct reconnaissance to update and verify current maps of the urban area and determine routes, obstacles such as damaged and destroyed buildings, and intact and destroyed bridges and other significant infrastructure. Increased UAS proliferation will require greater attention to airspace command and control procedures as well as the movement and protection of ground control stations.

5-36. As the media will likely be present in the urban environment, commanders may consider devoting intelligence resources (to include linguists) to monitoring the various local, national and international sources, particularly the medium of television. The media coverage of threat forces may provide important intelligence and relevant information of immediate tactical value. Commanders, however, should always be cognizant that threats may recognize that Army forces will be using this as a source of intelligence and, feeling no compulsion to be truthful, may use it for their own deception means.

Electronic Capabilities

5-37. The physical attributes of the urban area also diminish the effectiveness of electronic ISR capabilities. Buildings and other structures significantly disrupt radio communications in an urban area. Buildings not only make tactical radio communications difficult for the user, they also make them difficult to locate, intercept, and jam. The range and clarity of frequency modulation (FM) signals significantly diminish when antennas are located inside buildings or when buildings block line of sight between the source and receiving station. To mitigate this effect on signals intelligence means, detection capabilities often move closer to the transmission source. Thus, the threat's vulnerability to compromise by means of his FM and other wireless communications in an urban environment is much less than in many other environments.

5-38. Without losing tactical surprise and increasing risk, units cannot effectively use many electronic detection and surveillance capabilities until urban combat is imminent or perhaps already begun. Ground surveillance radars will have limited uses in urban areas because of the lack of wide-open spaces in which they best operate, but can be used along the periphery of urban areas to detect infiltration and exfiltration of threat forces. However, remote systems such as the remotely monitored battlefield sensor system and the platoon early warning device may have great utility in monitoring many of the numerous avenues of approach that cannot be kept under human surveillance due to manpower constraints. Among many urban applications, remote sensors and early warning devices can be used to monitor subsurface passageways, the entrances and exits to buildings, fire escapes, the perimeters of sustainment bases, and covert foot trails between urban and rural areas.

Human Capabilities

5-39. The limits on imagery and electronic ISR capabilities place a premium on human-based visual reconnaissance. Commanders have three types of human reconnaissance assets to augment electronic reconnaissance resources: special reconnaissance, conventional combat reconnaissance, and human intelligence (HUMINT) gathered by military intelligence from individuals and multimedia. The urban environment poses several challenges to these capabilities.

5-40. The urban area challenges special reconnaissance in several ways. First is the access to the urban area. Although avenues of approach may be numerous, concealed avenues of approach into a defended urban area may be limited and thoroughly covered. Air access is also more difficult because aircraft are detected more easily, airspace is smaller, drop and landing zones are limited or not secure, and more air defense systems probably exist. Still, special reconnaissance efforts to penetrate the urban area can be successful using unconventional techniques including high-altitude low-opening parachutes or underwater penetration.

5-41. Special reconnaissance then faces a second challenge: moving in and identifying targets in the urban area. Stealth movement in an occupied urban area is exceptionally difficult. Repositioning to new or alternate positions is also dangerous. The soldiers' ability to conceal themselves among the civil population can mitigate some of these challenges but includes inherent risks of a different nature. Also difficult is establishing observation positions that provide a field of view of several targets.

5-42. Lastly, special reconnaissance may face navigational and reporting challenges. Special reconnaissance's ability to locate themselves and communicate critical locations and routes are challenged by—

- Differences in language and numbering systems. Road signs, if they exist at all, may be removed or deliberately altered to confuse.
- Irregular street patterns.
- Outdated maps.
- Intervening structures that impede communications and global positioning systems.
- Changes to the landscape due to the effects of UO or natural disasters.
- Featureless shantytowns.

5-43. Conventional reconnaissance faces many of the same challenges as special reconnaissance. Conventional reconnaissance also may lack the advantage of surprise and the special equipment and training that provides special reconnaissance stealth capability. Conventional reconnaissance is not likely to operate undetected by the civilian population. Given the constraints discussed above on other sources, conventional reconnaissance units will likely begin their mission with much less information than they would have on threat dispositions in a less complex environment. Commanders may choose to have their reconnaissance elements fight for information in the urban area. While this high-risk option is more favorable under fluid conditions, it can be used at any time. It requires careful planning, rehearsal, and formulation of information requirements.

5-44. As previously indicated, human intelligence will be one of the most valuable sources for information regarding the situation inside an urban area. HUMINT may take advantage of the proximity and large numbers of potential informants to gather information about threat activities and capabilities. It is especially valuable because it can address all elements of the environment. HUMINT sources can describe political and religious nuances that may be significant to commanders. Such information is useful for insights regarding the human dimension but extremely difficult to obtain from other means. This intelligence also can describe the infrastructure uncovering essential details of how the infrastructure functions. Obtaining good HUMINT requires skilled interrogators and linguists—often in large numbers. Commanders must know and account for some of the possible shortcomings of HUMINT:

- It is susceptible to the influence of the threat; the threat can threaten and influence the source as well as local interpreters. Commander may need to consider that people may not want to talk through a local interpreter because they are providing information they might not want others to know; anonymity may be essential. On the other hand, a local interpreter may provide community access. Yet, those trusted by the locals may not be truthful with Army forces. Conversely, those truthful with Army forces may not be trusted by the indigenous population particularly if these interpreters come from another country, region, or demographic group. In such cases, commanders may need more than one interpreter.
- It is limited by the accuracy of the source and interpreter's perceptions. A local interpreter may have difficulty understanding and translating military topics or the intricacies of urban infrastructure. Some local translators may want to please both participants, so they may shade the conversation, telling each side what they want to hear. Finally, accuracy may be affected by the collector's own perceptions if not firmly grounded in an understanding of social, cultural, and other local conditions.
- Multiple echelons collecting from a given area may inadvertently collect from the same sources and thereby give the false impression of independent sources confirming the same data. To help mitigate this possibility, commanders should ensure that all HUMINT sources are properly recorded on intelligence source registers.
- It may not be timely. The process of identifying and cultivating a source (particularly in an environment where most civilians support threat forces), gathering information, analyzing the information, and providing the intelligence to commanders can be extremely time consuming.
- Some informants may come from unscrupulous or sordid elements of the urban society and may also have their own agenda. They may attempt to use protection afforded them by their relationship with Army forces to conduct activities (even atrocities) that will compromise

political and military objectives. In some cases, a local interpreter's own agenda may influence his translation. (For example, a local schoolteacher used as a translator may report to the collector that the source believes that repairing schools in their urban area is critical, when actually the source may have said that potable water is most needed.)

Conducting Urban ISR

5-45. To be successful, ISR efforts (national to tactical level) are exceptionally comprehensive and synchronized. Moreover, success necessitates the integration of all ISR sources into operational and tactical planning. This requires that ISR assets deploy and execute early, diversify, properly focus, and integrate under a comprehensive ISR plan. Successful ISR also requires flexibility to adapt to the operational and tactical needs of the commander (see Figure 5-2). Commanders must ensure that the appropriate intelligence architecture (including robust links to joint and multinational elements) is established and tested prior to execution of ISR effort.

Early Deployment

5-46. One of the first requirements for effective urban ISR is the early deployment and employment of assets. The complex urban terrain presents a significant challenge. Commanders should consider that ISR assets will normally take longer to gather data amid the complexity.

5-47. Limited national, strategic, and operational imagery intelligence (IMINT) and signals intelligence (SIGINT) capabilities should be considered and quickly requested. If they are approved, they are tasked and deployed or repositioned to begin urban ISR operations. This takes time. Fortunately, spacing the ISR effort over time permits the analysis of the information or data as it is received. Such time also permits subsequently refining the ISR effort before all assets are committed.

5-48. SOF or conventional units will require significantly more time to execute reconnaissance missions and maintain an acceptable survivability rate. Urban reconnaissance operations require additional time for stealthy insertion into the urban area. IMINT and SIGINT capabilities are used to identify possible locations of high-value targets and corresponding observation positions; this helps minimize time-consuming and high-risk repositioning in the urban area. Again, reconnaissance units may require extensive time to observe from observation positions for indicators of threat activity and disposition and identify patterns.

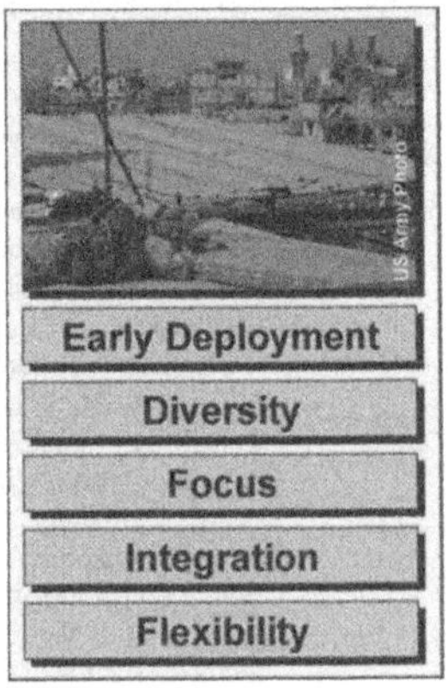

Figure 5-2. Urban ISR considerations

5-49. As conventional combat forces prepare to commit to the urban area, conventional reconnaissance precedes their actions. Conventional reconnaissance will often be a slow and methodical effort. Such forces may need time to reconnoiter the interior of structures for snipers and other small threat teams. They may also need time to deploy and destroy snipers and small delaying elements and to breach harassing obstacles. If necessary, they may need time to mass the combat power necessary to fight through enemy security and continue the reconnaissance.

Diversity

5-50. No single ISR capability can solve the riddle of the urban defense. In order to gain an accurate common operational picture of the complex urban terrain necessary to focus combat power on decisive points, commanders must employ diverse ISR capabilities. These capabilities will each contribute pieces of relevant information to permit identifying operational objectives and leveraging tactical combat power to achieve those objectives quickly. Higher-level commanders should understand that tactical reconnaissance capabilities alone often cannot provide all the tactical information required for success at lower echelons.

5-51. Using diverse capabilities challenges the threat's ability to conduct deception and otherwise defeat the friendly ISR effort. A threat that focuses on minimizing his vulnerability to overhead imagery (for example, satellites) may increase his reliance on communications and thus his vulnerability to SIGINT. At the same time, he may decrease his ability to detect the actions of ground reconnaissance units. Conversely, a threat that actively campaigns to detect ground reconnaissance may make himself more vulnerable to SIGINT and IMINT. Moreover, as Army forces enter urban areas, threat activity and susceptibility to SIGNINT often increases.

5-52. Diverse capabilities also facilitate the tactical ISR effort. Tactical reconnaissance units often consist of small dismounted teams and small combined arms teams with a dismounted element and an armor-protected mounted element. Combat engineers with their technical expertise and breaching capability are essential to the combined arms reconnaissance effort. The teams' movements are synchronized and coordinated with other assets, such as UAS and air cavalry reconnaissance. These teams use several movement techniques including infiltration, with the primary objective of conducting zone reconnaissance along key axes that support brigade and battalion actions against decisive points. To accomplish this mission, reconnaissance reconnoiters the proposed routes and alternate approaches. This supports deception and contingency planning. (Reconnaissance of alternate routes and objectives also applies to aerial reconnaissance; helicopters and UAS are not invisible. They can alert a threat to impending operations.) Infiltration of dismounted reconnaissance is made easier when a threat focuses on combined arms reconnaissance teams. Aerial reconnaissance, such as air cavalry and UAS, provides early warning of threat elements to ground reconnaissance, identifies obstacles and ambush sites, and helps select the routes for ground reconnaissance. Air elements may also reduce the mobility of counterreconnaissance forces.

Focus

5-53. Another key to successful ISR is the ability to focus the assets on commander's critical information requirements (CCIR). This focus begins with mission analysis and the commander's initial planning guidance. It is incrementally refined throughout planning and execution as each ISR effort provides information and permits more specific focus in subsequent efforts. The size and complexity of the urban environment require that the ISR effort center strictly on decisive points or centers of gravity (COGs). Therefore, the overall ISR effort will have two major focuses. The first is to uncover and confirm information on the decisive points and COG. The second is to determine the approaches (physical, psychological, or both) leading to the decisive points. The first focus will likely drive ISR in support of major operations. The second focus will likely provide the impetus for tactical ISR efforts. For example, special operations reconnaissance might focus on a major command center that controls the entire urban area and that is one of the higher-echelon command's CCIR. Tactical reconnaissance might focus on the nature of the defense along a particular avenue of approach to the objective.

Integration

5-54. Another important aspect of urban ISR is integration. All reconnaissance capabilities provide both distinctive information as well as information that confirms and adds to that coming from other sources. Essential to urban ISR is the link between all of these sources, either directly or through an integrating headquarters.

5-55. ISR operations must be vertically and horizontally linked. Vertical links ensure that ISR operations among the various levels of command are complementary and that the information flow between these levels is rapid. Horizontal links ensure that forces operating in close proximity (particularly adjacent units), where areas of interest overlap, can rapidly share results of their individual ISR efforts. Together, this helps ensure that all Army forces share a common operational picture and permits the greatest flexibility and survivability of ISR resources.

5-56. ISR operations also are integrated into the planning system, especially the targeting process. As part of targeting, positioned reconnaissance and surveillance elements may become the trigger and terminal control for applying precision fires when appropriate and after considering the risks of compromise of the position or platform. ISR operations and plans must also be synchronized with the air tasking order, as many surveillance platforms are coordinated on the same timeline.

Flexibility

5-57. The urban ISR effort must be more flexible than in other operations. This flexibility permits the ISR effort to meet unforeseen circumstances and to deal with the challenges of the urban environment. As indicated previously, the urban environment is particularly difficult to penetrate. The practical effects of this characteristic are that—

- The initial ISR effort may not be as successful as in other operations.
- More intelligence requirements may be discovered later while executing ISR operations than otherwise.
- The threat may be more successful in active counterreconnaissance because of the concealment advantages of the urban environment (hiding in structures as well as among the urban population).

Therefore, tactical and operational commanders should consider requesting greater than usual ISR support from higher headquarters. Higher headquarters should be proactive in augmenting units conducting urban operations with additional ISR assets. Additionally, ISR assets remaining under the control of the higher headquarters must respond more quickly to the CCIR of supported commanders. Sequencing reconnaissance missions over time provides flexibility by creating uncommitted reconnaissance assets.

5-58. Time sequencing of ISR assets is essential to flexibility. It makes ISR assets more survivable and allows the intelligence cycle to mature the CCIR. It also creates a ready ISR capability to augment committed forces in critical areas if required or diverts them around centers of threat resistance. If not required, the original ISR tasks can be executed as envisioned in planning. Cueing allows a high-value ISR asset to be able to respond to multiple targets based on an ongoing assessment of the overall reconnaissance effort and the changing CCIR. Redundancy permits the effort to overcome line of sight restrictions, the destruction of an ISR asset, and the ability to combine ISR resources to create combat power if required. Maximizing the ISR effort requires applying all available ISR assets to support the urban operation. Additionally, assets—such as air defense artillery and field artillery radars and engineer reconnaissance teams—are integrated into the ISR effort. In urban operations, units will also commit infantry and armor elements (plus their organic reconnaissance elements) into the tactical reconnaissance effort. These units increase the dismount capability and the ability of reconnaissance elements to fight for information and fight through security zones.

INFORMATION OPERATIONS

5-59. Information operations are an integral part of all Army operations and a critical component in creating and maintaining information superiority. The information environment is the sum of individuals, organizations, or systems that collect, process, and disseminate information; it also includes the information itself. In UO, the information environment is extremely dense due to the proliferation of INFOSYS and widespread access to those systems. In urban operations, commanders must consider how the urban environment, particularly the human component, uniquely relates to executing IO. They must also anticipate the threat's information operation campaign and preclude successful operations on its part. Overall in regard to the urban society, commanders must convince the urban inhabitants of the inevitable success of UO, the legitimacy of Army actions, and the beneficial effects that Army success will eventually generate.

5-60. Because urban operations are likely to be joint, interagency, and multinational, commanders must guard against various forms of information fratricide as a multitude of actors plan and conduct IO in the urban area. For example, the lead governmental agency's key IO objective may be to establish the legitimacy of newly formed civilian authorities. In an attempt to develop close relationships with the civilian populace, Army commanders may continue to work closely with traditional, informal leaders to the exclusion of the new authority. These actions, while they are often conducted out of practical and immediate necessity, may run counter to the lead agency's goal. Overall, Army commanders must nest their IO campaign objectives and themes within those of the lead agency, aggressively coordinate with other governmental agencies and coalition partners, and synchronize activities down to the tactical level to prevent working at odds and avoid information fratricide. However, the process must not be so centralized and rigid that subordinate units lose the flexibility necessary to develop their own products and themes essential to addressing their own AO and the particular situation that confronts them.

5-61. IO are executed using core and supporting elements and related activities (see Figure 5-3 and FM 3-13). The elements of IO are employed in either an offensive or defensive role. Many aspects of IO are not affected differently in an urban environment from any other environment. The following sections outline some IO considerations unique to urban operations.

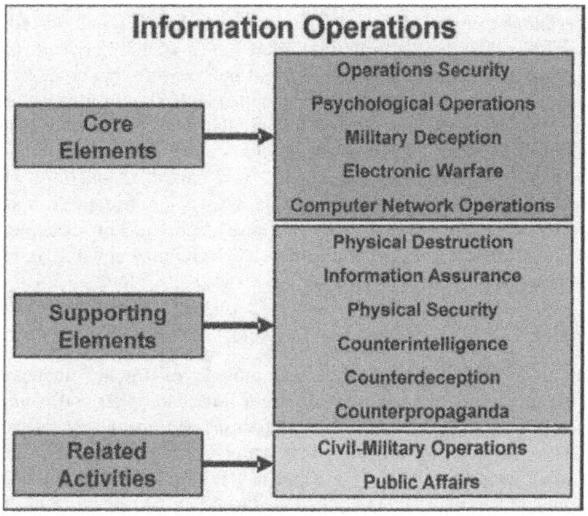

Figure 5-3. IO elements and related activities

Operations Security

5-62. In the urban environment, Army forces can leverage existing urban infrastructure, including the communications and information infrastructure, to enhance Army operations. The danger in integrating these systems is violating OPSEC. Commands ensure that Army forces use only approved systems and proper safeguards exist. Commands also supervise subordinate units for inadvertent breaches of OPSEC policies when using existing urban systems. Finally, established OPSEC procedures must be reasonable and practical. Overly complicated and stringent procedures increase the likelihood that unintentional OPSEC violations may occur as Army forces seek to rapidly accomplish the mission.

5-63. Of particular concern are computers (see also the discussion of Computer Network Operations below). During longer-term UO, commanders will constantly be upgrading and improving the quality of life for their soldiers through the establishment of well-resourced forward operating and sustainment bases. This will likely include the provision of unclassified access to the Internet and e-mail allowing Soldiers to communicate with friends and families around the world. E-mail addresses, facsimile numbers, cell phone numbers, photographs, and other sensitive (but unclassified) information can be valuable sources of threat information. Soldiers must be periodically trained and reminded of everyday procedures to combat potential threat exploitation of these unclassified sources (such as not opening e-mails from unknown sources as they may harbor Trojan programs and viruses).

5-64. To compound concerns, commanders will also establish protected systems to disseminate tactical lessons learned to individual soldiers. This creates the potential that Soldiers can inadvertently release classified information about ongoing operations to threat intelligence systems. In addition to OPSEC training and persistent awareness, commanders must establish procedures to shutdown unclassified Internet and e-mail access immediately, both before and after critical events, to mitigate potential OPSEC violations. (Commanders may also need to randomly shutdown Internet and e-mail access as part of their overall OPSEC measures to preclude such information blackouts serving as a potential warning of impending operations.) Technical surveillance countermeasures—the identification of technical collection activities conducted by threat intelligence entities—will be an important CI service to identify this and other technical OPSEC concerns.

5-65. The close proximity of Army operations to a civil population, particularly in stability and civil support operations, makes Army activities themselves an additional OPSEC concern. Hostile civilians or other threats integrated into the urban population may have more chances to observe Army activities closely. Such observations can provide insight into tactics, techniques, and procedures (TTP) and expose operational vulnerabilities. However, threats may coerce even friendly civilians to provide a threat with EEFI. Therefore, commanders in an urban environment must ensure that civilians cannot observe critical TTP. Any observable patterns and TTP vary and are supplemented with deception efforts. Physical security is increasingly important in urban areas to control civilians' access. Although many urban operations require close coordination with NGOs, commanders should screen information provided to these organizations to protect EEFI. Release of EEFI to NGOs is controlled and done with full recognition and understanding of potential consequences—the benefits must far outweigh the risks involved. (Even the best-intentioned NGO might inadvertently compromise security. For example, by revealing that a commander has closed a particular route the next day, the NGO may unwittingly reveal the objective of a pending offensive operation.)

Psychological Operations

5-66. PSYOP aim to influence the behavior and attitude of foreign audiences, both military and noncombatant, in the urban environment. PSYOP are a force multiplier and contribute in many ways to mission success (see FM 3-05.30). Effective PSYOP focuses on transmitting selected messages to specific individuals and groups in order to influence their actions. Their ability to influence the attitudes and disposition of the urban population cannot be overstated. While the complexity of the societal component of the urban environment can make PSYOP challenging, the urban society also offers many options and resources. Potentially, PSYOP (with other political and economic actions) may help limit or preclude the use of military force in urban areas. In some circumstances, UO may be relevant to the major operation only in terms of their psychological effect.

5-67. The positive influence created by PSYOP is often useful in developing an effective HUMINT capability particularly in an urban area where many civilians actively or passively support the threat. Persuading and influencing a few to support friendly forces may pay great dividends. These few supporters may allow Army forces to penetrate the urban area and obtain essential information. Such information can apply to threat capabilities, threat intentions, and even the urban environment itself.

5-68. PSYOP, combined with other elements of offensive IO, aid in isolation of a threat—a critical shaping action for any urban operation. For example, commanders may use PSYOP to inform civilians about new food distribution points located away from urban combat operations. This action supports the UO fundamental of separating combatants from noncombatants and helps to further isolate the threat (both physically and psychologically) from the civilian populace. Aside from projecting a positive image of friendly forces over threat forces, PSYOP can also isolate the threat by identifying and exploiting ethnic, cultural, religious, and economic differences between the elements of the civilian populace and threat forces as well as the differences among supportive and unsupportive civilian factions. The complexity of the urban environment enables quick changes in opinion or attitude. Commanders must continually evaluate the results of PSYOP for mission relevance.

Military Deception

5-69. Urban operations present numerous challenges to tactical commanders; however, higher-level commanders may help to mitigate some challenges through effective military deception Commanders can use military deception efforts designed to mislead threat decision makers as to friendly force disposition, capabilities, vulnerabilities, and intentions. Military deception actions may allow commanders to achieve tactical surprise or improve relative combat power at a selected location. For example, allowing the threat to observe certain activities on a selected avenue of approach may cause the threat to shift his forces (and effort) to the area perceived to be threatened. (This movement may also aid in determining the overall disposition of threat forces and intentions.) Repositioned forces or effort to activities or locations that are not decisive to the achievement of friendly objectives, combined with other IO designed to overwhelm threat information and intelligence systems, may create the force and tempo differential necessary to achieve success. Commanders should tailor urban deception plans to the specific urban area, paying close attention to the societal characteristics of the target population.

Electronic Warfare

5-70. Electronic warfare (EW) includes all actions that use electromagnetic or directed energy weapons to control the electromagnetic spectrum or to attack a threat. Conducting EW in urban areas seeks to achieve much the same results as in other environments. A major consideration in urban areas is collateral effects on portions of the urban infrastructure that rely on the electromagnetic spectrum for service. Thus, precision is a major factor in planning for EW operations. For example, EW attacking a threat's television broadcasts avoids affecting the television broadcasts of neutral or friendly television. Likewise, EW attacking military communications in a large urban area avoids adversely affecting the area's police and other emergency service communications. Urban offensive and defensive operations will have the least restrictions on EW operations while urban stability may have significant constraints on using EW capabilities.

Computer Network Operations

5-71. Computer network operations (CNO) include computer network attack (CNA), computer network defense (CND), and computer network exploitation (CNE). CNO are not applicable to units at corps and below. Echelons above corps (EAC) units will conduct CNA and CNE. If tactical units require either of these network supports, they will request it of EAC units.

Computer Network Defense

5-72. In urban operations, CND will require extreme measures to protect and defend the computers and networks from disruption, denial, degradation, or destruction. The nature of the urban environment and configuration of computer networks provides the threat with many opportunities to interdict local area networks (LANs) unless monitored by military forces. LANs controlled by military forces are normally more secure than the civilian infrastructure. Commanders should prepare for opportunities by the threat to insert misinformation.

Computer Network Attack

5-73. Considerations regarding the execution of CNA in urban operations are similar to those of EW: CNAs that do not discriminate can disrupt vital civilian systems. However, possible adverse effects on the civilian infrastructure can be much larger—potentially on a global scale. In the short term, CNAs may serve to enhance immediate combat operations but have a debilitating effect on the efficiency of follow-on urban stability operations. Because of these far-reaching effects, tactical units do not execute CNA. CNA is requested of EAC units. EAC units will receive all requests from lower echelons, carefully consider second- and third-order effects of CNA, and work to ensure its precise application.

Computer Network Exploitation

5-74. CNE consists of enabling operations and intelligence collection to gather data from target or adversary automated INFOSYS or networks. Tactical units do not have the capability for CNE. CNE contributes to intelligence collection at EAC. In UO, CNE will be centrally controlled.

Physical Destruction

5-75. Physical destruction includes those actions—including direct and indirect fires from air, land, sea, space, and Special Forces—taken with, to augment, or supplement IO actions. Like many other IO elements, major concerns with employing physical destruction in UO are precision and follow-on effects. Thus, commanders using physical destruction to support IO must adhere to the same constraints as all other urban fires.

Information Assurance

5-76. Information assurance in UO takes on an added dimension. As with other operations, availability of information means timely, reliable access to data and services by authorized users. In UO, the timeliness of information may be restricted because structures block the transmission waves. The need for retransmission facilities may overwhelm the signal community. The reliability can be questioned because of the blockage between units and communications nodes. Unauthorized users may intercept the communications and input misinformation or disinformation. Commanders must protect the integrity of all information from unauthorized changes, including destruction. INFOSYS with integrity operate correctly, consistently, and accurately. The authentication of information may be accomplished by sophisticated electronic means. However, it is more likely that communications-electronics operating instructions authentication tables will authenticate the information. Commanders should consider the confidential nature of all information in UO and establish procedures to protect the information from unauthorized disclosure. Of additional concern, the density of the infrastructure in urban areas may inhibit receipt by the intended individual or unit. Therefore, commanders must develop alternative means to determine if the message was received.

Physical Security

5-77. Physical security consists of those actions and measures to safeguard personnel; to prevent unauthorized access to equipment, installations, material, and documents; and to safeguard them against espionage, sabotage, damage, and theft. The joint, interagency, and multinational nature of urban operations increases the need for physical security measures, especially with organizations that do not emphasize physical and operational security within their own organizations. The potential for locating sustainment bases in or near urban areas and the use of indigenous personnel on these facilities to accomplish a multitude of tasks mandates that commanders carefully consider the effectiveness of their physical security measures. In some stability operations where Army forces must recruit and train security or police forces, and then conduct joint operations and patrols with them, commanders must examine how best to establish close working relations while maintaining essential physical (and operations) security.

Counterintelligence

5-78. The urban environment, particularly in stability operations, is ideal for espionage, other intelligence activities, sabotage, or assassination. Threats can approach, conduct reconnaissance, and escape under the concealment offered by the urban environment. The dense environment and electromagnetic clutter serves to hide technological collection means. To counter threat intelligence collection activities, the Army employs CI forces. They provide analysis of the threat's human, imagery, signals, and measurement and signature intelligence capabilities to develop countermeasures against them (see FM 34-60).

5-79. In conjunction with military police, engineers, medical personnel, and other as required, CI help create urban threat and vulnerability assessments necessary in developing comprehensive force protection measures and evaluating their effectiveness. CI personnel can also supplement the screening of linguists and other critical host-nation personnel employed by Army forces. As Soldiers operate in close proximity to the urban society, threats will attempt to use three, time-tested tools to influence them: money, sex, and drugs. Periodic CI refresher training remains crucial to decrease the effectiveness of these threat tactics.

Counterdeception

5-80. In UO, threat forces can easily accomplish deception operations. The force that controls the area above and below ground will have freedom of movement. Deception aimed at friendly commanders will cause them to deploy combat power at the wrong place and the wrong time. Counterdeception by friendly commanders will identify and exploit threat attempts to mislead friendly forces. Counterdeception is difficult. Cultures of certain rhetoric and actions are more predisposed to deception than others. Knowing a threat's previous deception methods is important. Dismissing tactical indicators because they conflict with preconceptions may allow a hostile deception operation that plays on the preconception to succeed.

Counterpropaganda

Four hostile newspapers are more to be feared than a thousand bayonets.

Napoleon Bonaparte

5-81. Because propaganda is aimed at both combatants and noncombatants, UO are especially concerned with its use. Propaganda can rapidly and dramatically affect the attitudes of the urban population and will probably occur (or increase significantly) after urban operations have begun. Thus, it can create situations in the human dimension of the environment quite different from those discovered in the pre-operations assessment. Counterpropaganda is, therefore, essential to urban operations. To negate, deflect, or destroy the threat's propaganda capability, counterpropaganda requires—

- Monitoring and evaluating the effectiveness the threat's propaganda efforts.
- Determining innovative methods using all Army force capabilities, especially PSYOP and PA units.

5-82. In particular, humanitarian assistance, when coupled with an effective information operations campaign, can be a powerful counterpropaganda tool to sway public support. For example, media coverage of Army forces directly providing or coordinating NGO humanitarian efforts to feed, shelter, and provide medical care to the urban population can help counter threat efforts to discredit Army forces as an evil occupying force. This type of coverage can be used to turn opinion against the threat and could also prove an effective tool in gaining favorable international support.

Public Affairs

5-83. A closely related activity to IO is public affairs (PA). PA influences urban operations by transmitting information through the media to internal (urban Army forces as well as the urban civilian populace) and external audiences. At higher levels of command, PA can help maintain popular national support for the urban operation by clarifying the links between strategic goals and operational objectives. At both the operational and tactical levels, it links Army units, the urban inhabitants, the U.S. and international public, and the media. The media will often focus on the negative actions that are ongoing in an AO; hence, the urban commander's challenge will be to publicize his unit's good news stories. Commanders should not wait for media inquiries and interview requests. Instead, they should aggressively plan public affairs to get the Army's story out to the media and, ultimately, the public.

5-84. PA can help determine potential media issues that may influence planned UO. It can also aid commanders in understanding the impact of UO on the environment (particularly its citizens) and other agencies and organizations operating in the urban area. PA also helps to counter rumors, uncertainty, fear, loneliness, confusion, and other factors that cause stress (to both soldiers and civilians) and undermine effective UO. If the populace does not understand the mission or the Army forces' capabilities, false expectations may be created that Army forces may not be able to meet. PA can help prepare the American public for the possibility of high casualty rates. Overall, PA supports urban commanders in their goals to achieve information superiority and preserve public support.

5-85. The density of information sources and reporters in UO ensures that all Army activities will be subject to media and public scrutiny. Many reporters will congregate in cities for their own comfort and take advantage of established communications networks. Urban areas are densely populated and, together with Army forces and NGOs operating there, will present the greatest number of human-interest stories.

The local urban or host-nation media, however, will often have their own agendas developed over a longer period of time. This local media may also have a greater influence over the urban population than the international media. The indigenous media may not follow international norms. Commanders are responsible for understanding the media (particularly the local media), its role, and its potential for negative or positive influence.

5-86. Commanders cannot allow themselves to be intimidated by the media. Media sources are going to get and report their story one way or another. Therefore, commanders should support open and independent reporting and grant access to their units as early and as far forward as the situation permits. Successful units often go beyond simply accommodating the media to actually integrating them into their units. In the future, embedding of media may become the standard for reporting Army operations. Other considerations for improving urban media relations include:

- Encouraging the media to report on all ongoing aspects of the unit's urban operations—not exclusively on combat operations. This may create a unit responsibility to protect journalists that participate in covering suggested activities.
- Hosting media luncheons or other events for key journalists. This provides commanders the ability to get the whole story out to the press from their unit's perspective. To gain media interest and unit credibility, this will often require key leadership participation.
- Identifying Soldiers that speak the local and international journalists' native language. Even if journalists speak English as a second language, they may overlook or misunderstand important points due to language subtleties. Additionally, people tend to be friendlier to those that speak their own language.
- Identifying selected Soldiers at lower tactical levels to act as PA representatives for their unit as an additional duty. However, commanders must provide them with appropriate training and resources to accomplish specific PA tasks.
- Writing articles for local newspapers that emphasize the unit's good news. Depending on the situation, subordinate units may be tasked to write stories about the good accomplishments in their AO. Commanders, with appropriate OPSEC considerations, may hire locals to assist in this effort. However, commanders must take extra precautions to ensure that these efforts are not perceived as covert propaganda or as undermining free and independent news reporting.

5-87. PA does not distort, direct, or manipulate information. Its effectiveness stems directly from establishing and maintaining credibility with the urban population and media. PA protects the overall integrity of the Army as an institution by ensuring that the Army is recognized as the most reliable source for official information among all other competing sources. Commanders must synchronize PA with the core and supporting elements of IO (particularly PSYOP and counterpropaganda) to ensure that all Army sources send only one message. Urban commanders must plan for the media and integrate PA into their decision-making and (through IO) targeting processes. Similar to lethal targeting, commanders should consider target audiences and coverage areas for the various new agencies and mediums. They also pay attention to print media circulation statistics and television and radio broadcast times. For example, an urban populace may tend to read morning newspapers vice late editions and watch or listen to nightly news instead of morning or daytime programming. Additionally, low literacy rates among the urban population may indicate that television and radio are better mediums to use to influence the urban population than print media such as newspapers, magazines, leaflets, and posters.

5-88. The PA principles listed in Figure 5-4 and addressed in FM 46-1 summarize PA. They serve as useful guides toward planning and executing PA operations regardless of the environment. However, the principles of "practice security at the source," and "truth is paramount" particularly apply to the urban environment. The compartmented nature of most UO impede commanders' and their PA officers' ability to be at all places where the media will likely be. Friendly inaccuracies, intentional or not, will help fuel the threat information campaign. Therefore, all soldiers are trained, provided with clear and understandable PA guidance, and prepared to accurately communicate to the civilian media. The keys at all levels are understanding, prepared acceptance, and truthfulness tempered with an essential concern for OPSEC.

- Truth is Paramount
- If News is Out, It is Out
- Public Affairs Must be Deployed Early
- Not All News is Good News
- Practice Security at the Source
- Media are Not the Enemy
- Telling Our Story is Good for the Army
- Soldiers and Families Come First

Figure 5-4. Public affairs principles

Civil-Military Operations

5-89. Civil-military operations (CMO) are critical and will be an essential, implied task of any urban operation. They are included here as a closely related activity of IO. Commanders use CMO to establish, maintain, influence, or exploit relations to achieve operational objectives. These relations are among military forces, civilian authorities and organizations (both governmental and nongovernmental), and the civilian population. CMO range from support to combat operations to assisting in establishing political, economic, and social stability. In some instances, commanders may need to be prepared to assume temporary responsibility for the functions and capacities of the urban government. Like PA, effective CMO is based on establishing and maintaining credibility and trust with the urban populace and civilian organizations operating in the urban environment. Because of the potential importance of CMO, commanders may consider placing an assistant commander, executive officer, or some other senior leader to monitor, guide, liaise with other units and organizations, and, overall, ensure that the appropriate command emphasis is placed on these activities.

Civil Affairs

5-90. Civil affairs (CA) units and teams will be critical during UO. While any military force can conduct CMO, CA units are specifically organized, trained, and equipped to conduct activities in support of CMO. They have experience in planning and conducting CMO, a regional focus (which includes enhanced cultural awareness and language training), civilian technical expertise that parallel common urban government and administration functions, and experience in negotiation and mediation. Such experience ensures relevant support to commanders conducting urban operations. (As it is likely that there will not be enough CA forces to cover all requirements, commanders should survey their Soldiers to determine the whether any possess civilian expertise that may be beneficial to CMO.) CA units organize their skills into six functional specialty areas (see Figure 5-5). Commanders should use these skills, often unfamiliar to most military personnel, to—

- Develop their situational understanding of the urban environment (particularly the infrastructure and society).
- Plan CMO to support UO.
- Achieve many of the fundamentals of UO.

Note: To ensure optimal integration, commanders must establish clear command and support relationships between CA and maneuver forces.

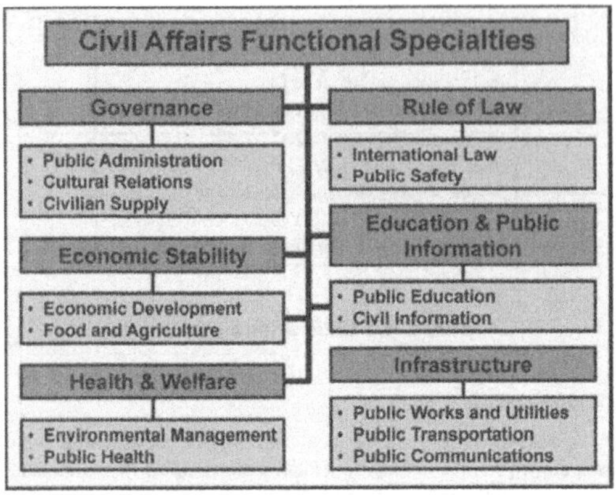

Figure 5-5. Civil affairs functional specialties

5-91. In addition to providing essential information for understanding the urban environment, CA personnel and activities help shape the battlefield, engage a civil problem, consolidate gains, and transition to a legitimate civil authority. Specifically, CA teams and CMO help urban commanders—

- Minimize civilian interference with UO and the impact of urban operations on the populace and infrastructure. CA personnel can help establish and run a civil-military operations center (CMOC), discussed later in this chapter, to coordinate UO with civilian agencies (both governmental and nongovernmental), other services, and multinational partners.
- Provide advice and assistance to restore or rehabilitate portions of the infrastructure.
- Plan, supervise, and execute necessary populace and resources control measures (in close coordination with military police units) until no longer required or the urban operation is completed.
- When requested or when military necessity or legitimate directives require, establish all or portions of the urban civil administration.
- Determine available supplies and services in the urban area and if necessary assist in negotiating their acquisition. CA also help commanders assess the capability, dependability, and willingness of urban sources to provide and sustain identified needs as well as to calculate the impact of using them on other aspects of the urban environment.
- In conjunction with the SJA, fulfill the Army's responsibilities toward the urban population under international, host-nation, and U.S. law.
- Plan and conduct the transition of control for the urban area or operation to another military or civilian governmental or nongovernmental organization or agency.

5-92. As part of CMO, stability or civil support operations will be critical as commanders begin to restore essential services to meet the emergency needs of a potentially devastated urban population. Besides repairing or restoring vital services and other critical infrastructure, these and other projects may also stimulate the urban area's economy and provide work for the unemployed. (Employed workers are less likely to be co-opted by terrorists, insurgents, and criminal gangs.) As such, they can be a critical stabilizing tool for commanders as they attempt to transition control and responsibility to other agencies or to legitimate civilian control. These projects may include—

- Providing clean water.
- Restoring waste management functions (less-extensive sewer repairs and trash collection and disposal activities).
- Improving medical care.
- Rehabilitating schools.
- Assisting in the establishment of local governments.
- Improving security and emergency services (police, fire, and rescue).
- Providing electrical generators for hospitals, clinics, and factories.
- Restoring irrigation to surrounding farmland.
- Repairing important cultural infrastructure.

5-93. As part of the initial planning process, CA units conduct an *area assessment*, which can provide commanders with essential information about the environment (see FM 41-10). Commanders should integrate this initial assessment into the overall urban-focused IPB process. To help analyze civil considerations in any environment, commanders and staffs can consider many characteristics such as areas, structures, capabilities, organizations, people, and events (ASCOPE). These characteristics easily align with an urban area's three main characteristics of terrain, society, and infrastructure; and, like them, they are overlapped and interdependent (see FM 6-0). Of these, the category of events is a critical addition to urban characteristics previously categorized and discussed. Significant spikes (and lulls) to threat and civilian activity can be expected to occur in conjunction with important religious, historic, and political events (for example elections). Therefore, commanders should determine, track, and pay particular attention to these events when planning operations and managing transitions (see also the Key Event Chart discussion in Appendix B, paragraph B-43).

5-94. Similar to PA operations, CMO are related to information operations. The nature of CMO and the need for CA personnel to develop and maintain close relationships with the urban population put CA personnel in a favorable position to collect information. CA personnel work daily with civilians, their equipment, and their records that may be prime sources of information. If used correctly, CA personnel can complement the intelligence collection process necessary to understand the dynamic societal component of the urban environment and detect significant changes. However, CA personnel are not, and cannot appear as, intelligence agents; otherwise, it will undermine their ability to interact with the civilian community. Examples of information available to CA units include government documents, libraries, and archives; files of newspapers and periodicals; industrial and commercial records; and technical equipment, blueprints, and plans.

5-95. Overall, CA expertise is critical to a commander's understanding of the complexities of the infrastructure and societal components of the urban area. These components (together with the terrain or physical component of the urban area) interconnect. CA forces help identify and understand the relationships and interactions between these urban components. From this understanding, commanders can anticipate how specific military actions affect the urban environment and the subsequent reactions. CA personnel help commanders predict and consider the second- and third-order effects and reactions as well as the long-term consequences. Understanding these long-term consequences helps in achieving stability and ensuring a smoother and speedier transition of the urban area back to civilian responsibility and control. Oppositely, an unconsidered short-term solution may, in the end, turn out to be the solution requiring the greatest amount of time. Poor solutions may exacerbate the situation. Others may contribute little to conflict resolution and waste precious resources (particularly time) until a more reasoned approach is formulated—one that should have been the first solution.

Civil-Military Reconstruction Efforts

5-96. Reconstruction efforts and activities, an integral part of CMO, should complement NGO efforts and are conducted in concert with larger urban area and nationwide projects often headed by the U.S. Army Corps of Engineers (USACE) and the U.S. Agency for International Development (USAID). For example, extensive, large-scale sewer projects or major repairs to the urban area's electrical infrastructure would be left to the USACE and USAID. When planning and conducting urban reconstruction, commanders should—

- **Keep expectations realistic.** With so much to do, it is easy for Army forces to become overcommitted. Commanders must focus on the achievable, build off successes, and create constructive forward process.
- **Seek opportunities.** While commanders think long-term, they should look for catalytic investment of resources. Minimal resources provided at the right time can generate significant activities resulting in increased momentum toward the desired end state.
- **Align resources with local needs.** In analyzing the merit of a project, commanders must check to determine if it is supported and desired by the urban populace. Local participation in the process reinforces ownership and acceptance and eventual transition of responsibility.
- **Balance speed and quality.** Commanders must balance the achievement of rapid response with quality construction and repairs.
- **Ensure efforts do not undermine locals.** Army reconstruction efforts should not undermine the growth and legitimacy of local institutions and authorities.
- **Integrate in the targeting process.** Commanders should routinely include approval of civil-military reconstruction products as part of the information operation's contribution to the overall targeting process.

INTEGRATION OF CONVENTIONAL AND SPECIAL OPERATIONS FORCES

5-97. One important Army and joint resource that commanders of a major operation can use to influence urban operations is SOF. Several types of these forces exist (including the CA forces discussed above), each with unique and complementary capabilities. They can be extremely valuable in UO for their ability to execute discrete missions with a higher degree of precision than conventional forces, to provide information, and to enhance cultural understanding. However, the challenges of using SOF include command and control; integration; coordination with conventional forces that will normally command, control, and conduct the bulk of UO tasks; and the imprudent inclination to use SOF forces for conventional purposes. The density and complexity of UO make close coordination and synchronization of conventional forces and SOF essential to mission success. The nature of the environment dictates that both forces will work in close proximity to each other; the separation in space and time between SOF and conventional forces will often be much less in urban areas than in other environments. Overall, the nature of the environment demands a synergistic combination of capabilities to achieve effects on the threat and mission success.

5-98. Successfully integrating SOF occurs with proper integration into, or coordination with, the command structure of the force conducting the UO. SOF within a theater (less CA and tactical PSYOP) ordinarily falls under joint command and control. Therefore, the commander of the major operation responsible for an urban area, if he is not a JFC, will have to coordinate through the JFC to integrate SOF capabilities into the UO. Examples of critical coordination elements include:

- Boundaries.
- No-fire areas. (If assigned, SOF liaison officers (LNOs) must be routinely included in targeting meetings.)
- Coordination points.
- The exchange of key intelligence and information.
- Requirements to support personnel recovery contingencies.

> **Conventional and Special Forces Integration**
>
> On 20 April 2003 during OPERATION IRAQI FREEDOM (OIF), an Army Special Forces (SF) operational detachment alpha (ODA) was patrolling through villages and secondary roads east-southeast of An Najaf. A local Iraqi approached the team and volunteered the location of a senior Ba'ath Party official in the nearby town of Ghamas. Unfortunately, he did not provide an exact location but rather an approximate location and the name of the family dwelling. The ODA coordinated with their company commander and nearby elements of the 101st Airborne Division to conduct a raid on the residence at dawn the next day. Joint planning, preparation, and rehearsals among the units occurred throughout the night.
>
> At 0430, the raid convoy departed its base en route to Ghamas. The small task force consisted of an assault team, a security team, a command and control element, and three blocking forces composed of Soldiers from a scout platoon and antiarmor company of the 101st. At 0515, the convoy hit the release point outside Ghamas. The 101st vehicles moved to their blocking positions at three bridges surrounding the town. The remainder of the task force moved into the town. An interpreter quickly found a local guide who knew where the house was located. He took the SF team to a walled two-story dwelling with a courtyard. The security team isolated the objective and provided overwatch while the assault team forcibly seized the house and apprehended three adult males. Among them was Abd Hamden, the target of the raid and a senior Ba'ath Party official from Baghdad.
>
> After a tactical interrogation, one of the men provided the location of a Fedayeen major nearby. The SF and conventional force raided this house minutes later but only found the major's relatives. As the Soldiers left the town with their captives, they were cheered by the locals throughout the town. Although only one small example, the extensive coordination between SOF and conventional forces during urban operations was a tremendous success during OIF.

SYNCHRONIZATION OF ACTIVITIES WITH OTHER AGENCIES

5-99. The population density of the urban environment, its economic and political importance, and its life-supporting infrastructure attracts many types of organizations. These organizations include—

- Other U.S. governmental agencies.
- International governmental organizations.
- Allied and neutral national governments.
- Allied and coalition forces.
- Local governmental agencies and politicians.
- NGOs.

5-100. Even in a major operation or campaign, many organizations operate in the area as long as possible before combat or as soon as possible after combat. Therefore, coordination with these organizations sharing the urban AO will be essential to achieve synchronization; however, effective synchronization is challenging, time consuming, and manpower intensive. The staffs of larger headquarters (divisions or higher) normally have the breadth of resources and experience to best conduct the necessary synchronization. They can effectively use or manage the organizations interested in the urban area and mitigate their potential adverse effects on UO. By taking on as much of the synchronization requirements as possible, the operational headquarters permits its tactical subordinates to remain focused on accomplishing their tactical missions. The higher headquarters should assume as much of the burden of synchronization as possible. However, the density of the urban environment will often require that smaller tactical units coordinate and synchronize their activities with other agencies and the local civilian leadership (formal and informal) simply because of their physical presence in the units' AOs. In urban stability and civil support operations, mission accomplishment will require effective civil-military coordination and synchronization activities and measures at all levels as either a specified or implied task.

Civil-Military Operations Centers

5-101. To coordinate activities among the varied agencies and organizations operating in an urban area and the local population, urban commanders can establish a CMOC. The CMOC synchronizes Army activities and resources with the efforts and resources of all others involved (see FM 41-10). This can be particularly important in stability and civil support operations where combat operations are not the dominant characteristic of the major operation. CMOCs can be established at all levels of command. Hence, more than one CMOC may exist in an AO, particularly large urban areas. CMOCs may be organized in various ways and include representatives from as many agencies as required to facilitate the flow of information among all concerned parties. Commanders must still ensure that force protection and OPSEC requirements are not compromised. Effective CMOCs can serve as clearinghouses for the receipt and validation of all civilian requests for support, can aid in prioritizing efforts and eliminating redundancy, can decrease the potential for inappropriate displays of wealth by one or more of the participating organizations, and, most importantly, can reduce wasting the urban commander's scarce resources.

Liaison Officers

5-102. LNOs—sufficiently experienced and adequately trained in liaison duties and functions—are necessary to deal with the other agencies that have interests in the urban area. Army LNOs work with the lead agency or other organizations (including local civilian agencies such as police) that the commander has identified as critical to mission success. Together they work to rapidly establish unity of effort and maintain coordination, often before a CMOC is established. The additional coordination afforded by the physical presence of LNOs within these organizations may be required even after the CMOC is fully functional. When commanders lack enough LNOs to meet requirements, they should prioritize and often assign a single LNO to several organizations. That LNO will then share his time and presence to those organizations based on the situation and his commander's guidance.

Commander's Personal Involvement

5-103. Overall, establishing a close relationship with other agencies and the urban civilian population will often be a major, positive factor in successful mission accomplishment, particularly in urban stability operations. Despite internally established command and control relationships, commanders that develop a direct and personal relationship with the leaders and staff of other agencies can often avoid conflict, win support, foster trust, and help eliminate the "us versus them" mentality that can frustrate cooperation among Army forces and civilian organizations.

Chapter 6
Foundations for Urban Operations

Utilities such as electricity and water are as much weapons of war as rifles, artillery pieces or fighter aircraft. . . . In the case of Manila, where there was a noncombatant, civilian population of one million in place, it was the attacker's aim to capture the utilities which the defender planned to destroy.

The Battle for Manila

Commanders conducting major urban operations (UO) use their ability to visualize how doctrine and military capabilities are applied within the context of the urban environment. An operational framework is the basic foundation for this visualization. In turn, this visualization forms the basis of operational design and decision making. To accurately visualize, describe, and direct the conduct of UO, commanders and their staffs must understand the basic fundamentals applicable to most UO.

URBAN OPERATIONAL FRAMEWORK

6-1. Army leaders who have an urban area in their area of operations (AO) or are assigned missions in an urban area follow an urban operational framework. They identify the portion of the urban area essential to mission success, shape the area, precisely mass the effects of combat power to rapidly dominate the area, protect and strengthen initial gains without losing momentum, and then transition control of the area to another agency. This framework divides into five essential components: understand, shape, engage, consolidate, and transition. These five components provide a means for conceptualizing the application of Army combat power and capabilities in the urban environment.

6-2. The urban operational framework assists commanders in visualizing urban operations. This framework is simply an aid to the commander. It is not sequential, nor is it a planner's tool for phasing an operation. Commanders should combine the urban operational framework with—

- The principles of war.
- The tenets of Army operations.
- The components of operational design.
- Considerations for stability operations and civil support operations.
- Sustainment characteristics.
- Running estimates.
- Commander's critical information requirements (CCIR).
- Each commander's experience.

The framework contributes to the visualizing, describing, and directing aspects of leadership that make commanders the catalysts of the operations process (see Figure 6-1). In the same manner, the urban operational framework contributes to the overall operations process (see FM 3-0).

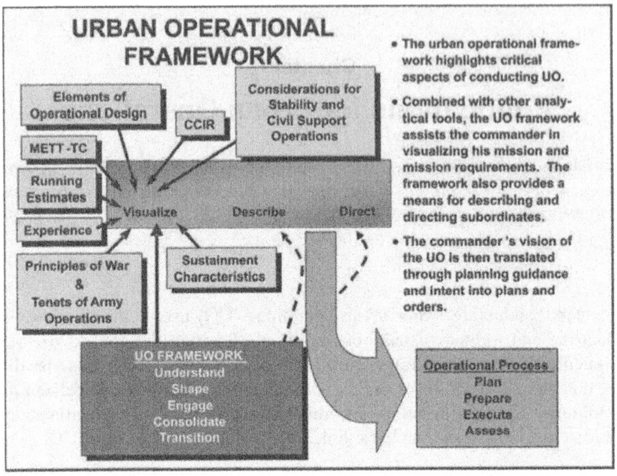

Figure 6-1. The urban operational framework and battle command

UNDERSTAND

6-3. Understanding requires continuous assessment—throughout planning, preparation, and execution—of the current situation and progress of an operation, and the evaluation of it against measures of effectiveness to make decisions and adjustments. Commanders use visualization, staff officers use running estimates, and all use the intelligence preparation of the battlefield (IPB) process to assess and understand the urban environment. Commanders and staffs begin to understand by observing and then collecting information about the situation. They observe and learn about the urban environment (terrain, society, and infrastructure), and other factors of METT-TC—mission, enemy, weather, troops and support available, and time available. They use intelligence, surveillance, and reconnaissance means; information systems (INFOSYS); and reports from other headquarters, services, organizations, and agencies. Then they orient themselves to the situation and achieve situational understanding based on a common operational picture (COP) and continuously updated CCIR. Largely, the ability to rapidly and accurately achieve a holistic understanding of the urban environment contributes to the commanders' abilities to seize, retain, and exploit the initiative during UO.

Disproportionately Critical

6-4. The Army operations process requires continuous assessment and understanding; it precedes and guides every activity. In UO, however, in-depth understanding is disproportionately critical for several reasons. First, each urban environment is unique. Other environments can be studied and their characteristics quantified in a general manner with accuracy. This is fundamentally not true of different urban areas. The characteristics and experience in one urban area often have limited value and application to an urban area elsewhere. This characteristic sets UO apart from operations in other environments.

Extremely Dynamic

6-5. The urban environment is also extremely dynamic. Either deliberate destruction or collateral damage can quickly alter physical aspects of the urban environment. The human aspect is even more dynamic and potentially volatile. A friendly civil population, for example, can become hostile almost instantaneously.

These dynamics (combined with initial difficulty of understanding and describing this unique environment) make it difficult for commanders and staffs to initially develop and maintain a COP and establish situational understanding. Furthermore, public reaction to media coverage of the urban operation and political changes influence national security strategy and objectives. Such changes can affect the basic nature of an operation, especially after it has commenced. Anticipating these potential effects and developing appropriate branches and sequels based on an accurate understanding often determines how quickly commanders can achieve the desired end state.

Risk Assessment

6-6. As in any environment, UO pose both tactical and accident risks. However, the level of uncertainty, ambiguity, and friction can often be higher than that of many other environments. Such challenges increase the probability and severity of a potential loss due to the presence of the enemy, a hostile civilian group, or some other hazardous condition within the urban environment (see Determining the Necessity of Urban Operations in Chapter 5). Therefore, commanders must—

- Identify, assess, and understand hazards that may be encountered in executing their missions.
- Develop and implement clear and practical control measures to eliminate unnecessary risk.
- Continuously supervise and assess to ensure measures are properly understood, executed, and remain appropriate as the situation changes.

6-7. Risk decisions are commanders' business. However, staffs, subordinate leaders, and even individual Soldiers must also understand the risk management process and continuously look for hazards at their level or within their area of expertise. Any risks identified (with recommended risk reduction measures) must be quickly elevated to the appropriate level within the chain of command (see FM 3-100.12 and FM 5-19).

Complex and Resource Intensive

6-8. The urban environment is the most complex of all the environments in which the Army conducts operations. It is often comprised of a diverse civil population and complex, ill-defined physical components. A sophisticated structure of functional, social, cultural, economic, political, and informational institutions unites it. Thus, the analysis to understand the environment is also complex and time and resource intensive. The nuances of the urban environment can take years to uncover. Hence, constant analysis of the environment requires greater command attention and resources. Accurately understanding the environment is a prerequisite to shaping it, and both understanding and shaping activities are crucial to effectively engage the elements of the urban environment critical to success.

SHAPE

6-9. Shaping operations, part of all Army operations, are essential to successful UO. They set the conditions for decisive operations at the tactical level in the urban area. Rapid action, minimum friendly casualties, and acceptable collateral damage distinguish this success when the AO is properly shaped. Failure to adequately shape the urban AO creates unacceptable risk. The commander of a major urban operation has several resources with which to begin shaping the AO. Important capabilities include—

- Fires.
- Information operations.
- Special operations capabilities.
- The maneuver of major subordinate units.

6-10. Critical urban shaping operations may include actions taken to achieve or prevent isolation, understand the environment, maintain freedom of action, protect the force, develop cooperative relationships with the urban population, and train Army forces for sustained UO.

Isolation

6-11. Isolation of an urban environment is often the most critical component of shaping operations. Commanders who's AO includes operationally significant urban areas often conduct many shaping operations to isolate, or prevent isolation of, those areas from other parts of the AO. Likewise, commanders operating in the urban area focus on isolating decisive points and objectives in the urban area or averting isolation of points that are critical to maintaining their own freedom of action. Isolation is usually the key shaping action that affects UO. It applies across full spectrum operations. Most successful UO have effectively isolated the urban area. Failure to do so often contributed to difficult or failed UO. In fact, the relationship between successful isolation and successful UO is so great that the threat often opposes external isolation actions more strongly than operations executed in the urban area (or critical areas within). In some situations, the success of isolation efforts has been decisive. This occurs when the isolation of the urban area compels a defending enemy to withdraw or to surrender before beginning or completing decisive operations. In UO that are opposed, Army forces attempt to isolate the threat three ways: physically, electronically, and, as a resultant combination of these first two, psychologically (see Figure 6-2).

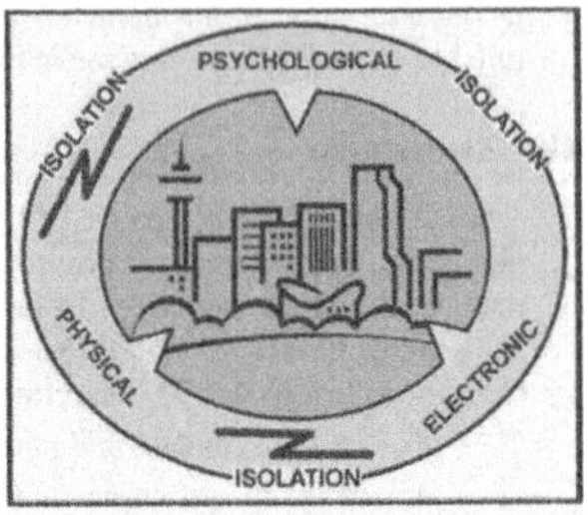

Figure 6-2. Urban isolation

Physical Isolation

6-12. In offensive UO, physical isolation keeps the threat from receiving information, supplies, and reinforcement while preventing him from withdrawing or breaking out. Conversely, a defending Army force attempts to avoid its own physical isolation. Simultaneously, this force conducts operations to isolate the threat outside, as they enter, or at selected locations in the urban area. Physical isolation can occur at all levels. In many situations, the commander of a major combat operation may attempt to isolate the entire urban area and all enemy forces defending or attacking it. At the tactical level, forces isolate and attack individual decisive points often using a cordon technique. In stability operations, physical isolation may be more subtly focused on isolating less obvious decisive points, such as a hostile civilian group's individual leaders. In many operations, isolation may be temporary and synchronized to facilitate a decisive operation elsewhere. To effectively isolate an urban area, air, space, and sea forces are often necessary additions to the capabilities of ground forces.

Electronic Isolation

6-13. Electronic isolation is achieved through offensive information operations (IO). Electronic warfare (particularly two of its components: electronic warfare support and electronic attack) and computer network attack are critical to electronic isolation (see FM 3-13 and the Information Operations discussion in Chapter 5). At the operational level, offensive IO aims to quickly and effectively control the information flow into and out of an urban area. This isolation separates the threat's command and control (C2) system in the urban area from its operational and strategic leadership outside the urban area. Offensive IO also focuses on preventing the threat from communicating with civilians through television, radio, telephone, and computer systems. At the tactical level, IO aim to isolate the threat's combat capability from its C2 and leadership within the urban area, thus preventing unity of effort within the urban area. Defensive IO can prevent isolation of friendly forces defending in an urban area.

Psychological Isolation

6-14. Psychological isolation is a function of physical actions, electronic warfare, and other forms of IO, especially military deception and psychological operations. Psychological isolation denies the threat political and military allies. It separates the enemy or hostile civilian group from the friendly population, nongovernmental organizations (NGOs) operating in the urban area, and from political leaders who may consider supporting Army forces. Psychological isolation destroys the morale of individual enemy soldiers or hostile civilians. It creates a feeling of isolation and hopelessness in the mind of the threat. It undermines the threat leadership's ability to think, plan, decide, and act. As such, it undermines the confidence of the threat in their leadership. On the other hand, IO, as well as the disciplined conduct of Army personnel, can help to forge legitimacy for Army operations. In stability operations, psychologically isolating the threat can result in the friendly urban population and NGOs positively supporting Army operations.

Other Shaping Actions

6-15. Other shaping actions can include the proper sequencing and deployment of forces, reconnaissance operations, and force protection. These actions contribute equally to the success of any urban operation. Commanders must understand how the urban environment affects their ability to accomplish these shaping actions. However, civil-military operations (CMO), another closely related activity of IO, are important to shaping the urban operational environment for decisive operations. The specific civil-military task can vary greatly and may include affecting a cooperative relationship with the civil political system, protecting portions of the civil population or infrastructure, or establishing refugee camps or safe areas for noncombatants. This is most true in stability and civil support operations. Successful CMO also can contribute to the psychological isolation of the threat. (See Civil-Military Operations in Chapter 5 for a more detailed discussion.)

Training and Education

6-16. Finally, Army commanders must consider that critical shaping actions often occur prior to the urban operation in the form of professional education, and home-station and in-theater training. Commanders can enhance training through joint, interagency, multinational, and combined arms exercises and effective rehearsals.

ENGAGE

6-17. In UO, Army forces engage by appropriately applying their full range of capabilities against decisive points leading to centers of gravity. Successful engagement takes advantage of the Army force's superior training, leadership, and, within the constraints of the environment, equipment and technology. Successful engagement also requires the establishment of necessary levels of control and influence over all or portions of the urban environment until responsibilities can be transferred to other legitimate military or civilian control. Engagement may range from the overwhelming, yet precise, application of combat power to defeat a threat, to large-scale humanitarian assistance operations, to unobtrusive advice and assistance to the urban security forces. It may require brief engagement in a relatively small portion of an urban area or with small segments of the population to long-term engagement throughout multiple, large urban areas and with vast sectors of the population. The commander's ability to engage in a specific urban operation depends, as always, on the situation and the assigned mission.

Offense: Attack Decisive Points

6-18. In offensive UO, forces successfully engage by striking at the enemy's center of gravity using multiple offensive actions from unexpected directions and throughout all dimensions. Army forces aim to engage identifiable decisive points. Successful efforts against decisive points lead to effects on the center of gravity. The center of gravity will differ in each offensive situation. It may be an individual enemy leader, the enemy's combat power, the enemy's communications capability, or a physical structure of cultural, political, or economic significance.

Defense: Deny Vital Functions and Critical Infrastructure

6-19. In defensive UO, engage translates into denying the enemy control of the vital functions and critical infrastructure of the urban area. Forces achieve this by leveraging the defensive advantages of the urban terrain, defending essential areas in depth, using economy of force in nonessential areas, controlling the enemy direction of attack with natural and man-made obstacles, and retaining the initiative through counterattacks.

Stability and Civil Support: Apply Innovation and Imagination

6-20. The ability to engage in urban stability operations hinges on the type of stability operation commanders execute. In a noncombatant evacuation operation, forces limit engagement to finite geographic areas and times. In contrast, a peace operation may require engagement of a large urban area for an extended time. In this operation, engage is defined as using the array of Army capabilities to create specific conditions among the belligerents. Thus, the techniques used for engagement in stability operations vary according to the situation and as situations mature during long-term operations.

6-21. In urban civil support operations, engagement is accomplished by innovative and subtle application of Army capabilities. Since Army forces usually support other agencies that lead the operation, effective engagement results from carefully and discretely applying Army capabilities to the tasks assigned by the lead agency. In a humanitarian relief situation, Army forces may be tasked to transport supplies in the urban area. Successful engagement in this activity then becomes the goal of Army forces and may be achieved by providing, managing, and protecting transportation assets and clearing and maintaining sufficient lines of communications (LOCs).

CONSOLIDATE

6-22. Army forces consolidate to protect and strengthen initial gains and ensure retention of the initiative. Consolidation includes actions taken to eliminate or neutralize isolated or bypassed threat forces (including the processing of enemy prisoners and civilian detainees) to increase security and protect LOCs. It includes the rapid repositioning and, as necessary, reorganization of forces and intelligence, reconnaissance, and surveillance (ISR) assets. Consolidation may also include activities in support of the civilian population such as relocating displaced civilians, reestablishing law and order, humanitarian assistance and relief operations, and restoration of key urban infrastructure. A significant consolidation effort may be the reduction or elimination numerous explosive hazards resulting from previous urban combat operations. During consolidation, commanders often coordinate with NGOs to establish unity of effort in the execution of stability or civil support operations. Consolidation begins with the initiation of the urban operation and follows each critical activity.

TRANSITION

6-23. When planning UO, commanders ensure that they plan, prepare for, and manage transitions. Transitions are movements from one phase of an operation to another or a change in responsibility for all or portions of the urban environment from one unit, organization, or agency to another. They involve significant changes in the type of operation, concept of the operation, mission, situation, task organization, forces, resource allocation and support arrangements, or C2. Transitions occur in all operations, but in UO they occur with greater frequency and intensity, are more complex, and often involve agencies other than U.S. military organizations. For example, a successful attack may transition to a defend mission that includes not only defense tasks but also stability tasks. All operations will likely include a transition of responsibility for some aspect of the urban environment to (or back to) a legitimate civilian authority. Unless planned and executed effectively, transitions can reduce the tempo of the urban operation, slow its momentum, and cede the initiative to the threat.

Mental and Physical Preparation

6-24. Transitions occur as conditions warrant. They can be carefully planned and controlled, or they can be quick and dramatic, such as the swift transformation of a stability operation into offense or defense—and back again. Units prepare mentally and physically to address rapid transitions. Accordingly, plans include branches and sequels that address anticipated or possible transition points. When the dominant type of operation changes from an offense to stability or civil support, the types of units originally conducting the UO may no longer be appropriate. A large mobile reserve may permit increased flexibility to react to unplanned transition requirements. Operations or units in one part of an urban area may transition before operations or units in a different part of the same urban area. This will require commanders to exercise careful command and control and execute various types of operations and associated tasks simultaneously.

Transition to Legitimate Civilian Authorities or Agencies

6-25. Combat operations are not an end to themselves. They are inevitably a means of transition, an activity for moving from an unsatisfactory state of affairs to an improved end (as determined by national objectives). In UO, a distinct aspect of transition is the requirement to quickly and efficiently transition the major portions of Army responsibilities to civil agencies. Some tasks to which units will transition are not traditional combat tasks but rather stability tasks more closely associated with CMO. In stability or civil support operations this is often a near-term critical mission objective. In these operations, commanders aim to alleviate the circumstances requiring Army forces and ensure that other civilian agencies assume the functions provided by Army forces. Following combat operations, civilian agencies are assisted and encouraged to quickly resume specific support activities—such as providing sanitary services, food services, law enforcement, and health services—because of their high demand on Army resources.

6-26. Transitioning responsibility for the urban environment (or selected portions) to legitimate civilian agencies or authorities requires commanders to make an accurate assessment and develop an in-depth understanding of the civilian organization's ability to accept and handle the responsibility. Commanders consider not only resources available but whether the organization has the appropriate leadership and technical expertise as well as a vision in consonance with U.S. objectives. Commanders should avoid forcing a transition only to alleviate the burden on Army forces. On the other hand, they may need to encourage acceptance of responsibility while continuing to provide "behind-the-scenes" advice, assistance, and resources. Transitioning responsibility prematurely may alleviate the commander's immediate resource concerns but may, in the longer term, negatively impact the achievement of strategic objectives.

Clearly Visualize and Describe the End State

6-27. Army UO conclude when Army forces depart and have no further mission requirements in the urban area. At the outset, commanders visualize and describe the intended end state of a unit's execution of UO. Commanders then clarify and update this visualization as the political or strategic situation is refined or changes. Importantly, commanders must create a command climate and procedures that allow their staffs and subordinate commanders the ability to ask questions that will enable all to share in an accurate perception of the commander's visualization. This enables subordinate units to identify likely transitions and ensures that current operational planning takes into account second- and third-order effects. As long as an active Army AO contains an urban area, some type of urban operation will exist. After urban combat successfully ends, combat forces may move on. Support forces conducting sustaining operations may then occupy the area and continue to conduct a different form of UO.

Applying the Urban Operational Framework: Panama – 1989

The U.S. conducted OPERATION JUST CAUSE in December 1989 to remove the illegal ruler of Panama, Manuel Noriega, and to restore that country to a democracy. It also conducted the operation to ensure the safety of a substantial number of U.S. personnel as well as the security of U.S. interests in Panama. The major focus of JUST CAUSE was in Panama City, the country's capital. Most operations occurred in this large urban area, one of the numerous smaller urban areas, or the urban-like military bases. These bases increased the AO and were directly linked to operations in the capital city. This successful operation illustrates how commanders can apply the urban operational framework to visualize, describe, and direct the critical aspects of urban operations.

Understand

The synchronization achieved during the operation may have obscured the challenges faced in the initial assessment process in Panama. However, it was not as simple as it may have seemed. Using the framework of the urban environment, U.S. forces required details of the physical characteristics of the environment, the infrastructure, and the human dimension including the capabilities of the Panamanian military.

Because Army forces had a long history in Panama, commanders clearly understood the physical challenges and layout of critical urban areas (see Figure 6-3), particularly Panama City. They also understood how the infrastructure in each urban area functioned and which parts was a key element to success. Examples of key portions of the infrastructure included the Madden Dam, which controlled the water flow through the Panama Canal, and the Cerro Azul television tower, which was the main Panamanian broadcast tower.

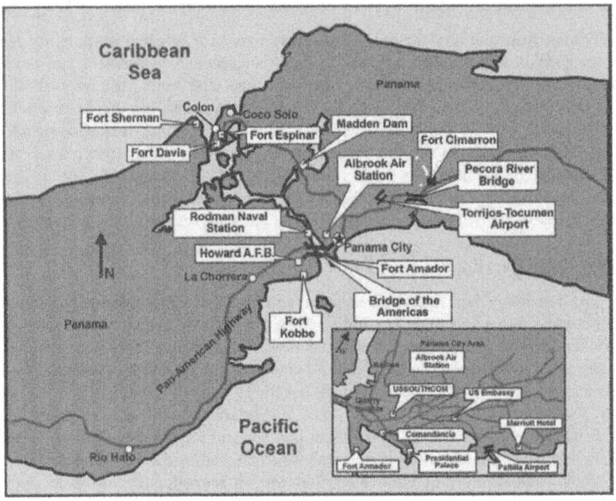

Figure 6-3. Panama

Collecting information and developing intelligence on the human elements of the urban environment was critical to operational success and a challenge. Because the Panamanian Defense Force (PDF) had traditionally been an ally of the United States, Army forces did not have a systemic database that adequately depicted their order of battle and their true capabilities. Additionally, much of the situation in Panama was colored in political terms making it more difficult for traditional military sources to evaluate the status of PDF forces. For example, Army planners needed to know if PDF military units (when faced with a formidable U.S. force) would fight at all for Noriega and if they did fight, how hard and long would they fight. The answers depended largely on their political loyalty to Noriega and on the individual loyalty of the unit officers to the Panamanian president. Thus, Army commanders needed to understand the military characteristics of PDF units and their political affiliations and tendencies.

Because transition from combat to noncombat tasks would be critical to achieving all objectives, particularly the restoration of democracy, Army forces also needed an accurate assessment of the political opposition to Noriega—including that opposition's capabilities and vulnerabilities. Again, Army forces were required to make assessments outside those needed solely for combat operations. Ultimately, assessing the political opposition's vulnerabilities led to assigning Army units to protect them throughout the operation so that they could serve as a foundation for a new democratic government.

Finally, the commander's assessment included an evaluation (often subjective) of the attitudes and disposition of the Panamanian people. Human intelligence (HUMINT) was the primary source of information on the population. Army forces had good access to the population because of their close proximity and historical ties to Panama. Many Soldiers were married to Panamanians, and the Army had total access to local media and to prominent individuals.

National imagery and special operations forces (SOF) also contributed to the ability of Army forces to assess the urban environments of JUST CAUSE. All units executing operations had detailed satellite photos of objective areas. Additionally, key objectives were placed under SOF surveillance well in advance. This surveillance revealed unit readiness, vulnerabilities, detailed disposition, and other patterns critical to mission success. The combination of the two capabilities allowed units to plan and achieve the synchronization necessary for such a complex urban operation.

Shape

During OPERATION JUST CAUSE, commanders conducted numerous shaping operations to establish the conditions for the decisive operations. Many operations were designed to control information, such as an assault on the Azul television tower identified during the assessment of the infrastructure. Planners designed many shaping operations to isolate various garrisons and PDF units. An example of tactical isolation was the plan for the Pacora River Bridge to prevent reinforcements from reaching the garrison at Torrijos-Tocumen Airport.

Operational isolation was achieved through the Ranger Regiment's and 82nd Airborne Division's assault on targets at Rio Hato in the west and Fort Cimarron in the east. These actions in conjunction with the securing of Maddam Dam had the primary objective of isolating Panama City. They were also the largest of the major actions occurring during OPERATION JUST CAUSE. The airborne assault was also the largest airborne operation conducted by U.S. forces since World War II. This large-scale shaping operation demonstrates that shaping operations are critical to mission success and can be more resource intensive than the actual operations that achieve domination.

Engage

U.S. Army forces achieved successful engagement in OPERATION JUST CAUSE by establishing unchallenged military control over Panama City and eliminating Noriega's capability to challenge that control. Toward this end, the operation attacked two decisive points. The first was the assault on the PDF headquarters located in Panama City: the Comandancia. The second was the operation undertaken to locate and seize Noriega himself.

Three battalions of task force (TF) Bayonet (5-87th Infantry, 1-508 Infantry [Airborne], and 4-6th Infantry [Mechanized]) executed the attack on the Comandancia and Fort Amador. They were also tasked to protect the American Embassy in downtown Panama City. To execute these missions, they moved from various staging areas located throughout the city to their assigned objectives using air assault, mounted, and dismounted approaches. The ground movement through the city proved to be the most difficult and hazardous part of the mission due to the vulnerability of the troops in their armored personnel carriers and trucks. The dismounted movement was slower than the mounted movement but allowed the Soldiers greater cover and concealment.

The strongest opposition to TF Bayonet occurred at the Comandancia. Elements of three PDF companies and two public order companies held out for three hours. The troops moving to Comandancia were subject to a large volume of sniper fire, and in the assault, unidentified indirect fire caused significant casualties among the mechanized forces. TF Bayonet forces, supported by airborne armored reconnaissance vehicles and Hellfire missiles from Apache helicopters, captured the

Comandancia. Commanders noted in particular the precision of the supporting fires from attack helicopters. The assault by fire from supporting AC 130 gunships severely damaged much of the reinforced Comandancia building.

Simultaneously, SOF attacked several targets where Noriega might be located. These initial attacks were unsuccessful. However, many subsequent actions neutralized Noriega's influence and eventually resulted in his apprehension on 3 January 1990. These actions included the well-organized and relentless manhunt conducted by SOF units, the isolation of Panama City itself, population control efforts, sophisticated IO, and cooperation with other U.S. agencies.

Consolidate and Transition

OPERATION JUST CAUSE demonstrated the vital need for a thought-out plan that adequately addresses consolidation and the transition from combat to noncombat before commanders initiate operations. Normally in complex UO, commanders cannot leave the details of transition until after the operation has begun without unacceptable risk to overall mission accomplishment. The follow-on stability operation, OPERATION BLIND LOGIC (later renamed OPERATION PROMOTE LIBERTY) began 24 hours after the initial assault and thus both operations were occurring simultaneously. This simultaneity of different types of operations is typical in major operations conducted in a large urban area. The stability operation involved more time than the combat operation and continued well after the close of OPERATION JUST CAUSE and after most of the major combat units had redeployed. It involved significant resources without the same level of risk to U.S. forces as the combat operations.

Civil affairs (CA) were a dominant part of the transition from combat to stability operations. The 96th Civil Affairs battalion was central to this operation. CA forces established a civil police force, emergency food distribution, property protection, production and distribution of newspapers, cleanup of the city, and building support for a new civil government. Most tasks were coordinated through Army CA forces and executed by other Army forces under the supervision of CA.

IO were also a major aspect of affecting a stable transition and successful postcombat operations. These operations built support for the U.S. operation among the population. They emphasized that the U.S. conflict was with Noriega and not the Panamanian people and that the U.S. forces would depart as soon as a new Panamanian government could take over.

Other U.S. agencies played critical roles in stability operations in Panama. The U.S. Drug Enforcement Agency and Justice Department were important to the negotiations that led to Noriega's capture. The U.S. State Department helped to negotiate for Noriega and develop military policies and plans during stability operations. The American Embassy advised commanders regarding the large diplomatic community that existed in Panama City.

FUNDAMENTALS OF URBAN OPERATIONS

6-28. UO often differ from one operation to the next. However, some fundamentals apply to UO regardless of the mission, geographical location, or level of command. Some of these fundamentals are not exclusive to urban environments. Yet, they are particularly relevant to an environment dominated by man-made structures and a dense noncombatant population (see figure 6-4). Appendix A provides an historical example of how many of these fundamentals applied to an actual conflict situation. Vitally, these fundamentals help to ensure that every action taken by a commander operating in an urban environment contributes to the desired end-state of the major operation.

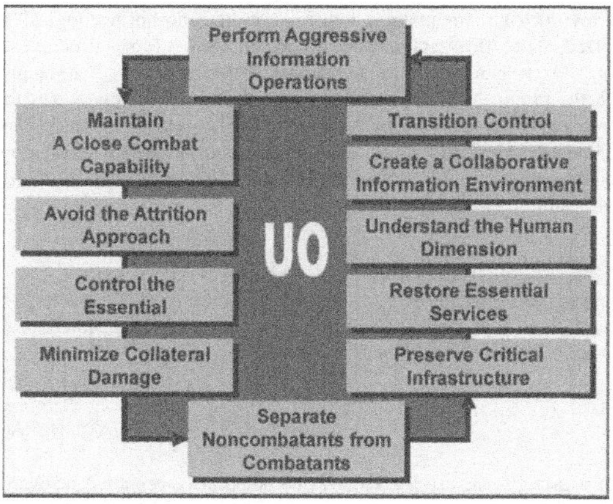

Figure 6-4. Fundamentals of urban operations

PERFORM AGGRESSIVE INFORMATION OPERATIONS

6-29. Information operations aimed at influencing non-Army sources of information are critical in UO across the spectrum of operations. Because of the density of noncombatants and information sources, the media, the public, allies, coalition partners, neutral nations, and strategic leadership will likely scrutinize how Army forces participate in UO. The proliferation of cell phones, video cameras, Internet capability, and media outlets ensure close observation of the activities of Army forces. With information sources rapidly expanding, public information of Army operations will often be available faster than the internal military INFOSYS can process it. Army forces should aggressively integrate IO into every facet and at all levels of the operation to manage perceptions and mitigate unintended consequences. Under media scrutiny, the actions of one Soldier may have significant strategic implications. IO aim to make the information accurate; placed in the proper context of the Army's mission; and available to all interested parties: the public, the media, and other agencies. Ensuring that the urban population understands how they and their urban area fit within the commander's vision, intent, and end state will be a critical task.

MAINTAIN A CLOSE COMBAT CAPABILITY

6-30. Close combat is required in all offensive and defensive UO. This core capability must also be present and visible in urban stability operations and may be required in urban civil support operations. Close combat in any urban operation is resource intensive, requires properly trained and equipped forces, and has the potential for high casualties. However, the ability to decisively close with and destroy enemy forces as a combined arms team remains essential. Potential threats respect strength and, instead, seek to attack ill-prepared or complacent Army forces—not necessarily forces that are better protected or armored, but those less likely or capable of inflicting casualties against them. In stability and civil support operations, a lack of respect and fear of Army forces can hinder recovery as much as the ill-advised use of forces. Hence, all Soldiers, regardless of branch or military occupational specialty, must be properly equipped and trained to fight in an urban environment. This ability allows Army forces to deter aggression, compel compliance, morally and physically dominate a threat and destroy his means to resist, and terminate urban conflicts on the Army commander's terms.

AVOID THE ATTRITION APPROACH

6-31. Previous Army doctrine was inclined towards a systematic linear approach to urban combat. This approach emphasized standoff weapons and firepower. Army force structure does not support this approach towards UO. It can result in significant collateral damage, a lengthy operation, and an inconsistency with the political situation and strategic objectives. Enemy forces that defend urban areas want Army forces to adopt this approach because of the likely costs in resources. Commanders should only consider this approach to urban combat as an exception and justified by unique circumstances. Instead, commanders should seek to achieve precise, intended effects against multiple decisive points that overwhelm a threat's ability to react effectively.

CONTROL THE ESSENTIAL

I'm talking about attacking those things from which the regime draws power but being very careful about it so that we don't get large bodies of young Americans caught up in house-to-house Berlin, World War II-type scenario.

Lt. Gen. William C. Wallace
Commander, V Corps
OPERATION IRAQI FREEDOM

6-32. Many modern urban areas are too large to be completely occupied or even effectively controlled without an enormous force. Therefore, Army forces focus their efforts on controlling only the essentials to mission accomplishment. At a minimum, this requires control of key terrain. Key terrain is terrain whose possession or control provides a marked advantage to one side or another. In the urban environment, commanders determine key terrain based on its functional, political, economic, or social significance. A power station or a place of worship may be key terrain.

6-33. All principles of war can apply to UO. The principle of mass and the principle of economy of force (in addition to the principle of unity of command discussed in the previous chapter) are particularly important in guiding UO and providing mission focus. Army forces mass combat power only to control those requirements essential for mission success. This permits conservation of combat power. It also implies economy of force and associated risk in those areas where Army forces choose not to exercise control.

MINIMIZE COLLATERAL DAMAGE

6-34. Forces should integrate precision fires, IO, and nonlethal tactical systems consistent with mission accomplishment while decreasing the potential for collateral damage. Commanders develop unique rules of engagement (ROE) for each urban operation and provide necessary firepower constraints. IO and nonlethal systems may compensate for some restrictions, especially in stability or civil support operations. Commanders continually assess the short- and long-term effects of operations and firepower on the population, infrastructure, subsequent missions, and national and strategic objectives. They must also consider what, if any, provisions should be made to amend or address potential collateral damage. Overall, commanders must balance restraint and precision with speed and overwhelming combat power. By avoiding unnecessary harm to all elements of the urban environment, commanders retain the moral high ground and help sustain legitimacy for their operations. Minimization of collateral damage allows civilians to continue to provide for their own needs or the rapid return of the urban area to civilian self-sufficiency.

SEPARATE NONCOMBATANTS FROM COMBATANTS

6-35. Promptly separating noncombatants from combatants (psychologically and physically) may make the operation more efficient and diminish some of the threat's potential asymmetrical advantages. This separation also may reduce restrictions on the use of firepower, enhance force protection, and strip the threat from its popular support base. This important task becomes more difficult when the threat is an unconventional force that can mix with civilians.

6-36. In more recent operations, threats have sought to integrate their military capabilities as closely as possible into the civilian population and infrastructure. In these conditions, commanders must increase their efforts to discriminate between the two. Soldiers managing violence in this setting require the highest level of individual and organizational discipline and judgment. Soldiers will require the mental agility to separate their aggression toward threats from the noncombatant civilian population. The training, effort, and command emphasis in this area is as important as fully successful results. Such efforts strongly impact national and international perceptions of the operation.

PRESERVE CRITICAL INFRASTRUCTURE

6-37. Commanders should analyze the urban area to identify critical infrastructure. They should attempt to preserve and protect the critical elements for postcombat sustainment operations, stability or civil support operations, or the overall health and well-being of the indigenous population. Urban areas remain in the AO after combat operations have ceased. Postcombat UO are unavoidable. Different from simply avoiding collateral damage, Army forces may have to initiate actions to prevent an enemy or a hostile civilian group from removing or destroying critical infrastructure and assets. This may include cultural infrastructure such as religious and historical places. In some cases, preserving the infrastructure and the urban society's sources of economic and cultural wealth may be the assigned objective of the urban operation.

RESTORE ESSENTIAL SERVICES

6-38. Army forces plan to restore essential services that may fail to function before or during an operation. Essential services include power, food, water, sewage, medical care, and security and law enforcement. When planning for and conducting Army UO, units can use nonlethal and less destructive munitions and capabilities to keep potentially vital infrastructure intact. Initially, Army forces may be the only force able to restore or provide essential services and commanders must plan accordingly. Failure to do so can result in serious health problems for the civilians, which can affect the health of Army forces and negatively impact overall mission success. Army forces seek to transfer responsibility for providing essential services to other agencies, NGOs, or the local government as quickly and effectively as possible. Despite the potential causes for the failure or destruction of essential services, commanders also seek to ensure that civilians continually perceive Army restoration activities as assistance rather than an Army requirement. Otherwise, the civilian populace may be slow to accept or resume responsibility for their urban area.

UNDERSTAND THE HUMAN DIMENSION

> *Do not try to do too much with your own hands. Better the Arabs do it tolerably than that you do it perfectly. It is their war, and you are to help them, not to win it for them. Actually, also, under the very odd conditions of Arabia, your practical work will not be as good as, perhaps, you think it is. It may take them longer and it may not be as good as you think, but if it is theirs, it will be better.*
>
> T. E. Lawrence
> "Twenty-Seven Articles"

6-39. Commanders carefully consider and manage the perceptions, allegiance, and morale of the civilians. Their assessment of the environment needs to accurately identify the attitudes of the people toward Army forces. Operational guidance to subordinates—including ROE, protection, sustainment operations, and fraternization—is based on this assessment. Commanders expect and consider the demographic variance in the attitudes of an urban population. They cannot inadvertently apply Western cultural norms to a non-Western urban population. Commanders can only make reliable assessments based on a thorough understanding and appreciation of the local society and their culture. Developing this ability also requires the additional ability to effectively share information among all echelons of command.

6-40. Sound policies, proper discipline, adequate consideration for local culture, and rapid engagement of local urban leaders will positively affect the attitudes of the population toward Army forces. Additionally, well-conceived and executed IO will enhance the position of Army forces relative to the urban population. Even during high-intensity urban combat, heightened awareness of and sensitivity toward the civilians can lead to a better postcombat situation than if civil considerations were unobserved or diminished in importance. An improved postcombat situation enhances transition. As the environment of conflict becomes more complex, the human dimension (and associated moral aspects) takes on greater importance and may have the greatest potential for affecting the successful outcome of UO. Therefore, the human aspect and development of cultural acuity creates a discrete planning factor.

Create a Collaborative Information Environment

6-41. The complexity of the urban environment, particularly the human dimension, requires rapid information sharing—providing and receiving—from the national level to the tactical level, among Army headquarters at each echelon, with sister services and coalition partners, and with participating governmental and (at appropriate times) nongovernmental agencies. The analysis of urban information into the relevant intelligence necessary to refine and deepen a commander's understanding of the urban environment and its infrastructure of systems also demands collaboration among the various information sources and consumers. Therefore, commanders seek to establish streamlined procedures, develop commonality among databases, and make best use of existing information systems to disseminate and receive the necessary intelligence and relevant information for subordinates and partner organizations and agencies to exercise effective leadership and decision making and establish synergistic unity of effort in this multifaceted environment.

Transition Control

6-42. Because UO are resource intensive, commanders should plan to end them as quickly as possible, yet consistently with successful mission accomplishment. (Depending on the factors of METT-TC, successful transition may take a few days or many years to achieve.) The end state of all UO transfers control of the urban area to another agency or returns it to legitimate civilian control and responsibility. Rapid transition releases Army resources for use elsewhere and improves the civilian morale and disposition toward Army forces. This requires the successful completion of the Army force mission and a thorough transition plan. The transition plan may include returning control of the urban area to another agency a portion at a time as conditions permit. A successful transition plan should consider early alignment of military capabilities with existing urban governmental and administrative organizations, agencies, structures, and districts. Transition planning must be conducted before the onset of operations and continually adjusted as the situation develops.

Chapter 7
Urban Offensive Operations

... Capture Suez City "provided it does not become a Stalingrad situation."

Order to the Adan Armored Division
prior to its 1973 attack on Suez City
On the Banks of the Suez

Offensive urban operations (UO) are one of the most challenging operations that military forces can undertake. Campaigns and wars have sometimes hinged on their success or failure. Costly in resources, even when successful, they are not lightly entered into. Once engaged, they are executed rapidly and decisively. For reasons already discussed, threat forces defending in UO may gain advantages from the environment while Army force capabilities may diminish. Despite the challenges, Army forces conduct successful urban offensive operations by combining the Army's existing offensive doctrine with a thorough understanding of the environment.

PURPOSE OF URBAN OFFENSIVE OPERATIONS

7-1. Like all offensive operations, urban offensive operations are designed to impose the will of commanders on the threat. The urban offense often aims to destroy, defeat, or neutralize a threat force. However, the purpose may be to achieve some effect relating to the population or infrastructure of the urban area. Army forces may conduct offensive operations to secure a port or a communications center, to eliminate a threat to a friendly government or the urban population, or to deny the threat use of urban infrastructure. No matter the purpose, commanders must use a combined arms approach for successful urban offensive operations.

OFFENSIVE CHARACTERISTICS APPLIED TO URBAN OPERATIONS

7-2. All offensive operations, including those in urban areas, contain the characteristics of surprise, concentration, tempo, and audacity (see FM 3-0). Commanders consider and incorporate these characteristics in their offensive UO plans.

SURPRISE

7-3. Army forces can achieve offensive surprise at two levels: operational and tactical. In urban offensive operations, operational surprise can be decisive. The goal is to attack the urban area before the threat expects it, from a direction he doesn't expect, or in a manner he doesn't expect. In UO, this requires an attack against urban areas that the threat believes will provide sanctuary from the technological advantages of Army forces. Usually, urban areas that meet this criterion are not easily accessible. Army forces launch an attack against these urban complexes in different ways: through a vertical assault using airborne or air assault forces, through an amphibious assault, or through a penetration followed by a rapid and deep advance. All three attacks aim to achieve surprise and to deny the threat time to prepare and establish a defense. Surprise in a major urban operation prevents a threat from falling back to occupy prepared positions in and around an urban area.

7-4. At lower tactical levels, forces achieve surprise by attacking asymmetrically. An asymmetric method attacks the threat so he cannot respond effectively. This may be achieved by using special operations forces

(SOF) against a threat prepared for a conventional attack, by attacking decisively with heavy forces when the threat expects an effort by light forces or SOF, or by leveraging Army forces' extensive information operations (IO) capability. Offensive IO—primarily using IO elements of deception, electronic warfare, and operations security (OPSEC)—can help achieve surprise at all levels (see Chapter 4). Attacking at night surprises the threat and maximizes the Army forces' training, command and control (C2), and technological advantages. Attacking from unexpected or multiple directions achieves surprise by leveraging Army information systems (INFOSYS) and superior synchronization of combat power and capabilities.

CONCENTRATION

7-5. In UO, the attacking force creates a major advantage by concentrating the effects of combat power at the point and time of its choosing. The area and its compartmented effects naturally disperse and dissipate combat capability. Often commanders must position considerable amounts of stay-behind forces to protect hard-fought gains. The environment also hinders repositioning forces rapidly. Such effects can work equally against defending and attacking forces. However, in a well-prepared defense, the defender often has the advantage of interior lines. The defender can reinforce or reposition forces more quickly using covered and concealed routes (such as, sewers, tunnels, or prepared holes made in walls). Successful UO need synchronized air and ground maneuver with overwhelming effects from fires at decisive points on the urban battlefield. To achieve proper synchronization and precise effects, commanders should consider the unique time and distance relationships set by the environment.

TEMPO

7-6. Tempo is the rate of military action; it differs between UO and operations in more open terrain. Often, the primary purpose of the threat's urban defense is to take advantage of that differential to disrupt the rapidity and overall tempo of the Army's major operation. Additionally, the complexity and the potential for increased risk in an urban environment may invoke a cautious and methodical response on the part of commanders and their staffs thereby exacerbating tempo differences. While careful and meticulous preparation and planning are required in UO, commanders conducting major operations that include urban areas must strive to maintain an active tempo in offensive operations through synchronization of combat power and anticipation of threat reactions. A high tempo allows Army forces to achieve surprise and quickly gain positions of advantage. Controlling operational tempo and not allowing the different tempo of UO to adversely affect other operations is a challenge that commanders of major operations must overcome.

Operational Context of Urban Operations: Brittany Ports – 1944

The plan for the invasion of Normandy, France, in June 1944 was meticulously developed. The plan not only addressed the invasion itself, but also contained detailed planning for the campaign to follow. A major concern of the detailed campaign planning was logistics. To address this critical concern, and specifically the problem of ports to supply the allied armies once ashore, the pre-invasion planning called for the major ports of the French province of Brittany—Brest, Lorient, and Saint Nazaire—to be objectives of General Patton's Third Army, once it was activated.

Early August 1944, almost two months after the successful Normandy invasion, the operational situation significantly differed from that envisioned by the D-Day planners. General Montgomery's Twenty-first Army Group was still fighting in the Bocage of Normandy. In contrast, General Bradley's Twelfth Army Group had just achieved a major breakthrough at Saint Lo, secured the Cotentin Peninsula, and reached the city of Avranches. Here was a decision point. Bradley and Eisenhower had to decide whether to adhere to the original plan and turn west with Patton's forces to secure the peninsula or to take advantage of the breakout at Saint Lo and turn east to exploit the disruption of the German defenses.

Ultimately they reached a compromise. General Middleton's VIII Corps was tasked to secure the peninsula, and the bulk of Patton's Army, three Army corps, was turned northeast to exploit the operational collapse of the main German defenses. (See figure 7-1.)

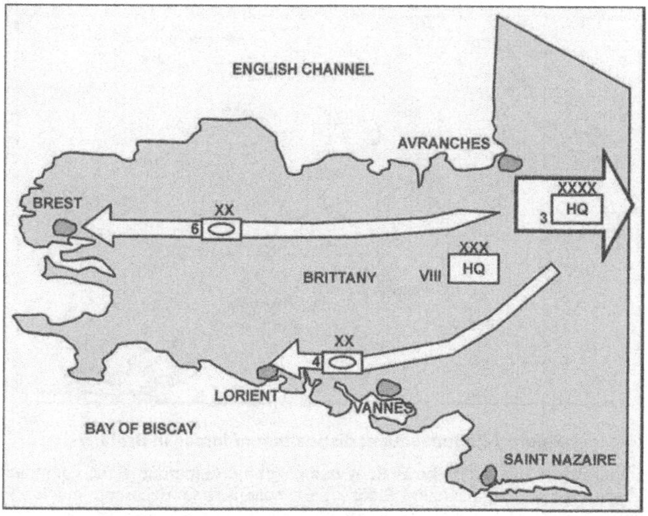

Figure 7-1. Initial attack in Brittany

Middleton's corps sprinted into the peninsula with the 4th and 6th Armored Divisions leading the way. However, poor communications, disagreements between commands, and contradictory orders caused the corps to hesitate before pushing the two divisions to continue to exploit toward the ports. The result: the 6th Armored Division missed an opportunity to seize Brest against light resistance by one day. The 4th Armored Division, after capturing the smaller port of Vannes, was also frustrated on the approaches to Lorient. The American reaction to the inability to rapidly seize the ports demonstrated an understanding of changing circumstances. The 6th Armored Division turned the attack at Brest to the 8th Infantry Division and then relieved the 4th Armored Division at Lorient. The 4th Armored was moved to rejoin the rest of Third Army exploiting to the east and north. Ultimately Brest fell to VIII Corps on 19 September after a 43-day siege by three infantry divisions. The victory yielded 36,000 German prisoners of war (POWs). However, the German defense and demolitions of the port left the port without an impact on the logistic situation of the allies. Brest cost the U.S. Army almost 10,000 casualties and the commitment of significant supplies. The experience convinced commanders to surround and bypass the other major Brittany ports. Lorient and Saint Nazaire remained under German control, deep in allied territory, until the war ended ten months later (see figure 7-2).

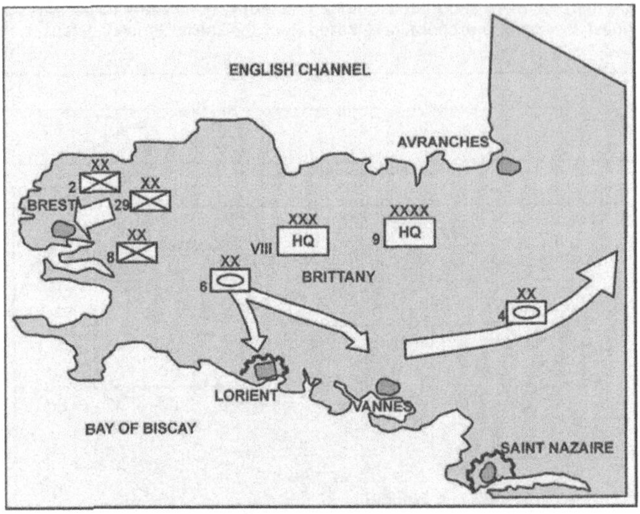

Figure 7-2. Subsequent disposition of forces in Brittany

The operational lessons of the Brittany campaign are numerous. First, commanders are responsible to continually assess assumptions and decisions made during planning based on the changing circumstances of the battlefield. This includes the planning decision to conduct urban offensive operations. When the allies arrived at the Brittany Peninsula, the focus of the operational maneuver was no longer securing logistics facilities but exploiting the breakthrough at Saint Lo and the disintegrating the German defense. The bulk of Third Army then was turned to the north and east rather than west into the peninsula.

The Brest experience also demonstrates that the costs of urban offensive operations are continually assessed against the operational value of the objective. This lesson was applied to the cities of Lorient and Saint Nazaire. The cities were never seized from the Germans because their logistic value failed to warrant the required resources to take them. German retention of the ports had no major adverse effect on the overall campaign.

Another lesson is that commanders cannot allow urban operations to disrupt the tempo of other offensive operations. One German goal of defending the ports was to disrupt the rapid tempo of the U.S. exploitation. They failed to achieve this goal because General Bradley continued the exploitation with the bulk of Third Army and executed the original plan with only a single corps.

Finally, commanders cannot allow emotion to color their decision to conduct or continue UO. The failure of 6th Armored Division to seize Brest rapidly caused some commanders to believe that Brest had to be captured because the prestige of the Army was committed to the battle. Costs of the continuing combat operations to seize Brest were significant. These resources might have been better committed elsewhere in the theater.

7-7. Tactical tempo is also important in urban combat. Because of the complex terrain, defending forces can rapidly occupy and defend from a position of strength. Once Army forces initiate tactical offensive operations, they cannot allow the threat to set the tempo of the operation. Instead, attacking forces seek to maintain a high tempo of operations. However, the tactical tempo of urban operations differs from operations in other terrain. Not necessarily slow, it requires a careful balance of preparation, speed, and security. In terms of unit fatigue, resource consumption, and contact with the threat, the tempo of most urban offensive operations may be rated as very high. On the other hand, in distances traveled and time consumed to achieve objectives, the tempo of many urban offensive operations might be rated as slow. The urban battlefield's density concentrates activity and consumes resources in a relatively small area. The lack of terrain seized or secured is not to be construed to mean a low tempo in the battle. Although the tempo may seem excruciatingly slow at higher levels of command and exceeding fast at lower, tactical levels, in reality, the natural tempo of urban operations is not faster or slower than other types of operations, merely different. Creating and operating at a tempo faster than an opponent can maintain, however, can favor forces which are better led, trained, prepared, and resourced.

7-8. A high tactical tempo in urban offensive operations challenges logisticians to provide for the increased consumption of munitions and degrades soldiers' physical capabilities. Commanders must anticipate these challenges and develop the means and abilities to overcome them. In the past, these challenges forced commanders to conduct urban offensives cyclically. They used night and other periods of limited visibility to resupply, rest, and refit forces. The environment influenced the tempo of their operations. This type of "battle rhythm" resulted in the forces spending each new day attacking a rested threat that was in a well-prepared position.

7-9. Army forces must maintain the tempo. Offensive operations continue even during darkness. Moreover, Army forces increase the tempo of operations at night to leverage the limited visibility capabilities, increased situational understanding, training, and INFOSYS that give an advantage to Army forces in all environments. To overcome the physical impact of the environment on soldiers, commanders can retain a large reserve to rotate, continuing offensive operations at night. The force that fights in daylight becomes the reserve, rests, and conducts sustaining operations while another force fights at night. Army forces can then maintain the tempo of operations and leverage technological advantages in urban offensive combat.

7-10. Tempo in UO does not necessarily mean speed. Offensive operations balance speed, security, and adequate firepower. Commanders must plan for the complex tactical environment and the requirements to secure flanks and airspace as the operation progresses. Mission orders allow subordinate units to make the most of tactical advantages and fleeting opportunities.

AUDACITY

7-11. Audacity is a simple plan of action, boldly executed. Superb execution and calculated risk exemplify it. In an urban attack, a thorough understanding of the physical terrain can mitigate risk. The terrain's complexity can be studied to reveal advantages to the attacker. Audacity can also be embodied in an operation by inventively integrating and coordinating the direct action tasks of SOF throughout the operation. Combining SOF actions with conventional attacks can asymmetrically unhinge a threat's defensive plan. Well trained Soldiers confident in their ability to execute urban offensive operations foster audacity.

URBAN OFFENSIVE OPERATIONS AND BATTLEFIELD ORGANIZATION

7-12. Urban offensive operations, like all operations, are arranged using the overall battlefield organization of decisive, shaping, and sustaining operations. Each operation is essential to the success of an urban offensive, and usually two or more of these operations occur simultaneously. Decisive operations are attacks that conclusively determine the outcome of UO. These attacks strike at a series of decisive points and directly lead to neutralizing the threat's center of gravity. Shaping operations in urban offensive operations create the conditions for decisive operations. In UO, much of the shaping effort focuses on isolation, which is critical in both major operations and tactical battles and engagements. Sustaining operations in urban offensive operations ensure freedom of action. They occur throughout the area of operations (AO) and for the duration of the operation.

DECISIVE OPERATIONS

7-13. A tactical commander fights decisive urban combat, whereas commanders conducting a larger major operation influence urban combat by setting the conditions for tactical success. Higher commanders may directly influence urban offensive operations by operational maneuver, by coordinating joint fires, by closely coordinating conventional forces, or with SOF.

7-14. Tactical urban offensive operations quickly devolve into small-unit tactics of squads, platoons, and companies seizing their objectives. The compartmented effect of the terrain and the obstacles to command and control of small units, especially once they enter close combat inside buildings or underground, often restricts the higher commander's ability to influence operations. Commanders can influence the actions of subordinates by clearly identifying the decisive points leading to the center of gravity; using mission orders (as discussed in Chapter 5); developing effective task organizations; synchronizing their decisive, shaping, and sustaining operations; and managing transitions.

7-15. Like all operations, successful decisive operations in UO depend on identifying the decisive points so the forces can destroy or neutralize the threat's center of gravity. Seizing a key structure or system that makes the threat's defense untenable; interdicting a key resupply route that effectively isolates the threat force from his primary source of support; or isolating the threat so that his force can no longer influence friendly activity may be more effective than his outright destruction.

7-16. Commanders must select the right subordinate force for the mission and balance it with appropriate attachments. Higher commanders do not direct how to organize the small tactical combined arms teams, but they must ensure that subordinates have the proper balance of forces from which to form these teams. Successful urban offensive operations require small tactical combined arms teams. Urban offensive operations require abundant infantry as the base of this force. However, successful urban combat requires a combined arms approach adjusted for the conditions of the environment. In urban offensive operations, it will normally not be a question of whether include armor and mechanized elements but, rather, how best to accomplish this task organization. Precision-capable artillery systems generally support urban operations better than unguided rocket artillery.

7-17. Divisions entering urban combat may require additional resources. These resources include military intelligence support in the form of linguists, human intelligence (HUMINT) specialists, and unmanned aircraft systems (UAS). Engineering assets will be at a premium; the task organization of a task force executing the decisive operation may require a one-to-one ratio of engineer units to combat units. Corps and higher engineering support may be necessary to meet these requirements and to repair vital and specialized infrastructure. A tailored and dedicated support battalion or group may need to assist in providing anticipated support to a displaced and stressed civil population. Finally, divisional CA units may require augmentation to deal with NGOs and civilian government issues.

7-18. Successfully conducting decisive operations in the urban environment requires properly synchronizing the application of all available combat power. Army forces have a major advantage in the command and control of operations. Commanders can use this advantage to attack numerous decisive points simultaneously or in rapid succession. They also use it to attack each individual decisive point from as many directions and with as many different complementary capabilities as possible. Commanders must completely understand urban environmental effects on warfighting functions to envision and execute the bold and imaginative operations required. Significantly, these operations require that C2 systems account for the mitigating effects of the environment as execution occurs.

7-19. Properly synchronized actions considerably enhance the relative value of the combat power applied at the decisive points. They present to the threat more requirements than he has resources with which to respond. Synchronized IO and multiple maneuver actions paralyze the threat's decision-making capacity with information overload combined with attacks on his C2 systems. Additionally, well-synchronized actions limit the time the threat has to make decisions and forces him into bad decisions. In the urban environment, these effects are enhanced because C2 systems are already strained, poor decisions are harder to retrieve, and units that do not react effectively are more easily isolated and destroyed.

SHAPING OPERATIONS

7-20. Shaping operations that support the urban attack separate into those focused on isolating the threat and all others. Army forces isolate the threat to ensure successful urban offensive operations. Depending on the threat reaction to isolation efforts and the nature of the threat center of gravity, this task may become decisive. Other shaping operations include those common to all offensive operations and others unique to urban operations. Unique urban shaping operations may include securing a foothold in a well-fortified defensive sector, securing key infrastructure, or protecting noncombatants. Because of the nature of UO, shaping operations may consume a much larger proportion of the force than during other operations and may take place both inside and outside the urban area (see Applying the Urban Operational Framework: Panama in Chapter 5). By successfully isolating a threat force, the force needed to conduct the decisive operation may be relatively small.

SUSTAINING OPERATIONS

7-21. Commanders conducting urban offensive operations must ensure security of the sustaining operation and bases; in many situations, protecting lines of communications (LOCs) and sustaining operations may be the greatest vulnerability of the attacking force. Those supporting an urban offensive are tailored to the urban environment and are well forward. Ideally, the supporting forces closely follow the combat forces and move within or just outside the urban area as soon as they secure an area. Operating in the urban area during offensive operations allows the sustaining operation to take advantage of the defensive attributes of the environment for security purposes.

7-22. Counterattacks against sustaining operations may take the form of special operations activities aimed at LOCs leading to or within the urban area. Choke points—such as bridges, tunnels, and mountain passes—are vulnerable to these attacks and may require combat forces to protect them. Threat forces attack the LOC to blunt the Army's combat power advantage in the urban area.

7-23. Attacks against the LOC into the urban area may also attempt to isolate the attacking Army forces from its sustainment base. Isolated forces in an urban area are greatly disadvantaged. Commanders must plan and aggressively execute strong measures to protect their LOC, even if it requires reduced combat power to execute their offensive operation.

7-24. Sustaining operations anticipate the volume and unique logistics requirements of urban operations. Specialized individual equipment—such as grappling hooks, ladders, and pads—is identified and provided to troops in quantity before they are needed. Forces stockpile and distribute their attacking units' special munitions requirements including small arms, explosives, and grenades of all types, precision artillery munitions, and mortar ammunition. Forces also supply transport to move the resources rapidly forward, both to and through the urban environment. Sustaining operations cannot rely on "operational pauses" to execute their tasks. Commanders must plan to continuously supply resources and capabilities to the most forward combatants as offensive operations advance.

7-25. Sustaining operations also anticipate the growth of sustainment requirements as Army forces secure and take responsibility for large portions of the urban area. The success of Army urban offensive operations will often uncover the civil population in former threat occupied areas. It may attract the civil population from sections of the urban area where the Army is not operating to areas occupied by Army forces. Rural populations may migrate to the urban area as the result of successful Army offensive operations.

7-26. Army forces may be required to take initial responsibility to provide for the urban population. This consideration is integrated into logistics planning and organization from the start of the planning process. To be successful and efficient in such a situation, sustainment planning includes Army civil affairs (CA) specialists and local government representatives. It also integrates and consults with the international community and nongovernmental organizations (NGOs) that might augment or supplement Army sustainment capabilities.

FORMS AND TYPES OF URBAN OFFENSE

7-27. Traditional forms of offensive maneuver include envelopment, turning movement, infiltration, penetration, and frontal attack. Traditional types of offensive operations are movement to contact, attack, exploitation, and pursuit. These traditional forms and types listed apply to urban combat. Some have greater application to an urban environment than others do. Moreover, success will belong to commanders who imaginatively combine and sequence these forms and types throughout the depth, breadth, and height of the urban battlefield. This is true at the lowest tactical level and in major operations.

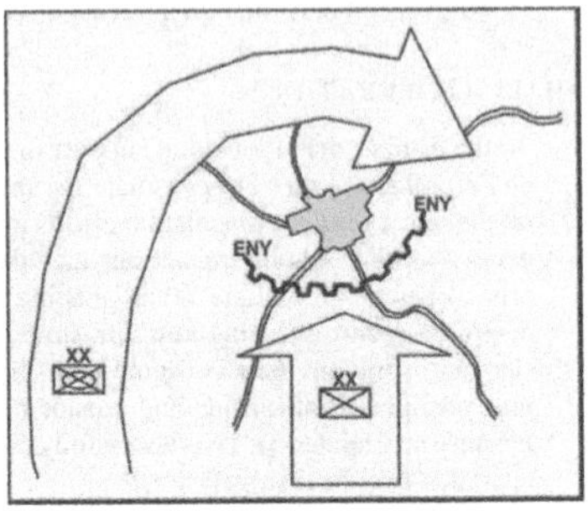

Figure 7-3. Envelopment isolates an urban area

FORMS OF OFFENSIVE MANEUVER

Envelopment

7-28. The envelopment is the ideal maneuver for isolating threat elements in the urban area or isolating the area itself. A deep envelopment effectively isolates the defending forces and sets the conditions for attacking the urban area from the flank or rear. Yet, enveloping an objective or threat force in the urban area is often harder since achieving speed of maneuver in the environment is so difficult (see figure 7-3). Vertical envelopment, however, works effectively if Army fires can effectively suppress or neutralize the threat air defense.

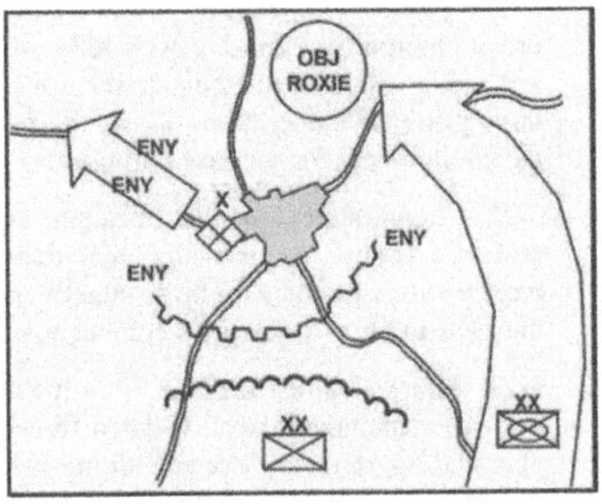

Figure 7-4. Turning movement

Turning Movement

7-29. Turning movements can also be extremely effective in major operations (see figure 7-4). By controlling key LOCs into the urban area, Army forces can force the threat to abandon the urban area entirely. These movements may also force the threat to fight in the open to regain control of LOCs.

Infiltration

7-30. Infiltration secures key objectives in the urban area while avoiding unnecessary combat with threat defensive forces on conditions favorable to them (see figure 7-5). This technique seeks to avoid the threat's defense using stealthy, clandestine movement through all dimensions of the urban area to occupy positions of

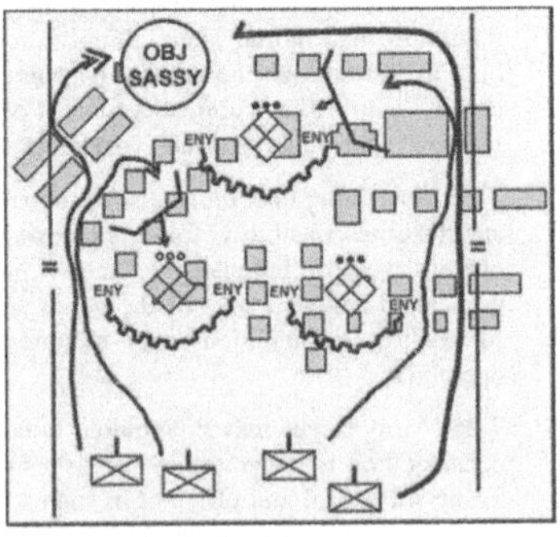

F

igure 7-5. Infiltration

advantage in the threat's rear (or elsewhere).It depends on the careful selection of objectives that threaten the integrity of the threat's defense and a superior common operational picture (COP). Well-planned and resourced deception operations may potentially play a critical role in masking the movement of infiltrating forces. The difficulty of infiltration attacks increases with the size and number of units involved. It is also more difficult when Army forces face a hostile civilian population. Under such circumstances, infiltration by conventional forces may be impossible. Armored forces are generally inappropriate for infiltration operations. However, they may infiltrate large urban areas if the threat is not established in strength and had insufficient time to prepare defenses.

Penetration

7-31. Penetration is often the most useful form of attack against a prepared and comprehensive urban defense (see figure 7-6). It focuses on successfully attacking a decisive point or on segmenting or fragmenting the defense—thereby weakening it and allowing for piecemeal destruction. (The decisive point may be a relatively weak or undefended area that allows Army forces to establish a foothold for attacks on the remainder of the urban area.) Ideally in urban combat, multiple penetrations in all dimensions are focused at the same decisive point or on several decisive points simultaneously. In urban combat, the flanks of a penetration attack are secure, and resources are positioned to exploit the penetration once achieved. Although always a combined arms team, rapid penetrations are enhanced by the potential speed, firepower, and shock action of armored and mechanized forces.

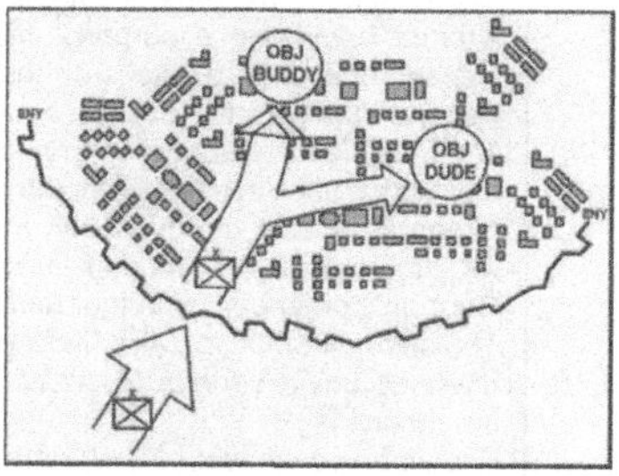

Figure 7-6. Penetration

7-32. Importantly, commanders must consider required actions and resources that must be applied following a successful (or unsuccessful) penetration. A penetration may result in the rapid collapse and defeat of the threat defense and complete capitulation. On the other hand, success may cause threat forces to withdraw but leave significant stay-behind forces (or disperse into the urban population as an insurgent-type force) which may necessitate methodical room-to-room clearance operations by significant dismounted forces. Additionally, securing portions (or all) of an urban area requires occupation by Army forces to prevent re-infiltration of threat forces thereby further increasing manpower requirements. Based on the factors of the situation, commanders may conclude that over time methodical clearance operations conducted from the outset, while frontloading time and resource requirements, will be less costly than a penetration followed by systematic clearing operations.

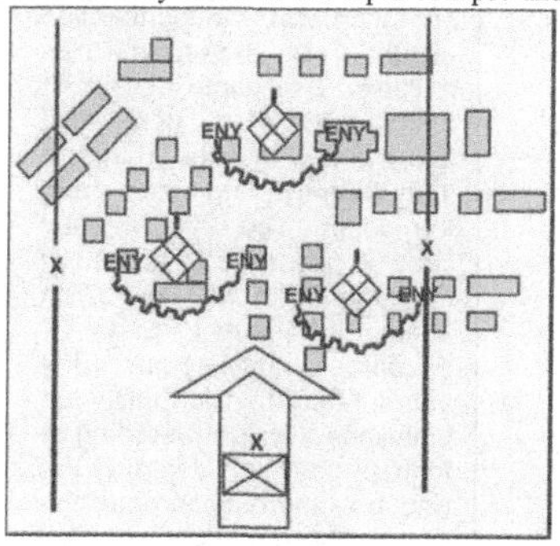

Figure 7-7. Frontal attack

Frontal Attack

7-33. For the commander of a major operation, the frontal attack is generally the least favorable form of maneuver against an urban area unless the threat is at an obvious disadvantage in organization; numbers, training, weapons capabilities, and overall combat power (see figure 7-7). Frontal attacks require many resources to execute properly, risk dispersing combat power into nonessential portions of the area, and risk exposing more of the force than necessary to threat fires. In urban offensive combat, forces most effectively use the frontal attack at the lowest tactical level once they set conditions to ensure that they have achieved overwhelming combat power. Then the force of the frontal attack overwhelms the threat with speed and coordinated and synchronized combat power at the point of attack. The assigned frontage for units conducting an attack on an urban area depends upon the size and type of the buildings and the anticipated threat disposition. Generally, a company attacks on a one- to two-block front and a battalion on a two- to four-block front (based on city blocks averaging 175 meters in width).

Forms of Attack in the Urban Offense: Metz – 1944

In November 1944, the U.S. Third Army launched its final effort to take the French city of Metz from the defending Germans. This was the Army's third attempt. The first attempt had been a surprise, mounted attack. This was followed by a series of piecemeal infantry assaults on the surrounding fortresses. Finally, a deliberate effort was made to take the city in a coordinated effort by XX Corps.

The initial failures stemmed from a shortage of resources on the U.S. side, to include fuel and units—especially Infantry. This added to the ad-hoc nature of the first two efforts. The third effort, though more deliberate, was still constrained by resources. XX Corps could only muster three nearly full-strength infantry divisions to attack the German's defending with four under-strength divisions.

The third attempt to take Metz demonstrates how a corps operates with multiple divisions using various forms of attack to achieve its objective in urban offensive operations.

The opening phase of the Metz battle had attacks by the 90th and 5th Infantry Divisions to envelop the city from the north and south (see figure 7-8). This isolated the city and ensured the garrison could not escape nor be reinforced. The garrison was under orders from Hitler not to conduct a breakout and to resist to the last man; thus, German forces strongly counterattacked both wings of the envelopment to prevent isolation.

The second phase of the operation was the penetration of the city defenses from multiple directions. The 5th Infantry Division penetrated into the city with the 10th and 11th Infantry Regiments from the south. The 95th Infantry Division penetrated into the city from the north with the 377th Infantry Regiment and TF Bacon. Simultaneously, the 95th Infantry Division infiltrated battalions through the string of fortress positions guarding the western approaches into the city, isolated, and bypassed these positions with its other two regiments.

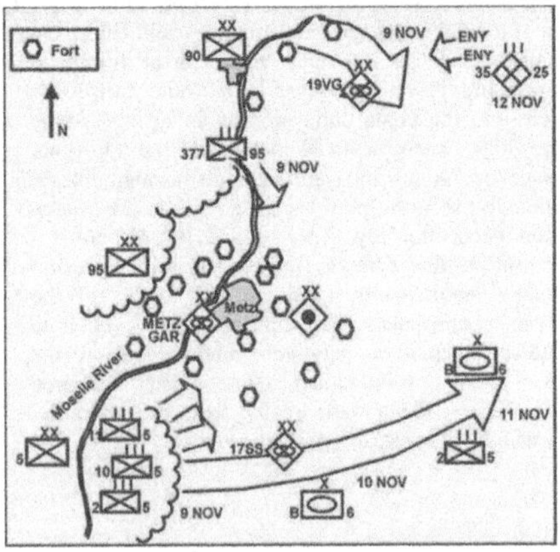

Figure 7-8. Metz envelopment

The final reduction of the defense was a series of battalion frontal attacks, which took place against the last remaining strongholds within the city. Even in these final engagements, however, the infantry battalions isolated, bypassed, and then attacked from the flanks and rear whenever possible (see figure 7-9).

The city was declared secured on 19 November. However, at that point more than a half-dozen of the fortresses had yet to be reduced. The 95th Infantry Division, after a four-day rest, quickly moved forward to rejoin the still rapidly advancing corps forward elements. Elements of the 5th Infantry Division remained in siege posture around the remaining strong points, the last of which surrendered on 19 December 1944 when it ran out of food. U.S. forces made no efforts to attempt to assault these bypassed fortresses although extensive psychological operations (PSYOP) were used.

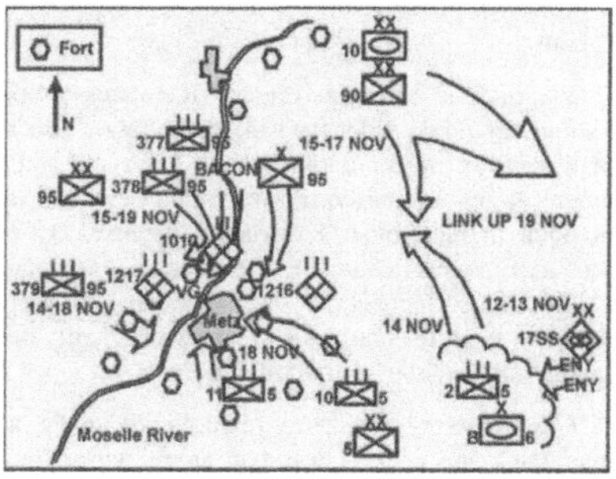

Figure 7-9. Metz final assault

TYPES OF OFFENSIVE OPERATIONS

Movement to Contact

7-34. In an urban area where the threat situation is vague, Army forces will often conduct a movement to contact to establish or regain threat contact and develop the situation. A movement to contact in an urban area occurs as both sides try to establish their influence or control over a contested urban area. The situation determines whether the movement to contact or its specific technique, the search and attack, is appropriate. A conventional force-oriented movement to contact will likely take place when friendly and threat conventional forces attempt to establish control simultaneously. Initially, neither side is defensive. The friendly force aims to quickly locate and fix the threat while establishing control of the urban area and its key infrastructure. The search and attack technique works well when a smaller threat has established a noncontiguous defense in an urban area. This operation is characterized by the friendly point defense of key infrastructure, robust reconnaissance, and rapidly concentrated combat power to fix and defeat or destroy threat resistance once located.

7-35. A meeting engagement often results from the movement to contact. It occurs when a moving force that is partially deployed for battle collides with and engages a threat at an unexpected time and place. In a meeting engagement in an urban area, the unit that reacts most quickly and decisively will likely win. Rapid and accurate decision making depends heavily on understanding the nature of the urban area and its impact on operations. Thus, in a meeting engagement, commanders should quickly assess the impact and role of all components of the urban environment (terrain, infrastructure, and society) on the operation. To this end, responsive reconnaissance is important. This permits accurate decision making regarding where to attack, where to defend, and how to allocate resources. Situational understanding enhanced by robust digital INFOSYS that provide an accurate COP also facilitates the rapid reaction of Army units and a synchronized response. This reaction and response allow Army forces to seize the initiative and dominate the threat.

Attack

7-36. The attack is the most common and likely offensive operation that Army forces will conduct in an urban environment. Commanders conducting major operations and commanders of large tactical units usually execute deliberate attacks. In the urban environment, units larger than battalion-size rarely conduct hasty attacks. Hasty attacks are common below company level as units use their initiative to take advantage of tactical opportunities. However, larger units will conduct hasty attacks when threat defenses are disrupted or unprepared, to take advantage of an unexpected situation, and to prevent the threat from establishing or re-establishing a coherent defense.

Exploitation

7-37. Exploitation follows a successful attack to disrupt the threat in depth. Commanders of major operations should consider focusing exploitation attacks on urban areas. A threat defeated in an attack will attempt to rally units, reinforce with reserves, and reorganize his defense. With its information and communications capability, transportation network, and defensive attributes, the urban area is the natural focal point to reestablish a disrupted defense. By establishing urban centers as the objectives of the exploitation, commanders deny the threat the sanctuary needed to reorganize and reestablish his defense. The exploitation focuses on the urban area as well as on the remnants of the threat. A successful exploitation to seize an urban area works efficiently because the attack preempts the defense and denies the threat the full advantages of urban terrain.

7-38. Commanders conducting exploitation must acknowledge the vulnerability of their forces to counterattack and ambush in urban areas. An urban area provides ideal cover and concealment to hide threat reserves, reinforcements, or reorganized forces. Constrictions of routes into and through the urban area make exploitation forces a potentially dense target and limit maneuver options. Robust and well-coordinated reconnaissance, tactical dispersal, and use of advance guard security forces protect against this threat (see Defensive Combat Power: Suez City vignette in Chapter 8).

Pursuit

7-39. The pursuit is designed to destroy threat forces attempting to escape. It focuses on the threat and not on urban areas. When conducting a pursuit, Army forces move through undefended urban areas and, if possible, bypass those in which threat forces successfully take refuge. The threat will likely attempt to use urban areas to disrupt the pursuit and permit the threat main body to escape. Commanders can prevent escape by denying the threat the time to establish forces in urban areas that cannot be bypassed. The agility of Army aviation forces for attack, reconnaissance, and transportation is essential to execute a successful pursuit around and through urban areas.

URBAN OFFENSIVE CONSIDERATIONS

7-40. The urban operational framework (understand, shape, engage, consolidate, and transition) provides a structure for developing considerations unique to urban offensive operations. The considerations vary depending on the situation and scale of the operation. Some considerations applicable to major operations that include an urban area will also be considerations at the tactical level focused in the urban area. However, no set rules exist. All urban operations are unique. Issues addressed at the operational level in one situation may be addressed in a new situation only at the tactical level. Under the right circumstances, a consideration may become an operational issue, a tactical issue, or a combination of the two. The following identifies some planning and execution issues that commanders conducting major operations should address.

UNDERSTAND

7-41. The first requirement, and a continuing requirement throughout the conduct of urban operations, is the assessment and understanding of the situation. Commanders should base this understanding on detailed information regarding the particular urban area. Since the threat will likely dominate or control most of the urban area during the planning phase of offensive operations, achieving an accurate understanding of the urban environment will be difficult. A comprehensive intelligence, surveillance, and reconnaissance (ISR) effort in support of a rigorous intelligence preparation of the battlefield (IPB) process overcomes this obstacle.

Integrated Intelligence, Surveillance, and Reconnaissance

7-42. The commander of a major operation that includes an urban area, unlike his subordinate commanders, can target reconnaissance deep into his AO and area of interest. This begins the application of ISR resources against the urban area that may lead to decisive ground operations. This ISR effort and the understanding it supports continue as long as the urban area remains in the AO. Commanders of major operations initially direct ISR assets on those information requirements that support determining whether or not to conduct urban offensive operations. Once decided, ISR resources shift to support the planning and execution of the operation in the urban area.

7-43. The first resources that a senior commander can use are national and strategic sensors. He requests them through the appropriate joint force commander. The commander aggressively pursues full use of these systems to begin building an initial database for analyzing the significant aspects of the terrain; key infrastructure considerations; the status and disposition of the population; and the size, type, disposition, and intentions of threat forces in the area.

7-44. Simultaneously, multiple intelligence sources contribute to the database. The sources collect, process, store, display, and disseminate the relevant information on large urban areas through open and classified resources. These information sources include—

- Historical research.
- Travel brochures that include cultural information and recent maps.
- Classified debriefings of diplomats, businesses, DOD personnel, and allies.
- Military maps and special geospatial products of the urban area.
- Previous intelligence assessments of the country, government, and population.
- Reachback to appropriate economic, political, cultural, and infrastructure subject matter experts not in the commanders AO.

The gathering and analysis of HUMINT plays a critical part of this process and assists commanders in understanding ethnic, cultural, religious, economic, political, and other societal and infrastructural facets of the environment.

7-45. As the intelligence and the national reconnaissance and surveillance efforts progress, commanders will insert, if available and feasible, Special Forces reconnaissance assets into the urban environment. These elements will seek to confirm or deny the information received from imagery intelligence (IMINT), signals intelligence (SIGINT), and HUMINT sources. Among many factors, using SOF depends on their availability, the particular urban area, the area's ethnic composition, and the relationship between the urban population and the threat. Other joint operational reconnaissance and surveillance assets that higher-echelon commanders may have available might include the Joint Surveillance Target Attack Radar System, Guard Rail targeting aircraft, UAS, and space-based systems.

7-46. The commander's staff will use all sources of information—IMINT and SIGINT sensors, HUMINT, historical research, and reconnaissance—to refine his ability to understand the urban environment. Digitally linking subordinate commanders with information sources helps to develop a COP essential to their situational understanding of the urban environment. The IPB process guides this assessment. As operations progress, additional reconnaissance and surveillance assets may become available. These may include UAS, long-range reconnaissance and surveillance units, counterfire radar, and air and ground cavalry. As these assets are employed, they are linked into the net of sources sharing information and further refine a common situational understanding of the environment.

Focused Assessment Efforts

7-47. In urban offensive operations, the tactical commander's assessment focuses on defeating the threat in the urban area within the constraints of the environment. Toward this end, identifying and assessing decisive points to attack is a commander's priority assessment task. Some unique aspects of the urban environment also require the focus of the commander's assessment efforts. These include the character of the urban defense, collateral damage considerations, and the effects of the environment on friendly and threat courses of action.

Character of the Urban Defense

7-48. To be both efficient and effective, Army urban offensive operations focus on what is decisive. Decisive points for an urban attack depend primarily on the mission within the urban area. They can vary widely in composition and size. Since commanders only focus on the essential, they may determine the decisive point to be a single building or a limited sector of an urban area. It could be an entire system within the urban infrastructure such as communications and information, or a limited subsystem of the transportation and distribution infrastructure such as a single airfield. Sometimes what is decisive in the urban area is the threat military capability, but even this large an objective, when carefully analyzed, may not require destruction of all threat forces or control of the entire urban area. Decisive points relate directly to the threat's center of gravity and to mission success. Some decisive points related to the urban threat's center of gravity may be physically located outside the urban area.

7-49. To gain specifics on threat dispositions within the urban area requires reconnaissance capability to see into the depths of the area and the intelligence capability to determine the threat's likely defensive course of action. With this information, commanders can determine decisive points and apply Army combat power discretely against them. Effective urban offensive operations require detailed situational understanding of an area of interest that extends well beyond the perimeter of the urban area.

7-50. Commanders see throughout the depth of the urban area using several actions (see figure 7-10). First, they evaluate sensor data and imagery. This guides targeting of special reconnaissance. Simultaneously, HUMINT is conducted using any persons who might know the urban area and threat. This includes civilians (allies, aides, neutrals, obstacles, and hostiles) and POWs. Finally, tactical conventional reconnaissance assets including reconnaissance forces, aviation, artillery radar, signals intelligence, and UAS are directed at the urban area. All these sources and data are linked through digital INFOSYS to provide commanders and their subordinates with improved situational understanding and a COP.

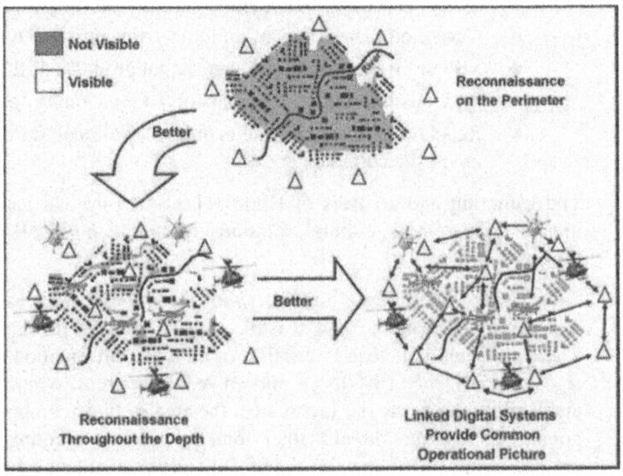

Figure 7-10. Required urban reconnaissance capabilities

Collateral Damage Considerations

7-51. Commanders must also assess the collateral damage risks that their operations may include. This assessment helps to initially determine the viability of a course of action. However, commanders should reassess their courses of action at frequent intervals in urban offensive operations based on known information to determine if the original evaluations remain valid. This reassessment minimizes potential collateral damage from a change in mission or a change in a course of action. Many aspects of the environment can change during mission execution.

The Environment's Effects on Courses of Action

7-52. The urban environment's unique aspects can significantly impact the course of action chosen by Army forces and the threat. Commanders must assess these effects in planning, but they must also verify and monitor these effects as forces execute offensive missions. In particular, commanders will want to confirm the civilian population's locations, beliefs, and actions and to monitor any changes. They will need to validate terrain considerations and monitor the effects of any changes due to rubble and other damage to structures. In urban terrain, dead space, cover, and concealment can only be identified physically and may change considerably as operations are executed.

SHAPE

7-53. Commanders of major operations have a primary contribution to urban operations: the planning and conduct of effective shaping operations that set the conditions for subordinate tactical success. In urban operations, isolation will be a critical condition. Effective isolation will require persistent, continuous surveillance and reconnaissance, innovative use of fires and maneuver (including effective force allocation decisions), and well-established sensor-to-shooter links. These efforts—combined and synchronized with SOF direct actions, IO that minimize noncombatant influences, and necessary shaping attacks (particularly the seizure of a foothold)—establish the conditions necessary for the subsequent offensive domination of the area.

Isolation is Essential

7-54. One key to success in the history of urban operations has been the effective isolation of the threat force (see figure 7-11). This applies today and equally well to major urban offensive operations as it does to smaller-unit attacks. This isolation not only denies access to the urban area from outside but also contains threat forces within. In a modern metropolis or megalopolis, this can appear a daunting task. Operational isolation requires dominating all physical and electronic contact between the threat in the urban area and supporting threat forces outside the urban area. This does not necessarily require physically encircling the urban area, but it does require that Army forces be able to exert control over the area's entire perimeter, as well as decisive points within. For a sprawling urban area, successful isolation may require the commitment of a large amount of resources.

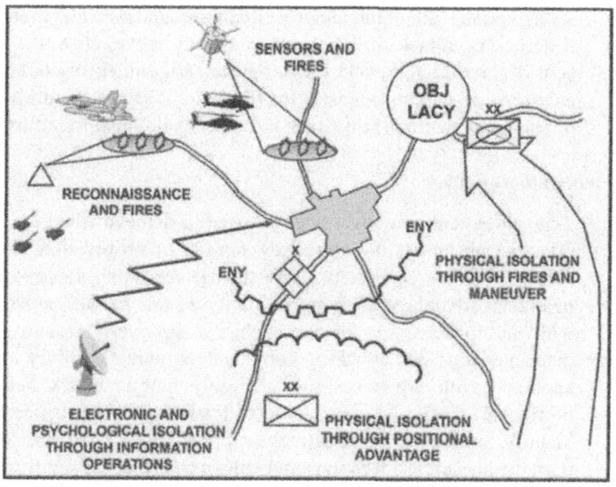

Figure 7-11. Shaping through isolation

7-55. Successful isolation of the urban area depends as much on the nature of the threat as it does on any other factor. A conventional threat in a large urban area may be much easier to isolate than an insurgent threat in a much smaller urban area. The forces needed in the former situation may be less than those needed in the latter. The more the characteristics of the threat are conventional in nature generally the easier it will be to isolate him using standard combat methods and equipment. Isolating a more unconventional force requires many of the same techniques as used against conventional forces. However, it also requires a much greater ability to simultaneously conduct offensive IO, to integrate CA units and civil-military operations (CMO), and to work with allies, nongovernmental organizations, and local authorities. Fundamentally, isolating a less conventional threat puts increased emphasis on separating combatants from noncombatants.

Offensive Isolation Objectives

7-56. Isolation seeks to achieve two primary objectives with respect to defeating a threat's urban defense:

- Weaken the overall coherence of his defense.
- Limit or manipulate his maneuver options.

Isolating the threat in the urban area from external support, as well as isolating him from sources of support within the urban area, weakens his overall defense. The defense is weakened through a combination of attrition (the threat cannot replace his losses) and the diversion of his combat power from the defense to operations to counter the isolation effort. Isolation can also prevent the threat from shifting his forces to reinforce decisive points in the urban area or to conduct counterattacks.

7-57. Commanders may choose not to isolate the urban area completely—or at least make it appear so to threat forces. Instead, they may afford the threat an apparent means of escape, create the conditions for its use through effective fire and maneuver against the defenders, and then destroy the threat through various ambush methods. While friendly forces may be able to move undetected to appropriate ambush sites, it is more likely that this technique will necessitate rapidly mobile air and ground forces moving along carefully chosen routes through the urban area. Commanders must consider maintaining the ability to complete the isolation of the urban area to prevent reinforcement and escape of urban threat forces—particularly if the ambush attack does not achieve desired effects.

Persistent Surveillance

7-58. Persistent surveillance of the urban area is essential to all types of actions used to isolate an urban area and as complete as resources will allow. Surveillance of the urban area relies on either reconnaissance forces or sensors continuously observing or monitoring urban avenues of approach. This network of ISR assets updates the commander's situational understanding and provides the means to quickly identify and, if necessary, attack threat elements as they move. However, particularly with sensors, commanders must consider that each detection is not necessarily an enemy to be attacked. Noncombatant activity clutters the environment making it easier for threats to disguise themselves and increases the burden (and the number of resources required) on Army forces to distinguish friend from foe.

Fires and Maneuver

7-59. Fires and maneuver may be used to achieve isolation, either singly or in combination. (As always, effective obstacles, monitored by sensors or observation, are integral to any isolation technique.) First, attacking forces can pre-position themselves along avenues of approach to deny entry and exit through positional advantage. Relying primarily on this method of isolation, particularly around a large urban area with multiple avenues of approach, can be overly resource intensive. Instead, the pairing of fires and maneuver provides attacking commanders more flexibility and allows them to isolate several avenues of approach with fewer resources. Highly mobile attack helicopters, operating outside threat-controlled portions of the urban area, are ideal for this purpose. Inside threat-controlled areas, it is more difficult to identify, eliminate, or effectively suppress the air defense threat. The threats may have numerous man-portable air defense weapons and enhanced effects of small arms used for air defense. Therefore, the risk to using this equipment may outweigh the potential benefits. However, mobile ground units—such as an air assault (subject to the same air defense threat considerations as attack aviation), armored, or mechanized forces—can also rapidly move to attack and destroy a threat moving in or out of an urban area. Potential disadvantages of the combined, fires and maneuver, option are that—

- Critical assets, on standby and dedicated to isolation efforts, may be unavailable for other missions.
- The attacking force may not locate the threat in time to complete its mission (an inherent risk to any attack).

7-60. Another alternative relies on indirect or joint fires alone to destroy the threat force. Its disadvantage is that fires alone often cannot completely destroy or stop a determined force from moving into or out of an urban area. Although targets and avenues of approach will require continual surveillance, it is usually a less resource-intensive option than those that include maneuver. It also does not normally require fires assets to remain on standby to accomplish the mission. However, fires must be able to reliably and quickly respond. For Army field artillery units and naval gunfire, the units must also be in range, which requires careful positioning. A skilled threat can avoid interdiction fires by using the geometry of the area to identify gaps due to obstructing terrain or the firing unit's range limitations. It can also use concealment and weather to avoid observation. However, effective sensor-to-shooter links throughout the urban battlefield will reduce the threat's ability to hide (see figure 7-12). A resolute threat may risk significant losses to fires to prevent isolation or may attempt to use noncombatants as a shield. Ultimately, commanders should use innovative combinations of all techniques discussed. Some units will physically block key avenues of approach. Surveillance will monitor less important routes and avenues. Artillery fires, joint fires, and maneuver units will then respond to the results of surveillance depending on the circumstances.

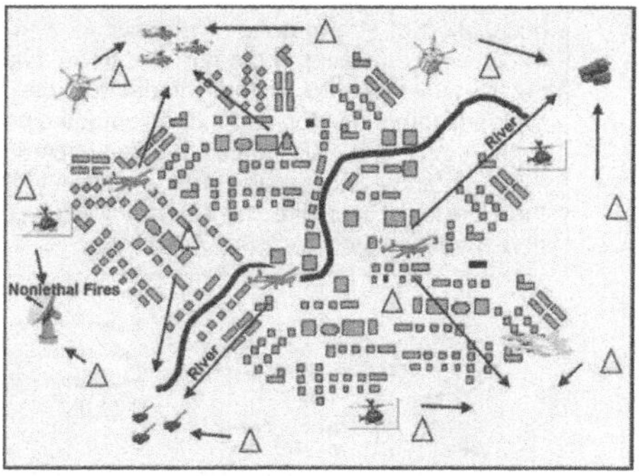

Figure 7-12. Critical sensor-to-shooter links

Threat Reactions

7-61. The reaction of the threat to the effects of isolation will depend on his mission, morale, force structure, and overall campaign plan. The threat may recognize isolation actions early and withdraw from the urban area before isolation is completed instead of risking destruction. On the other hand, the threat, based on a different or flawed assessment (perhaps a perception shaped by the Army force commander), may choose to—

- Continue to defend (or hide) and conduct local ambushes and counterattacks.
- Attack to break into the urban area or infiltrate forces and supplies in.
- Attack to break out of the urban area or infiltrate forces out.
- Execute any combination of the above (see figure 7-13).

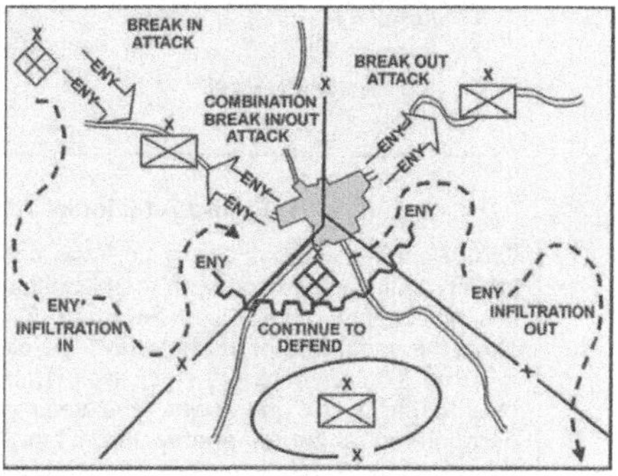

Figure 7-13. Reactions to isolation

7-62. Attacking commanders must consider how the threat leadership's subsequent actions will affect the continuance of overall offensive operations. They deliberate many considerations, to include—

- The allocation of more forces to the shaping operations to isolate the urban area.
- The allocation of more combat power to achieve rapid penetration and seizure of objectives to take advantage of developing threat dispositions in the urban area.

Isolating an Urban Area: Hue, Vietnam – 1968

On 31 January 1968, the 4th and 6th North Vietnamese Army (NVA) regiments and attached NVA and Vietcong (VC) sapper battalions attempted to seize control of Hue from the South Vietnamese Army's (ARVN) 1st Division as part of the North Vietnamese Tet Offensive. The attack, which was launched with complete surprise, successfully established temporary control over most of the city and facilitated the occupation of strong defensive positions within the city's ancient fortress known as the Citadel. However, it did not lead to the capture of the headquarters of the ARVN 1st Division located in the Citadel nor the U.S. military assistance command—Vietnam compound located in the southeast part of the city.

Both ARVN and U.S. forces swiftly reacted to the NVA attack. Within 24 hours, ARVN infantry and cavalry units counterattacked to recapture the Citadel. South of the Perfume River, the 1st Marine Division began attacking to clear the southern portion of the city (see figure 7-14).

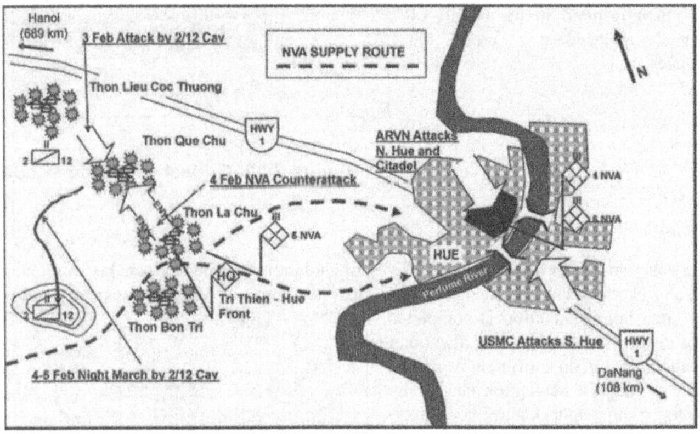

Figure 7-14. Initial attack to isolate Hue

On 2 February, the first Army unit was committed to Hue. The 2-12th Cavalry (an infantry battalion) was ordered to attack southeast along Highway 1 and clear enemy interdicting that route. On 3 February, 2-12th Cavalry began their attack moving along the south side of the highway. The plan was to clear a series of small hamlets in succession en route to the city itself. The first hamlet, Thon Lieu Coc Thuong, was cleared easily. Moving toward the second hamlet, Thon Que Chu, the battalion encountered a strong enemy force that was well entrenched. Air and artillery supported the battalion as it attacked and captured the hamlet against strong resistance. The battalion dug in for the night and prepared to resume the attack against the third hamlet, Thon La Chu, on 4 February.

Unknown to the 2-12th Cavalry, Thon La Chu was the headquarters of the Tri Thien-Hue Front, and it was defended by the NVA's 5th Infantry Regiment. Also important, the hamlet sat astride the NVA's primary supply route to the regiments fighting in Hue. The NVA was determined that 2-12th Cavalry's attack would not succeed; the NVA was determined to destroy the threat to its LOCs.

As the 2-12th Cavalry prepared to resume the attack on 4 February, the NVA fiercely counterattacked with all three battalions of the 5th NVA Regiment. It soon became apparent that 2-12th Cavalry could not continue to attack. As the day continued and the NVA pressure steadily increased, the battalion began to lose its position in Thon Que Chu. To avoid destruction, the battalion broke contact.

As darkness fell on 4 February, 2-12th Cavalry finally broke contact with the NVA. However, instead of retreating north back to its start line, the battalion moved west and then south to good defensive high ground deeper in the NVA AO. Eleven hours later the battalion was set in its new defensive position.

At dawn on 5 February, 2-12th Cavalry was established 5 kilometers west of Hue. The battalion soon observed enemy forces and supplies moving toward Hue. From its high ground position, the battalion directed artillery and air strikes against the NVA forces. By its bold move to bypass the 5th NVA Regiment, the battalion now held excellent position to direct fires on the primary NVA supply line into Hue. These fires were the first step toward isolating the NVA in Hue.

The fires controlled by the 2-12th Cavalry shut down the NVA LOCs into Hue during the daytime. However, under the cover of darkness supplies and reinforcements were still entering the city (see figure 7-15). It became apparent that the isolation of the NVA in Hue would require the capture of Thong La Chu. The problem facing American forces was concentrating combat power against the NVA. All U.S. units at this time were actively engaged against the numerous NVA attacks that constituted the NVA's 31 January Tet Offensive.

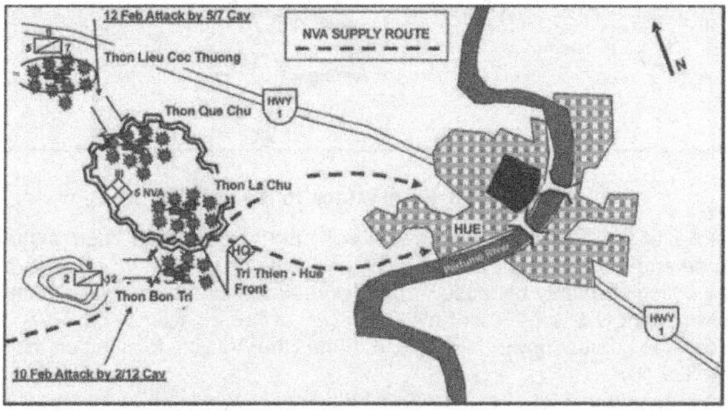

Figure 7-15. Subsequent attack to isolate Hue

The first additional American unit was not available until 12 February when the 5 7th Cavalry attacked Thong Que Chu much like the 2-12th Cavalry had attacked previously. The 5-7th Cavalry had even less success against the alert 5th NVA. The 5-7th Cavalry was forced to occupy defensive positions in Thon Lieu Coc Thuong and await the build up of combat power before it could continue to attack. In the interval, 2-12th Cavalry had moved off the high ground and captured the hamlet of Thon Bon Tri, south of the 5th NVA Regiment.

On 21 February, the 1st Cavalry Division had moved enough resources to the area to launch an effective attack to isolate Hue (see figure 7-16). In addition to the 5-7th and 2-12th Cavalry, the 1-7th Cavalry arrived in the AO and the 2-501st Airborne Infantry of the 101st Airborne Division was attached. On 21 February, after a combined artillery, air, and naval gunfire bombardment, the four battalions attacked the Thon La Chu stronghold. Elements of the 5th NVA Regiment were either destroyed in place or fled northeast. The next day resistance in Hue was noticeably lighter. U.S. Marine and ARVN units began the last phase of fighting to recapture the Imperial Palace. On 26 February, the North Vietnamese flag was removed from the Citadel and the ARVN I Corps declared the city secured.

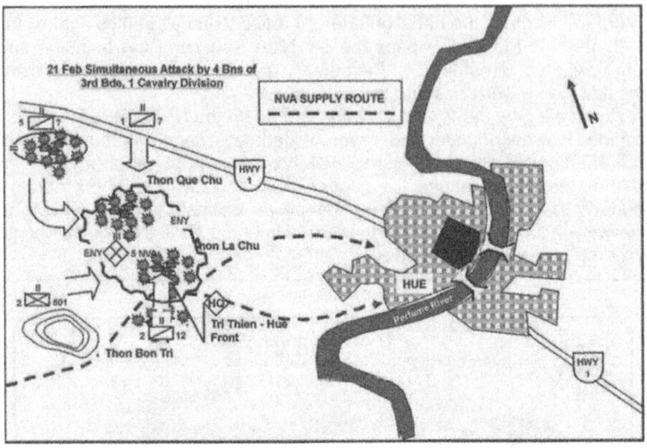

Figure 7-16. Final attack to isolate Hue

> The actions of the 1st Cavalry Division forces northwest of Hue demonstrated the importance and the difficulty of isolating an enemy fighting in an urban area. Isolating Hue was difficult not only because of the dispersion and surprise with which the Tet Offensive caught U.S. forces, but also because of the tenacity of the NVA. At least one-third of the combat power of the NVA in the Hue AO was focused on maintaining access to the city.
>
> Although only one aspect of this urban battle, Hue's isolation had an immediate and important, if not decisive, impact on operations. It not only resulted in restriction and then elimination of supplies and reinforcements, but it also immediately impacted the conduct of the defending NVA forces. Isolation caused an immediate drop in NVA morale and changed the nature of the defense. Once the enemy was isolated from external support and retreat, the objective of the NVA in the city changed from defending to avoiding destruction and attempting to infiltrate out of the city.

Civilian Reactions

7-63. Commanders must also consider the potential effects on and reactions and perceptions of the population living in the urban areas that they choose to isolate and bypass—either as a direct effect or as a response of the threat force being isolated. Isolation to reduce the threat's ability to sustain itself will likely have similar (and worse) effects on the civilian population remaining in the isolated area. (If food and water are in short supply, threat forces may take from noncombatants to satisfy their needs, leaving civilians to starve.) Isolation may also create a collapse of civilian authority within an urban area as it becomes apparent that the military arm of their government is suffering defeat. Due to their isolation, elements of the population may completely usurp the governmental and administrative functions of the former regime and establish their own local control, or the population may lapse into lawlessness. Returning later, Army commanders may find that these self-governing residents are proud of their accomplishments and, in some instances, less willing to allow Army forces to assume control since they may be perceived as having done nothing to earn that privilege. Alternatively, as witnessed in some urban areas during OPERATION IRAQI FREEDOM in 2003, a power vacuum may lead to intra-urban conflicts among rival factions coupled with general public disorder, looting, and destruction of the infrastructure (see also the discussion of Competing Power Structures in Chapter 3).

Direct Action by Special Operations Forces

7-64. Although SOF in urban offensive operations will likely conduct essential reconnaissance, they also have a direct action capability to shape the offensive operation (see figure 7-17). Special Forces and Rangers can use direct action capabilities to attack targets to help isolate the urban area or to directly support decisive actions subsequently or simultaneously executed by conventional forces. Successful attacks against urban infrastructure, such as transportation or communications centers, further the area's physical and electronic isolation. Direct action against command centers, logistics bases, and air defense assets can contribute to the success of conventional attacks by destroying or disrupting key threat capabilities. Direct action can also secure key targets such as airports, power stations, and television stations necessary for subsequent operations. Direct action by Special Forces and Rangers in these operations can help achieve precision and reduce potential damage to the target or noncombatant casualties.

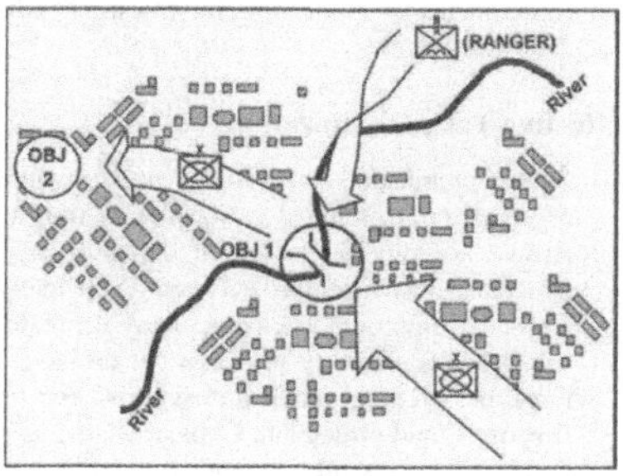

Figure 7-17. Coordination of SOF and conventional capabilities

Information Operations

7-65. Regardless of how Army forces physically isolate the urban area, they combine physical isolation with offensive IO to electronically isolate the threat and undermine his morale. Electronic isolation will cut off communications between forces in the urban area from their higher command to deny both from knowing the other's status. IO combined with isolation may persuade the threat's higher command or leadership that its forces located in the urban area are defeated. Thus, the command or leadership's intentions to break through to the besieged threat forces may be affected. PSYOP can undermine the morale of the threat in the area and reinforce electronic isolation and perceptions of abandonment. IO can be used to reduce any loyalty the civil population may have to the threat. IO can also ensure that civilians have the information that minimizes their exposure to combat and, as a result, overall noncombatant casualties. In addition, IO aim to deceive the threat regarding the time and place of Army force operations and intentions.

Detailed Leader Reconnaissance

7-66. Army commanders must clearly see the urban environment to understand the challenges facing their brigades, battalions, companies, platoons, and squads. Urban terrain can be deceptive until viewed from the soldier's perspective. Commanders are responsible to intimately know the conditions to allocate resources effectively to subordinate units. Often, particularly at battalion level and above, commanders will not be able to command and control dispersed forces from positions forward, but be forced by the terrain to rely on semifixed command posts. Detailed leader reconnaissance of the AO by commanders, their staff, and their subordinates before the mission can compensate for this challenge. This reconnaissance will give commanders a personal feel for the challenges of the terrain and will facilitate more accurate planning and better decision making during operations.

Mission Orders

7-67. Before contact, commanders can mitigate some terrain challenges to effective C2 using mission orders. Subordinates have mission orders to take advantage of opportunities before C2 systems can adversely impact the environment. To see the battle and provide effective and timely direction, tactical leaders will follow closely behind units as they assault buildings, floors, and rooms. Thus, only the most mobile INFOSYS can accompany tactical leaders into combat, and they will suffer the degrading effects of the environment. Mission orders permit rapid and decisive execution without commanders intervening at battalion level and above. Higher-level commanders can facilitate mission orders through their subordinates by articulating their desired end state, clearly stating their intent, and building flexibility into the overall plan.

Effective Task Organization

7-68. Commanders can shape urban offensive operations through effective and innovative task organization. Combined arms, often starting with an infantry base, are essential to success and may be the Army's asymmetric means of defeating an urban threat. Urban attacks will quickly break down into noncontiguous firefights between small units. To achieve the tactical agility for mission success in this nonlinear environment, many Army capabilities are task organized down to the company, platoon, and squad levels. Infantry provides the decisive capability to enter buildings and other structures to ensure threat destruction. Tanks, gun systems, and fighting vehicles provide additional speed and mobility, direct firepower, and protection. Combat engineers provide specialized breaching and reconnaissance capability. Field artillery provides the indirect (and if necessary, direct) firepower. Such mobility and firepower create the conditions necessary for the dismounted infantry to close with and destroy a covered threat in an urban defense. When a threat skillfully uses the urban area to limit ground maneuver, vertical envelopment or aerial attack using precision-guided munitions from Army aviation may circumvent his defenses and achieve necessary effects. Generally, ground systems used within the urban area will not be able to operate independently from dismounted infantry. The infantry will be required to protect armor and mechanized systems from close antiarmor weapons, particularly when those weapons are in well-prepared positions located throughout the urban area but especially on rooftops and in basements.

7-69. In urban offensive operations, direct fire support can be critical. Armor vehicle munitions types do not always achieve decisive effects against some urban structures. In some cases, field artillery high explosive munitions work better than armor for direct fire support of infantry. Large caliber (105 or 155mm) high explosives directly fired at a structure often produce a more severe shock effect than tank and fighting vehicle cannon and machine guns produce. Artillery is also able to achieve higher elevation than armor and engage threats located at greater heights.

7-70. However, commanders must view artillery as not just a weapon but a weapon system. As such, artillery should normally be placed under tactical control (TACON) of maneuver commanders, such as a platoon of three guns TACON to a company or a battery to a battalion, not just one gun to a company or other maneuver unit. Self-propelled artillery has some of the mobility characteristics of armor; however, it provides minimal ballistic protection from fragmentation for the crew. Although these systems seem formidable, they provide less crew protection than a Bradley fighting vehicle, for example, and contain large amounts of onboard ammunition and propellant. They are susceptible to catastrophic destruction by heavy automatic weapons, light cannon, and antitank fire. Therefore, infantry units must carefully secure and protect these systems (even more so than armored vehicles) when employed in urban offensive operations, particularly when forward in the direct fire role.

Creative Task Organization: Using Artillery in the Direct Fire Role

Task organizing artillery to permit its use in a direct fire role demonstrates the innovative task organization required for urban operations. The following provides three historical examples of task organizing and using field artillery for a direct fire role.

In 1944, U.S. Army units of the 1st Infantry Division were assigned to attack and seize the German city of Aachen. The city's internal defense included bunkers designed to serve as air raid shelters. These positions, buildings of stone, seemed impervious to direct fire tank weapons, demolitions, and small arms. To reduce the positions, the 1st Infantry Division relied on the artillery's direct fire.

Field artillery used this way had physical and psychological effects on the defenders. The 26th Regimental Combat Team's history of the battle describes the German reaction to the artillery pieces:

The chief shock to the defenders, Colonel Wilck (Aachen defense commander) said, came from the self-propelled 155s and tanks. The colonel spoke with considerable consternation of the 155mm self-propelled rifles. A shell from one of the guns, he said, pierced three houses completely before exploding and wrecking a fourth.

The 26th Infantry Regiment also described how the artillery, one piece attached to each assaulting infantry battalion, helped the infantry to penetrate buildings.

With solid blocks of buildings comprising most of the city, there wasn't any easy way to get at the Germans in the buildings. The eight-inch gun solved the problem. Beginning on the eastern outskirts the gun would plow a round into the side of the built up block of buildings at about ground level. One shell would usually open an entrance into the first tier of floors, i.e. the first building. Then several more shells were fired through the first hole. Thus a tunnel would be rapidly made all the way to the next cross street. Soldiers could then rush the newly formed entrance, clear the upper floors with hand grenades and rifles and then move on to the next building to repeat the process. When a block or square, was thus completely cleared of Germans—soldiers, skulkers, or even snipers—the next square was treated in the same way, working forward square by square, right and left, thereby avoiding nearly all exposure in the streets.

In 1982, Israeli forces invaded southern Lebanon to destroy base camps of the Palestine Liberation Organization (PLO). This operation involved significant fighting in urban areas including major operations in Beirut. Artillery, firing in a direct fire role, played a major part of the tactical solution. Artillery was particularly effective in the 33-day siege of Beirut. During this siege, Israeli forces used artillery in its traditional role as well as in the direct fire role.

The Israeli army was committed to a policy of disproportionate response during the Beirut siege. When fired on with small arms, crew-served weapons, tanks, or indirect artillery, the Israeli forces responded with intense, high-caliber direct and indirect fire from tanks and artillery positioned around the city. Many firing positions were on heights to the south and southwest that dominated much of the city. These positions had almost unrestricted fields of view. Israeli artillery fired from these positions directly into high-rise buildings concealing PLO gunners and snipers. The artillery, using direct fire, destroyed entire floors, collapsed floors on top of each other, and completely removed some upper floors. Such a response, as in Aachen in 1944, had as much a psychological impact as it did a physical impact on the PLO defenders.

In the early hours of 20 December 1989, the United States launched OPERATION JUST CAUSE. One of this operation's objectives was removing the Panamanian dictator, Manuel Noriega. U.S. forces carefully planned using all fires before the operation to minimize casualties and collateral damage. Part of this detailed fire planning called for applying artillery in a direct fire role.

The Panamanian Defense Force's (PDF) 5th Rifle Company based at Fort Amador was one of the key objectives of U.S. forces at the start of hostilities. This unit was high priority because it was the closest PDF unit to Noriega's headquarters, the Comandancia. U.S. forces expected the 500-man company to react first to

OPERATION JUST CAUSE by reinforcing the defense of the Comandancia. It also posed a threat to U.S. military dependents housed at Fort Amador.

To quickly neutralize this force, the United States assembled a three-company force composed of A, B, and headquarters elements of 1-508th Infantry (Airborne), supported by 105mm towed howitzers of 320th Field Artillery and M113 armored personnel carriers. The howitzers and the personnel carriers were covertly pre-positioned at the fort. At approximately 0100, helicopters transported the two airborne rifle companies into position. The howitzers then suppressed any personnel in the PDF-controlled buildings on Fort Amador while demonstrating the firepower of the U.S. task force. They used direct fire into the PDF barracks. The impact of the 105mm high explosives and .50-caliber fire from the M113s convinced the PDF infantry to give up after token resistance. Following the direct fire, U.S. infantry assaulted and cleared the dozen PDF buildings, finding that most occupants had fled or surrendered. For more details of OPERATION JUST CAUSE, see Applying the Urban Operational Framework: Panama in Chapter 5.

The three examples cited indicate the importance of the situation-dependent and innovative task organization of artillery and its use in the direct fire role. Using artillery helps overcome some challenges of offensive operations in the urban environment, and it has an important psychological effect on a defending threat. Such task organization takes a traditional tool of a higher-level tactical commander and uses it to directly influence the company-level battle. This philosophy of task organization can be applied to other types of forces—not just artillery. PSYOP teams, interpreters, CA specialists, armor, engineers, and reconnaissance teams may require task organization different from traditional organization. The compartmented urban environment drives the requirements for these assets lower in the tactical scheme than in open operations. Consequently, commanders must understand and account for more of these assets for UO than for operations in less restrictive environments.

7-71. Army aviation may also be inventively task organized. It can support urban operations with lift, attack, and reconnaissance capabilities. Tactical commanders down to company may use all these capabilities to positively influence ground close combat. Army attack and reconnaissance aircraft can provide flank security for attacking ground forces. Attack aircraft may also provide direct fire support to individual platoons or squads. Lift may move entire battalions as part of brigade operations, or it may move single squads to a position of advantage (such as a roof) as part of a small unit assault. Army aviation can assist with C2 by providing airborne retransmission capability, airborne command posts, and the confirmed status and position of friendly forces. However, Army aviation is a limited and high-value asset; commanders should review its use in innovative task organizations. It is particularly vulnerable to urban air defense threats unless used over terrain secured by ground forces. From these positions, aircraft can use enhanced sensors to conduct reconnaissance and use precision weapons with standoff capability.

Shaping Attacks

7-72. In a large urban area, the defending threat cannot be strong everywhere. Shaping operations can also take the form of attacks against vulnerable positions to force the threat to maneuver and redeploy his forces in the urban area. This prevents him from merely defending from prepared positions. Forcing the threat to move negates many of the defensive advantages of urban terrain, confirms his dispositions, exposes vulnerable flanks, and permits target acquisition and engagement with precision standoff fires.

7-73. A critical shaping operation in urban offensive operations is usually an initial attack to seize a foothold. Once Army forces establish this foothold, they accrue some of the defensive advantages of urban terrain. From this protected location, Army forces continue offensive operations and have a position of advantage against neighboring threat defensive positions.

ENGAGE

7-74. Commanders may employ several methods to decisively engage elements of the urban area during offensive operations. These include—

- Rapid maneuver.
- Appropriate use of SOF.
- Precise application of fires and effects.
- Proper balance of speed and security.

7-75. None is unique to UO. Their effective execution, however, allows Army commanders to dominate in this challenging environment by effectively using resources with the least amount of collateral damage. Overall, decisive engagement results from urban offensive operations when forces achieve the objective of the assigned mission and establish preeminent control over the necessary terrain, population, and infrastructure. Largely, the Army commander's ability to engage is based on superior situational understanding and the correct application of unit strengths to the challenges found in the urban environment.

Rapid and Bold Maneuver

7-76. Commanders of major operations may have or create the opportunity to seize an urban area with rapid and bold maneuver. Such maneuver requires striking while the area remains relatively undefended—essentially preempting an effective defense. This opportunity occurs when the urban area is well to the rear of defending threat forces or before the onset of hostilities. Under such conditions, an attack requires striking deep behind threat forces or striking quickly with little time for the threat to make deliberate preparations. Attacks under such conditions may entail significant risk; however, the potential benefit of audacious offensive operations may be well worth possible losses. Three potential ways to accomplish such attacks (and their combinations) are:

- Airborne or air assault.
- Amphibious assault.
- Rapid penetration followed by an exceptionally aggressive exploitation, for example, a heavy force using shock, armor protection, and mobility.

7-77. Commanders should analyze all potential urban operations to seek an opportunity or advantage to apply rapid and bold operational maneuver to the task. Using operational maneuver to avoid urban combat against an established threat defense potentially marks a significant operational achievement and can have decisive strategic consequences. Just influencing the threat's morale can positively affect all future operations. However, commanders must evaluate the challenges of such a course of action. These challenges may include—

- Sustaining the operation.
- Avoiding isolation and piecemeal destruction.
- Successfully conducting shaping attacks.
- Achieving the necessary tactical, operational, and strategic surprise.

7-78. Commanders can also build on the shaping effects of isolating the urban area internally and externally by attacking urban decisive points from multiple directions. They can attack multiple decisive points either simultaneously or in a systematic, synchronized manner. This complicates the threat's situational understanding of the urban environment, further impedes his decision making, and allows Army commanders to dictate the tempo.

Bold Operational Maneuver to Seize an Urban Area: Inchon and Seoul, Korea – 1950

In August 1950, UN forces in Korea were desperately attempting to stave off defeat by establishing a viable defense in southern Korea. This defense, known as the Pusan Perimeter, was the focus of the world's attention. The commander of U.S. Forces Far East, General Douglas MacArthur, was not focused on the U.S. Eighth Army in Pusan. Instead, he focused on how to dramatically and decisively transition to the offense—the recapture of the South Korean capital of Seoul.

Seoul was more than just the South Korean political capital, although that was an important consideration. Seoul was the key to the Korean transportation system, the hub of the national communications system, and the center of the country's economy and culture. It was also strategically placed astride maneuver corridors to the north, south, and northeast. It had been Korea's capital since 1394 when King Taejo Yi Sung-Gye established it as the center of the Chosun dynasty. It had been captured by the surprise communist invasion of the south in June 1950, and the key to success in Korea was its recapture. The question was how.

The obvious military solution to the Korean situation in August 1950 was to build military strength within the Pusan Perimeter and then, when sufficiently strong, to breakout and counterattack north and recapture Seoul. The disadvantage of this course of action was that the counterattack would be through hundreds of miles of mountainous terrain, through several other major cities, and across numerous rivers. The North Korean People's Army (NKPA) would be able to fall back from one mountain defense and one river line to another and would have weeks or months to prepare the defenses of Seoul.

MacArthur's solution was to delay a conventional counterattack, boldly maneuver by sea deep into the flanks of the NKPA, land on the Korean west coast at Inchon, and rapidly seize Seoul before the enemy could react (see figure 7-18). This course of action had numerous disadvantages: achieving surprise; assembling a landing force trained in amphibious operations; few good landing sites; supplying a large force once landed; and needing to simultaneously continue to defend and prepare a counterattack with the Eighth Army from Pusan. MacArthur understood the challenges of the operation but believed that the potential value compensated for the risk.

The potential value of a rapid assault on Seoul through a deep amphibious operation was immense. The operation would trap the bulk of the NKPA in South Korea and facilitate its destruction. It would also capture Seoul before the NKPA could prepare a defense. The quick recapture would immeasurably impact the morale of UN forces and South Korea. MacArthur felt the payoff of success justified the risks and challenges of the operation.

On 15 September, elements of the 1st Marine Division landed in Inchon taking the NKPA completely by surprise. On D+1 they were joined by elements of the 7th Infantry Division, both divisions under the control of U.S. X Corps. Naval gunfire and Marine close air support flying from Navy aircraft carriers supported the landings. Seven days later, the 5th Marine Regiment had battled 25 miles to the outskirts of Seoul and began attacking the city from the north. Three days later, on 25 September, the 1st Marine Division and elements of the 7th Infantry Division had defeated the 10,000 defenders of the NKPA's 18th Rifle Division.

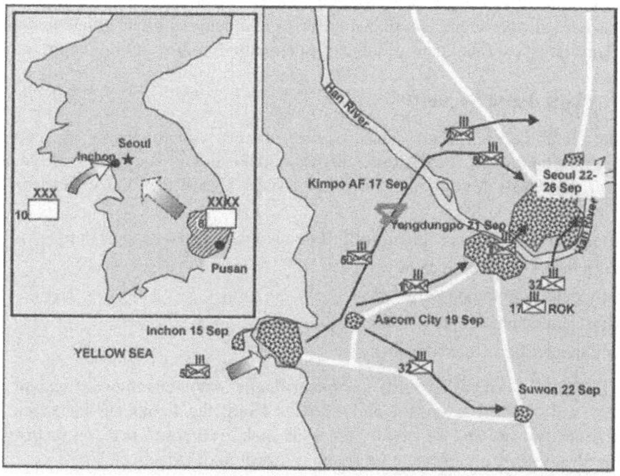

Figure 7-18. Inchon-Seoul campaign, September 1950

> Most of MacArthur's predictions for the Inchon-Seoul campaign were accurate. Superb staff work, excellent logistics operations, and unmatched training permitted the UN forces to assemble a trained landing force, land it over one of the most difficult shores in the history of military amphibious operations, and keep it supported. More importantly, the operation achieved total strategic and operational surprise. Seoul was recaptured quickly and, although resistance was fierce, the NKPA could not react fast enough to influence the outcome. In conjunction with the Eighth Army counterattack from Pusan, 75 percent of the NKPA was destroyed. The Inchon-Seoul campaign was an important turning point in the war, and had not the Chinese People's Army intervened two months later, it would have been the decisive campaign of the war.

Appropriate Use of Special Operations Forces

7-79. Sometimes Army forces can decisively engage (instead of simply shape) the urban area using the direct action capability of SOF. When the threat fails to develop a comprehensive defense and does not possess large, capable conventional forces, then Army forces can achieve operational surprise. Commanders, by synchronizing conventional and SOF effects, may actively control offensive operations to dominate the area. Importantly, however, conventional ground forces must be available to quickly assume the mission because SOF units acting as the primary striking force have limited logistic capability to sustain long-term operations. OPERATION JUST CAUSE offers several examples of this type of synchronization (see Applying the Urban Operational Framework: Panama in Chapter 6).

Precise Application of Fires and Effects

7-80. Precisely applied fires and the massed effects of combat power characterize successful urban attacks. The fires can be direct fire from combined heavy or light ground teams; direct or indirect fires from supporting Army aviation standing off from the target and any possible air defense threat; precision indirect fires from conventional tube artillery; or direct and indirect fires from supporting joint assets including Air Force and naval assets. All efforts strive to reduce collateral damage around the point of attack, consistent with mission success. Forces use fires to deny the threat the ability to maneuver in the urban area and to destroy the threat when he attempts to maneuver. When the threat exposes himself by moving, the environment no longer protects him, and fires can effectively engage him. Overall, precise fires and effects demonstrate the power of Army forces and help the urban population understand that only legitimate military targets are the focus of attacks, potentially building public support of Army UO.

Proper Balance of Speed and Security

7-81. Attacking units balance speed and security. Forces secure flanks as units advance, control dominating terrain (buildings), evacuate civilians, and keep the integrity and synchronization of the combined arms team. Obstacles are anticipated and rapidly breached. Commanders can choose avenues of approach to—

- Provide cover and concealment for following aviation and support units.
- Permit travel by all classes of vehicles.
- Easily defend from counterattack.
- Avoid nonessential centers of threat resistance.
- Avoid population concentrations.

7-82. Army aviation is a critical resource to protect flanks. Another important resource is engineers who seal off surface and subsurface entries and avenues along the flanks of the attack. Finally, as in all offensive operations, ground and air cavalry are ideal mobile forces to perform security in an economy of force role along flanks allowing decisive forces more freedom of maneuver.

CONSOLIDATE

7-83. Commanders, at all levels, must consolidate to strengthen their position during urban offensive operations without loss of momentum. They take the steps necessary to make permanent any temporary battlefield successes while maintaining relentless pressure on threat forces. Consolidation provides security and protection, facilitates reorganization, and allows forces to prepare for counterattack.

Repositioning of Forces

7-84. Following seizure of the objective, commanders normally consolidate by adjusting and repositioning forces. While UO are likely to be noncontiguous, commanders reposition ISR assets (including observation posts and reconnaissance patrols) to maintain contact with the threat, to establish contact with nearby friendly units and ensure that there are no exploitable gaps or seams, and to help maintain freedom of action. Physical occupation of the terrain (as well as continued reconnaissance) will provide commanders with a fuller understanding of the urban environment. With this enhanced understanding, commanders may need to adjust boundaries and other control measures to better adapt to the effects of urban terrain features such as canals, subway tunnels, raised roadways, and tall buildings. As necessary, commanders may also need to reposition communications assets and C2 facilities to enable subsequent operations.

7-85. Immediately after the conduct of successful urban operations, units must also be alert to the potential for rapid and violent counterattacks. Defenders may launch a quick counterattack to regain terrain before the offensive forces have consolidated and fully assumed the defensive advantages of the urban terrain. Delaying a counterattack in UO, even for a few minutes, permits the environment's advantages to shift to the successful attacker. Thus, attacking units anticipate this reaction and reposition forces such that they are prepared to defeat it.

7-86. As necessary, commanders should reposition armor and artillery (and other fire support assets) to account for the changing situation and battlefield geometry. Commanders should also consider consolidating and repositioning armored and artillery forces in positions—either inside or outside the urban area—to add significant combat power to a hasty defense, to defeat threat counterattacks, or to allow for quick resumption of the attack. If integrated into a hasty defense inside the urban area, these forces will require continued infantry protection. As a mobile counterattack force positioned inside the urban area, armored forces will also require careful selection of attack positions and counterattack routes. Damage to buildings and infrastructure may limit maneuver and the use of direct and indirect weapon systems.

Repositioning forces outside the urban area may contribute to strengthening or reestablishing the isolation of the urban area and will make better use of range and standoff capability, allow them to take full advantage of their speed and mobility, and decrease their need for additional protective support.

Eliminating Pockets of Resistance

7-87. Focusing on controlling the essential and attacking decisive points means that attacking units will have bypassed some elements of the threat's defense. Within the bounds of the initial plan and current situation, commanders should consider whether it is necessary to devote the time and resources to immediately clear and eliminate remaining threat forces and pockets of resistance with all or parts of the attacking force or leave the task to follow-on forces. In their deliberations, commanders should determine if remaining threat forces are capable of consolidating and mounting effective counterattacks before follow-on units can engage the remaining threat and take advantage of their dispersion and disunity. As another part of their deliberations, commanders should also consider if remaining threat forces will likely (and immediately) endanger—

- The urban inhabitants.
- Lines of communications.
- Critical resources within the urban area necessary to support the inhabitants or subsequent operations.

Reorganization

7-88. Reorganization includes all measures taken by the commander to maintain combat effectiveness or return to a specified level of combat capability. As necessary, these actions include—

- Redistributing or cross-leveling supplies and equipment until resupply can be accomplished.
- Replacing key personnel and combining units or crews to form mission-capable ones.
- Integrating replacement soldiers and systems into the unit and matching operational systems with available crews.
- Recovering, treating, and evacuating casualties, enemy prisoners of war, and damaged equipment.
- Conducting training and disseminating critical lessons learned.
- Other actions to reestablish unit cohesion.

TRANSITION

7-89. Effective transitioning allows commanders to continue full spectrum operations in the urban area and elsewhere in the AO without unnecessary delays. Commanders can accomplish this by thorough planning—including appropriate branches and sequels (revised as the situation changes)—that gives adequate consideration to post-offensive organizational, training, psychological, and civilian requirements. If properly prepared, commanders can anticipate rather than react to potential mission changes.

Early and Concurrent Transition Planning

7-90. Commanders can ensure smooth transitions of urban offensive operations by planning for post-offensive operations early. Based on the mission envisioned, they determine which subordinates and what type of force structure to use. Post-offensive missions, like all urban operations, encompass full spectrum operations. At the successful conclusion of offensive operations, Army forces transition to some type of stability or civil support operation. Commanders may leave the subordinate unit in place to execute the new mission, may reorganize the subordinate unit for the mission, or relieve the unit that just completed offensive operations with a new unit.

Changes to Task Organization

7-91. Commanders should consider the organization of forces following offensive operations. Hostile civilians may require significant combat forces or military police forces to maintain stability. On the other hand, friendly civilians may require a minimum of military police or combat forces, but significant logistic support. Commanders must carefully consider the urban situation before deciding how to use combat forces that recently participated in a high-intensity offensive operation.

Training and Psychological Considerations

7-92. Many Army combat tasks may not support follow-on stability or civil support operations without considerable modification (if at all). Often, noncombat skills—not normally part of a unit's mission essential task list such as negotiating or mediating skills—will be required. However, the greatest modification required applies to each soldier's mental outlook. Forces that transition directly from combat to stability operations may not be psychologically prepared for a rapid and drastic change of mission. Commanders cannot expect troops who have just completed high-intensity offensive operations to rapidly adjust and exercise the sensitivity and judgment required in most stability operations. This especially applies if the population is hostile to Army forces. If possible, combat forces assisting in stability operations, particularly in hostile civilian situations, should not have had recent experience in high-intensity urban operations and they should have trained for the mission.

Return to Civil Agencies

7-93. Commanders of major operations also have the critical role of transitioning aspects of the urban offensive operation to civil, allied, NGOs, and other agencies as appropriate. Transition planning is detailed and aims to quickly return as much civil control of the area as is feasible after the attack. Beyond local civil control, outside civil agencies and NGOs assume tasks as completely and as rapidly as possible. These organizations are consulted and integrated into the planning process as early as possible. Commanders should begin planning for transition simultaneously with planning for offensive operations. They consider the feasibility of relinquishing control of urban areas to civil government, law enforcement, or NGOs even before completing offensive operations. During the conduct of urban operations, these transition operations are closely synchronized with the execution of the attack.

Transition to a New Mission

7-94. In urban offensive operations, like other offensive missions, the change in mission after a successful urban attack may be to a hasty defense or a continuation of offensive operations outside the area. However, in urban offensive operations the mission will just as likely rapidly change to a civil support or stability mission. This is particularly true if the unit has had special training and is task organized for urban operations. Transition to stability or civil support operations is often accompanied by a transition in roles from supported to supporting.

7-95. Even more challenging than transition at the end of the mission is transition during the conduct of the mission. Soldiers may have a difficultly transitioning from stability or civil support to offense and defense, and back again multiple times during an urban offensive operation. Soldiers may apply the tactics, techniques, and procedures of urban offensive operations directly to the stability or civil support missions with potentially disastrous results. Commanders need to segregate missions in time and space, and, if sufficient forces exist, by unit. To this end, commanders should consider permanently designating specific units to conduct civil-military and humanitarian support tasks and avoid rapid mission changes that rotate units (particularly at company level and below) between violent and nonviolent tasks. However, commanders may not have that luxury and may need to rely heavily on preparatory training (including the inculcation of Army values) and strong unit leadership to mitigate potential difficulties.

Chapter 8
Urban Defensive Operations

Generally, a modern city magnifies the power of the defender and robs the attacker of his advantages in firepower and mobility. A city can ingest an invading army, paralyze it for weeks on end, and grind it down to a state of ineffectiveness.

"Military Operations on Urbanized Terrain:
The 2d Battalion, 26th Infantry, at Aachen, October 1944"

The skillful defense of an urban area can decisively affect a campaign. The urban area offers many advantages to defending forces. An adroit defender can use the advantages of the urban environment to negate combat power disparities, blunt the tempo of an attack, attrit threat forces, and sap the morale of attacking troops. The defender gains an opportunity to concentrate resources, reconstitute attrited units, and transition to the offense. A successful defense of an urban area can also deny the threat vital resources. Defense in the urban environment is an essential Army capability and can significantly affect the outcome of entire campaigns and the achievement of national objectives.

PURPOSE OF URBAN DEFENSIVE OPERATIONS

8-1. Army forces defend urban areas for various reasons: defeating a threat attack, buying time, economizing forces, protecting an ally's political institutions and economic infrastructure, protecting an urban population, shaping conditions for decisive offensive operations, and shaping conditions for executing stability or civil support operations. During force projection operations, urban areas may be used as initial lodgment areas that Army commanders may need to defend at the outset until they build sufficient combat power. Usually two or more of these purposes apply to the urban defense. Urban defensive operations provide commanders great opportunities to turn the environment's characteristics to the advantage of Army forces. Urban areas are ideal for defensive operations and greatly enhance the combat power of defending units.

DEFENSIVE CHARACTERISTICS APPLIED TO URBAN OPERATIONS

8-2. There are five general characteristics of the successful defense: preparation, security, disruption, massing effects, and flexibility. All apply to the successful urban defense and to the higher commander supporting a subordinate defending in the urban area.

PREPARATION

8-3. The urban area suits the defense since the area's physical characteristics naturally enhance the combat power of defending units. These characteristics include protection, obstacles, and concealment. Urban terrain provides superb defensive positions with minimum preparation. With deliberate preparation, urban defensive positions can rapidly become strong points.

8-4. One primary characteristic of urban terrain that enhances the defense is protection. With little or no advance preparation, buildings, subsurface structures, and walls protect Soldiers from direct and indirect fire, interdict indirect fire, limit observation, and limit engagement ranges (requiring Soldiers to be skilled at combat in close quarters and quick-fire techniques). Nearly all buildings provide some ballistic protection from direct and indirect fire. Mason and stone buildings with basements and cellars can protect Soldiers from most fires except the largest caliber or tonnage bomb. Minimal additional preparation turns them into formidable, defensive strong points.

8-5. Buildings in urban areas, because of their height and close proximity, also can protect Soldiers by masking them from indirect fire. The height of a building may interdict the flight path of an artillery round, rocket, missile, or bomb at a point short of the intended target. Masking protects static defending forces and protects forces moving along routes bordered with tall buildings that form urban "canyons". These protected routes can be used for sustainment, counterattacks, and maneuver.

8-6. Structurally significant buildings in an urban area can create major obstacles to maneuver. These obstacles immediately canalize maneuver into existing streets and routes without any preparation by the defense. These obstacles then become kill zones for well-positioned and sited defensive forces. Minimal obstacle construction as point obstacles blocking streets and routes can further restrict the maneuver options of the attacking force. Rubble from structures collapsing into streets after fires (intentional or unintentional) can also block routes.

8-7. Buildings also conceal the location, disposition, and intent of the defense. They limit visual observation to the external perimeter of the urban area. They degrade radar and electronic position identifiers and decrease the utility of overhead imagery. The physical aspect of the urban environment greatly enhances the defense by degrading the opposition's intelligence, surveillance, and reconnaissance (ISR) capabilities. Buildings can conceal static defensive positions and the maneuver of defensive forces in the urban area. Although the environment constrains defensive mobility in much the same manner as offensive mobility, the defender has the time and opportunity to conduct careful reconnaissance and select and prepare routes. This gives the defender the ability to move reserves, maneuver counterattack forces, and plan sustainment without observation. Careful preparation provides the defender a mobility advantage over attacking forces.

Security

8-8. The urban area can be an advantage or a disadvantage to the security of defending forces. This largely depends on the nature of the human dimension of the environment. If the population is evacuated or allied with Army forces, then the environment may assist in the security of defending Army forces. However, if the population is present and hostile, then the environment may make security difficult.

8-9. The physical aspects of the urban environment, uninfluenced by the human dimension, may assist in the security of defending Army forces. The combat power of small security forces manning observation posts is greatly enhanced. Forces can more easily restrict and monitor avenues of approach for threat reconnaissance. Defending forces positioned mostly in structures are difficult to locate.

8-10. The physical aspects of the environment may also present some security challenges, primarily with observation. The compartmented terrain limits the field of observation from any one point. The defense may require more security forces to adequately observe the mounted and dismounted avenues to prevent infiltration. Threat forces that successfully infiltrate will be more difficult to locate. These forces will gain numerous hide positions for small reconnaissance units in complex terrain and the effect the terrain has in masking electronic signatures.

8-11. Friendly civilians in the urban area can help identify threat forces attempting to conduct reconnaissance. Civilian activity will also help to mask defense preparations. However, a hostile element of the population may pass intelligence information to the threat. They may assist threat reconnaissance to infiltrate the urban area or provide guides, manpower, or resource support for threat forces. Commanders should take measures to ensure strict control of hostile populations. If resources permit, commanders may consider removing potentially hostile civilians from the area.

Disruption

8-12. The urban environment's attributes assist defending Army forces to disrupt the attacker. It does this through compartmentalization, inhibiting command and control (C2), and facilitating counterattacks.

8-13. The physical aspects of the urban area force the attacking threat into compartmented urban canyons that make mutual support between attacking threat columns difficult. Shifting resources from one portion of the threat attack to another is also difficult. Physically, the urban area disrupts tactical communications making synchronization of combat power difficult.

8-14. The urban terrain hinders the mobility capabilities of the defense. However, careful planning, preparation, and rehearsals can facilitate more rapid movement of larger forces. Defending forces can assemble counterattacks undetected, move them along covered and concealed routes, and achieve surprise at the point of the counterattack. Attacking forces, using the compartmented terrain, often leave forward elements in position to be isolated or expose long and vulnerable flanks to friendly counterattack and interdiction.

MASSING EFFECTS

8-15. The urban environment allows defenders to better protect their centers of gravity and decisive points. The restrictive terrain reduces the attacker's maneuver options. Defenders can position forces in protected and mutually supportive positions oriented on deadly engagement areas. Relatively few well-positioned defenders can generate significant combat power. Without the positional advantage and the corresponding protective effects of the terrain, attacking forces often mass numbers to achieve the necessary combat power.

8-16. Knowing the complex terrain permits defending forces to plan engagement areas that maximize the effects of their combat power. Defending forces can remove fences, walls, rooftops, and even entire buildings to facilitate fields of fire and unmask indirect fire flight paths. Forces carefully choose firing positions for indirect fire systems so that flight paths travel between buildings into engagement areas. By leveraging this knowledge of the terrain, numerically inferior defenders can synchronize devastating fires on offensive forces that are forced by terrain and reinforcing obstacles to mass in confined spaces where fires can have the greatest effect.

FLEXIBILITY

8-17. Defensive flexibility results from detailed planning and mental agility. Commanders develop defensive flexibility by ensuring that plans adequately address branches and sequels that include alternate and subsequent positions and emphasize counterattack options. The urban area facilitates defensive flexibility because the urban terrain can be quickly adapted for defensive operations with little or no preparation. The effect is similar to having multiple, prepared positions on nearly every possible approach. The urban area can also permit rapid, covered movement on interior lines. This permits swift movement to and occupation of strong defensive positions with little or no preparation. The defense also has more flexibility since defenders often know and better understand the urban terrain's effects on operations. Normally, defenders will not get lost as easily, will know complex lines of sight and masking effects, and will best understand the ballistic characteristics of individual structures.

8-18. Mental agility allows commanders to see that the best urban defense may actually be to defend outside of the area. Such a defense mitigates the danger to the urban population and potentially reduces collateral damage. It takes advantage of Army long-range engagement capabilities and denies the threat the opportunity to "hug" Army forces or noncombatants as protection from fires. This defense may be appropriate when Army forces have enough resources to defend more open terrain, when time permits deploying extensive obstacles and constructing protected positions, and when natural terrain such as river obstacles aids the defense.

URBAN DEFENSIVE OPERATIONS AND BATTLEFIELD ORGANIZATION

8-19. Urban defensive operations are organized within the overall battlefield organization of sustaining, shaping, and decisive operations. The success of urban defense depends on each operation, but commanders must synchronize these simultaneous operations as one action. Sustaining operations in defensive urban operations (UO) ensure freedom of action. Critically, urban sustaining operations ensure security of the lines of communications and establish effective movement control. Shaping operations in defensive UO create the conditions for decisive operations. Shaping operations vary greatly depending on the type of defense. For example, in a mobile defense the shaping operation may be the fixing force. In contrast, in an area defense the fixed defense may be the decisive operation. In the urban defense, decisive operations focus on accomplishing the commander's mission. The decisive operation may not defeat the threat's main effort, and it may not prevent threat occupation of large portions of the urban area if those tasks are not essential to mission accomplishment. For example, if the defense's objective is to protect a critical communications node, then, depending on the commander's overall intent, threat actions to secure an airfield elsewhere may not be important.

TYPES OF URBAN DEFENSE

8-20. Commanders can view urban area defensive operations two ways: as conducting a major defensive operation with an urban area in their area of operations (AO) and as defending entirely in an urban area.

AREA DEFENSE

8-21. At the operational level, an area defense may include both urban areas and open maneuver areas. The most common defense in an urban area and the most suitable for the characteristics of this distinct environment is the area defense. As a defensive operation, the area defense concentrates on denying threat forces access to designated terrain for a specific time rather than destroying the threat outright. Although an area defense in an urban area does not directly seek to destroy or defeat attacking threat forces, as an objective it does aim to force culmination of the threat's attack. The urban area defense often works effectively to exhaust threat resources and shape conditions for a transition to offensive operations. The urban area may also be used as a strong point to force threat movement in a different direction or to fix threat forces as part of a large, mobile defense taking place in the AO outside the urban area.

MOBILE DEFENSE

8-22. A mobile defense can operate in an urban area but only under specific conditions. It focuses on destroying or defeating the threat through a decisive attack by a striking force. It requires the defender to have greater mobility than the attacker. To shape a mobility advantage, the urban defender effectively uses the terrain and correctly task organizes his forces' mobility. The principles of applying the mobile defense in the urban area remain the same: a small fixing force stops the threat and limits any ability to maneuver while a striking force quickly maneuvers and counterattacks to destroy him.

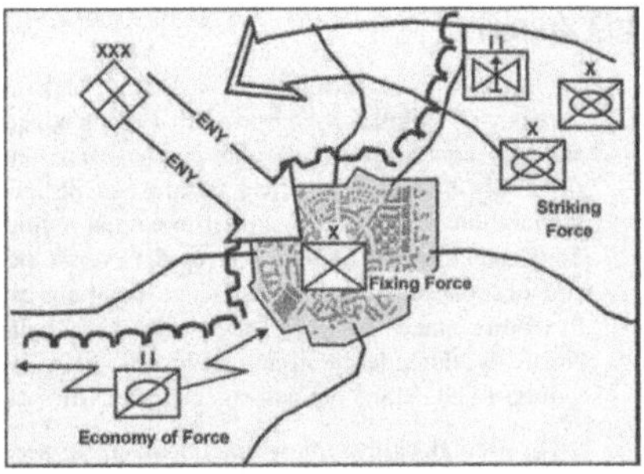

Figure 8-1. An urban area incorporated into a larger mobile defense

8-23. One key to executing a mobile defense in the urban area is to entice a threat force into the depths of the urban area where it begins to lose mobility options. A well-placed fixing force augmented with man-made obstacles and taking advantage of the naturally constrictive terrain can stop a much larger force. If the attacking force is largely mounted and armored, its mobility in the urban area may be reduced to less than that of dismounted infantry. In addition, if the attacking force's movement into the urban area is mounted and rapid, the commander's situational understanding also diminishes. Then the striking force, consisting of dismounted infantry forces, can execute the counterattack with surprise from multiple directions and dimensions (subsurface, surface, supersurface, and airspace). Man-portable antiarmor weapons—firing from flanks and top down and supported by precision indirect fires from both organic and joint systems—can rapidly destroy the threat.

8-24. From the perspective of commanders of the major operation, the urban environment can help defending forces achieve a mobility advantage over an attacker in a broader sense. Defending commanders can attempt to shape the battlefield so that the attacker commits significant resources into an urban area, where his maneuver capabilities are reduced (see figure 8-1). A disproportionately small defending force, which relies on the defensive combat power advantages of the urban environment, reduces and fixes the attacker's maneuver capabilities. Other defending forces mass outside the urban area then strike the threat with a combined mobility and firepower advantage.

Urban Defense in a Major Operation: Stalingrad – 1942 to 1943

The German and Soviet battle for Stalingrad in late 1942 and into early 1943 illustrates how a tactical urban area defense integrates into a larger mobile defense. Stalingrad was a battle fought on a huge scale: it involved army groups on both sides and thousands of square kilometers. Though the city was relatively small, it remained the focus of both German and Soviet offensive and defensive operations during the six-month battle.

In the summer of 1942, the Germans launched a strategic offensive in southern Russia. Its goal was the valuable oil fields of the Caucasus. German forces turning south into the Caucasus exposed a vulnerable flank to Soviet forces positioned between the Don and Volga Rivers. For the German Caucasus operation to succeed, it had to destroy Soviet forces between the Don and Volga, establish a good defensive line, and capture Stalingrad. This city would anchor the German defense and interdict the critical flow of supplies from the Caspian Sea via the Volga River into central Russia. Stalingrad, by virtue of its name, also had important political and cultural value to the Germans and Soviets.

The opening phases of the German offensive were successful; German forces—the 6th Army and 4th Panzer Army—entered the outskirts of Stalingrad in late August 1942 (see figure 8-2).

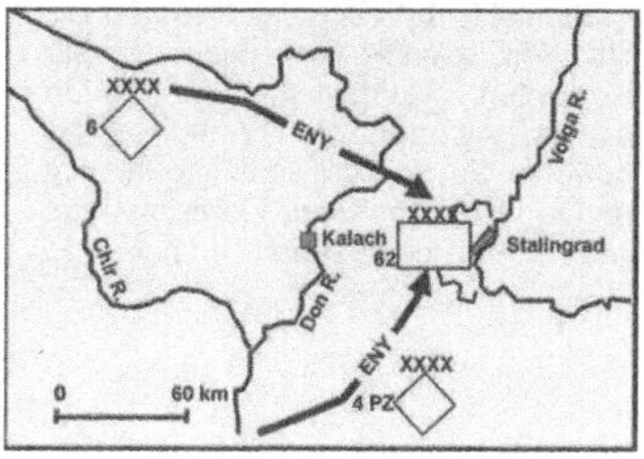

Figure 8-2. German attacks to seize Stalingrad

After a month of intense fighting, the Germans possessed nearly 90 percent of the city. At this point, the 6th Army commanded all German forces in and around Stalingrad. The Soviet 62nd Army's defense was reduced to a front only a few hundred meters deep and a couple of kilometers long on the banks of the Volga. The Soviet defenses hinged on fortress-like concrete industrial buildings and the fanatical bravery and tenacity of soldiers and civilians fighting in the city's remains (see figure 8-3 on page 8-6).

Beginning in mid-September, the Soviet command began looking at how to convert the defense of Stalingrad into an operational opportunity. During October and November, the 62nd Army held on to its toehold in Stalingrad. While maintaining the defense of the 62nd Army, the Soviets secretly began to build up strength on both flanks of the German 6th Army. The Germans increased their vulnerability by securing the German 6th Army's flanks with less capable Romanian, Hungarian, and Italian armies. Also, the 6th Army moved powerful German divisions into the city and rotated with German divisions that were exhausted by urban combat.

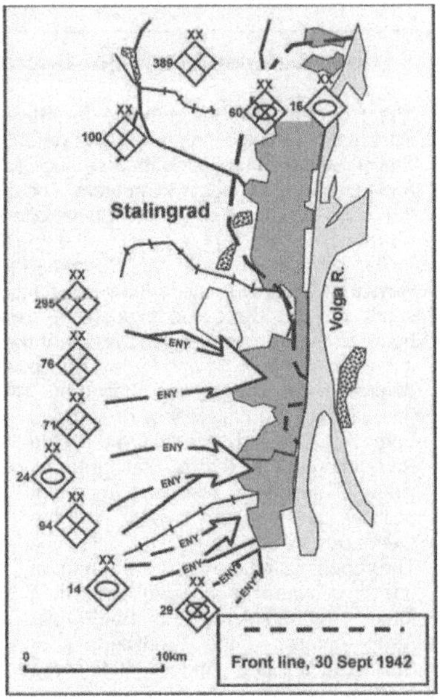

Figure 8-3. German attacks to seize Stalingrad, September 1942

On 19 November, the Soviets launched Operation Uranus that attacked two Romanian armies with seven Soviet armies. Simultaneously, the 8th Russian Army attacked to aid the 62nd Army in further fixing the German 6th Army. Within five days, the Soviet armies of the Don Front, Southwest Front, and Stalingrad Front met near the city of Kalach and sealed the fate of the German 6th Army's 300,000 troops in Stalingrad (see figure 8-4).

On the third day of the Soviet offensive, when encirclement seemed inevitable but not yet complete, the 6th Army commander asked permission to withdraw from the trap. The German high command denied permission believing that the Army could be supplied by air and then a renewed offensive could break through to the city. On 12 December, the German LVII Panzer Corps launched an offensive north to break through to Stalingrad. This offensive made progress until another Soviet offensive on 16 December forced its cancellation. This ended any hope of recovering Stalingrad and the 6th Army. On 31 January 1943, the 6th Army surrendered after sustaining losses of almost two-thirds of its strength. The Soviets took over 100,000 prisoners.

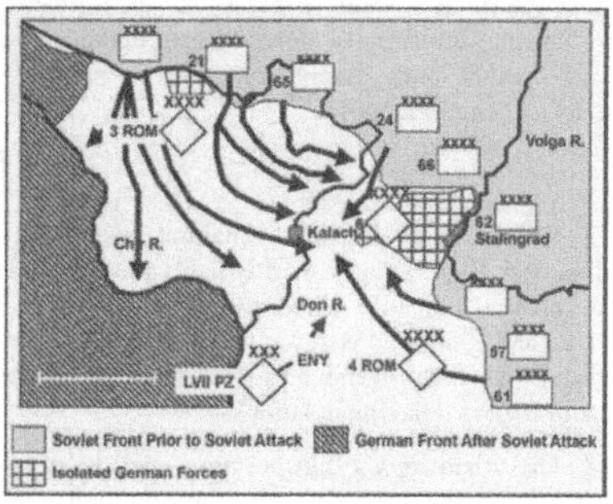

Figure 8-4. Soviet attacks trap German 6th Army

Many lessons emerge from the successful defense of Stalingrad. Tactically, the defense showed how using the terrain of a modern industrial city wisely could increase the combat power of an inferior, defending force and reduce the maneuver options of a mobile, modern attacking force. Another element in the Soviet's tactical success was the Germans' inability to isolate the defenders. The Germans never threatened the Soviet supply bases east of the Volga and, despite German air superiority, the Soviets continuously supplied and reinforced the 62nd Army across the Volga River. Also, Soviet artillery west of the river was able to fire in support of Soviet forces and was never threatened with ground attack.

At the operational level, the Soviets demonstrated a keen understanding of using an urban area within the context of a mobile defense. The 62nd Army's stubborn area defense of Stalingrad drew the bulk of the German combat power into the urban area where they were fixed by a smaller and quantitatively inferior defending force. This allowed the Soviets to build combat power outside the urban area. The Soviets set the conditions for a mobile defense by positioning powerful Soviet armor forces in open terrain outside the urban area against quantitatively inferior German allied forces. In OPERATION URANUS, the mobile defense's strike force destroyed the enemy outside the urban area and trapped the greater part of the best enemy formations inside the urban area. The trapped units were then subjected to dwindling resources and extensive psychological operations, further isolated into pockets, and defeated in detail.

Retrograde

8-25. A retrograde involves organized movement away from the threat. Retrograde operations include withdrawals, delays, and retirements. These defensive operations often occur in an urban environment. The urban environment enhances the defending force's ability to conduct retrograde operations successfully (see figure 8-5).

8-26. The cover and concealment afforded by the urban environment facilitates withdrawals where friendly forces attempt to break contact with the threat and move away. The environment also restricts threat reconnaissance, which is less able to detect friendly forces moving out of position, and presents excellent opportunities for deception actions. Finally, a small security force's ability to remain concealed until contact in the urban environment significantly slows threat attempts to regain contact once Army forces have broken contact and begun to move.

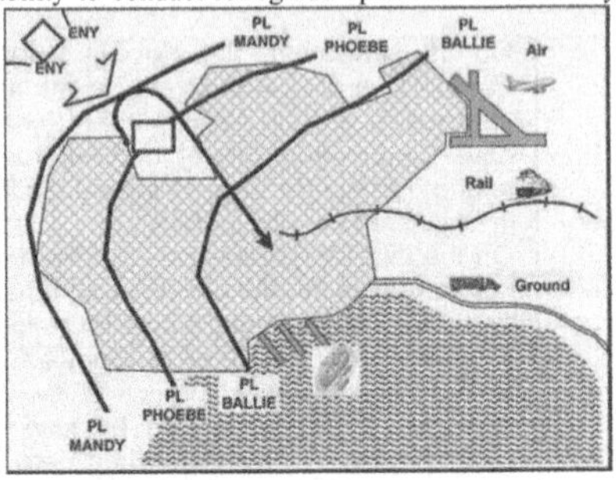

Figure 8-5. Retrograde through an urban area

8-27. The urban environment's natural cover and concealment, as well as the compartmented effects, facilitates delays. Delays can effectively draw the threat into the urban area for subsequent counterattack or as an integral part of a withdrawal under threat pressure. Delaying units can quickly displace from one covered and concealed position to another; the repositioning options are vast. Compartmented effects force the attacking threat to move on well-defined and easily interdicted routes and limit the threat's ability to flank or bypass delaying positions.

8-28. The urban area's transportation and distribution network facilitates retiring forces that are not in contact. Properly used, the urban transportation system can quickly move large forces and associated resources, using port facilities, airfields, railheads, and well-developed road networks.

URBAN DEFENSIVE CONSIDERATIONS

What is the position about London? I have a very clear view that we should fight every inch of it, and that it would devour quite a large invading army.

Winston Churchill
War in the Streets

8-29. The urban operational framework—understand, shape, engage, consolidate, and transition—provides structure to developing considerations for defensive operations. The considerations can vary depending on the level of war at which the operation is conducted, the type of defense, and the situation. Most issues discussed may, in the right circumstances, apply to both commanders conducting major UO and commanders at lower tactical levels of command.

UNDERSTAND

8-30. The commander defending in the urban area must assess many factors. His mission statement and guidance from higher commanders help him focus his assessment. If the mission is to deny a threat access to port facilities in an urban area, the commander's assessment will be focused much differently than if the mission is to deny the threat control over the entire urban area. The mission, enemy, terrain and weather, troops and support available, time available, civil considerations (METT TC) structure guides the commander's assessment. Of these, the impacts of the threat and environment—to include the terrain, weather, and civil considerations—are significant to the commander's understanding of urban defensive operations.

The Threat

8-31. In the urban defense, a key element is the commander's understanding of the threat. One of his primary concerns is to determine the attacker's general scheme, methodology, or concept. Overall, the attacker may take one of two approaches. The most obvious would be a direct approach aimed at seizing the objectives in the area by a frontal attack. A more sophisticated approach would be indirect and begin by isolating Army forces defending the urban area. Innumerable combinations of these two extremes exist, but the threat's intentions toward the urban area will favor one approach over another. The defending Army commander (whose AO includes but is not limited to the urban area) conducts defensive planning, particularly his allocation of forces, based on this initial assessment of threat intentions. This assessment determines whether the commander's primary concern is preventing isolation by defeating threat efforts outside the area or defeating a threat attacking the urban area directly. For the higher commander, this assessment determines how he allocates forces in and outside the urban area. For the commander in the urban area, this assessment clarifies threats to sustainment operations and helps shape how he arrays his forces.

The Environment's Defensive Characteristics

8-32. A second key assessment is the defensive qualities of the urban environment. This understanding, as in any defensive scenario, is based on mission requirements and on a systemic analysis of the terrain in terms of observation and fields of fire, avenues of approach, key terrain, obstacles, and cover and concealment (OAKOC). It is also based on potential chemical, biological, radiological, nuclear and fire hazards that may be present in the urban area. This understanding accounts for the unique characteristics of urban terrain, population, and infrastructure as discussed in Chapter 2.

8-33. Generally, units occupy less terrain in urban areas than more open areas. For example, an infantry company, which might occupy 1,500 to 2,000 meters in open terrain, is usually restricted to a frontage of 300 to 800 meters in urban areas. The density of building in the urban area, building sizes and heights, construction materials, rubble, and street patterns will dictate the actual frontage of units; however, for initial planning purposes, figure 8-6 provides approximate frontages and depths for units defending in an urban area.

UNIT	Frontage (Blocks*)	Depth (Blocks*)
Battalion	4 – 8	3 – 6
Company	3 – 4	2 – 3
Platoon	1 – 2	1
*Average block is 175 meters		

Figure 8-6. Approximate defensive frontages and depths

SHAPE

8-34. Commanders of a major operation shape the urban battle according to the type of defense they are attempting to conduct. If conducting an area defense or retrograde, they use shaping actions like those for any defensive action. Important shaping actions that apply to all defensive UO include—

- Preventing or defeating isolation.
- Separating attacking forces from supporting resources.
- Creating a mobility advantage.
- Applying economy of force measures.
- Effectively managing the urban population.
- Planning counterattacks.

Preventing or Defeating Isolation

8-35. Failure to prevent isolation of the urban area can rapidly lead to the failure of the entire urban defense. Its importance cannot be overstated. In planning the defense, commanders must anticipate that the threat will attempt to isolate the urban area. Defensive planning addresses in detail defeating threat attacks aimed at isolation of the urban area. Commanders may defeat this effort by allocating sufficient defending forces outside the urban area to prevent its isolation. Defensive information operations (IO) based on deception can also be used to mislead the threat regarding the defensive array in and outside the urban area. Such information can convince the threat that a direct attack against the urban area is the most favorable approach.

8-36. If the threat has successfully isolated the urban area, commanders of a major operation have several courses of actions. Three options are an exfiltration, a breakout attack by forces defending the urban area, or an attack by forces outside the urban area to relieve the siege. A fourth option combines the last two: counterattacks from both inside and outside the urban area to rupture the isolation (see breakout operations in FM 3-90). Time is critical to the success of either operation. Commanders should plan for both contingencies to ensure rapid execution if necessary. Delay permits threat forces surrounding the urban area to prepare defenses, reorganize their attacking force, retain the initiative, and continue offensive operations. The passage of time also reduces the resources of defending forces and their ability to breakout. Therefore, commanders and staff of a major operation must vigilantly avoid isolation when Army forces are defending urban areas in their AO.

Separating Attacking Forces from Supporting Resources

8-37. Commanders of the major operation primarily use fires and IO for separating in space and time threat forces attacking the urban area from echelons and resources in support. The purpose of this shaping action is the same as for any conventional area defense. It aims to allow the defending forces to defeat the threat piecemeal as they arrive in the urban area without support and already disrupted by deep fires and IO against information systems. This separation and disruption of the threat also sets the conditions for a mobile defense if commanders choose to execute that type of defense. These operations also prevent the threat commander from synchronizing and massing his combat power at the decisive point in the close battle.

8-38. If the urban area is part of a major mobile defense operation, the urban defense becomes the fixing force. Commanders can shape the defense to encourage the threat to attack into the urban area. They can lure the threat using a combination of techniques depending on the situation. They may make the urban area appear only lightly defended while other alternative courses of action appear strongly defended by friendly forces. Placing the bulk of the defending forces in concealed positions well within the urban area and positioning security forces on the periphery of the urban area portray a weak defense. In other situations, the opposite is true. If the urban area is an important objective to the threat, friendly forces can make the urban area appear heavily defended, thus ensuring that he commits sufficient combat power to the urban area to overwhelm the defense. Both cases have the same objective: to cause a major commitment of threat forces in the urban area. Once this commitment is made, the mobile defense striking force attacks and defeats the threat outside the urban area. This isolates the threat in the urban area and facilitates its destruction.

8-39. In the urban tactical battle, many shaping actions mirror those in all defensive operations. The size and complexity of the urban area prevent defending forces from being strong everywhere; shaping operations designed to engage the threat on terms advantageous to the defense have particular importance. Shaping actions include reconnaissance and security operations, passages of lines, and movement of reserve forces prior to their commitment. In addition, shaping operations critical to urban defense include mobility and countermobility operations, offensive IO, economy of force operations, and population management operations.

Creating a Mobility Advantage

8-40. In urban terrain, countermobility operations can greatly influence bringing the threat into the engagement areas of defending forces. Countermobility operations—based on understanding the urban transportation system, design, and construction characteristics—can be unusually effective (see Chapter 2). Demolitions can have important implications for creating impassable obstacles in urban canyons as well as for clearing fields of fire where necessary. Careful engineer planning can make the already constrictive terrain virtually impassable to mounted forces where appropriate, thus denying the threat combined arms capabilities. Countermobility operations in urban terrain drastically increase the defense's ability to shape the attacker's approach and to increase the combat power ratios in favor of the defense. As with all aspects of UO, countermobility considers collateral damage and the second- and third-order effects of obstacle construction.

8-41. Well-conceived mobility operations in urban terrain can provide defending forces mobility superiority over attacking forces. This is achieved by carefully selecting routes between primary, alternate, and subsequent positions, and for moving reserves and counterattack forces. These routes are reconnoitered, cleared, and marked before the operation. They maximize the cover and concealment characteristics of the terrain. Using demolitions, lanes, and innovative obstacles denies the defense of these same routes.

Applying Economy of Force Measures

8-42. Economy of force is extremely important to effective tactical urban defense. A megalopolis is too large and too easily accessible for defending forces to be strong everywhere. Economy of force enables the defending force to mass effects at decisive points. Forces used in an economy of force role execute security missions and take advantage of obstacles, mobility, and firepower to portray greater combat power than they actually possess. They prevent the threat from determining the actual disposition and strength of the friendly defense. If, contrary to expectations, they are strongly attacked, their mobility—stemming from a mounted maneuver capability, planning, and an intimate knowledge of the terrain—allows them to delay until reserves can meet the threat. Security forces in an economy of force role take position in parts of the urban area where the threat is less likely to attack.

Defensive Combat Power: Suez – 1973

At the end of October, the Israeli Army was in the midst of effective counterattack against the Egyptian Army. The Israelis had success attacking west across the Suez Canal. Their armored divisions were attempting to achieve several objectives, to include destroying Egyptian air defense sites and completing the encirclement of the Egyptian 3rd Army, which was trapped on the canal's east side.

To completely encircle the Egyptian 3rd Army, the Israelis had to seize all possible crossing sites to it from the canal's west bank and the Red Sea. Also, as international negotiations towards a cease-fire progressed, the Israeli government wanted to capture as much Egyptian territory as possible to improve their negotiating position after hostilities.

Consequently, the Israeli Adan Armored Division was tasked to seize the Egyptian Red Sea port of Suez on the morning of 24 October. A cease-fire was to begin at 0700, and the Israeli intent was to be decisively engaged in the city by that time and then consolidate their position as part of the cease-fire compliance.

The Adan Division plan to seize Suez was a two-part operation. Each of the division's armored brigades would have a role. The 460th Brigade would attack west of the city and complete the city's encirclement. Simultaneously, the 217th Brigade would attack in columns of battalions through the city to seize three key intersections in the city. This was in accordance with standard Israeli armored doctrine for fighting in an urban area. The 217th Brigade would seize its objectives through speed, firepower, and shock action. Once the objectives were seized, infantry and armored teams would continue attacking from the secured objectives to mop up and destroy pockets of resistance. The Israeli commanders expected to demoralize the defending Egyptians—two infantry battalions and one antitank company—by this rapid attack. The armored division commander was specifically advised by his commander to avoid a "Stalingrad" situation.

The attack got off to an ominous beginning as mist greatly inhibited a scheduled aerial bombardment in support of the attack. The 217th Brigade began its attack without infantry support and was quickly stopped by antitank missiles and antitank fire. Infantry was quickly integrated into the brigade and the attack resumed.

At the first objective, the Israelis encountered their first problems. A withering barrage of small arms, antitank missiles, and antitank fire hit the lead tank battalion, including direct fire from SU-23 anti-aircraft guns. Virtually all the officers and tank commanders in the tank battalion were killed or wounded, and several tanks were destroyed. Disabled vehicles blocked portions of the road, and vehicles that turned on to secondary roads were ambushed and destroyed. The battalion, however, successfully fought its way through the first brigade objective and on to the final brigade objective.

Hastily attached paratroop infantry in company strength were next in column following the tanks. They were traveling in buses and trucks. As the lead tank battalion took fire, the paratroopers dismounted, and attempted to secure adjacent buildings. The tank battalion's action of fighting through the objective caused the paratroopers to mount up and also attempt to move through the objective. Because of their soft skinned vehicles the paratroopers were unable to remain mounted and again dismounted, assaulted, and secured several buildings that they could defend. Once inside the buildings, the paratroopers found they were cut off, pinned down, and unable to evacuate their considerable casualties, which included the battalion commander. The paratroopers were on the initial brigade objective but were unable to maneuver and were taking casualties.

A second paratroop company also dismounted and quickly became stalled in house-to-house fighting. The brigade reconnaissance company in M113 personnel carriers brought up the rear of the brigade column and lost several vehicles and was also unable to advance.

By 1100 the Israeli attack culminated. Elements of the 217th Brigade were on all three of the brigade's objectives in the city. However, the armored battalion, which had achieved the deepest penetration, was without infantry support and under severe antitank fire. Both paratroop companies were isolated and pinned down. In addition, an attempt to link up with the paratroopers had failed. At the same time, the civilian population of the city began to react. They erected impromptu barriers, ambushed isolated Israeli troops, and carried supplies and information to Egyptian forces.

The Israeli division commander ordered the brigade to break contact and fight its way out of the city. The armored battalion was able to fight its way out in daylight. The paratroop companies were forced to wait until darkness and then infiltrated out of the city carrying their wounded with them. Israeli casualties totaled 88 killed and hundreds wounded in addition to 28 combat vehicles destroyed. Egyptian casualties were unknown but not believed to be significant.

The fight for Suez effectively demonstrates numerous urban defensive techniques. It also vividly demonstrates the significant effect on defensive combat power of the urban environment.

The Egyptian defense demonstrates how the compartmented urban terrain restricts the mobility and the massing of firepower of armored forces. Trapped in column on the road, the Israelis were unable to mass fire on particular targets nor effectively synchronize and coordinate their fires. The short-range engagement, also a characteristic of urban combat, reduced the Israeli armor protection and eliminated the Israeli armor's ability to keep out of small arms range. Thus, hand-held antiarmor weapons were more effective in an urban area. Additionally, Egyptian small arms and sniper fire critically affected Israeli C2 by successfully targeting leaders.

The Egyptian defenders effectively isolated the mounted Israelis by defending and planning engagement areas in depth. The Egyptians synchronized so that they engaged the entire Israeli force simultaneously. This forced the Israelis to fight in multiple directions. It also separated the Israeli infantry from the armor and prevented the formation of combined arms teams necessary for effective urban offensive operations.

Suez also demonstrated how civilians come to the advantage of the defense. After the battle was joined, the population—by threatening isolated pockets of Israelis and building barricades—helped prevent the Israelis from reorganizing while in contact and hindered the Israelis breaking contact. The population was also a valuable source of intelligence for the Egyptians and precluded covert Israeli movement in daytime.

Suez shows the ability of a well-placed defense in depth to fix a superior force in an urban area. Despite the Israeli commander's caution to avoid a "Stalingrad," the Israeli division, brigade, and battalion commanders were quickly trapped and unable to easily break contact. Even a successful defense on the perimeter of the city would not have been nearly as effective, as the Israelis would have easily broken contact once the strength of the defense was recognized.

Another key to the success of the Egyptian defense was the Israelis' inadequate reconnaissance. While the Israelis knew the approximate size of the defending forces, they had no idea of the Egyptian dispositions. In this case, time prevented adequate reconnaissance. Key to a successful defense is adequate security to obscure defense dispositions, which permits surprise and shock effect.

The Suez defense was a decisive defeat of elite Israeli forces by regular infantry units inferior in training, morale, and numbers. Total disaster was averted only because of the professionalism of the Israeli armored forces and paratroopers that permitted them to continue to fight and eventually exfiltrate the urban trap. The Israeli forces thus escaped total destruction. Suez strongly demonstrates how the enhancing effects of the urban environment on defensive combat power are significant enough to permit inferior regular forces to defeat elite formations. Since the 1973 Suez battle, U.S. forces in Mogadishu, Somalia, and Russian forces in Grozny, Chechnya have faced similar urban defensive ambushes.

Effectively Managing the Urban Population

8-43. Another way to shape the urban defensive battle is population management. In most cases, defending force commanders are in the urban area before combat. This time gives them the chance to manage civilians. Consequently, they can better manage and protect the population (a legal requirement) and gain more freedom of action for their forces.

8-44. Managing the civilians during the defense is a function of the size, disposition, and needs of the population and the resources available to the commander. Requesting higher support or coordinating with nongovernmental organizations and the local civil leadership for support may make up shortages of resources. Resources devoted to population management are carefully weighed against availability, military mission requirements, and possible collateral damage affecting tactical, operational, or strategic success. It may prove impractical to evacuate an urban area's population; still, commanders should attempt to create and move most civilians to protected areas. Moving the population allows defending forces to more liberally apply fires, emplace obstacles, and relieve combat units and support units of requirements to continue life support for civilians while executing combat operations. Overall, effective civil-military operations can turn a friendly (or a neutral) population into an effective force multiplier providing support to every warfighting function.

Planning Counterattacks

8-45. Counterattacks are also an important tool in shaping the battlefield for defensive success. Counterattacks as a shaping tool have two applications: retaining the initiative and separating forces. However, opportunity for effective counterattacks will be brief and, therefore, timing will be critical. If conducted too soon, the counterattack may expend resources required later; if conducted too late, it may not be effective. Commanders should understand the effect of the urban environment on time-distance relationships; otherwise, the timing of the attack may be upset and the operation desynchronized. Additionally, commanders should develop plans beyond the counterattack to exploit potential success.

ENGAGE

8-46. Engaging the urban area in a defensive operation requires decisively defeating the threat's attacks. Defensive forces use the terrain to their advantage, employ precision supporting fires, and use direct fire from protected positions aligned against carefully selected avenues of approaches and kill zones. The combat power of the defense augmented by shaping actions and the characteristics of urban terrain force culmination of the threat attack. Like urban offensive operations, effective engagement in urban defensive operations typically results from successful actions at the tactical level of war. These actions include—

- Performing aggressive ISR.
- Creating depth.
- Executing an effective obstacle plan.
- Conducting coordinated counterattacks.

Performing Aggressive ISR

8-47. ISR efforts of the defender are focused initially on identifying relevant information about the location and nature of the threat's main effort. Once identified, the defender's ISR focus shifts to assessing the rate at which the threat attack moves to its culminating point. Indicators of culmination may be physical fatigue of Soldiers, a breakdown in C2 capability, difficulty providing logistics support, or the increasing time required to reorganize small units to attack. When that culmination is achieved, friendly forces counterattack before the threat has a chance to transition to a hasty defense.

Creating Depth

8-48. Depth in the defense is the key to forcing the threat to culminate. The urban defense cannot allow itself to be penetrated nor permit forward elements to be destroyed. The defense is designed with the greatest depth possible. Defending forces weaken the threat to the fullest extent possible by attack from each position but not permit themselves to be destroyed by fires or close assault. Instead, as threat combat power builds up against individual positions, the use of mission orders permits subordinate leaders to disengage on their own initiative and move on preplanned routes to subsequent positions. Positions are designed to be mutually supporting—withdrawing from one position to a subsequent one while supporting positions cover by fires. The attacker is constantly forced to deploy and reorganize without being able to achieve decisive effects against the defender.

Executing an Effective Obstacle Plan

8-49. Obstacles in the urban defense are designed to break up the threat's combined arms capability. Separating dismounted forces from mounted forces disrupts the cohesion of the attacker and reduces his combat power. It also exposes his individual elements to the effects of asymmetric counterattack. The leading threat dismounted force can be effectively counterattacked by a friendly combined arms element while the threat armored force remains vulnerable to antiarmor attack by dismounted forces.

Conducting Coordinated Counterattacks

8-50. The counterattack is one of the key actions of the urban defense. However, commanders do not counterattack unless there is a reasonable chance of success. As the attacker moves into the depth of the urban area, his forces become fatigued, attrited, and increasingly disorganized. He likely also creates an increasingly long and exposed flank. At all levels, forces defending in urban terrain look for opportunities to counterattack. As the offensive force reaches the culmination point where it can no longer continue to attack with the available forces, the defensive commander executes a planned and coordinated counterattack. The counterattack aims to regain the initiative and to make the threat fight in multiple directions. Infiltration using superior knowledge of the terrain (including supersurface and subsurface capabilities) and the skillful use of stay-behind forces permits attacking the threat throughout the depth of his formations. Small-scale counterattacks focus on C2 and sustainment capabilities. These counterattacks can set the conditions for a deliberate attack leading to the ultimate destruction of the attacking threat force.

CONSOLIDATE

8-51. Consolidation is as important to urban defensive operations as to offensive operations. Many of the same consolidation considerations for the urban offense apply equally well to the defense. Commanders should reinforce or reposition maneuver forces and fire support assets on the urban battlefield based on weaknesses uncovered during rehearsals and opportunities discovered during actual execution. While maximizing the many advantages of the urban defense, the commander of the urban defense aggressively seeks ways of attriting and weakening enemy forces before they enter the urban area and close combat is initiated. Commanders should combine the static and mobile elements of the defense to strengthen their positions in relation to the enemy while seeking every opportunity to transition to urban offensive operations. As in urban offensive operations, commanders should conduct any necessary reorganization actions that they were unable to accomplish during execution.

TRANSITION

8-52. Transitions in urban defensive operations occur at all levels. As with offensive operations, commanders of major operations should address which units are assigned to continue to operate in the area after defensive operations have ceased. In defensive UO, this task is not as challenging as an occupation mission during urban offensive operations. The psychology of troops defending an urban area differs from those attacking into it. Defending forces become accustomed to the environment, having experience in the environment before combat. In terms of training, it is easier for follow-on missions to be assigned to a unit that has successfully defended the urban area. This course of action takes advantage of the defending unit's experience in the area and its relationships with other agencies—agencies that were operating alongside the units before and possibly during the defense. In defensive operations, regardless of the civilians' attitudes, policies regarding that population are established before the successful defense, and the command likely has experience executing operations with civil authorities and other agencies. Thus, these relationships are neither new nor as significant an issue as in offensive operations. Therefore, commanders must be prepared to execute various stability or civil support operations or use a successful defense to springboard into more decisive offensive operations elsewhere in the commanders' AO.

Transition Emphasis to Stability or Civil Support Operations

8-53. At the end of a successful urban defense, operational commanders should generally expect civil authority, control, and jurisdiction to increase. Additionally, the civil population may be anxious to return. Defensive combat will require virtually complete military control of the urban area; however, after the successful defense, a rapid transition will occur from military control to civil or joint military and civil control afterward. Although the full spectrum of Army operations will often be conducted simultaneously in UO, commanders will transition emphasis to stability or civil support operations. Important transition tasks will include demilitarizing munitions, clearing obstacles, and searching for isolated threat pockets of resistance. Conclusion of the defensive operations will also require transition to joint civil-military tasks, such as evaluating structures for safety, restoring essential services, and possibly creating joint law enforcement. Commanders of major operations, using a civil-military operations center and G9, should anticipate these requirements and begin early preparations to ensure a smooth, successful transition.

Transition to Offensive Operations

8-54. Units that have successfully defended the urban area may then transition to offensive operations. A rapid transition to offensive operations will require identification, preparation, and training of units designated to assume missions as the defending units leave the urban area. This preparation emphasizes continuity of policies and relationships already established. A relief in place occurs. The new occupying units provide not only a continuity of policy, but also a continuity of attitude toward the urban area, its population, and its institutions.

Chapter 9

Urban Stability and Civil Support Operations

The fundamental shared aims between stability and civil support operations and any operation conducted in an urban environment are the increased significance and influence of the civil population and nonmilitary organizations. Often, no military victory is to be achieved. The center of gravity for these operations normally cannot be attacked through military means alone; the Army (and the military in general) is often but one tool supporting a larger, civil-focused effort. When conducted, military defeat of an armed adversary is only a phase in a larger campaign, the overall objectives of which will be to bring about a more satisfactory political, social, diplomatic, economic, and military situation than existed prior to the conflict. Without a tightly coordinated civil-military effort, overall success will be difficult or impossible to achieve. Commanders who can understand and cope with the complexities of stability or civil support operations gain insights that directly apply to executing any urban operation. Urban stability or civil support operations may complement urban offensive and defensive operations, or may dominate the overall operation. Army forces may need to conduct offensive and defensive operations to defend themselves or destroy urban threats seeking to prevent the decisive stability or civil support mission. During hostilities, urban stability operations may keep armed conflict from spreading, encourage coalition and alliance partners, and secure the civilian population's support in unstable urban areas (and the surrounding rural areas under their influence). Following hostilities, urban stability operations may provide a secure environment for civil authorities to rebuild. Urban civil support operations can range from transporting, feeding, and sheltering the population made homeless as a result of man-made or natural disasters to conducting civil search and rescue and providing urgent medical care.

PURPOSE OF URBAN STABILITY AND CIVIL SUPPORT OPERATIONS

9-1. Army forces conduct stability and civil support operations to deter war, resolve conflict, promote peace, strengthen democratic processes, retain U.S. influence or access abroad, assist U.S. civil authorities, and support moral and legal imperatives. Stability operations promote and sustain regional and global stability. Civil support operations meet the urgent needs of designated U.S. civilians until civil authorities can accomplish these tasks without military assistance. Nearly every urban operation will involve some type or form of stability operation or civil support operation combined and sequenced with offensive and defensive operations.

CHARACTERISTICS OF URBAN STABILITY AND CIVIL SUPPORT OPERATIONS

9-2. Worldwide urbanization, migration trends from rural to urban areas, and more centralized populations in urban areas increase the chance that Army forces will conduct stability or civil support operations in or near urban areas. Simply put, many people live in urban areas, and their welfare will be the primary reason for conducting these operations. Urban areas that serve as economic and government centers (the ideal location for U.S. and allied embassies) are often the focal point for many threat activities. Therefore, Army forces may need to conduct stability operations in these foreign cities to counter those threats. Additionally, urban areas may contain the resources and infrastructure to support both types of operations, regardless of whether the overall focus is in urban or rural areas. Repairing or restoring the infrastructure may be a critical task in accomplishing either type of operation. Supported governmental and nongovernmental agencies are not as logistically self-sufficient as the Army. As such, these agencies will often need to center their operations in and around urban areas to use the area's infrastructure to support themselves and their objectives. These agencies may require military protection to accomplish their missions. Figure 9-1 lists some defining characteristics of these wide-ranging operations.

- Long or Short Duration
- Joint and Interagency
- Unilateral or Multinational
- Increased Civil-Military and Legal Considerations
- Greater Potential for Ambiguity
- Increased Constraints Necessitating More Restrictive ROE
- Amplified Need for Cultural and Political Sensitivity

Figure 9-1. Characteristics of stability and reconstruction and support operations

9-3. Stability and civil support operations are diverse, varied in duration, and unilateral or multinational. Stability operations are conducted overseas as part of a campaign or major operation, while civil support operations are conducted within the United States as part of homeland security. Like all urban operations (UO), they are usually joint. Unlike urban offensive and defensive operations, they are more often interagency operations and require more restrictive rules of engagement (ROE). The multiplicity of actors involved usually increases the scope and scale of required coordination and communication. In urban stability or civil support operations, adverse conditions arising from natural or man-made disasters or other endemic conditions—such as human suffering, disease, violations of human rights, or privation—will significantly modify the urban environment. Unresolved political issues and tenuous agreements, difficulties discriminating combatants from noncombatants or between parties of a dispute, and the absence of basic law and order all serve to complicate an already complex and uncertain environment. Civil-military and legal considerations take on added significance in all urban operations (see Civil-Military Operations and Legal Support in Chapters 5 and 10 respectively), but even more so in urban stability and civil support operations. Finally, recognizing, defining, and achieving the desired end state is often more difficult than in offensive and defensive operations.

9-4. Overall, commanders of major operations involving urban stability and civil support operations should not expect clear guidance. They must learn, adapt, and live with ambiguity. They cannot expect to operate in a political vacuum (even commanders at the tactical level); rather, they should expect to work alongside both governmental and nongovernmental leaders and organizations. Commanders conducting UO should not expect an easily identifiable enemy located across a clearly demarcated line. In fact in many peace operations, they and their Soldiers must resist the need to have an enemy—difficult at best when one side or another (or both) may be sniping at them. Commanders should also expect changing and additional missions and tasks, without being allowed to use every means at their disposal to carry out those missions. Many tasks required may be ones for which their units have never, or rarely, trained. Finally, commanders should expect to show restraint with a keen sensitivity to political considerations and to alien cultures, either or both of which they might find confusing or even repugnant.

URBAN STABILITY AND CIVIL SUPPORT OPERATIONS AND BATTLEFIELD ORGANIZATION

9-5. Each urban stability operation or civil support operation will be distinct from one another. These operations differ even more when applied to a specific urban area. Due to the complexity of the environment, commanders must carefully arrange their forces and operations according to purpose, time, and space to accomplish the mission. In most UO the terrain, the dense population (military and civilian), and the participating organizations will further complicate this arrangement.

9-6. The support and assistance that Army forces will provide during these operations is only temporary (although it may be of long duration) and this facet requires particular emphasis. Commanders should plan and execute both of these types of operations with that essential consideration always in mind. Eventually, the government and administration—either foreign or domestic—must secure and support their population by themselves. Therefore, commanders must envision and set the conditions that allow for the transition of control and responsibility to shift to legitimate civilian authorities. While commanders may only provide assistance and support based on specific and well planned civilian requests, more often commanders will need to determine requirements in collaboration with competent civilian authorities and agencies or, in some cases, with little or no initial civilian assistance at all. Ultimately, transition planning should occur as an integral part of the overall operational planning and include collaboration with appropriate civilian agencies and organization as early as possible to ensure a seamless transition to civilian control without major setbacks and loss of forward momentum.

STABILITY OPERATIONS

9-7. Stability operations establish, sustain, and exploit security and control over foreign areas, populations, and resources. Urban areas will be decisive to accomplishing many types of stability operations because urban areas are the centers of population, culture, economy, and government. Much of the support provided by Army forces will aim to assist local, regional, or national governments to restore essential services and infrastructure and reestablish civil order and authority. The location of civilian authorities in urban areas will, by necessity, be a dominating factor in accomplishing the mission. As importantly, many stability operations—enforcing peace in Bosnia for example—will require interacting with, influencing, controlling, or protecting all or parts of the civilian population. Assessing, understanding, and gaining the support of civilians in key economic, cultural, or political urban areas may influence surrounding regions (smaller urban areas and the rural countryside) and may be decisive to achieving overall stability objectives.

9-8. Stability operations may involve both coercive and cooperative actions. They are conducted in situations in which legitimate civil authority is unable to provide the necessary security and control for the urban population as a result of—

- Deliberate operations to influence a regime change.
- Offensive or defensive operations or natural disasters resulting in ineffective civil authorities, conditions beyond the capabilities of the foreign urban government, or both.

CIVIL SUPPORT OPERATIONS

9-9. Civil support operations are an integral part of homeland security which aim to protect U.S. citizens and infrastructure from conventional and unconventional threats and mitigate the effects of man-made and natural disasters. (The other part to homeland security is homeland defense.) Army forces conduct civil support operations by providing Army resources, expertise, and capabilities in support of a lead agency—which may be to Department of Defense—to address the consequences of man-made or natural accidents beyond the capabilities of local or regional U.S. civilian authorities. Such a situation could result from a chemical, biological, radiological, nuclear, and high-yield explosive (CBRNE) incident, hurricane, flood, or civil disturbance in a domestic urban area.

CBRNE INCIDENTS

9-10. CBRNE incidents can be disastrous and are of particular concern for both stability and civil support UO. In urban areas, the potential for catastrophic loss of life and property is enormous. The Army categorizes CBRNE incidents separately from other natural and man-made disasters because it has specific expertise with these weapons. CBRNE incidents usually result from a military or terrorist threat (adding a law enforcement dimension to the disaster). Other urban CBRNE considerations include the following:

- Subways and other subsurface areas offer ideal areas for limited chemical or biological attacks. Nuclear attack (and high-yield explosives) can produce tragic results due to the effects of collapsing structures, flying debris, and fires. Dispersion patterns are affected by the urban terrain and are more difficult to predict and monitor. Large-scale incidents may produce hundreds of thousands of casualties, but even a limited attack may require evacuating and screening large numbers of civilians. Requirements for medical support, basic life support, and, if necessary, decontamination may quickly overwhelm the Army force's capabilities even with augmentation.
- Similar to natural disasters, panic and disorder may accompany the CBRNE event. Fleeing civilians may clog elements of the transportation and distribution infrastructure. Physical destruction may also affect other components of the infrastructure of critical and immediate concern, such as energy and administration and human services (water, sanitation, medical, fire fighting, and law enforcement). Because all elements of the infrastructure may be affected, the overall recovery time may be lengthened and the effects broadened to include much of the surrounding area. The effects of a single urban CBRNE event potentially could be felt nationally or globally.

DECISIVE OPERATIONS

9-11. In urban stability operations, decisive operations may take many years and include multiple actions before achieving the desired end state. This particularly applies to the strategic and operational levels. Oppositely, decisive operations involved in an urban civil support operation for mitigating or reducing disease, hunger, privation, and the effects of disasters normally achieve faster results. However, any operation that attacks the underlying cause and seeks to prevent or relieve such conditions will usually take longer. In urban areas, establishing law and order to protect critical infrastructure and the inhabitants from lawlessness and violence is often critical and also the decisive operation.

SHAPING OPERATIONS

9-12. Shaping operations establish and maintain the conditions for executing decisive operations. In urban stability or civil support operations, shaping operations always include information operations (IO) that influence perceptions and maintain legitimacy. Often, various participants, and their potentially divergent missions and methods, are involved. Army commanders must coordinate their planning and efforts (early and continuously) to ensure that their decisive, shaping, or sustaining operations are not working against other agencies' efforts and operations—agencies that may have the lead role in the operation. Thus, a critical shaping operation may be to establish the coordination to help develop a common purpose and direction among agencies, particularly those that may experience continuous personnel turnover during the conduct of a lengthy operation. In some instances and with some organizations and agencies, particularly nongovernmental organizations (NGOs), genuine unity of effort may not be achievable; however, recognizing the differences in aims and goals will allow Army commanders to conduct operations with less friction. Commanders should actively request and include NGOs and appropriate governmental agencies in mission readiness exercises or any other training for stability or civil support operations.

SUSTAINING OPERATIONS

9-13. Sustaining operations enable decisive and shaping operations and include logistics, rear area and base security, movement control, terrain management, and infrastructure development. Sustainment bases, especially those located in urban areas, become an attractive target for hostile civilians; therefore, commanders must actively and aggressively protect these bases as well as lines of communications (see Chapter 10).

CONSIDERATIONS FOR URBAN STABILITY AND CIVIL SUPPORT OPERATIONS

In wars of intervention the essentials are to secure a general who is both a statesman and a soldier; to have clear stipulations with the allies as to the part to be taken by each in the principal operations; finally, to agree upon an objective point which shall be in harmony with the common interests.

Lieutenant General Antoine-Henri,
Baron de Jomini

9-14. The urban operational framework (understand, shape, engage, consolidate, and transition) provides a structure for developing considerations unique to urban stability and civil support operations. Many considerations presented in urban offensive and defensive operations apply to urban stability and civil support operations, particularly those that address how to understand the urban and overall operational environment. Because the situations in which stability and civil support operations normally occur share strong similarities with any urban environment, many of these considerations are closely linked to the urban fundamentals presented in Chapter 5. Taken together, commanders will often find them useful in conducting UO throughout full spectrum operations. Appendix C has a historical example of how to apply the urban operational framework to a stability operation.

UNDERSTAND

9-15. In urban stability or civil support operations, commanders must carefully assess and understand the political dimension of the operational environment, as well as their role and the media's part in managing information. These operations are inherently tied to the exercise of diplomatic power. All operations in urban areas are often the focus of the media and thus gain considerable public and political attention. Therefore, military objectives in urban stability or civil support operations are more directly linked with political objectives. The relationship between the levels of war—strategic, operational, and tactical—is often closer than in urban offensive and defensive operations. Military objectives must be carefully nested within political objectives. Commanders must ensure that the ways and means to accomplish their objectives, to include security and force protection measures, will hold up to media scrutiny and are appropriate for the situation and environment. All levels of command must understand the link between political and military objectives, to include a basic understanding at the Soldier level. One uncoordinated, undisciplined, or inappropriate action, even at the lowest level, could negate months or years of previous, disciplined effort. Commanders must balance security and force protection measures with mission accomplishment. Ineffective measures can put Soldiers at too great a risk and jeopardize the mission. Conversely, overly stringent measures may make it difficult for forces to interact with the population closely—essential in many of these operations. Finally, commanders will need a thorough assessment of the governmental and nongovernmental organizations and agencies that will be operating in or near urban areas that fall within their area of operations.

Political and Military Objectives

9-16. Commanders must translate political objectives into military objectives that are clear and achievable (clear tasks and purposes) and can lead to the desired end state. (In most stability operations, defeat of a threat is not the ultimate desired end state.) Political objectives may be vague making it difficult for commanders to conduct their mission analysis. This applies to tactical- and even operational-level commanders, unskilled at higher level, strategic political-military assessments. Each type of stability operation or civil support operation is distinct, often unfamiliar to the executing unit, and unique to the specific situation. These factors often make it difficult to confidently determine the specific tasks that will lead to mission success. Therefore, commanders must also establish measures of effectiveness that aid in understanding and evaluating progress and help gauge mission accomplishment.

Note: Commanders should consult the US Agency for International Development's (USAID) Field Operations Guide for Disaster Assessment and Response when conducting their assessments and developing measures of effectiveness for many urban relief operations.

9-17. These measures should be quantifiable (in some circumstances, a qualitative assessment may be most appropriate) and link cause with effect. They help determine the changes required and are essential to the assessment cycle required for urban stability or civil support operations. In a humanitarian relief operation to aid the starving, commanders could determine that the decisive effort is delivering safe food to the urban area. To judge success or effectiveness, they could determine that the appropriate measure is the number of food trucks dispatched daily to each distribution site; the more trucks, the more effective the efforts. However, this measure must correlate with the overarching measure of effectiveness: decline in the mortality rate. If no significant decrease in deaths due to starvation occurs, they may need to reassess and modify the tasks or measures of effectiveness. A better measure may be to track the amount of food consumed by those in need instead of simply counting the number of trucks dispatched. Areas around which measures of effectiveness can be formed for many stability or civil support operations (including the example above) and which will help return most societies to some degree of normalcy and self-sufficiency include:

- Restoring law and order.
- Decreasing morbidity and mortality rates.
- Securing safe food and water.
- Restoring critical infrastructure.
- Resettling the population.
- Reestablishing economic activity.

Commanders can often develop measures of effectiveness to address these areas in terms of establishing or restoring security and providing logistics. (Figure 9-2 provides example measures of effectiveness from the strategic to tactical levels for a possible stability operation.)

9-18. Political objectives are fluid and modified in response to new domestic and international events or circumstances. Thus, assessment is continuous, and commanders must adjust their own objectives and subsequent missions accordingly. In urban stability or civil support operations, commanders often develop military objectives that support or align with the objectives of another agency that has overall responsibility for the urban operation. In this supporting role, commanders may receive numerous requests for Soldier and materiel assistance from the supported agency and other supporting agencies operating in the urban area (to include elements of the urban population). With such unclear lines of authority and areas of responsibility, they must ensure that the tasks, missions, or requested Army resources fall clearly in the intended scope and purpose of the Army's participation in the operation. They must not develop or execute missions based on inadequate or false assumptions, misinterpreted intent, or well-meaning but erroneously interpreted laws or regulations by any organization, to include even the lead agency. When missions appear outside their scope, commanders must quickly relay their assessment to their higher headquarters for immediate resolution. The commander's goal is not to limit or slow military participation but to contribute as intended and in consonance with political objectives and the law.

Security and Force Protection Measures

9-19. Commanders must plan for and continually assess the security of their forces operating in an urban area as well as constantly review protection measures. Establishing a robust intelligence—particularly human intelligence (HUMINT)—network that can determine the intentions and capabilities of the threat and the urban populace is the basis for establishing force protection for Army forces operating in the urban environment. However, many such operations, particularly stability operations, require extra time to forge a lasting change. Over time, and particularly in peacetime when objectives center on helping others and avoiding violence, even the complex urban environment may seem benign. Without continued, aggressive command emphasis, Soldiers may become lulled into complacency. Also, during periods of transition or transfer of authority from one unit or organization to another, departing Soldiers' focus often shifts to redeployment activities and away from force protection concerns. It is usually then that Army forces are most vulnerable to terrorist and insurgent tactics, such as bombings, kidnappings, ambushes, raids, and other forms of urban violence.

Measures of Effectiveness: The Need to Measure Progress		
Strategic Level Criteria	**Operational Level Criteria**	**Tactical Level Criteria**
Accountable to the American people for defining and measuring progress towards defeating terrorism and meeting national security goals.	Accountable to the strategic level for measuring operational success and providing linkage to strategic goals.	Responsible to the operational level for measuring tactical success and providing linkage to operational goals.
Examples: • Prevention of the insurgency from receiving aid or resources from other international groups. • A functioning national government. • Amount of international support and aid to reconstruction. • Number of nations contributing manpower to coalition forces.	Examples: • Host-nation security forces trained and equipped. • Denial of the merging of insurgent forces with terrorist groups. • Amount and distribution of— • Electricity. • Liquid propane gas. • Gasoline. • Functioning provincial governments.	Examples: • Number of insurgent forces in the AO neutralized. • Reduced attacks on coalition forces in the AO. • Reduced civilian-on-civilian violence in the AO. • Host-nation security force recruitment goals met. • Host-nation security force training goals met. • Number of reliable human intelligence walk-ins. • Amount of unexploded explosive ordnance cleared. • Functioning neighborhood and district advisory councils.

Figure 9-2. Example measures of effectiveness

9-20. Although force protection will not ensure successful urban stability or civil support operations, improper assessment and inadequate force protection measures can cause the operation to fail. In either operation, keeping a neutral attitude toward all elements of the urban population, while maintaining the appropriate defensive posture, enhances security. One threat principle discussed in Chapter 3 was that threats would seek to cause politically unacceptable casualties. An improper threat assessment and a lapse in security at the tactical level could result in casualties. That result could affect strategy by influencing domestic popular support and subsequently national leadership decisions and policy.

9-21. Emphasizing security and force protection measures does not mean isolating Soldiers from contact with the urban population. On the contrary, commanders should balance survivability with mobility according to the factors of METT-TC—mission, enemy, terrain and weather, troops and support available, time available, civil considerations. Survivability measures—such as sandbagging, hardening, or fortifying buildings and installations, particularly where large numbers of Soldiers are fed and billeted—will be necessary and require considerable command attention. On the other hand, mobility operations are essential in preserving freedom of action and denying a threat the opportunity to observe, plan, and attack urban forces. Continual Army presence in the urban area may provide the urban population a sense of security and allow Soldiers to develop a detailed knowledge of the "patterns of life" in their assigned area of operation (AO). Armed with this knowledge, they can detect the absence of the normal or the presence of the abnormal that might indicate a potential threat. Overall, mission degradation and increased risk to the force can result if force protection measures prevent Army forces from conducting prudent missions and establishing an active and capable presence.

Assessment of Security and Force Protection: Belfast, Northern Ireland

Since 1969, Belfast has significantly affected the British military campaign for stabilizing the area. British operations in Belfast illustrate the difficulty of balancing the security and protection of forces with maintaining the stabilizing presence necessary to uphold law and order, minimize violence, and control the urban population.

British successes in protecting Belfast's infrastructure and government facilities from terrorist attacks compelled various terrorist cells, especially the Irish Republican Army and the Provisional Irish Republican Army, to attack more military targets.

At the time, British soldiers and bases presented relatively unprotected targets to these factions, and attacks against them solidified their legitimacy as an "army." In response, British commanders implemented extreme security and force protection measures—from ballistic protection vests and helmets to fortress-like operational bases and large unit patrols. These protection measures successfully decreased the violence against British soldiers in Belfast. However, they also decreased the soldiers' interaction with the population and their ability to stabilize the city. The large patrols protected the soldiers, but also inhibited effective saturation of neighborhoods. These patrols, coupled with fortress-like bases and bulky protective clothing, created an "us-versus-them" mentality among civilians and soldiers.

As force protection increased and stabilizing effects decreased, the terrorists were provided more targets of opportunity among the civilians and infrastructure. British commanders reassessed the situation, identified this "see-saw" effect, and adapted to strike a better balance between force protection and effective presence patrols. For example, British forces switched to four-man patrols to enable greater mobility and wore berets instead of helmets to appear less aggressive.

While the above historical example illustrates a critical aspect of stability and civil support operations, it subtly reveals another important lesson for Army commanders. Many of our future coalition partners and assisting governmental and nongovernmental organizations will bring key insights to UO (particularly dealing with urban societies) based on their historical and operational experience. Army commanders and their staffs should seek out and remain open to others' unique understandings that are based on hard-earned and well-analyzed lessons learned.

Participating Organizations and Agencies

9-22. Across the spectrum of urban operations, but more so in these operations, numerous NGOs may be involved in relieving adverse humanitarian conditions. Dense populations and infrastructure make an urban area a likely headquarters location for them. (In 1994 during OPERATION UPHOLD DEMOCRACY, for example, over 400 civilian agencies and relief organizations were operating in Haiti.) Therefore, commanders should assess all significant NGOs and governmental agencies operating (or likely to operate) in or near the urban area to include their—

- Functions, purposes, or agendas.
- Known headquarters and operating locations.
- Leadership or senior points of contact (including telephone numbers).
- Communications capabilities.
- Potential as a source for critical information.
- Financial abilities and constraints.
- Logistic resources: transportation, energy and fuel, food and water, clothing and shelter, and emergency medical and health care services.
- Law enforcement, fire fighting, and search and rescue capabilities.
- Refugee services.
- Engineering and construction capabilities.
- Other unique capabilities or expertise.
- Previous military, multinational, and interagency coordination experience and training.
- Rapport with the urban population.
- Relationship with the media.
- Biases or prejudices (especially towards participating U.S. or coalition forces, other civilian organizations, or elements of the urban society).

Commanders can then seek to determine the resources and capabilities that these organizations may bring and the possible problem areas to include resources or assistance they will likely need or request from Army forces. These organizations will be critical to meeting the population's immediate needs and minimizing the effects of collateral damage or disaster. However, commanders should consider whether a close relationship with any of these organizations will compromise the organization's appearance of neutrality (particularly threat perceptions) and adversely affect their ability to assist the population.

SHAPE

9-23. Commanders conduct many activities to shape the conditions for successful decisive operations. In urban stability and civil support operations, two activities rise to the forefront of importance: aggressive IO and security operations.

Vigorous Information Operations

9-24. IO, particularly psychological operations (PSYOP) and the related activities of civil affairs (CA) and public affairs, are essential to shape the urban environment for the successful conduct of stability or civil support operations. Vigorous IO can influence the perceptions, decisions, and will of the threat, the urban population, and other groups in support of the commander's mission. IO objectives are translated to IO tasks that are then executed to create the commander's desired effects in shaping the battlefield. These operations can isolate an urban threat from his sources of support; neutralize hostile urban populations or gain the support of neutral populations; and mitigate the effects of threat IO, misinformation, rumors, confusion, and apprehension. Developing effective measures of effectiveness is essential to a good IO campaign strategy. One of the most valuable methods for obtaining data for use in this process is face-to-face encounters with targeted audiences by unit patrols and HUMINT, PSYOP, and CA teams. A valuable technique may be to conduct periodic, unbiased surveys or opinion polls of the civilian population to determine changes in their perceptions and attitudes.

Security Operations

Protecting Civilians and Critical Infrastructure

9-25. Security for NGOs and civilians may also be an important shaping operation, particularly for stability and civil support operations. Commanders may need to provide security to civil agencies and NGOs located near or operating in the urban area so that these agencies can focus their relief efforts directly to the emergency. Commanders may also need to protect the urban population and critical infrastructure to maintain law and order if the urban area's security or police forces are nonexistent or incapacitated, or the urban area's security situation has undergone drastic change (such as the result of a natural disaster) and security or police forces require additional augmentation. (See also the discussion of the legal aspects of Civilians Accompanying the Force in Chapter 10.)

Preserving Resources

9-26. Just as forces may be at risk during urban stability or civil support operations, so may their resources. In urban areas of great need, supplies and equipment are extremely valuable. Criminal elements, insurgent forces, and people in need may try to steal weapons, ammunition, food, construction material, medical supplies, and fuel. Protecting these resources may become a critical shaping operation. Otherwise, Army forces and supporting agencies may lack the resources to accomplish their primary objectives or overall mission.

Prioritize Resources and Efforts

9-27. During UO, commanders will always face limited resources with which to shape the battlefield, conduct their decisive operations, and accomplish their objectives. They must continually prioritize, allocate, and apply those resources to achieve the desired end state. To this end, they may develop an order of merit list for proposed projects and constantly update it over time. To some degree, the local urban population will usually need to be protected and sustained by Army forces. Hence, commanders must tailor their objectives and shape their operations to achieve the greatest good for the largest number. Commanders first apply the urban fundamental of preserving critical infrastructure to reduce the disruption to the residents' health and welfare. Second, they apply the urban fundamental of restoring essential services, which includes prioritizing their efforts to provide vital services for the greatest number of inhabitants possible. In operations that include efforts to alleviate human suffering, the criticism for any participating organization is likely to be there is not enough being done or Army forces are not being responsive enough. Therefore, commanders must develop clear measures of effectiveness not only to determine necessary improvements to operational plans but also to demonstrate their Soldiers' hard work and sacrifice and U.S. commitment to the operation.

Improve the Urban Economy

9-28. When conducting reconstruction and infrastructure repair, commanders should consider using such activities to simultaneously improve the urban economy. Hiring local civilians and organizations to do reconstruction work helps satisfies urban job requirements, may inspire critical elements of the urban society to assume responsibility for the success or failure of urban restoration efforts, and may potentially reduce threat influence by diminishing their civilian sources of aid. Hiring indigenous personnel for short-term projects does not replace the need for long-term economic planning and the development of stable jobs. The overall reconstruction effort must be guided by the commander's vision of the end state. However, the commander's guidance and intent should be broad and expansive enough to allow responsive decision making by subordinate commanders based on their analysis of the urban population's needs in their assigned AO.

ENGAGE

If there is any lesson to be derived from the work of the regular troops in San Francisco, it is that nothing can take the place of training and discipline, and that self-control and patience are as important as courage.

Brigadier General Frederick Funston
commenting on the Army's assistance following
the 1906 San Francisco earthquake and fire

9-29. The focus of the Army is warfighting. Therefore, when Army commanders conduct many urban stability and civil support operations, they must adjust their concept of success. Commanders will most often find themselves in a supporting role and less often responsible for conducting decisive operations. They must accept this supporting function and capitalize on the professional values they have instilled in each Soldier, particularly the sense of duty to do what needs to be done despite difficulty, danger, and personal hardship. Commanders must also put accomplishing the overall mission ahead of individual desires to take the lead—desire often fulfilled by being the supported rather than supporting commander. In many stability and civil support operations, success may be described as settlement and compromise rather than victory. Yet, the Army's professionalism and values—combined with inherent adaptability, aggressive coordination, perseverance, and reasonable restraint—will allow Army forces to engage purposefully and dominate during complex urban stability or civil support operations.

Adaptability

9-30. Adaptability is particularly critical to urban stability or civil support operations because these operations relentlessly present complex challenges to commanders for which no prescribed solutions exist. Commanders often lack the experience and training that provide the basis for creating the unique solutions required for these operations. Since the primary purpose for the Army is to fight and win the nation's wars, the challenge then is to adapt urban warfighting skills to the unique stability or civil support situation.

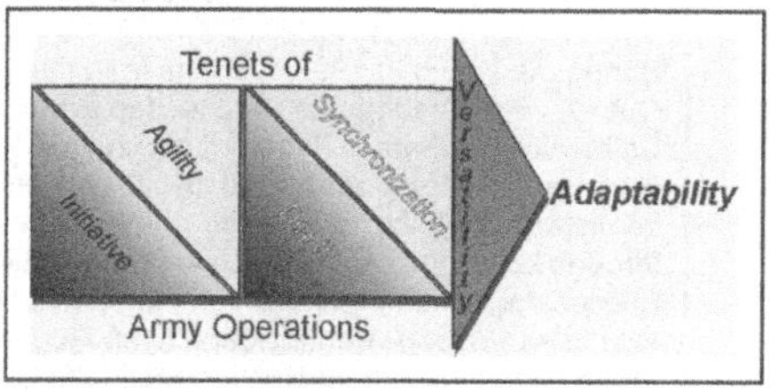

Figure 9-3. Adaptability

9-31. Doctrine (joint and Army) provides an inherent cohesion among the leaders of the Army and other services. Still, Army commanders conducting urban stability or civil support operations will often work with and support other agencies that have dissimilar purposes, methods, and professional languages. Army commanders must then capitalize on three of the five doctrinal tenets of Army operations: initiative, agility, and versatility (see figure 9-3 and FM 3-0). Commanders must bend as each situation and the urban environment demand without losing their orientation. They must thoroughly embrace the mission command philosophy of command and control addressed in FM 6-0 and Chapter 5 to encourage and allow subordinates to exercise creative and critical thinking required for planning and executing these UO. Commanders must also be alert to and recognize "good ideas," and effective tactics, techniques, and procedures, regardless of their source—other services, coalition partners, governmental and nongovernmental organizations, and even the threat—and adapt them for their own purposes in UO.

9-32. Adaptability also springs from detailed planning that carefully considers and realistically accounts for the extent of stability and civil support operations. Although no plan can account for every contingency and completely eliminate the unexpected, good plans—which include detailed civil considerations—provide platforms from which to adjust course more readily. Adequate planning allows commanders not only to react more effectively, but also to be forward-thinking and take actions that favorably steer the course of events.

Aggressive Coordination and Synchronization

9-33. In urban stability and civil support operations, the increased number of participants (both military and nonmilitary) and divergent missions and methods create a significant coordination and synchronization challenge. Significant potential for duplicated effort and working at cross-purposes exists. The success of UO often depends on establishing a successful working relationship with all groups operating in the urban area. The absence of unity of command among civil and military organizations does not prevent commanders from influencing other participants not under his direct command through persuasion, good leadership, and innovative ideas.

9-34. Commanders may consider establishing, as necessary, separate organizations for combat operations and for stability operations or civil support operations to increase coordination and enhance local, NGO, and international support. Further, aligning the unit or subordinate units with NGOs may contribute to establishing popular legitimacy for the operation and place greater pressure on threat forces. In some instances, commanders may consider organizing part of their staff around government, administrative, and infrastructure functions that mirror the urban area in which their forces are operating. Development of a mirror urban area organization may give greater legitimacy to the urban government or administration and ease transition of responsibility once the end state is achieved. Commanders must be mindful, however, that local groups seen as allying themselves with Army or coalition authorities will likely experience pressure to demonstrate their independence as established dates for redeployment or other critical events approach. In some instances that demonstration of independence may be violent.

Civil Support and Coordination with Civilian Authorities: Los Angeles – 1992

During the spring of 1992, Soldiers from the 40th Infantry Division, California National Guard were among the forces deployed to Los Angeles County to assist the California Highway Patrol, Los Angeles County Sheriffs, and civilian law enforcement. They worked to quell the riots that were sparked by the "not guilty" verdicts concerning four police officers who, following a lengthy high-speed chase through Los Angeles, were accused of brutally beating Rodney King.

Successful accomplishment of this civil support operation was attributed to the exercise of strong Army leadership and judgment at lower tactical levels, particularly among the unit's noncommissioned officers. An essential component of combat power, it was especially critical in executing noncontiguous and decentralized operations in the compartmented terrain of Los Angeles. As important, however, was the clear understanding that Army forces were to support civilian law enforcement—and not the other way around. The 40th Infantry Division aligned its area of operations with local law enforcement boundaries and relied heavily on police recommendations for the level at which Soldiers be armed (the need for magazines to be locked in weapons or rounds chambered).

One incident emphasized the need for coordination of command and control measures with civilian agencies even at the lowest tactical levels. To civilian law enforcement and Army forces, the command "Cover me" was interpreted the same: be prepared to shoot if necessary. However, when a police officer responding to a complaint of domestic abuse issued that command to an accompanying squad of Marines, they responded by immediately providing a supporting base of fire that narrowly missed some children at home. However, the Marines responded as they had been trained. This command meant something entirely different to them than for Army Soldiers and civilian law enforcement. Again, coordination at all levels is critical to the success of the operation (see also the vignette in Appendix B).

9-35. In the constraints imposed by METT-TC and operations security (OPSEC), commanders should seek to coordinate all tactical stability operations with other agencies and forces that share the urban environment. Importantly, they should seek to coordinate appropriate information and intelligence sharing with participating organizations. Commanders must strive to overcome difficulties, such as mutual suspicion, different values and motivations, and varying methods of organization and execution. Frequently, they must initiate cooperative efforts with participating civilian agencies and determine where their objectives and plans complement or conflict with those agencies. Commanders can then match Army force capabilities to the needs of the supported agencies. In situations leading to many urban civil support operations, confusion may initially make it difficult to ascertain specific priority requirements. Reconnaissance and liaison elements—heavily weighted with CA, engineers, and health support personnel—may need to be deployed immediately to determine what type of support Army forces provide. Overall, aggressive coordination will foster trust and make unity of effort possible in urban stability or civil support operations where unity of command is difficult or impossible to achieve.

Perseverance

9-36. The society is a major factor responsible for increasing the overall duration of urban operations. This particularly applies to urban stability operations where success often depends on changing people's fundamental beliefs and subsequent actions. Modifying behavior requires influence, sometimes with coercion or control, and perseverance. The urban population often must be convinced or persuaded to accept change. Necessary change may take as long as or longer than the evolution of the conflict. Decades of problems and their consequences cannot be immediately corrected. Frequently, the affected segments of the urban society must see that change is lasting and basic problems are being effectively addressed.

9-37. In most stability operations, success will not occur unless the host nation, not Army forces, ultimately prevails. The host urban administration must address the underlying problem or revise its policies toward the disaffected portions of the urban population. Otherwise, apparent successes will be short lived. The UO fundamental of understanding the human dimension is of paramount importance in applying this consideration. After all Army forces, particularly commanders and staff of major operations, understand the society's history and culture, they can begin to accurately identify the problem, understand root causes, quickly engage and assist key civilian leadership, and, overall, plan and execute successful Army UO.

Reasonable Restraint

9-38. Unlike offensive and defensive operations where commanders seek to apply overwhelming combat power at decisive points, restraint is more essential to success in urban stability and civil support operations. It involves employing combat power selectively, discriminately, and precisely (yet still at decisive points) in accordance with assigned missions and prescribed legal and policy limitations. Similar to the UO fundamentals of minimizing collateral damage and preserving critical infrastructure, restraint entails restrictions on using force. Commanders of major operations should issue or supplement ROE to guide the tactical application of combat power. Excessively or arbitrarily using force is never justified or tolerated by Army forces. Even unintentionally injuring or killing inhabitants and inadvertently destroying their property and infrastructure lessens legitimacy and the urban population's sympathy and support. Collateral damage may even cause some inhabitants to become hostile. In urban stability or civil support operations, even force against a violent opponent is minimized. Undue force often leads to commanders applying ever-increasing force to achieve the same results.

9-39. Although restraint is essential, Army forces, primarily during urban stability operations, must always be capable of decisive (albeit in some circumstances, limited) combat operations. This is in accordance with the UO fundamental of maintaining a close combat capability. This capability must be present and visible, yet displayed in a nonthreatening manner. A commander's intent normally includes demonstrating strength and resolve without provoking an unintended response. Army forces must be capable of moving quickly through the urban area and available on short notice. When necessary, Army forces must be prepared to apply combat power rapidly, forcefully, and decisively to prevent, end, or deter urban confrontations. Keeping this deterrent viable requires readiness, constant training, and rehearsals. It also requires active reconnaissance, superb OPSEC, a combined arms team, and timely and accurate intelligence.

Restraint: An Najaf, Iraq – 2003

On 31 March 2003, during OPERATION IRAQI FREEDOM, The 101st Airborne (Air Assault) Division transitioned from isolating An Najaf—among the holiest places for Shia Muslims in Iraq and the home of one of their leading holy men, the Grand Ayatollah Ali Hussein Sistani—to an attack to clear Iraqi fighters from the town. The decision to attack was made in response to increasing attacks against 101st forces situated near An Najaf and due to alarming reports that the Fedayeen were killing Iraqi family members to force adult males to fight coalition forces. Since An Najaf had an airfield, control of the town would also allow the 101st to obtain needed hardstands for their aircraft.

Ultimately, the division used two brigades to attack and clear the town. The 2nd Brigade Combat Team (BCT) attacked from the north and the 1st BCT—responsible for the initial penetration—attacked from the southwest. Using tanks, infantry fighting vehicles, and light infantry, the commander of the 1st BCT formed effective combined arms teams supported by artillery and air that successfully fought their way into the city.

On 3 April 2003, Soldiers of TF 2-327 Infantry turned a street corner to face a group of civilian men blocking their way and shouting in Arabic, "God is great." The crowd quickly grew into hundreds as they mistakenly thought the Soldiers were trying to capture Sistani and seize the Imam Ali Mosque. Someone in the crowd threw a rock at the Soldiers, which started a hailstorm of rocks; even the battalion commander was hit on the head, chest, and the corner of his sunglasses.

Based on information from a Free-Iraqi Fighter accompanying the battalion and his own observations during the attack, the commander of TF 2-327 believed that most of the people in An Najaf neither wanted to fight nor obstruct his unit's efforts. Consequently, he ordered his Soldiers to position themselves on one knee, smile, and point their weapons toward the ground. At this gesture, many of the Iraqis backed off and sat down, enabling the commander to identify the true troublemakers. He identified eight. In case these agitators were to produce weapons and start to shoot, the commander wanted to make sure that the remainder of the crowd would know from where the shooting would originate. Next, he ordered his Soldiers to withdraw to allow the tension to subside. With his own rifle pointed toward the ground, the battalion commander bowed to the crowd and led his Soldiers away. When tempers had calmed, the Grand Ayatollah Sistani issued a decree (fatwa) calling on the people of Najaf to welcome the American Soldiers.

The attack on An Najaf by the 101st provides an excellent example of well-led Soldiers capable of understanding cultural differences and possessing the discipline and mental flexibility to transition from the aggressive mindset required for high-intensity urban combat to the restraint essential for the initiation of stability operations. In his weekly radio address, President George W. Bush commented: "This gesture of respect helped defuse a dangerous situation and made our peaceful intentions clear."

CONSOLIDATE

9-40. As urban operations will often be full spectrum, many of the consolidation activities necessary to secure gains in urban offensive and defensive operations will be applicable to urban stability and civil support operations. However, the greatest obstacles to attaining strategic objectives will come after major urban combat operations. Therefore, emphasis will appropriately shift from actions to ensure the defeat of threat forces to those measures that address the needs of the urban population, manage their perceptions, and allow responsibility to shift from Army forces to legitimate indigenous civilian control (or the intermediate step to other military forces, governmental agencies, and organizations).

Continued Civilian and Infrastructure Protection

9-41. Following urban offensive or defensive operations, there will likely be a need to secure and protect the civilian population and much of the civilian infrastructure from the civilians themselves. After having minimized collateral damage and preserved critical infrastructure, commanders must implement measures to preclude looting and destruction of critical and essential infrastructure by the urban population and civilian-on-civilian violence. This may be as relatively simple as allowing the urban police force to return to work or may be as difficult as hiring, vetting, and training an indigenous police force. In the latter case commanders may need to determine—

- Number and operability of police stations.
- Responsibility for recruiting, hiring, training, and equipping the urban security or police force. The accountable unit, organization, or agency will need to consider a vetting process, suitable salaries and wages, and appropriate training standards.
- The appropriate responses toward those civilians who threaten, oppose, or harm the new police force.

9-42. Alternatively, civilian security firms (from inside or outside the urban area or country) can provide supplemental protection until indigenous police forces are fully functional. Further, the commander of the major urban operation may plan to manage expected instability primarily with Army forces. Often, this will require larger numbers of infantry, military police, and dismounted forces. Other populace and resource control measures such as curfews may also assist in civilian (including NGOs) and infrastructure protection. Previous shaping operations aimed at improving the local economy can also assist in this regard.

Resolute Legitimacy

9-43. Closely linked to the restraint described previously is legitimacy or the proper exercise of authority for reasonable purposes. Achieving or maintaining legitimacy during urban stability or civil support operations is essential in gaining and maintaining the support of the urban population. Commanders can ensure legitimacy by building consent among the population, projecting a credible force, and appropriately using that force. Perceptions play a key role in legitimacy, and skillful IO can shape perceptions. Commanders must send messages that are consistent with the actions of their forces. Generally, the urban population will accept violence for proper purposes if that force is used impartially. Perceptions that force is excessive or that certain groups are being favored over others can erode legitimacy and generate resentment, resistance, and, in some situations, violent acts of revenge.

9-44. In stability operations against an elusive insurgent threat, commanders may need to explain to urban residents why damage was necessary, apologize, or make near-instant restitution for some unsuccessful Army operations that may have been planned based on inaccurate or incomplete intelligence. (Soldiers may even make it a point to thank homeowners for allowing the search of their homes during cordon and search operations.) During urban operations, a single Soldier's misbehavior can significantly degrade a commander's ability to project an image of impartiality and legitimacy. Fortunately, disciplined Soldiers can contribute immeasurably to gaining and maintaining legitimacy, mitigating ill will, or otherwise winning the urban population's trust and confidence. In stability operations, the greater fight will often be to win the battle of perceptions and ideas instead of one to seize terrain and triumph over an enemy. Inconsistencies in message and behavior will provide threats with raw material for their propaganda and precipitate doubt in the minds of the urban populace who might otherwise support Army objectives.

TRANSITION

9-45. Commanders of major operations are the focal point for synchronizing tactical stability operations or civil support operations with strategic diplomatic and political issues. They are also the critical links between national intelligence resources and the tactical commander. Because strategic, diplomatic, and political changes can quickly transition the type of urban operation, they must keep subordinate tactical commanders abreast of changes in intelligence, policy, and higher decisions. The potential to rapidly transition to urban combat operations emphasizes the need to maintain the capability to conduct close, urban combat. Failure to recognize changes and transition points may lead to UO that do not support the attainment of the overall objective and needlessly use resources, particularly Soldiers' lives. Therefore, Army forces on the ground in an urban stability operation must be more aware of the strategic environment than the threat and the civilian population, each of whom will have their own means of monitoring the national and international situation.

Legitimate and Capable Civilian Control

> *I met immediately with the [special operations forces liaison officer] to discuss how we would establish a new government that would be able to make instant progress. We decided to select a city council with the following departments: commerce, public works, social services, health, emergency services, education, public relations, and agriculture. A city council director would lead the council, and each department would also have a director in charge of its respective services. A critical step in selecting the city council was ensuring that there was a balance on the council between Altun Kupri's three major ethnic groups. An ethnic imbalance on the council or perceived favoritism would have decreased the legitimacy of the Coalition forces' efforts in the city and served as fuel for the fire of competition for power between the Turkomen and Kurds.*
>
> Captain Jeffrey B. Van Sickle
> "Stability Operations in Northern Iraq"

9-46. Commanders must maintain or enhance the credibility and legitimacy of the government and police of the urban area and, in the case of stability operations, of the host nation's military forces operating there. In accordance with the urban fundamental of transitioning control, urban commanders must conclude UO quickly and successfully, often to use assets elsewhere in their area of operations. This entails returning the control of the urban area back to civilian responsibility as soon as feasible. The host nation's military and the urban area's leadership and police are integrated into all aspects of the urban stability or civil support operations to maintain their legitimacy. They must be allowed (or influenced) to take the lead in developing and implementing solutions to their own problems. This requires commanders to transition from "leading from the front" to "leading from behind" in an advisory and assistance role. Effective transition to civilian control and responsibility requires commanders at all echelons who understand the basic operation of civil governments and the administration and management of key urban infrastructure.

9-47. If the host nation's leadership, military, and police are not up to the task, commanders can take steps to increase its capabilities through training, advice, and assistance by CA units or by other nongovernmental or governmental organizations and agencies. Sometimes, new leadership and a restructured police force may be required, particularly when corrupt and no longer trusted by the population. This candid assessment of the urban leadership's ability to govern, protect, and support itself is made early in the planning process. Only then can commanders ensure that resources and a well thought-out and coordinated plan (particularly with civilian organizations and agencies) are available for a speedy transition. IO will be paramount in these instances to ensure that the urban population sees the training and rebuilding process itself as legitimate. Throughout urban stability or civil support operations, commanders shape the conditions to successfully hand over all activities to urban civilian authorities.

Longer-Term Commitment

Maintain the Focus

9-48. Many stability operations often require perseverance and a longer-term U.S. commitment requiring operational endurance and the establishment of a sustainable battle rhythm. Lengthy operations will also require a transitional rotation of Army units into the area of operations to continue the mission. Considerations for these transitions are similar to a relief in place and battle handover (see FM 3-90) combined with considerations for deployment and redeployment. FM 41-10 contains comprehensive guidance on transition planning and coordination activities applicable to UO. In addition to any threat considerations, planning for urban transitions between units will often include emphasis and understanding of—

- Formal and informal civilian leadership and relationships.
- Government institutions and administrative functions.
- On-going reconstruction projects.
- The urban economy.
- Participating governmental and nongovernmental organizations and established relationships and cooperation activities, particularly information sharing.
- Significant key events affecting (or likely to affect) operations.
- Significant cultural lessons learned as a result of the outgoing unit's operations.

9-49. The commander of the major operation must ensure that the incoming unit also understands the political and strategic objectives behind the tasks that they will be required to accomplish. Otherwise, the new unit may begin to plan operations that are similar to those conducted by the previous unit without achieving the desired end state or accomplishing the mission.

Transition Trust

9-50. If units are successful in establishing and cultivating relationships with the urban populace and their leadership, and have built an effective level of trust, their anticipated departure from the AO as part of a unit rotation schedule may have detrimental effects on working relationships. As a result, local leaders may develop deep anxieties about the replacement force and their ability to "be as good" as the current command. Therefore, incumbent forces must take steps to identify successful and ongoing efforts and activities in which the new unit can continue to achieve success. In some instances, the current unit may delay a project or activity whose scheduled completion is close to the date of relief or the transfer of authority so as to allow the new unit immediate success. In other instances, projects or activities can be backward planned to ensure that the completion date is scheduled soon after the new unit arrives. The goal for the outgoing unit is to ensure that the new unit is readily accepted and that both units execute a seamless relief in place.

Modify Objectives to Match the Current Environment

9-51. While it is important that incoming commanders create a seamless transition with their outgoing counterparts, they must make full use of the opportunity to review the current political, strategic, and local urban environment to determine potential modifications to logical lines of operations and unit objectives. Otherwise, units may fall into the trap of executing a series of six-month to one-year rotations that do not significantly contribute to solidifying the conditions required to ultimately transition the urban environment back to legitimate civilian control. In operations that involve longer-term commitments, the potential exists for the same unit to conduct two (or more) rotations back into the same AO. Commanders must resist the desire to simply continue with previous lines of operations that, at the time, were deemed successful. The urban environment will change in response to Army actions and this change will necessitate subsequent operational adaptations.

Provide an End State Not an End Date

9-52. Though it may not be feasible, due to political constraints, commanders of major urban operations should fight the public desire to establish a fixed transition timetable. Instead, they should plan urban stability operations based on an attainable end state and not on a specified withdrawal or end date. Progress toward strategic (and operational) goals is susceptible to many changes and delay, particularly in coalitions which will likely be a part of most future urban operations. As such, an end date will only serve to unduly constrain commanders in the attainment of their urban operational objectives. Having set and announced a firm timetable, American and coalition publics will expect withdrawal to commence as scheduled and will not easily be convinced of the need to stay longer even if the situation changes drastically. In general, the public can understand and accept a logical and clearly articulated end state as long as commanders can demonstrate adequate progress based on well developed measures of effectiveness. As important, a timetable allows threats to adjust their plans to the anticipated withdrawal date. Instead of yielding to relentless diplomatic, informational, military, and economic pressure backed by firm political resolve, threats will likely exercise strategic and operational patience.

This page intentionally left blank.

Chapter 10
Urban Sustainment

Sustainment activities exist to enable the Army to initiate and sustain full spectrum operations. Inseparable from decisive and shaping operations, logistics is the central component of sustaining operations and provides the means for commanders to generate and maintain combat power and extend the operational reach of the force. In offensive and defensive operations, sustainment activities are not by themselves likely to be decisive or shaping, yet significantly contribute to those operations. In stability and civil support operations, when the critical objectives may be restoring the infrastructure and the welfare of civilians, sustainment activities will often be decisive. Their success will allow Army forces to dominate this complex environment. However, like all urban operations (UO), sustainment activities affect and are affected by the environment. The urban terrain, infrastructure, and existing resources, coupled with supportive civilians, may facilitate sustainment activities. In contrast, a poorly designed or damaged infrastructure and a hostile population may severely hamper sustainment activities. In the latter case, critical Army resources required elsewhere in the area of operations (AO) may be diverted to repair facilities and control and support the inhabitants of the urban areas.

URBAN SUSTAINMENT CHARACTERISTICS

10-1. Fundamental sustainment characteristics (see figure 10-1) guide prudent sustainment planning regardless of the environment. They provide commanders an excellent framework to analyze and help develop urban sustainment requirements, assess the impact of the environment on the provision of logistics, and gauge the effectiveness of urban sustainment activities.

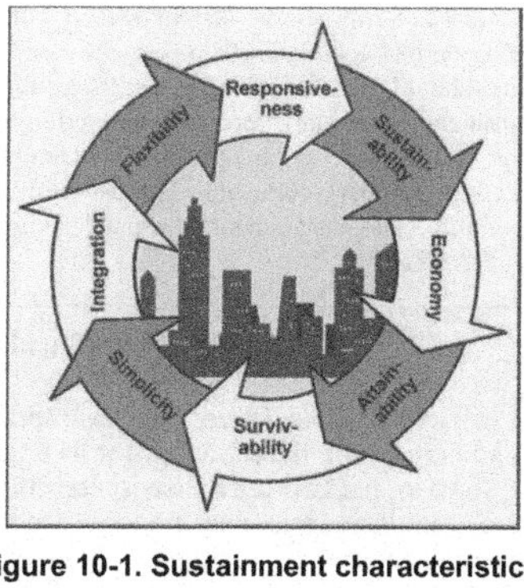

Figure 10-1. Sustainment characteristics

Responsiveness and Sustainability

10-2. UO require responsiveness and sustainability to establish and maintain the tempo necessary for success. Responsiveness—providing the right support in the right place at the right time—is the essential sustainment characteristic. It requires that sustainment commanders and planners accurately forecast urban operational requirements. Continuous urban operations will drain personnel, equipment, and supplies at rates vastly different than in other environments. Therefore, sustainability—the ability to maintain continuous support throughout all phases of the operation—will be a significant concern. Anticipation is critical to both responsiveness and sustainability. It requires that sustainment commanders and planners comprehend the potential effects that components of the urban environment (terrain, infrastructure, and society) may have on operations and sustainment, either benefiting or impeding UO. Effective urban operational and sustainment planning cannot be accomplished separately. Operational and sustainment planners, as well as sustainment operators, must be closely linked to aid in synchronizing and attaining responsiveness and sustainability. The key is often the right balance between efficiency and effectiveness.

Economy and Attainability

10-3. A thoughtful assessment and understanding of the urban environment can also help determine how specific urban areas can contribute to or frustrate the achievement of economy and attainability. Economy is providing the most efficient support at the least cost to accomplish the mission. Attainability means generating the minimum essential supplies and services necessary to begin operations. If available, obtaining support in the AO costs less than purchasing the supplies outside the area and then transporting them there. Critical resources may be available in urban areas to support the operation. However, relying on sources outside the established military sustainment system may create conflict with other sustainment characteristics. A strike by longshoremen, for example, may shut down port operations (at least temporarily) lowering responsiveness and sustainability.

Survivability

10-4. Survivability is being able to protect support functions from destruction or degradation. Commanders often choose to locate sustainment functions in an urban area because the buildings may better protect and conceal equipment, supplies, and people. Urban industrial areas and airports are frequently chosen as support areas because they offer this protection as well as sizeable warehouses, large parking areas, and materials handling equipment (MHE). Such areas facilitate the storage and movement of equipment and supplies. They may also provide readily available water, electricity, and other potentially useful urban resources and infrastructure. However, these areas may also contain toxic industrial materials (TIM) (see Chapter 2). These materials and chemicals in close proximity to support areas may unjustifiably increase the risk to survivability. Sustainment activities in any environment will always be targeted by threat forces. Furthermore, sustainment activities located in any type of confined urban area can offer lucrative targets for terrorists, insurgents, or even angry crowds and mobs. Therefore, no sustainment activity should be considered safe from attack during UO. (During OPERATION JUST CAUSE, Panamanian paramilitary forces and deserters even attacked marked ambulances). Although host-nation support may include assets to assist in defending sustainment units, bases, and lines of communications (LOCs), sustainment commanders must carefully consider if adequate protection measures can ensure survivability. The sustainment commander's greatest challenge to force protection may be complacency born of routine.

Base Security: Tan Son Nhut, Vietnam – 1968

Colonel Nam Truyen was the commander of the 9th Vietcong Division who planned and conducted the attack on the U.S. airbase at Tan Son Nhut during the 1968 Tet Offensive. He had previously entered the airbase during the 1967 Christmas cease-fire using forged identity papers to conduct his own personal and highly thorough reconnaissance.

> UO increase the likelihood that the indigenous population will be utilized as a workforce to supplement Army sustainment activities. Hence, strict operations security will be paramount. Commanders will need to carefully screen potential workers to determine their relative trustworthiness. Human intelligence (HUMINT) and counterintelligence (CI) assets will likely be required to assist in this effort. However, higher-level commanders should seek to take on as much of this burden as possible so that lower-level, tactical commanders can use their limited intelligence assets to accomplish other critical intelligence collection and analysis activities.
> All commanders with base protection responsibilities should implement random, changing force protection measures that focus on civilians—even those civilians who have, over time, earned a perceived measure of trust. (Civilians can be co-opted and coerced by threat forces.) As such, all civilian activity that seems out of the ordinary should be promptly investigated regardless of whether a specific force protection measure is in effect. Although leaders and Soldiers seek to establish mutual trust and respect between themselves and the local populace, commanders should normally err on the side of protecting their Soldiers.

SIMPLICITY

10-5. Simplicity is required in both planning and executing sustainment activities in this complex environment. Developing standard procedures among the Army, other services, and especially civilian governmental and nongovernmental agencies, liaison and open channels of communication, simple plans and orders, and extensive rehearsals contribute immeasurably to attaining this necessary characteristic.

INTEGRATION

10-6. The need for sustainment integration increases in UO due to its joint and multinational nature and greater numbers of other governmental and nongovernmental organizations (NGOs) and agencies operating in or near urban areas. More NGOs will likely exist because urban areas often contain most of a region's population. Most NGOs focus on people. Army forces and other military and nonmilitary groups must cooperate and coordinate their actions. Much of their coordination will revolve around sustainment activities. Cooperation and coordination will take advantage of each group's sustainment capabilities, help to avoid duplicated effort (contributing to economy), and create sustainment synergy. It will also help to curtail competition for the same urban resources and assist in developing a unified list of priorities. Such coordination will help ensure that other operations by one force or agency will not disrupt or destroy portions of the urban infrastructure critical to another's sustainment operations and the overall mission.

10-7. Success in UO requires combined arms and innovative task organizations. As such, UO will often result in the integration of heavy and light units necessitating a simultaneous integrated approach to sustainment. Gaining units must integrate forces into their maintenance systems and carefully consider potentially increased resource requirements. (For example, armor units consume greater amounts of Class III—bulk and petroleum—than a similar-sized light unit.) When integrating into a light unit, armored and mechanized units should bring as much of their critical parts and supplies as possible and immediately integrate into the gaining unit's logistical reporting system to ensure adequate resupply.

FLEXIBILITY

10-8. Lastly, commanders must develop flexibility. Although they and their staffs must thoroughly understand the urban environment essential to planning sustainment operations, they cannot possibly anticipate every eventuality. Urban commanders must possess the ability to exploit fleeting opportunities. Knowledge of the environment, particularly its infrastructure, can aid in developing innovative solutions to sustainment acquisition and distribution problems. Flexibility enables sustainment personnel to remain adaptive and responsive to the force commander's needs. Key to maintaining flexibility is the constant assessment of the sustainment situation and the readiness to modify or change procedures to adapt to the current conditions confronting Army forces operating within the urban environment.

10-9. Maneuver and sustainment commanders should consider and prioritize these characteristics as they visualize UO. Each characteristic does not affect every operation and urban area in the same way. The sustainment characteristics seldom exert equal influence, and their importance varies according to mission, enemy, terrain and weather, troops and support available, time available, civil considerations (METT-TC). Like the principles of war, commanders must not ignore the potential impact of sustainment characteristics and how their influence changes as the operation evolves (see FM 4-0).

LOGISTICS PREPARATION OF THE THEATER

[Sustainment] planning accounts for increased consumption, increased threats to lines of communications, and anticipated support to noncombatants...Urban operations place a premium on closely coordinated, combined arms teams and carefully protected [sustainment activities]. Urban operations are [logistic]-intensive, demanding large quantities of material and support for military forces and noncombatants displaced by operations.

FM 3-0

10-10. A thorough logistics preparation of the theater (LPT) is critical for an adaptable UO support plan. Sustainment planners conduct the LPT to understand the situation from a sustainment perspective and determine how best to support the maneuver commander's plan. However, decisions that impact the urban environment and the political situation require a combined operational and sustainment perspective. Sustainment planners must understand the urban environment, the fundamentals of UO, the proposed course of action, and the urban environment's effects on sustainment (as well as the other warfighting functions). Such knowledge allows the planners to develop a detailed estimate of support requirements. A thorough LPT helps commanders determine the most effective method of providing adequate, responsive support to meet support estimates while minimizing the sustainment footprint. Overall, it helps tie together UO requirements with acquisition and distribution. As with all operations, but particularly in a dynamic urban environment, this assessment process is continuous since requirements will change as the urban operation unfolds and matures.

SUPPORT TO INTELLIGENCE

10-11. The LPT resembles and runs parallel to the intelligence preparation of the battlefield (IPB). Products generated under IPB may be useful in the sustainment analysis. Conversely, the LPT may contribute to the IPB by identifying critical resources and infrastructure and assessing their potential to influence (positively or negatively) the operational plan. This information may warrant a course of action that includes offensive or defensive operations to seize, secure, or destroy those critical resources. In UO initially planned for other than sustainment reasons, the information may require altering the plan or imposing additional constraints to protect the identified resources. While these resources may or may not be critical to current operations, they are usually important to set or shape the conditions necessary for Army forces to consolidate and transition to subsequent missions or redeploy. This close relationship between IPB and LPT underscores the need to quickly and continuously involve sustainment personnel for their sustainment expertise and perspective in planning UO.

10-12. As sustainment units are among the most dispersed and omnipresent in any AO, sustainment Soldiers are potentially valuable intelligence collectors. Soldiers working in and around urban sustainment bases and executing frequent convoys through urban areas quickly become very familiar with the urban terrain and the routine associated with that terrain. As with all Soldiers, sustainment personnel must be incorporated into a comprehensive human intelligence collection system.

URBAN SUSTAINMENT INFORMATION

10-13. Figure 10-2 illustrates that a thorough analysis of the key components of urban areas in the commander's AO provides the data for an accurate LPT and subsequent UO sustainment plan (see Chapter 2 and Appendix B). Analyzing the urban terrain and infrastructure helps to determine—

- Geographic influences on consumption factors and on the provision of support (weather, climate, and topography).
- The availability of supplies, such as safe food, potable water, petroleum, electrical energy, barrier material, and compatible repair parts.
- The location of facilities, such as warehouses, cold-storage sites, water treatment facilities, manufacturing plants, hospitals, hotels for billeting, and waste treatment facilities.
- Transportation information, such as seaport and harbor facilities, airfields, rail and road networks, traffic flow patterns, choke points, and control problems.
- Locations and accessibility of maintenance facilities and equipment, and machine works for the possible fabrication of parts.
- The available general skills among the urban population, such as linguists, drivers, mechanics, MHE operators, longshoremen, and other vital trade skills.

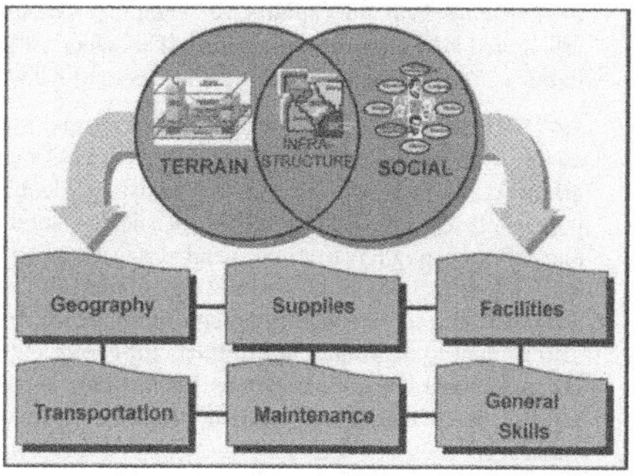

Figure 10-2. Urban environment and essential elements of sustainment information

POTENTIAL RESTRICTIONS

10-14. Commanders must be aware of restrictions that apply to the use of some non-U.S. resources. Security and requirements for U.S. national control dictate that only U.S. assets may perform certain services and functions. Therefore, some foreign urban-area capabilities, even if abundantly available, may not be used. These might include—

- Command and control of medical supply, service, maintenance, replacements, and communications.
- Triage of casualties for evacuation.
- Treatment of chemical, biological, radiological, nuclear, and nuclear (CBRN) casualties, as well as the decontamination of U.S. equipment, personnel, and remains.
- Identification and burial of deceased U.S. personnel.
- Veterinary subsistence inspection.
- Law and order operations over U.S. forces and U.S. military prisoner confinement operations, as well as accountability and security of enemy prisoners of war in U.S. custody.

URBAN SOCIETAL CONSIDERATIONS

10-15. As in all aspects of UO, the urban society is a critical element of the LPT analysis. Sustainment planners must assess whether they can acquire and use urban resources and infrastructure without overly disrupting the urban society and their environment. If the resources are only sufficient for the urban area's inhabitants (and dependent populations in outlying areas), and the facilities cannot increase production to accommodate the needs of Army forces, then commanders may not be able to rely on those resources to support their operations. In fact, the opposite may be true. The effects of UO on the inhabitants, particularly during offensive and defensive operations, may place increased burdens on the Army's resources. For example, commanders may require civilians to evacuate all or parts of an urban area to safeguard them from the effects of planned combat operations. This precautionary measure may require Army forces to support civilian evacuees with temporary housing, food, water, clothing, and medical and sanitary facilities. It will also require carefully planned evacuation routes to ensure that the movement does not interfere with sustainment of the fight.

10-16. Sustainment planners must take into account the potential effects that the provision of supplies to the urban population may have on the operational environment (provided by Army forces or not). For example, an embargo on petroleum products was lifted after Army forces entered Haiti in the 1990s and a less repressive regime assumed governmental control. Few civilian vehicles had moved on the streets of the Port-au-Prince, the capital city, when fuel was scarce and the movement of mounted forces through the urban area was virtually unimpeded. This changed almost overnight once fuel supplies became available, initially making it very difficult for forces to move where required (unless transported by helicopter).

10-17. Sustainment planners must also consider the urban society's ability to restore their own facilities and provide for themselves (if necessary with assistance from Army forces). Throughout this analysis, civil affairs (CA) units can advise and assist in identifying and assessing urban supply systems, services, personnel, resources, and facilities. Critically, commanders must understand that purchasing local goods and services may have the unintended negative consequence of financially sustaining the most disruptive and violent factions in the area. Army forces must seek to purchase urban products and services that will not contribute to prolonging the conflict or crisis. In many stability or civil support operations, they must also attempt to distribute the contracts for goods and services purchased locally as fairly as possible among urban factions and ethnic groups to maintain impartiality and legitimacy. As part of their coordination efforts, commanders should attempt to achieve the cooperation of relief agencies and other NGOs in this endeavor.

10-18. Finally, sustainment planners must also identify potential threats and increased protection requirements that the urban society (criminals, gangs, and riotous mobs) may present, particularly when support units and activities are positioned in urban areas. The disposition or allegiance of the urban population is also important to consider. The infrastructure of an urban area may exhibit great potential to support the sustainment efforts of Army forces, but if the population is hostile or unreliable, use of the facilities and resources may be unfeasible.

URBAN SUPPORT AREAS AND BASES

10-19. A major influence on the operation plan and its subsequent execution is often the proper identification and preparation of support areas and bases. The LPT helps commanders determine the need, advantages, and disadvantages of using urban areas in the AO as bases from which to provide support and conduct distribution operations. Ideally, these areas should support reception, staging, onward movement, and integration operations. They should also allow easy sea and air access, offer adequate protection and storage space, increase the throughput of supplies and equipment, and be accessible to multiple LOCs. Consequently, commanders often establish support areas and bases near seaports and airports that are part of a larger urban area. However, threats recognize the Army's need for ports and airfields and may devote substantial resources and combat power to defend them. Army commanders also face the challenge of integrating airfield and port defensive operations with air and naval component commanders to deny threats the ability to conduct stand-off attacks against aircraft and ships. Only an integrated, joint approach can ensure persistent air and sea support– essential not only to sustainment but to firepower as well—is not degraded by the threat. Therefore, planners may determine during the LPT that the risks of seizing or establishing urban lodgment areas may be too high (see Chapter 4). Instead, they may recommend building an airfield, conducting logistics over-the-shore operations, or constructing sustainment bases in more isolated—but more easily secured—locations.

OVERALL ASSESSMENT

10-20. As shown above, the LPT process and analysis help to determine if urban areas in the AO—

- Are suitable as areas and bases for support.
- Can contribute sufficient quantities of and are a dependable source for resources for the overall operation.
- May additionally drain the supported commander's resources.

10-21. The results of this process often serve as a basis for reviewing requirements for civilian contract support and host-nation support and for developing sustainment input into time-phased force and deployment data. This chapter focuses on the effects urban areas may have on accomplishing sustainment functions and related activities, particularly when sustainment units and activities are in urban areas.

ACHIEVING FORCE AGILITY

10-22. To maintain the responsiveness necessary to support full spectrum operations against a wide array of threats and operational environments, sustainment forces must remain agile, possessing the mental and physical agility to transition within or between types of operation with minimal augmentation, no break in contact, and no significant additional training. Responsiveness, flexibility, and economy are key characteristics that enable sustainment forces to support a dynamic combat force; they represent the capability of sustainment forces to provide the resources to initiate, sustain, and extend UO. Agile sustainment forces are characterized by modular force designs, the ability to tailor sustainment organizations for the supporting mission, and the ability to conduct split-based operations. These sustainment aspects are critical to achieving the agility necessary for Army forces to succeed in UO.

MODULAR DESIGN

10-23. Sustainment units are structured as modular organizations, or must possess the ability to modularize in order to conduct decentralized sustainment operations required in UO. At the company level, modularity enables each major company sub-element to assume a cross-section of the company's total capabilities, allowing commanders to employ individual modules to provide a support function, while the rest of the unit remains operational. Modularity enhances responsiveness while enabling support forces to project capabilities as far forward as possible into urban areas, yet with the minimum forward footprint necessary to sustain operations.

FORCE TAILORING

10-24. Sustainment force tailoring refers to the process of determining and deploying the right mix of capabilities to support the force or mission. Sustainment commanders may task organize specific capabilities into functional elements in order to maximize effectiveness and efficiency, and to minimize the sustainment footprint. During UO, the sustainment commander can tailor the support element required to accomplish a specific mission or task, thereby mitigating the risk associated with deploying a larger, more robust capability package forward into the urban area.

SPLIT-BASED OPERATIONS

10-25. Split-based operations refer to performing certain sustainment administrative and management functions outside the area of operations. During UO, Soldiers and civilians can perform personnel, materiel, and distribution management functions without deploying forward into the urban area if the information systems are adequate. This is essential to minimizing risk to support forces and the sustainment footprint in the urban area, and still ensures that all support requirements are met.

SUSTAINMENT FUNCTIONS

- Supply and Field Services
- Maintenance Support
- Transportation Support
- Force Health Protection
- Explosive Ordnance Disposal Support
- Human Resources Support
- Financial Management Support
- Religious Support
- Legal Support to Operations
- General Engineering Support

Figure 10-3. Sustainment functions

SUSTAINMENT FUNCTIONS

10-26. Sustainment consists of eleven functions that must be carefully planned, managed, and synchronized to provide responsive and efficient support to maneuver commanders (see figure 10-3).

Similar to the components of the urban environment, particularly its infrastructure, they overlap and are interdependent. The success of one function depends on the success of several others. Like urban infrastructure, they have two components: a physical component (supplies, equipment, and facilities) and a human component (the Soldiers and civilians who execute these functions). Like city mayors, commanders must plan, manage, and synchronize these functions to provide responsive and efficient sustainment to UO.

10-27. Commanders and planners should consider two essential aspects when addressing these sustainment functions. One aspect looks outward and one looks inward. The first aspect is how these functions can best support full-spectrum UO—the outward analysis. The second aspect is how the urban area affects the conduct of sustainment functions, particularly when those functions are located or performed in an urban area—the inward analysis.

SUPPLY AND FIELD SERVICES

10-28. The supply function involves acquiring, managing, receiving, storing, protecting, maintaining, salvaging, and distributing all classes of supply (except Class VIII) required to equip and sustain Army (and joint) forces. For UO, commanders may need to decide early in the planning cycle whether to stockpile resources forward or to rely on distribution-based sustainment management to satisfy requirements. Some specialized items identified below may not be available through the normal military supply system and may take sustainment personnel much longer to obtain or fabricate. Operation planners must quickly identify the special equipment and increased supply requirements for UO to give sustainment personnel the time necessary to acquire them.

10-29. Field services are those essential services required to enhance a Soldier's quality of life. They consist of clothing exchange, laundry and shower support, textile repair, mortuary affairs, preparation for aerial delivery, food services, billeting, and sanitation. The urban commander determines the need and priority of each service after careful METT-TC analysis. Some facilities such as shower, laundry, and cold storage may be available in the urban area through theater support contracting or provided by external support contracts such as the logistics civilian augmentation program (LOGCAP). Additionally, requirements to care for the urban population will increase requirements for field services immensely. In some circumstances, most notably urban stability or civil support operations, adequately protected field service units or activities will be critical and may be the only support provided.

Increased Urban Supply Requirements

Water

10-30. Water production and distribution will become more difficult for sustainment forces during UO. (Many parts of an urban area can become virtual deserts if public water supplies are cut off or contaminated due to deliberate military action or natural disasters.) Consequently, sustainment managers will be called upon to find more efficient and effective means of moving and distributing potable water, especially when bulk storage of water within the urban area may not be possible or feasible. In sustainment bases that do not have access to urban or other water sources, the transportation and storage of non-potable water may be as great a concern as potable water. Unless adequate storage is available for both potable and non-potable water, units may be forced to rely solely on potable water for all bulk requirements to mitigate health risk. Due to the compartmented nature of many urban operations, planners may need to ensure that additional containers for water (and fuel) are available to support dispersed stock at small-unit level, and that man-portable systems are available in sufficient quantities to ensure timely replenishment.

Class II

10-31. Clothing and individual equipment exposed to the urban environment (concrete, glass, and steel) experience wear and damage at accelerated rates. Units will also require items such as rope, grappling hooks, crowbars, ladders, chain saws, elbow- and kneepads, special vehicle and personnel body armor, fire-fighting equipment, packboards, carpentry tools, and other specialized items to conduct UO. Planners must recognize this and plan accordingly to ensure operations are not interrupted due to equipment shortfalls.

Class III (Bulk)

10-32. With the potential exception of aviation fuel, bulk Class III requirements for UO generally decrease at the maneuver unit level. However, increased fuel requirements for engineer and power-generation equipment attached to or operating with forward units may offset these decreases. Generally, fuel sources within the urban area (refinement facilities, gas stations, garages, and airfields) should not be utilized unless tested and approved for use by a fuel laboratory.

Class IV

10-33. Requirements for construction and barrier material (and the power tools to cut it) during UO will likely far exceed the capability of the sustainment force to provide. Commanders must prioritize distribution, defensive projects, and construction and repair efforts in order to maximize the efficient use of available Class IV. Every effort should be made to obtain supplemental Class IV material from the local economy, as long as that effort does not adversely affect operations or the restoration and improvement of critical urban infrastructure. Of note, specialized, prefabricated road barriers will also be in great demand to block roads, create checkpoints, reinforce defensive positions, and protect headquarters and sustainment activities.

Class V

10-34. Urban combat operations will often increase requirements for Class V. Ammunition consumption rates for urban environments have been as much as five to ten times greater than operations in other environments. Urban offensive and defensive operations will often increase overall requirements for certain types of ammunition (crew-served and small arms ammunition, mortars, antitank rounds, grenades, mines, and demolitions) while others may typically decrease (tank, artillery, attack helicopter). Since forces operating on urban terrain may not be capable of storing ammunition, an established system of distribution is vital to ensuring a constant flow of ammunition necessary to sustain continuous operations and retain the initiative.

Class IX

10-35. Requirements for repair parts generally decrease in UO, with the notable exception of those systems that experience increased ammunition consumption. The heavy use of crew-served and small arms, mortar, and antitank systems in the urban area causes wear and failure at significantly increased rates. Additionally, wheeled-vehicle tires and armored-vehicle track pads and road arms, road wheels, center guides, wedge bolts, and track blocks may wear rapidly on urban streets, rubble, and debris. Based on the tempo of operations, it may be more feasible to replace whole sets of tracks rather than have crews replace individual track pads. Hence, commanders may consider resourcing all wheeled vehicles with spare tires and creating containerized and rapidly transportable sets of track and track parts for speedy delivery to urban sustainment bases. On the whole, planners should consider the use of repair parts packages specifically configured for UO that can be transported by tailored, modular sustainment elements that directly support operations within the urban environment. During long-term stability operations, outgoing units may leave large portions of their prescribed load list in theater to ensure that relieving units have adequate supplies to maintain the tempo of operations.

Storage and Distribution

10-36. UO present challenges to storage and distribution that may not be experienced outside of the urban environment. Forces operating within these areas may not be static and, therefore, cannot store equipment and resources beyond what they can transport organically. Even forces conducting operations from secure locations within the urban area may not be able to store materiel, necessitating a continuous, coordinated distribution system to ensure forces are adequately resourced. During UO, sustainment forces may need to utilize unit or throughput distribution techniques to provide the level of support required to sustain the force.

10-37. Urban commanders must also consider the benefits of stockpiling resources forward against relying on the Army's distribution-based sustainment system. Stockpiling brings supplies close to urban forces and helps ensure supplies are immediately available to support potentially increased consumption rates. However, this method may burden the support structure that moves, handles, and protects large quantities of resources often on a repetitive basis. The Army's normal distribution-based sustainment system reduces this burden significantly; however, available transportation assets (sea, air, and ground) impact delivery response times. These transportation assets often combine military assets overlaid on the host nation's (and urban area's) transportation and distribution infrastructure. Although the civilian infrastructure may initially support the Army's distribution system, later effects of UO, such as destruction of equipment and facilities or loss of civilian workers, may degrade the system. Moreover, Army forces may have to share these assets with other military, multinational force, or civilian organizations participating in the urban operation, as well as with the indigenous civilian population at large. This shared system may put at risk the timely delivery of critical supplies to Army forces.

10-38. Sustainment planners must understand the urban environment and its effects on the proposed method of distribution. They must also understand how urban operations (to include sustainment UO) may affect the urban environment. Storing bulk fuel or ammunition in or near an urban area, for example, may increase the risk of fire and explosive hazard to civilians and Army forces. With this awareness, planners present the commander with an estimate that considers both risks and benefits. Depending on the particular area and other METT-TC factors, they may recommend one method of distribution then transition later to another or a combination of methods. Preconfigured resupply loads delivered as close as possible to where they are needed is vital to responsiveness. Only an agile distribution-based sustainment system will allow Army forces to be responsive and operationally effective across full spectrum operations in an urban environment.

Food Preparation

10-39. The daily feeding standard for Soldiers operating in urban areas remains the same: three quality meals per Soldier. UO creates an increased energy demand on Soldiers, requiring a caloric intake of about 5,000 calories each day. Since producing and delivering prepared meals to forward elements may be difficult or even impossible due to the urban environment, the innovative use of meal supplements can be critical to meeting the dietary requirements of the force. Energy drinks and bars, rich in electrolytes and carbohydrates, are an essential supplement to prepared meals and should be provided in quantities that support or exceed the requirements of the forces operating in the urban area.

10-40. Food (and water) may be available in the urban area; however, local sources must be tested, carefully monitored, and medically approved before consumption. Garbage disposal may be an important consideration in the urban area. Improper trash disposal may leave a signature trail (particularly during urban defensive operations) that may degrade sanitary conditions leading to increased disease and nonbattle injuries (DNBIs). Commanders must understand that food operations, if not properly positioned and secured, can become a focal point for the urban population. Strict policy regarding distribution and control of any Class I supplies (including waste products) will be enforced. Black marketers will be attracted to Army food service activities as well (even during offensive and defensive operations).

Water Purification

Vigilant Monitoring

10-41. Urban areas will often have a ready source of water to support the urban inhabitants and its infrastructure. However, this water may not be suitable for U.S. and allied forces (though the urban population may have developed immunity toward its microorganisms). The higher concentration of TIM in urban areas compounds this problem, as ground water is highly susceptible to chemical contamination, even supplies located miles away from the source of contamination. U.S. Soldiers should be trained and cautioned against using water from an urban area (to include ice and bottled water) until preventive medicine and veterinary personnel can determine its quality. When water quality is unknown, commanders ensure use of tactical water purification equipment that will upgrade it to the Army's water quality standards. Even if initial testing indicates that urban water sources are safe for Army forces, personnel must continuously monitor the water quality. However, Army water purification, storage, and transportation requirements for UO can be greatly reduced if the existing urban water supply can be integrated into sustainment operations. An early assessment of the feasibility of this course of action is critical to sustainment planning.

Potentially Greater Requirements

10-42. Offensive and defensive UO are often intense and can produce more casualties, including civilians. Consequently, medical facilities, already consumers of large volumes of water, may require even more water. Water purification, particularly in the urban areas of developing nations and during urban disaster relief operations, will be a critical and constant concern for Army forces.

Possible Key Terrain

10-43. Sites that can control the water of the urban area may be key terrain, providing not only a resource for Army forces, but also a means to control the threat, the civilians, or both. These sites may be the sources of the water—the river, lake, reservoir, or storage tanks—or the means that process and transport the water—pipelines, pumping stations, or treatment facilities. Many sites may be outside the urban area, as many large urban areas draw water from distant sources. The seizure of a pumping station or pipeline may make it possible for commanders to control water supplies without expending resources required to enter the urban area. To preserve critical infrastructure, commanders may increase security to protect these locations from contamination or destruction. Engineers also may need to restore, maintain, or operate existing water facilities damaged by the threat or disaster and to drill new wells and construct new water facilities.

Mortuary Affairs

10-44. Mortuary affairs provide the necessary care and disposition of deceased personnel. It supports the Army across the spectrum of operations. It may directly and suddenly impact (positively or negatively) the morale of the Soldiers and the American public and may influence relations with the civilian population in the AO. It can also affect the health of Soldiers and the urban populace. Overall, commanders must carefully plan evacuation routes and temporary collection and internment sites, adhering to local customs and traditions to lessen potentially negative consequences.

10-45. Units are responsible for recovering the remains of their own fatalities and evacuating them to the closest mortuary affairs collection point, usually located at the nearest support area. Urban governmental authorities, family, and representatives of appropriate nongovernmental organization are responsible (and should always be relied on to the greatest extent possible) for recovering and handling indigenous civilian remains. However, because of the sheer density of noncombatants in UO, commanders may also find themselves responsible for civilian remains. Deaths of civilians under Army control, such as urban evacuees and refugees at Army-operated sites, often obligate the Army to care for their remains including medical certification and records of death. High-intensity urban combat may result in civilian deaths, and health concerns will require Army forces to deal with civilian remains expeditiously. Commanders should consult local religious leaders, the Staff Judge Advocate (SJA), CA personnel, and chaplains to verify that they are abiding by law and customs. Overall, commanders must ensure that forces treat all deceased, including civilians, with dignity and respect (another important aspect of adhering to the urban fundamental of understanding the human dimension).

Aerial Delivery

10-46. Aerial delivery is the movement by fixed- or rotary-wing aircraft and delivery by the use of parachute or sling load of Soldiers, supplies, and equipment. As a vital link in the distribution system, it contributes to flexibility and provides the capability of supplying the force even when urban ground LOCs are disrupted. Forces use aerial delivery to deliver supplies and equipment when not possible or timely by other means. During urban stability or civil support operations, it is used extensively to move supplies to meet the urgent needs of a population in crisis. In all UO, delivery aircraft are highly vulnerable to small arms, rockets, and air defense systems. A threat may further decrease an already limited number of urban drop zones (aircraft may be able to avoid air defense systems, but ground forces may not be able to secure the drop zone and retrieve the cargo). Equipment and supplies transported by helicopter sling-load lessen the latter disadvantage. There are usually more available sites to deposit sling loads, such as rooftops (engineers may need to determine if they have sufficient structural integrity), parking lots, and athletic fields. However, load instability during flight may restrict a helicopter's airspeed and maneuver capabilities making it more vulnerable to small arms and man-portable air defense systems. For these reasons, aerial delivery of supplies in UO may be much less efficient than in many other environments.

MAINTENANCE SUPPORT

10-47. Maintenance entails keeping materiel in operating condition, returning it to service, or updating and upgrading its capability. It includes recovering and evacuating disabled equipment; replacing forward; performing preventive maintenance checks and services (PMCS); increasing battle damage assessment and repair (BDAR); and determining potential maintenance resources in the urban area.

Rapid Recovery Essential

10-48. Disabled vehicles produce obstacles to movement during UO, blocking narrow thoroughfares and creating vulnerable targets for threat forces. This makes rapid recovery operations essential to avoid clogging limited LOCs and mounted avenues with vehicle evacuation operations. Hence, maintenance personnel should be proficient in hastily securing unit maintenance collection sites near the damaged equipment, ensuring route security to and from the disabled vehicle, and methods of egress under fire. The increased potential for the task organization of armored units into smaller attachments during UO (often platoon-sized or smaller) will strain limited recovery assets and increase the need for sections to be well trained and resourced (to include adequate tow-bars and chains) for self-recovery operations. Units must maintain centralized and responsive control over these potentially critical recovery assets and position them as far forward as the situation allows.

Replace Forward

10-49. One of the guiding maintenance principles is to replace forward and fix rear. Maintenance activities, with a forward focus on system replacement, task and use the distribution and evacuation channels to push components and end items to the sustainment level for repair. However, the urban environment may make distribution and evacuation difficult or even impossible. Repairing equipment as far forward as possible will often be vital to maintaining the tempo necessary to succeed in UO. When recovery is required, equipment is moved only as far back as the point where repairs can be made. When selecting a maintenance site, commanders consider:

- Security and force protection measures.
- Sufficient area around equipment for lift or recovery vehicles.
- Availability of existing maintenance facilities or garages.

10-50. The unforgiving urban terrain will invariably increase damage to man-portable weapons and equipment, particularly electronic equipment sensitive to jarring. More frequent jarring can also translate into more frequent calibration requirements. Although by definition man-portable weapons and equipment are easier to transport, evacuating these systems may prove as difficult as evacuating vehicles and larger, heavier equipment. Therefore, field support maintenance support teams (MSTs) will frequently need to repair equipment at (or as near as possible to) the point where it was damaged. Equipment operators are responsible to properly diagnose the fault or damage. Such action ensures that the correct repair parts and maintenance personnel are sent forward to complete necessary repairs. In UO, particularly offensive and defensive operations, units may need to replace rather than repair equipment. Subsequently, sustainment personnel may need to plan for increased replacement of what might normally be repairable equipment, as well as plan for increased repair parts for man-portable items.

Add-On Protection Increases Wear

10-51. To increase protection against small arms, mines, rocket-propelled grenades, and lightweight antiarmor weapons, units may need to attach additional armor and improvised wire-mesh screens and cages to both wheeled and tracked vehicles operating in an urban environment. (Units can also sandbag vehicles to achieve a degree of increased protection.) These modifications, however, increase wear on brakes, springs, suspension, and tires, all of which are already vulnerable to the increased amount of debris present in the urban environment. Over time, this wear will begin to increase failure rates in major assemblies, such as engines, transmissions, and transfers. To offset the increase in wear and, subsequently repair parts requirements, commanders conducting UO should consider emphasizing these aspects during daily operator PMCS and increasing the frequency of scheduled services.

Increased Battle Damage Assessment and Repair

10-52. In UO, operators, crews, MSTs, and recovery teams execute BDAR far more than in other environments. BDAR quickly restores minimum essential combat capabilities for a specific mission (normally of short duration) or allows the equipment to self-recover by expediently fixing or bypassing components. Commanders may need to authorize supervised battlefield cannibalization and controlled exchange when units lack critical parts or cannot bring them forward. Sustainment commanders should consider the use of BDAR kits, specifically designed for the systems they support. The use of these kits, though providing only temporary repair, will extend the availability of combat systems until full repair capability is accessible.

Potential Urban Maintenance Resources

10-53. Although urban areas can complicate maintenance, they may contribute to this sustainment function. Analyzing the urban area in the commander's AO may reveal potential sources of parts, tools, equipment, and facilities, and the expertise necessary to fix equipment, repair components, and fabricate critical parts. Sites of potential value may include automotive repair shops, foundries, car dealerships, parts stores, junkyards, scrap-metal dealers, and machine shops. Urban areas may also serve as key sources for parts and facilities (and theater support contractor personnel) to repair automation and network communication equipment.

Host Nation Repair Operations: OPERATION IRAQI FREEDOM

During operations in Mosul, Iraq in the summer months of 2003, elements of the 2nd Brigade Combat Team (BCT), 101st Airborne Division (Air Assault) suffered from a decline in combat power that outpaced the ability of the distribution-based sustainment system to provide repair parts. As readiness across the BCT threatened to degrade operational capability, sustainment personnel from the brigade's forward support battalion (FSB)—drawing on lessons learned from the 1990-1991 Gulf War—scoured the local populace for repair parts, facilities, and expertise in an attempt to offset the losses suffered due to supply shortfalls.

Initially, local retailers provided limited repair parts and facilities to the FSB, but the effect was significant and readiness rates began to stabilize. As time passed, however, the continued wear on equipment operating in the urban environment, coupled with the harsh summer weather, began to take a toll on reparable components and major assemblies. With limited availability in theater, and order-ship times as long as 60 days, requisitions would not be filled before the effect on combat fleet readiness reached a critical point.

Recognizing a trend that threatened the success of the BCT's urban operations in Mosul, the FSB established a contract maintenance facility in the brigade support area, used its contingency contracting officer to hire a group of local mechanics certified through General Motors, and opened the first host nation repair facility in Iraq. The group repaired major assemblies and reparable components and systems, maintaining a failure rate consistent with major CONUS rebuild facilities. The effect was immediate: during the first four months of the program, a full two-thirds of the combat platforms repaired and returned to service operated with components or assemblies rebuilt in the host nation facility.

Over time, the program was expanded to support other divisional units in northern Iraq. This effort provided maneuver commanders with responsive maintenance support and demonstrated the effect of sustainment flexibility and sustainability in urban operations. Furthermore, this action met the division commander's intent of infusing much-needed funds into the local economy while strengthening American-Iraqi relations in the region.

TRANSPORTATION SUPPORT

10-54. Transportation supports the concept of the urban operation by moving and transferring units, Soldiers, equipment, and supplies. Transportation incorporates military, commercial, host-nation, and urban area capabilities to establish a flexible system that expands to meet the needs of the force. Transportation includes movement control, terminal operations, and mode operations.

10-55. Urban areas are often critical to transportation operations. These areas may serve as a lodgment or support area for entry of Army forces and sustainment supplies. The existing transportation and distribution infrastructure may be essential to reception, staging, and onward movement. Theater support contracts, as well as host-nation support agreements, may greatly increase the ability of Army forces to use the urban area's facilities, which may include docks, airfields, warehouses, and cargo handling equipment. Urban support may also include skilled urban workers, such as longshoremen, truck drivers, and MHE operators.

Urban Terminals

10-56. In addition to serving as major seaports and aerial ports of debarkation, urban areas may provide additional terminals in the AO. Forces may use these terminals for further staging, loading, discharging, and transferring the handling between various inland transportation modes and carriers (motor, air, rail, and water). These urban terminals—with synchronized movement control—permit commanders to rapidly shift transportation modes and carriers. This increases flexibility and ensures the continued forward movement of equipment and supplies to support ongoing operations. Movement control, particularly in urban areas, relies heavily on support from military police in their maneuver and mobility support role. Without this support, urban LOCs may become congested, hinder movement and maneuver, and degrade force effectiveness (see FM 3-19.4). Urban commanders may need to establish multiple checkpoints, roadblocks, and traffic control points; restrict selected roads to military traffic; and reroute movement to unaffected road networks when civil support and refugee control operations compete for available routes. Again, military police operations are critical in this regard and will require continuous, close coordination with urban civilian police (if available).

Obstacles to Ground Transportation

10-57. Although urban areas can contribute to transportation operations, rubble and other damage can become obstacles to ground movement. Even in an undamaged urban environment, road and bridge weight restrictions, barriers and medians, steep embankments, large drainage systems, and other urban structures may limit transportation operations. Urban route maintenance, to include reinforcing bridges and constructing bypasses, may become a priority task for engineer units. Bypassed pockets of resistance and ambushes pose a constant threat along urban supply routes and routes in close proximity to urban areas. Urban convoy and resupply operations will often require increased security in the form of continuous route security operations, regular (daily, if necessary) mine and improvised explosive device (IED) clearance operations, numerous observation posts, attack helicopter support, a larger, more mobile tactical combat force, and adequate communications capabilities among convoy vehicles and with supporting forces and higher headquarters. While the level of security and force protection may vary, convoy operations are never considered an "administrative" operation (see FM 4-01.45). Security needs may increase manpower requirements for sustaining operations and potentially reduce resources from decisive operations. Moving critical supplies may require heavily armed convoys and armored vehicles. Drivers must be well trained in convoy operations, rehearsed, and alert. They must recognize and avoid potential mines and IEDs (stationary and vehicle-borne) and be able to react rapidly to ambushes. Additionally, commanders may need to institute deception measures such as randomly closing and opening supply routes to confuse threat pattern analysis. As before, aerial resupply can alleviate problems due to ground obstacles, but the air defense threat and proximity of threat forces may preclude their routine use.

Population Effects

10-58. The ability of Army forces to use vital urban transportation facilities depends largely on the civilians and the threat. The civilian population can affect the transportation system if they do not support the goals of Army operations. Urban transportation systems—such as ports, railroads, and rivers—require many specialists to operate. Without these specialists, the system's utility is degraded and may not function at all. In urban stability or civil support operations, Army forces will share the system with civilians and other agencies. Civilian authorities may refuse to allow Army forces to use any portions of an urban area's transportation system. Negotiating for access to that system under these circumstances then becomes a command priority.

Threat Effects

10-59. The threat can significantly affect urban transportation systems. Many are composed of smaller subsystems. Each subsystem is vulnerable to attack, which in turn often shuts down the whole system. A large canal system, for example, may have entrance and exit facilities, the canal itself, a means to pull the vessel along such as a locomotive engine, and the civilians that run each of these subsystems. Both an attacker and defender understand the components of the particular transportation system. If important to current or subsequent operations, defending forces then develop plans and allocate forces to protect these subsystems. Attacking forces, on the other hand, often avoid collateral damage to the system, while simultaneously preventing enemy destruction of the facilities.

FORCE HEALTH PROTECTION

10-60. Force health protection (FHP) consists of all services performed, provided, or arranged to promote, improve, conserve, or restore the mental or physical well-being of personnel in the Army and, as directed, for other services, agencies, and organizations. FHP conserves the force by—

- Preventing DNBIs and controlling combat stress.
- Clearing the urban area of casualties.
- Providing forward medical treatment and hospitalization and en route care during medical evacuation.
- Providing required veterinary, dental, and laboratory services.
- Ensuring that adequate Class VIII supplies, medical equipment, and blood are available.

10-61. FHP operations minimize the effects of wounds, injuries, disease, urban environmental hazards, and psychological stresses on unit effectiveness, readiness, and morale. Effective UO require acclimated Soldiers trained in specific urban tactics, techniques, and procedures. FHP helps maintain the health of urban forces, thereby conserving that trained manpower. Historically, urban combat operations have generated three- to six-times greater casualty rates than operations in any other type environment. (Greater enemy wounded and a need to provide care for them will also exacerbate this situation.) FHP operations that maintain the health and readiness of Army forces reduce the strain on the replacement and evacuation systems. Such care allows Soldiers to concentrate on the task at hand instead of the increased risks associated with UO.

10-62. As part of the overall LPT, commanders and medical planners must analyze and continuously assess the urban area. They must identify medical threats, required medical resources, and the quality and availability of medical facilities and resources (to include civilian medical personnel). This assessment prevents duplicated services and permits more effectively and efficiently organized medical resources.

10-63. FHP analysis also identifies available hospitals, clinics, medical treatment facilities, and medical supplies and equipment (including production facilities) in the urban area. It may also identify NGOs capable of providing medical services and supplies. (However, stringent federal regulations, standards of medical care, and a need for unavailable advanced technologies may limit their use by Army forces. But, this analysis will still be crucial in determining whether urban medical facilities can support civilian requirements.) FHP personnel keep abreast of the operational situation and its impact on FHP and the urban population. Peace operations, for example, may rapidly transition to high-intensity offensive and defensive operations requiring medical support able to handle potential mass casualty scenarios which may include civilians.

Care of Civilians

10-64. During combat operations, the military normally does not provide injured civilians with medical care, instead relying on civilian medical personnel to provide this service. However, based on METT-TC and requirements under Geneva Conventions, commanders may need to recover, evacuate, and treat numerous civilians (particularly in urban civil support or stability operations) until local civilian medical personnel and facilities can be reconstituted and supplied. In urban civil support operations involving weapons of mass destruction, the primary focus of Army support may be FHP.

10-65. If commanders provide civilian medical support, they adhere to the UO fundamental of transitioning control and transferring responsibilities for medical care of civilians to another agency or into civilian hands as soon as is practical. In UO, commanders consider and address the medical treatment of civilians (enemy and friendly) early in the planning process. Any initial assessment or survey teams therefore contain medical representatives. Of significant concern, commanders may need to ensure that adequate supplies and equipment are on hand to treat the urban population's children as combat health service support is normally focused on the treatment of adult Soldiers. Additionally, follow-on military care for civilians after treatment at a combat support hospital is often nonexistent or extremely limited to a case-by-case basis necessitating the quick restoration of civilian hospital capabilities. Finally, military health practitioners must also assess and consider cultural factors related to civilian medical treatment. (For example, male doctors may not be able to examine or treat indigenous women.) To alleviate confusion and preclude avoidable negative repercussions and unrealistic expectations, commanders may need to develop specific medical policies, directives, and standing operating procedures. These "medical rules of engagement" (ROE) ensure that subordinates know how much medical care they may provide to—

- The urban population.
- Other host-nation and third-country civilians.
- Coalition and host-nation forces.
- Contractor personnel.

10-66. Commanders should also consider that medical aid to the local urban population and other civilians can be a powerful influential tool. Properly planned and executed, its potential benefits for Army forces include enhanced security and intelligence collection created by increasing the willingness of civilians to interact with Soldiers. These effects cannot be left to happenstance. They should be a conscious part of the information operations campaign nested within the overall UO plan. Consideration must also be given, however, to mitigating the potential negative consequences involved in the inevitable reduction or curtailing of medical assistance by Army FHP personnel. As with any aspect of UO, providing care without considering the longer-term consequences could result in far greater disadvantages than benefits.

Disease and Nonbattle Injury

10-67. DNBI is a major medical threat during all operations and presents additional risk during UO. Urban pollution hazards and potential exposure to TIM may increase the risk to soldier health. Some urban areas, particularly those in developing countries, are already large sources of communicable diseases, such as tuberculosis, cholera, typhus, hepatitis, malaria, dengue, and acquired immune deficiency syndrome (AIDS). Other factors that increase the number of disease vectors and the potential for the transmission of infectious diseases might include—

- Physical damage or deterioration of urban infrastructure—such as electricity, water, and sewage services.
- Industries that use or produce hazardous materials.
- Increased habitats for rodents and other pests due to rubble, debris, and uncollected trash.
- Inability for the threat or local authorities to recover and dispose of human remains and animal carcasses.
- Greater threat of animal diseases transmittable to humans due to larger numbers of animals wounded or injured by the effects of UO.
- Greater susceptibility of urban civilians to infectious diseases—increasing the potential of person-to-person disease transmission—due to declining public health capabilities and nutrition levels.

10-68. The density of the environment extends these urban medical risks to Army forces. Commanders should establish a medical epidemiological surveillance system early to assess the health of the force and promptly identify unusual or local occurrences that may signal preventive medicine problems or the influence of biological or chemical agents. These potential hazards, particularly the release of TIM, will influence the type of medical supplies needed by medical personnel and will also necessitate critical planning and preparation for potential mass casualties (civilian as well as military).

10-69. Adhering to the UO fundamental of preserving critical infrastructure will likely prevent surges in DNBI. Preventive medicine personnel must identify the diseases and recommend control and preventive measures. In urban areas, they may also conduct civilian health screening, health education, and immunization programs. Medical screening of military personnel, particularly multinational forces, may be required to prevent introducing new diseases (especially drug-resistant strains) into an urban area. A new disease may tax the medical system and introduce a new medical problem into an area already in crisis. Lastly, field sanitation training (to include a general emphasis on hand washing; training in the use of barrier protection such as latex gloves when rendering care to any person; and animal, rodent, and pest control), equipment, and supplies are part of overall preventive medicine measures and considerations.

Combat Stress

10-70. Stress occurs in every operation and type of environment; while limited stress levels are beneficial, too much is harmful. Controlled combat stress can invoke positive, adaptive reactions such as loyalty, selflessness, and heroism. On the other hand, uncontrolled combat stress can result in negative, harmful behavior and combat stress casualties. Such behaviors and casualties—battle fatigue, misconduct stress behaviors, and post-traumatic stress disorder—can interfere with the unit's mission. Physical and mental factors leading to combat stress result from the environment and the deliberate enemy actions aimed at killing, wounding, or demoralizing Soldiers. However, many stressors are also generated from the Soldier's own leaders and mission demands.

Combat Stress: Chechnya – 1994 to 1996

Russia's 1994-1996 conflict with Chechnya, a republic in the southwestern part of the Russian Federation, produced an increased number of psychological trauma and combat stress casualties. One medical survey found 72 percent of the soldiers screened had some sort of psychological disorder symptoms. Of that, 46 percent exhibited asthenic depression (a weak, apathetic, or retarded motor state). The other 26 percent exhibited psychotic reactions such as high states of anxiety, excitement, or aggressiveness and a deterioration of moral values or interpersonal relations. The statistics showed more troops experienced combat stress disorders than during their 1980s war in Afghanistan. One primary difference was that in Chechnya, Russian forces conducted combat mostly in cities rather than in mountains, valleys, and other rural areas.

Combat always invokes fear in soldiers. However, poor training and planning, uncertainty in their cause, and urban populations that resented their presence exacerbated the psychological climate for the Russian forces in Chechnya. Acts of subversion and terrorism by Chechen guerrillas kept the Russians in a constant, high state of readiness and anxiety; the soldiers viewed every civilian—young or old, male or female—as a potential enemy. This psychological pressure was not simply a by-product, but an objective of information operations and a prime reason for taking the fight into the close confines of Chechnya's urban areas. Torture and mutilation of prisoners; immediate execution of captured pilots; imitative electromagnetic deception (Chechens mimicked Russian radio transmissions and directed Russian close air support against their own forces); and propaganda to convince civilians that Russia's actions had a religious bias against Muslims and Islam were conducted to exert intense, unremitting psychological pressure on Russian forces—with great success.

The characteristics of urban areas combined with Chechen insurgent activities and information operations, civilians that did not welcome foreigners, and an unpopular and poorly supported conflict with an open-ended mission reinforced the need for a national will to strengthen and support the urban fight. These characteristics also reinforce the need for clear objectives, proper leadership and training, and available medical assets able to properly diagnose and treat combat stress casualties.

10-71. During UO, compartmented urban combat leads to physical isolation while difficulties in transmitting radio signals lead to communications isolation; together, they create an overwhelming sense of loneliness. Snipers, mines, and IEDs combined with the closeness and high intensity of urban combat contribute to an unremitting fear of attack from any quarter that further increases stress casualties. Additionally, seeing and perhaps accidentally inflicting casualties on civilians (especially women and children) increase battle fatigue. If civilians are hostile or a threat uses the population as cover and concealment, then the potential for misconduct stress behaviors often increases. Urban areas may provide temptations for looting, alcohol and substance abuse, black marketeering, and harmful social interactions; these temptations may increase misconduct stress behaviors. Leaders can prevent or rapidly identify, successfully manage, and treat stress-related casualties (see FM 22-51 and FM 6-22.5) and prevent misconduct stress behaviors as well as potential violations of the law of war. To reduce the number of stress casualties, commanders must—

- Provide tough, realistic training in urban combat (and combat stress management).
- Plan for rotation of forces and adequate rest and relaxation opportunities.
- Establish effective ROE.
- Create unit cohesion.
- Ensure adequate mental health support is available.
- Provide strong, capable leadership. Leaders, forward on the battlefield, will determine whether the necessary discipline perseveres.

Evacuation

10-72. Transportation restrictions may preclude the medical evacuation (MEDEVAC) of urban casualties. MEDEVAC by air ambulance remains the preferred method; however, it may not be feasible. The threat from small arms fire, rocket-propelled grenades, and other shoulder-fired weapons may prohibit the employment of helicopters. Aeromedical evacuation can also be hampered by the lack of landing zones due to narrow streets, heavily damaged buildings, and barricaded or booby-trapped roofs. If the employment of air ambulances is not feasible, a ground evacuation operation will be necessary. In UO, ground ambulances should often be—

- Hardened to improve the safety of the crew and patients as well as to increase the survivability of the evacuation platform.
- Field-sited as close to the operation as is tactically and physically possible.
- Maneuverable in tight spaces and narrow streets.

10-73. Evacuation by ground ambulance may have their own set of obstacles to surmount such as rubble; debris; barricades; mines, booby-traps, and IEDs; damaged or destroyed vehicles, and flooded or snow-filled streets. If enough ground ambulances are unavailable due to increased casualties or the tactical situation does not allow, units will need to execute casualty evacuation (CASEVAC) operations.

10-74. In UO, CASEVAC may entail evacuation by litter and the expedient use of a vehicle of opportunity or other military vehicle not dedicated for medical evacuation. Based on the threat situation, CASEVAC may also be limited to armored vehicles, conducted only during limited visibility, or both. For evacuation by litter, units may require more litter bearers to move the injured to a point where they can be further evacuated by ground or air ambulance. Lengthy evacuation routes require more litter bearers, as multiple litter relay teams will be necessary to conserve energy and expedite evacuation. However, unless augmented or relieved of this responsibility by another unit, these litter bearers will come from the casualties' own unit thereby diminishing the unit's strength levels necessary to accomplish its primary mission. Instead, tactical commanders may need to wait until a lull in or a cessation of the battle occurs. Depending on projected casualty rates, higher-level commanders may augment units with additional personnel to perform evacuation or may assign maneuver units this mission. Evacuation personnel will require specific training in urban evacuation techniques (moving casualties from subsurface and supersurface levels to and along the surface level).

10-75. Overall, urban commanders at all echelons must develop detailed casualty evacuation plans that include the establishment of treatment facilities as far forward as the situation allows. Engineers are critical to clear routes for medical resupply and evacuation. Based on the threat situation, additional armored vehicles may need to accompany ambulances in a multi-vehicle convoy not only to and from the patient pickup point but also through to a higher level of care. Army forces may also need to plan for and acquire unique capabilities, equipment, and skills now typically seen in civilian urban search-and-rescue teams to clear debris and search for urban casualties. For example, specially trained dogs may play a vital role in locating victims among urban rubble.

Treatment

10-76. First aid training will have increased significance in UO. The compartmented nature of UO, transportation restrictions, communications difficulties, and the low density of trauma specialists (combat medics) may limit the urban casualty's initial treatment administered by nonmedical personnel or to self-treatment measures. To mitigate this risk, units identify and train combat lifesavers (CLS) to perform in the absence of trauma specialists. Since the likelihood of Army forces performing UO continues to increase, commanders strive to meet or exceed Army standards for the number of fully trained CLS required for their specific unit. CLS (and trauma specialist) training must be realistic, hands-on, and conducted at appropriate intervals to ensure CLS can perform lifesaving tasks to include starting intravenous lines. This increase in self, buddy, and CLS care, as well as longer delays in evacuation, may also increase requirements for additional first aid, CLS, and trauma specialist-carried supplies. (Ultimately, commanders may have to consider weighting the main effort with greater overall FHP support for units conducting UO than for units conducting operations in other environments.)

10-77. The increased potential for delayed evacuation during UO mandates that Army trauma specialists be skilled in prolonged casualty care. During the 3-4 October 1993 battle in Somalia, for example, seven trauma specialists managed 39 casualties for more than 14 hours before they could be evacuated. Evacuation delays significantly increase the potential for infection; therefore, trauma specialists should also be trained to administer antibiotics on the battlefield.

10-78. FHP personnel must be capable of recognizing and treating injuries due to IEDs and incendiary or thermobaric weapons (fuel-air explosives)—favored, urban-oriented threat weapons (see Chapter 3). Thermobaric weapons explode; create a cloud of volatile gases, liquids, or powders; and then ignite, creating an immense fireball consuming oxygen and creating enormous overpressure. When employed in an urban structure, the blast wave or overpressure of any explosive is greatly amplified. Injuries resulting from these weapons are massive burns, broken or crushed bones, concussions and traumatic brain injuries, missile injuries, and internal injuries. To counter this threat, medical personnel must be aware of the capabilities of these blast weapons, wounding patterns, and treatment methods.

10-79. The increased use of body armor during UO will help prevent penetrating chest and abdomen wounds. As such, FHP personnel should expect more groin, pelvis, head, and extremity injuries. Furthermore, when fighting soldiers that are known to use body armor, a threat (particularly snipers) can be expected to target the head and face more often than other anatomic areas resulting in more head injuries. Medical personnel must also anticipate and be prepared to treat an increase in hearing loss due to the firing of weapons (particularly recoilless) in enclosed spaces with little ventilation. Medical personnel (and engineers) should also be trained on how to properly remove rubble that has fallen on a Soldier or civilian so as not to increase the extent of his injuries. Lastly, FHP personnel will need to research and develop unique treatments for injuries due to an ever-increasing arsenal of nonlethal weapons.

Explosive Ordnance Disposal Support

10-80. Explosive ordnance disposal (EOD) support provides the capability to neutralize domestic or foreign conventional explosive hazards which include unexploded explosive ordnance (UXO), booby traps, IEDs, captured enemy ammunition (CEA), and bulk explosives. Such material and devices threaten military operations and military and civilian facilities, equipment, and personnel. Explosive hazards create a much greater risk during UO than operations in any other environment. Confined spaces, hard surfaces, and more personnel (both Soldiers and civilians) in the vicinity may magnify the detonating effects of explosives. Dense terrain makes all types of explosive hazards more difficult to locate. Moreover, terrorists select urban settings for their booby traps and IEDs to potentially kill and destroy more, thus gaining greater visibility for their message.

10-81. 10-81. EOD units detect, mark, identify, render safe, and dispose of explosive hazards. Importantly, EOD specialists will work closely with unit intelligence personnel to conduct explosive forensics—technical intelligence collection and exploitation—to help identify the makers of the devices as well as their ever-evolving methods of construction, placement, concealment, and detonation. Chemical reconnaissance with their array of chemical sensors will also work closely with EOD units to reduce the uncertainty of the chemical threat. This integrated forensic approach is a proactive method to take the advantage away from a terrorist or insurgent threat that seeks to employ these devices as a primary tactic.

10-82. Urban operations will increase EOD specialists' role as advisors and instructors on explosive hazards, protection measures, and disposal techniques. They will advise and train other Army forces, other services, multinational partners, and civilian authorities. EOD specialists will also work closely with public affairs and psychological operations personnel to increase public awareness of the horrific effects of these munitions and devices. This education process is designed not only to teach the urban population to identify and avoid explosive hazards, especially UXO, booby traps, and IEDs, but to gain civilian assistance in reporting their locations. (In this regard, rewards for information leading to the discovery of unexploded ordnance may be beneficial. However, the largest reward should normally be reserved for information leading to the arrest or capture of bomb makers.)

10-83. In an urban environment where there are large amounts of UXO, IEDs, and CEA, the sheer amount of explosives may overwhelm the capabilities of available EOD personnel. Because of their training in demolitions and explosives, commanders will often use combat engineers to augment the destruction and clearance effort. Engineers are normally given responsibility for UXO clearance and disposal while EOD teams use their technical skills for IEDs. Commanders may also consider embedding EOD expertise within engineer units to provide the knowledge necessary to determine the correct amount and type of explosive necessary to detonate UXO and to understand potential risks and effects of detonation of various munitions. For example, EOD specialists would better understand the second-order effects and risk reduction measures necessary to detonate a large missile containing numerous submunitions.

10-84. Unfortunately, there is no single device or technique that can eliminate every explosive hazard encountered. Clearance operations are dangerous and time-consuming. Based on METT-TC, commanders may consider the following options when dealing with these explosive hazards:

- Report, mark, and bypass the hazard.
- Employ tactical breaching procedures.
- Self-extract to alternate routes or positions.
- Restrict routes or positions from further use.
- Clear specific areas or positions, or conduct an extensive route clearance.
- Lastly, accept risk of casualties and continue with the assigned mission.

10-85. Effective tactics, techniques, and procedures (TTP) for addressing explosive hazards must be captured and shared with friendly forces and coalition partners. Commanders recognize, however, that adaptive threats will continually modify their own techniques to counter the unit's response. See FM 3-34.119, FM 20-32, FM 3-100.38, FM 4-30.51, and FM 4-30.16.

HUMAN RESOURCES SUPPORT

10-86. Human resources support (HRS) encompasses the following functions: manning the force, personnel support, and personnel services. These activities include personnel readiness; replacement operations; personnel accounting; personnel information management; casualty operations; essential personnel services; postal operations; morale, welfare, and recreation; and band support provided to Soldiers, their families, Department of Army civilians, and contractors.

10-87. Successful UO require HRS functions. Clearly, HRS focuses on caring for the needs of people: Soldiers and civilians who support them. Since a critical component of the urban environment is the population, these activities, when required, may also support them. For example, HRS elements, in conjunction with mortuary affairs, military police, and CA, may provide administrative support to assist in the accountability of displaced persons and civilian internees and the establishment of other populace and resources control measures. This support may include the provision of identification and registration cards, travel permits, and other documents, and administrative support to casualty operations. Human resources managers may also assist with community activities and coordination with the Red Cross. If necessary, a postal platoon may assist urban officials in training and reestablishing civilian postal operations. Overall, this type of assistance and support helps to care for the immediate needs of the urban population and, ultimately, speed transition back to legitimate civilian responsibility and control.

10-88. In offensive and defensive UO, HRS personnel will need to account for more casualties and more frequent reconstitution. In all UO, success relies on training individual replacements. Urban combat requires Soldiers skilled in specific urban combat TTP. They understand the societal aspects of the urban population and have training in crowd control to avoid escalating potentially volatile situations. Training these replacements prior to their actual deployment or while in the replacement system frees urban commanders from having to do so. It also helps reduce Soldier (and supporting civilian) isolation, anxiety, and fear associated with urban areas and operations.

FINANCIAL MANAGEMENT SUPPORT

10-89. Money and its management are often critical to the success of UO. As such, financial management support makes resources available when and where they are needed and assist the urban commander in maintaining fiscal responsibilities. Financial management operations are necessary to conduct the contracting necessary to restore and rebuild infrastructure and provide real-time information, accounting, and services. Financial management operations ensure that urban operational policies and procedures adhere to laws and regulations, develop command resource requirements, and leverage appropriate fund sources to meet them.

10-90. UO will likely include other U.S. military services, governmental agencies, and contractors; multinational forces; NGOs; and various elements of the host nation. So many actors complicate financial accounting, resource management, and cost capturing. Yet, these activities are vital to accomplishing UO and maintaining legal requirements. Army forces will receive support from and provide support to these participants. Financial managers provide crucial expertise to the urban commander on the obligation authority. They also furnish the documentation necessary to obtain and pay for local goods and services using contract or commercial vendor services. However, commanders (with financial management assistance) must assess the economic impact of UO on the economy. An appropriate analysis includes how well the urban economics and commerce infrastructure can support the deployed force and how an influx of U.S. currency may affect the overall economy either positively or negatively.

Labor, Reward, and Solatia Programs

10-91. Financial management elements can also provide joint pay support and non-U.S. pay support for host-nation employees and day laborers supporting UO, as well as for civilian internees. These units can support reward programs and solatia (financial compensation). Commanders can use reward programs to purchase weapons, radios, information, and other items from the urban inhabitants. However, they must use caution when paying for critical human intelligence. First, they must ensure that multiple collection agencies in the command avoid paying the same source for the same information and interpreting these repeated inputs as validation that information is accurate and reliable. Second, they must establish a price or pay scale so subordinate units (particularly in multinational UO) equally compensate their sources for information. Commanders may also need to make solatia to alleviate grief, suffering, and anxiety resulting from injuries and property or personal loss resulting from some aspect of UO. At other times, commanders make solatia to meet cultural expectations of the urban population. These payments are nominal in amount and made without an admission of liability by the Army. Significantly, they may help maintain the legitimacy of the operation and preserve the support of the urban populace.

Financial Support to Civil-Military Projects

> *...[I]n an endeavor like that in Iraq, money is ammunition. In fact, depending on the situation, money can be more important than real ammunition—and that has often been the case in Iraq since early April 2003 when Saddam's regime collapsed and the focus rapidly shifted to reconstruction, economic revival, and restoration of basic services. Once money is available, the challenge is to spend it effectively and quickly to rapidly achieve measurable results. This leads to a related observation that the money needs to*

be provided as soon as possible to the organizations that have the capability and capacity to spend it in such a manner.

Lieutenant General David H. Petraeus
Commander, 101st Airborne Division (Air Assault) and, later,
Commander, Multi-National Security Transition Command-Iraq
during OPERATION IRAQI FREEDOM

10-92. Other financial resources separate from operational funds may be appropriated to support civil-military operations (CMO) projects during stability operations. (During OPERATION IRAQI FREEDOM, these funds were referred to as commander's emergency response program [CERP] funds.) These funds are typically disbursed as bulk currency to tactical units, where the funds are allocated against lower-cost, high-impact projects that support the immediate needs of civil authorities and their citizens. As such, commanders must be prepared to receive, account, secure, and properly disburse these resources in accordance with established procedures. Commanders have an inherent responsibility for ensuring the proper use and disbursement of these funds and must execute appropriate judgment, fiscal responsibility, and supervisory oversight of any resources appropriated in support of CMO. While commanders establish appropriate guidelines to accomplish this fiscal responsibility, they must provide their subordinate commanders—who have the need, capacity, and capability to put these funds to use—reasonable flexibility in how they spend at least a portion of the money, so that they can address emerging needs within their AO.

Contracting Expertise

10-93. Due to the nature of UO, tactical-level units will be likely required to plan and initiate contracts not only for reconstruction projects but to supplement their own sustainment needs. Therefore, higher-level commands must be prepared to assist tactical-level units with contracting expertise. This expertise is required to overcome difficulties in delineating the scope of work, defining construction standards, and establishing viable delivery dates. For small projects and purchases, a standard contractual format may be sufficient. However, for large projects, commanders must provide their subordinates with knowledgeable contracting personnel, who can write contracts with clearly-understood specifications, create purchase orders that specify quality and delivery times, and help monitor work for quality control and assurance. Payment guidelines and standards among all services and members of a coalition are essential to eliminate competition and potential overpayment. Translators will be also essential to accurately translate requirements and standards in addition to trusted local agents to assist in selecting quality engineering and construction firms and workers. Overall, subordinate commanders need adequate contracting support and expertise to meet civilian needs as well as their expectations for timeliness and quality. Without it, commanders may risk losing critical opportunities to positively influence civilian perceptions. (See also Legal Support to Operations below.)

Religious Support

10-94. Religious support entails providing and performing religious support operations for commanders to protect the free exercise of religion for Soldiers, family members, and authorized civilians. This includes personal delivery of rites, sacraments, ordinances, pastoral and spiritual care, and religious education. Such support also consists of:

- Advising the commander on matters of religion, morals, and morale as affected by religion.
- The impact of local religions on threat courses of action, the urban society, and, subsequently, the mission.
- The ethical impact of command decisions.

Moral and Ethical Climate

10-95. Chaplains can help commanders sense the moral and ethical climate in Army units and understand potential moral dilemmas associated with planned UO. The urban environment affects Soldiers' mental health and can increase combat stress casualties, especially misconduct stress behaviors. Chaplains are an important part of combat stress support. The chaplain's presence and faith can sustain Soldiers throughout periods of great trauma. Religious support contributes to the total well being of Soldiers and aids in their return to combat readiness. Chaplains help bolster Soldiers' moral and ethical behaviors through spiritual fitness training so that Soldiers may better cope with ambiguous moral and ethical situations. Chaplains can help identify ethical concerns before they become critical command problems. To this end, their observations help develop and modify ROE, which—due to ambiguity or overly restraining rules—may be part of the problem.

Influence of Local Religions

10-96. Chaplains advise commanders on matters of religion as it affects the Soldiers within their units. They also explain the influences of local religions on threat courses of action and the urban populace and, subsequently, their potential effects on Army forces and UO. Religion is a crucial aspect of assessing the societal component of the urban environment. Understanding the major tenets and concepts of the religions and the impact of faith on civilians' lives may help commanders understand what motivates the threat and the urban populace. (Historically, when religion was a critical factor in armed conflict, the level of lethality has been significantly higher.) To understand threat strategy and tactics, commanders understand their ideology, which, in many cases, involves strong religious beliefs. This understanding also helps commanders to appreciate the inhabitants' attitudes toward other races, religions, and cultures and to identify unacceptable kinds of social interaction (particularly between Soldiers and civilians). Failure to recognize and respect religious beliefs can rapidly erode the perceived legitimacy of the mission and increase the length of the operation. In addition to its influence on threat courses of action, a thorough analysis of the urban environment also includes the degree of influence religion and religious leaders have on the area's government and economy.

LEGAL SUPPORT TO OPERATIONS

10-97. Legal support provides operational law support in all legal disciplines (including military justice, international law, civil law—composed of contract, fiscal, and environmental law, claims, and legal assistance). This support assists in command and control, sustainment, and HRS of UO. Legal considerations are important in any operation; they take on added significance during UO. They form the foundation for establishing ROE and are critical in the targeting process. (For example, with the increased likelihood that the threat will use the media in an attempt to categorize Army attacks as unlawful, legal support will be critical in determining the protected status of targets.) Adequate legal instruction and training may help subordinate commanders and their Soldiers better understand some aspects ROE and, with this understanding, help them make the rapid decisions often required in a complex urban environment.

10-98. Legal considerations also affect how units acquire goods and services from urban areas and provide support to other agencies and organizations operating in an area. The environment's complex nature requires commanders and their staffs to review and closely consider applicable legal constraints when developing and executing friendly courses of action. Most urban areas have a highly developed legal system nested within county, provincial, state, and national systems. The SJA support to commanders addresses this urban system in conjunction with other legal systems and their potential to affect, positively or negatively, the success of UO.

10-99. International, host-nation, and U.S. law and other regulatory guidelines may vary in their applicability by time and place; actions permissible in one situation may be prohibited in another. These exceptions and complexities increase requirements for SJA, often working with CA personnel, to identify and resolve technical legal issues. Therefore, the SJA must actively advise and participate in all aspects of UO from predeployment training and initial planning through transition and redeployment. FM 27-100 contains detailed legal guidelines affecting UO.

International and Host-Nation Law

10-100. International law consists primarily of agreements, treaties, and customary law to include the law known as the law of war (see FM 27-10). The law of war consists of four general principles applicable when conducting any operation but requiring particular attention during UO. Figure 10-4 lists and describes the four principles: military necessity, discrimination (or distinction), unnecessary suffering (or humanity), and proportionality.

10-101. International law may affect urban operational issues, such as the right of entry, base operations, use of urban infrastructure, and overflight and landing rights. Status-of-forces agreements (SOFAs) exist or can be negotiated to resolve legal issues, such as the status of Soldiers (and, as necessary, contractors) operating in foreign areas to include criminal and civil jurisdictions, taxation concerns, and claims for damages and injuries. Unless a SOFA or other convention exists, Soldiers operating in foreign urban areas have the same legal status as tourists; they are subject to the laws and judicial process of the host nation. (During armed conflict, however, Soldiers always have the rights afforded to them by the Geneva Conventions). Commanders are responsible for understanding the international and host-nation agreements and laws that influence foreign UO. If local law hinders the operation, commanders may be able to inform the local U.S. diplomatic mission and request that it negotiate a solution.

10-102. Commanders may encounter civilian resistance groups whose actions may range from providing enemies with sustainment support to actively fighting Army and coalition forces. Members of such resistance groups must be dealt with in accordance with applicable provisions of the law of war. Commanders should seek legal guidance concerning detention and disposition of persons participating in various acts harmful to friendly forces and detrimental to the mission.

U.S. Law

10-103. Commanders conducting UO must also comply with U.S. law whether it is in the form of a statute, executive order, regulation, or other directive from a federal branch or agency. U.S. law influences UO by governing the acquisition of supplies and services for Army forces, regulating the assistance that can be rendered to foreign nations, and controlling intelligence activities. The Posse Comitatus Act, for example, makes it a crime for Army forces to enforce civil law. Similarly, portions of the Foreign Assistance Act prohibit Soldiers from performing law enforcement activities in foreign urban areas. However, circumstances—expressly authorized by the Constitution, acts of Congress, and other exceptions to these statutes—exist that allow the Army to support civilian law enforcement. Although not nearly all-inclusive, the above demonstrates how U.S. law further complicates urban operations, particularly stability operations, and increases the need for proactive SJA advice and counsel in all facets of UO.

Military Necessity: The principle that justifies the use of measures not forbidden by international or domestic law necessary to rapidly achieve military objectives.

Discrimination: The principle of distinguishing between combatants (who may be attacked) and noncombatants (who may not be attacked).

Unnecessary Suffering: The principle that prohibits the use of weapons, projectiles, or other materials in a manner calculated to cause superfluous injury and unnecessary suffering.

Proportionality: The principle that injury to persons and damage to property incidental to military action, in the circumstances ruling at the time, must not be excessive in relation to the concrete and direct military advantage anticipated.

Figure 10-4. General principles of the law of war

Legal Aspects of Nonlethal Force

10-104. Nonlethal capabilities can augment the means of deadly force and extend urban firepower options. They enhance the commander's ability to apply force in proportion to the threat and to allow discrimination in its use. The range of nonlethal capabilities includes offensive information operations, smoke and obscurants, irritants (such as chemical riot control agents), nonpenetrating projectiles, high-pressure water devices, and directed-energy weapons. These continually expanding capabilities give commanders more options to confront situations that do not warrant deadly force but require Soldiers to employ overwhelming decisive power. However, nonlethal capabilities are subject to the same legal constraints as lethal force (in fact, some nonlethal capabilities can cause serious injury and death, particularly if not employed properly) and undergo the same legal review. Like lethal force, nonlethal capabilities show military necessity, distinguish between combatants and noncombatants, distinguish between military objectives and protected property, are used proportionally, and do not result in unnecessary suffering.

10-105. Of special note, commanders cannot employ chemical herbicides or riot control agents as a method of warfare without prior Presidential approval. However, the Secretary of Defense and regional combatant commanders have authority to use herbicides and riot control agents in limited circumstances during armed conflict and in peacetime operations. Considering the complexity of this issue, commanders of major operations must streamline the approval process so that subordinate commanders do not discount use of these agents simply because of the high-level approval authority. Subordinate commanders must, however, be able to anticipate and clearly justify their potential use in UO. Also, any approval does not mean that these agents must be employed.

Civilians Accompanying the Force

10-106. LOGCAP personnel and other civilian support contractors provide a variety of sustainment functions for the Army. These functions may range from providing unskilled labor, transportation support, and health care to technical support of sophisticated equipment and weapons systems. Commanders must ensure that civilians providing support in their AO are not placed in positions of jeopardy, but if they are, these civilians must understand the risks that they assume when they engage in activities that might be misconstrued as direct or active participation in hostilities.

10-107. This last caution indicates an important legal consideration for commanders. They must try to ensure that threat forces can distinguish civilian support contractors from combatants and thereby remove any justification for them to be targeted or attacked. (As Army forces seek to work closely with the members of governmental and nongovernmental agencies and organizations in urban stability operations, commanders may need to take steps necessary to mitigate the risk of attack caused by their involvement with Army forces.) In the event that threat forces capture civilian support contractors, this same concern helps to ensure that, as lawful civilians accompanying the force, they will be afforded prisoner of war status. As the role of contractor support to Army operations continues to expand, commanders must stay abreast of changes in the law (to include evolving interpretative distinctions) as it affects the use of contractors on the urban battlefield. Other important considerations for commanders with civilian contractors supporting their force are—

- **Minimize Exposure to Harm**. Civilian contractors should be assigned duties at locations that minimize their exposure to harm. Commanders will need to periodically review these locations since they are likely to change as the urban operation evolves and transitions.
- **Prohibit the Performance of Combat Functions**. Civilian support personnel should not be working in situations that involve combat operations where they might be conceived as combatants. As an example, they should not perform force protection for Army forces such as fortification construction and checkpoint security.
- **Prohibit the Use of Arms**. Civilians should not be armed, unless approved by the combatant or service component commander on a limited, by-exception basis for personal defense.
- **Provide Identification Cards**. Civilians should be provided, and carry on their person, a Geneva Convention identification card identifying them as civilians authorized to accompany military forces and entitling them to be treated, if captured, as prisoners of war.
- **Restrict the Wear of Military Uniforms**. Civilians should not normally wear distinctive U.S. military uniforms unless the combatant or service component commander authorizes them to do so. Regardless of their clothing, however, civilians will wear a symbol that establishes their civilian status.

GENERAL ENGINEERING SUPPORT

10-108. General engineering support will be essential during UO. This support helps assess, construct, maintain, and restore essential LOCs and urban facilities to sustain Army forces, the urban population, or both. Using civilian resources and investing Army general engineering resources requires careful consideration by commanders and staff planners. Since all elements of the urban infrastructure interconnect, general engineering support touches each category to some degree. Figure 10-5 illustrates how urban-specific, general engineering tasks align primarily with the transportation and distribution, energy, and administration and human services components of the urban infrastructure. These engineering tasks are significant and readily apply to UO. The last two, providing fire fighting support and waste management, have not been previously addressed and require more specific consideration.

Construct, Maintain, or Restore

- Roads and Highways
- Over-the-Shore Facilities
- Ports
- Railroad Facilities
- Airports and Heliports
- Fixed Bridges
- Electric Power Facilities
- Petroleum Pipelines and Storage Facilities
- Water Facilities

and Provide

- Mobile Electric Power
- Construction Support
- Fire Fighting Support
- Waste Management

Figure 10-5. General engineering support tasks

The Use or Investment of Resources

10-109. During urban offensive and defensive operations, Army engineer units accomplish tasks to sustain or improve movement and mobility, protection, and sustainment of U.S. and allied forces. These units should maximize the existing urban facilities, host-nation support, civilian contractors, and joint engineer assets. However, commanders should consider how using urban facilities to support military forces may negatively affect the population. On the other hand, construction and repair may benefit both Army units and the urban inhabitants. Restoring the urban transportation network not only improves military LOCs, but may also allow needed commerce to resume. Repairing urban airfields or ports increases throughput capabilities for military supplies, facilitates medical evacuation operations to the support base, accelerates needed relief efforts, and allows international commerce to proceed. Commanders may first invest resources and conduct general engineering tasks to restore facilities for civilian use. Such actions stem future drains on operational resources or facilitate later transition of control back to civilian authorities. For example, repairing police stations, detention facilities, and marksmanship ranges may help urban governments reestablish law and order after completing urban offensive or defensive operations. During stability or civil support operations, the focus of general engineering will often be to support and assist the urban population rather than Army forces.

Fire Fighting Support

10-110. Fire protection and prevention, as well as fire fighting, takes on added significance during UO, particularly offensive and defensive operations. Most ordnance produces heat and flame. This, coupled with an abundance of combustible material (buildings, furniture, gasoline, oil, and propane), poses a serious risk to Soldiers, civilians, and the urban operation. Large shantytowns can exacerbate this problem. In highly combustible areas, commanders may even need to limit or preclude the use of small-arms tracer ammunition. Fire threats to urban areas can be categorized as—

- **Isolated Fires**. These are restricted to a single structure or a specific area within a structure.
- **Area Fires**. These fires consume two or more structures and may extend to encompass an entire block. Generally, streets will serve as firebreaks and help contain the fire within a single block.
- **Firestorms**. These are the most violent and dangerous fires capable of rapidly consuming large areas by creating windstorms and intense heat. Firestorms are often inextinguishable until they have consumed all available combustible materials.

- **Explosive Hazards**. These are present in areas containing fuels and chemicals as well as military explosive hazards.

10-111. When analyzing the administration and human services component of the infrastructure, commanders must determine the adequacy of existing civilian fire fighting support. A deteriorated or nonexistent infrastructure that cannot support the urban area will likely fail to handle the increased risk due to military operations. Commanders may need to provide fire fighting teams to support their own forces and civilians.

10-112. A military force task organized with multiple fire fighting teams (even with maximum use of available civilian fire-fighting assets) will only be able to fight some fires in the AO. Water distribution systems damaged during operations, chemical and other TIM, and hostile activities will further complicate and limit fire fighting capabilities. Commanders must develop priorities for equipment, facility, and infrastructure protection. All Soldiers need training in fire prevention and initial or immediate response fire fighting. Such training includes planning covered and concealed movement, withdrawal, and evacuation routes. Soldiers should be trained to identify and remove ignition and fuel sources and provided additional fire fighting material such as extinguishers, sand, and blankets (see FM 5-415 and the U.S. Department of Transportation's current version of the Emergency Response Guidebook).

Waste Management

10-113. Management of all forms of waste, particularly human, putrescible (such as food), and medical, may become a critical planning consideration for Army forces. This particularly applies if the urban waste management infrastructure was previously inadequate or damaged during UO or natural disasters, the Army force is operating in the urban area for an extended period of time, and a significant number of the urban population remains. Failure to adequately consider this aspect, possibly coupled with the decaying remains of humans and animals and an inadequate or tainted water supply (such as may be caused by flooding), may create unacceptable sanitary and hygiene conditions and subsequently increase DNBIs as well as civilian casualties.

Appendix A

Siege of Beirut: An Illustration of the Fundamentals of Urban Operations

The [Israeli Defense Forces] IDF had neither the strategy nor the experience nor the configuration of forces to fight and sustain a house-to-house campaign in Beirut.

Richard A. Gabriel
Operation Peace for Galilee: The Israeli-PLO War in Lebanon

OVERALL STRATEGIC SITUATION

A-1. In 1982, Israel launched OPERATION PEACE FOR GALILEE designed to destroy the Palestine Liberation Organization (PLO) presence in southern Lebanon. On 1 June, the IDF launched a massive assault across the border into southern Lebanon. The Israeli attack focused on the PLO, but the operations quickly involved major ground and air combat between Israel and Syrian forces.

A-2. In the first few weeks, Israeli forces quickly pushed back both the Syrians and the PLO. However, except for some PLO forces isolated in bypassed urban areas, such as Tyre and Sidon, most of the PLO fell back into Beirut (see Figure A-1). By 30 June, Israeli forces had reached the outskirts of southern Beirut, occupied East Beirut, isolated the city from Syria and the rest of Lebanon, and blockaded the sea approaches to the city. Even so, with most of the PLO intact inside and with significant military and political capability, the Israelis had yet to achieve the objective of OPERATION PEACE FOR GALILEE. The Israeli command had to make a decision. It had three choices: permit the PLO to operate in Beirut; execute a potentially costly assault of the PLO in the city; or lay siege to the city and use the siege to successfully achieve the objective. The Israelis opted for the latter.

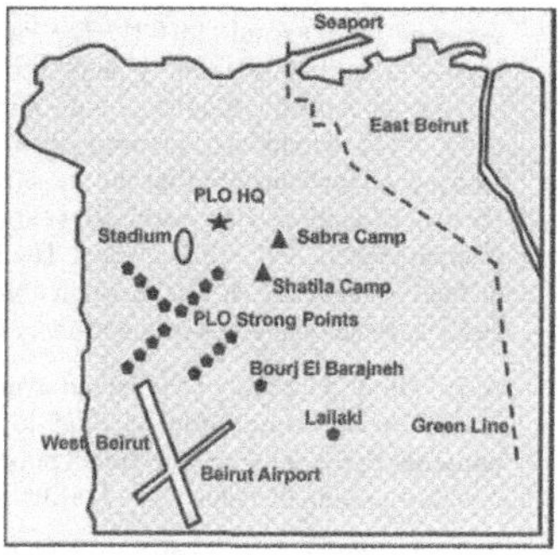

Figure A-1. The city of Beirut

Israeli Military Position

A-3. The Israelis had an excellent position around Beirut. They occupied high ground to the south and west, virtually dominating the entire city. Israeli naval forces controlled the seaward approaches to Beirut. The Israelis' position was also strong defensively, capable of defeating any attempt to break out of or into the city from northern Lebanon or Syria. The Israeli air force had total and complete air superiority. The Israelis controlled the water, fuel, and food sources of West Beirut. Although the PLO forces had stockpiles of food and supplies, the Israelis regulated the food, water, and generating power for the civil population.

A-4. Despite the superior positioning of Israeli forces, the IDF faced significant challenges to include the combat power of the PLO, Syria, and other threats in Beirut. Israeli doctrine and training did not emphasize urban operations. Additionally, Israel was constrained by its desire to limit collateral damage and friendly and noncombatant casualties. Organizationally, the Israeli army was not optimized to fight in urban terrain.

Armor and self-propelled artillery formations dominated the Israeli forces, and most Israeli infantry was mechanized. The Israeli forces had only a few elite formations of traditional dismounted infantry.

PLO MILITARY POSITION

A-5. Despite being surrounded and cut off from support, the PLO position in Beirut offered numerous advantages in addition to the characteristic advantages of urban defense. The PLO had long anticipated an Israeli invasion of southern Lebanon; it had had months to prepare bunkers, obstacles, and the defensive plan of Beirut and other urban areas. Approximately 14,000 Arab combatants in West Beirut readied to withstand the Israeli siege. This was done with the advice of Soviet, Syrian, and east European advisors. The preparation included stockpiling essential supplies in quantities sufficient to withstand a six-month siege. Also, the PLO fighters integrated into the civil populations of the urban areas. Often their families lived with them. The civil population itself was friendly and provided both information and concealment for PLO forces. PLO fighters were experienced in urban combat and knew the urban terrain intimately. PLO forces had been involved in urban fighting against Syrian conventional forces and Christian militias in Beirut several years prior to the Israeli invasion. Finally, the organization of the PLO—centered on small teams of fighters armed with machine-guns and antitank weapons, and trained in insurgent, hit-and-run tactics—was ideally suited to take maximum advantage of the urban environment.

ROLE OF CIVILIANS

A-6. Various ethnic and religious groups make up the civil population of southern Lebanon. However, West Beirut's population was heavily Palestinian and Lebanese. The civil population of West Beirut was between 350,000 and 500,000. The Palestinian population supported the PLO. The Lebanese population may be described as friendly neutral to the Israelis. Although unhappy under Palestinian dominance, this population was unwilling to actively support Israel. The civilian population was a logistic constraint on the PLO, which would have become significant had the siege lasted longer. The civilians in West Beirut were an even larger constraint on the Israelis. The presence of civilians significantly limited the ability of the Israelis to employ firepower. However, the Palestine refugee camps located in West Beirut were both civilian centers and military bases. The Israeli constraints on artillery and other systems against these parts of the city were much less restrictive than in other parts of West Beirut where the population was mostly Lebanese and where fewer key military targets existed.

A-7. The PLO knew of the Israeli aversion to causing civil casualties and purposely located key military centers, troop concentrations, and logistics and weapons systems in and amongst the population—particularly the refugee Palestinian population in the southern part of West Beirut. Tactically, they used the civilians to hide their forces and infiltrate Israeli positions.

A-8. The friendly Palestinian population provided intelligence to the PLO while the friendly Lebanese population provided intelligence for the IDF. Throughout the siege, the IDF maintained a policy of free passage out of Beirut for all civilians. This policy was strictly enforced and permitted no weapons to leave the city. Some estimates are that as many as 100,000 refugees took advantage of this policy.

INFORMATION OPERATIONS

A-9. The siege of Beirut involved using information operations (IO) to influence the media. PLO information operations were aimed at controlling the media and hence the international perception of the operation. This was done by carefully cultivating a select group of pro-PLO media years before hostilities even began. Once hostilities started, only these media sources were permitted to report from the besieged portions of the city, and they were only shown activities that portrayed the IDF negatively. The IDF did not vigorously counter the PLO plan. In fact, the IDF contributed to it by limiting media access to their activities. The PLO information operations had a successful impact. The international community was constantly pressuring the Israeli government to end hostilities. This put pressure on the IDF to conduct operations rapidly and to limit firepower and casualties.

CONDUCT OF THE URBAN OPERATIONS

A-10. The siege of Beirut began 1 July (see Figure A-2). By 4 July, Israeli forces occupied East Beirut, the Green Line separating East and West Beirut, and dominating positions south of the airport. IDF naval forces also controlled the sea west and north of Beirut. On 3 and 4 July, IDF artillery and naval fire began a regular campaign of firing on military targets throughout West Beirut. On 4 July, the IDF cut power and water to the city.

A-11. From 5 to 13 July, the Israeli fires continued to pound PLO targets in West Beirut. The PLO gave one significant response, firing on an Israeli position south of the city and causing several casualties. On 7 July, reacting to international pressure, the IDF returned power and water to West Beirut's civil population. On 11 July, the IDF launched its first attack, probing the southern portion of the airport with an armored task force (see Figure A-3). The PLO repulsed this attack and destroyed several IDF armored vehicles.

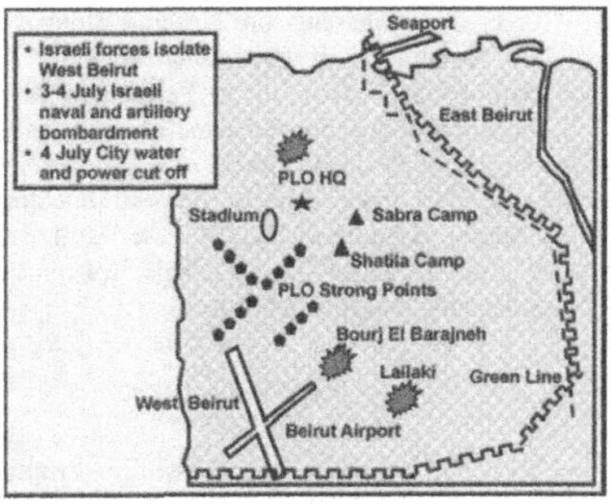

Figure A-2. Initial conduct of the urban operation

A-12. On 13 July, both sides entered into a cease-fire that lasted until 21 July. They began negotiations, mediated by international community, to end the siege. The PLO used this period to continue to fortify Beirut. The Israelis used the time to train their infantry and other arms in urban small unit tactics in Damour, a town the Israeli paratroopers had captured.

A-13. The cease-fire ended on 21 July as PLO forces launched three attacks on IDF rear areas. The Israelis responded with renewed and even more vigorous artillery, naval, and air bombardment of PLO positions in the city. The IDF attacks went on without respite until 30 July. On 28 July, the IDF renewed its ground attack in the south around the airport (see Figure A-4). This time IDF forces methodically advanced and captured a few hundred meters of ground establishing a toehold.

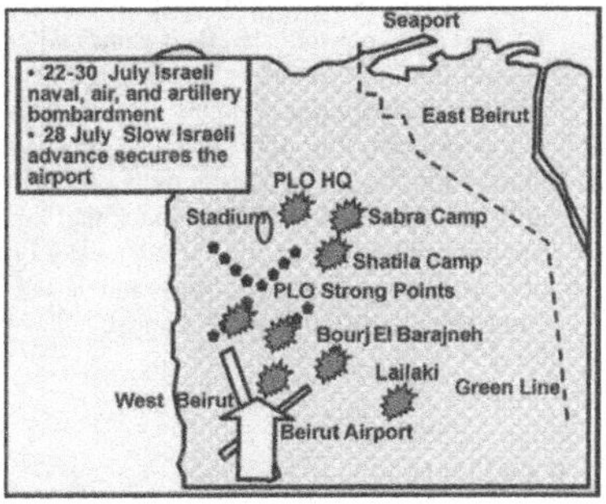

Figure A-3. Israeli probe of PLO defenses

A-14. The Israeli bombardment stopped on 31 July. However, on 1 August the IDF launched its first major ground attack, successfully seizing Beirut airport in the south (see Figure A-5). Israeli armored forces began massing on 2 August along the green line, simultaneously continuing the attack from the south to the outskirts of the Palestinian positions at Ouzai. On 3 August, the Israeli forces continued to reinforce both their southern attack forces and forces along the green line to prepare for continuing offensive operations. On 4 August, the IDF attacked at four different places. This was the much-anticipated major Israeli offensive.

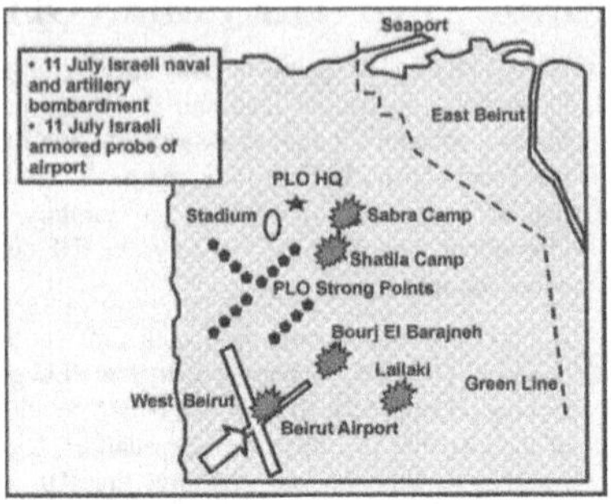

Figure A-4. Initial Israeli attack

A-15. The Israeli attack successfully disrupted the coherence of the PLO defense. The southern attack was the most successful: it pushed PLO forces back to their camps of Sabra and Shatila and threatened to overrun PLO headquarters. Along the green line the IDF attacked across three crossing points. All three attacks made modest gains against stiff resistance. For this day's offensive, the Israelis suffered 19 killed and 84 wounded, the highest single day total of the siege, bringing the total to 318 killed. Following the major attacks on 4 August, Israeli forces paused and, for four days, consolidated their gains and prepared to renew the offensive. Skirmishes and sniping continued, but without significant offensive action. On 9 August, the IDF renewed air and artillery attacks for four days. This activity culminated on 12 August with a massive aerial attack that killed over a hundred and wounded over 400—mostly civilians. A cease-fire started the next day and lasted until the PLO evacuated Beirut on 22 August.

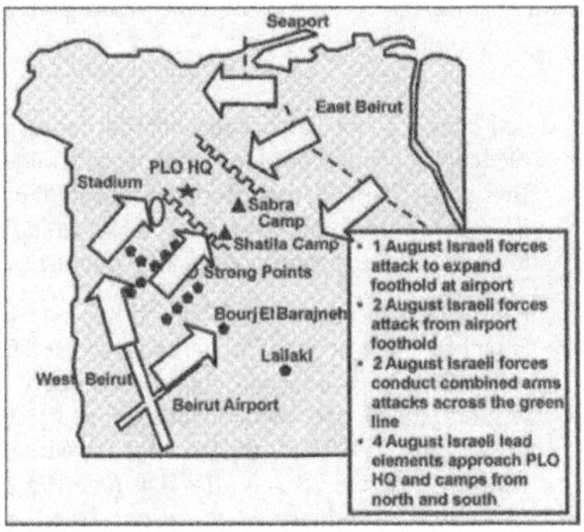

Figure A-5. Final Israeli attack

LESSONS

A-16. The Israeli siege of West Beirut was both a military and a political victory. However, the issue was in doubt until the last week of the siege. Military victory was never in question; the issue in doubt was whether the Israeli government could sustain military operations politically in the face of international and domestic opposition. On the other side, the PLO faced whether they could last militarily until a favorable political end could be negotiated. The answer was that the PLO's military situation became untenable before the Israeli political situation did.

A-17. This favorable military and political outcome stemmed from the careful balance of applying military force with political negotiation. The Israelis also balanced the type of tactics they employed against the domestic aversion to major friendly casualties and international concern with collateral damage.

PERFORM AGGRESSIVE INFORMATION OPERATIONS

A-18. The PLO devoted considerable resources and much planning on how to use IO to their best advantage. They chose to focus on media information sources as a means of influencing international and domestic opinion.

A-19. The PLO's carefully orchestrated misinformation and control of the media manipulated international sentiment. The major goal of this effort was to grossly exaggerate the claims of civilian casualties, damage, and number of refugees—and this was successfully accomplished. Actual casualties among the civilians were likely half of what the press reported during the battle. The failure of the IDF to present a believable and accurate account of operations to balance PLO efforts put tremendous pressure on the Israeli government to break off the siege. It was the PLO's primary hope for political victory.

A-20. In contrast to the overall weak performance in IO, the IDF excelled in psychological operations. IDF psychological operations attacked the morale of the PLO fighter and the Palestinian population. They were designed to wear down the will of the PLO to fight while convincing the PLO that the IDF would go to any extreme to win. Thus defeat was inevitable. The IDF used passive measures, such as leaflet drops and loudspeaker broadcasts. They used naval bombardment to emphasize the totality of the isolation of Beirut. To maintain high levels of stress, to deny sleep, and to emphasize their combat power, the IDF used constant naval, air, and artillery bombardment. They even employed sonic booms from low-flying aircraft to emphasize the IDF's dominance. These efforts helped to convince the PLO that the only alternative to negotiation on Israeli terms was complete destruction.

Maintain a Close Combat Capability

A-21. The ground combat during the siege of Beirut demonstrated that the lessons of tactical ground combat learned earlier in the twentieth century were still valid. Small combined arms teams built around infantry, but including armor and engineers, were the key to successful tactical combat. Artillery firing in direct fire support of infantry worked effectively as did the Vulcan air defense system. The Israeli tactical plan was sound. The Israelis attacked from multiple directions, segmented West Beirut into pieces, and then destroyed each individually. The plan's success strongly influenced the PLO willingness to negotiate. Tactical patience based on steady though slow progress toward decisive points limited both friendly and noncombatant casualties. In this case, the decisive points were PLO camps, strong points, and the PLO headquarters.

A-22. The Israeli willingness to execute close combat demonstrated throughout the siege, but especially in the attacks of 4 August, was decisive. Decisive ground combat was used sparingly, was successful and aimed at decisive points, and was timed carefully to impact on achieving the political objectives in negotiations. The PLO had hoped that their elaborate defensive preparations would have made Israeli assaults so costly as to convince the Israelis not to attack. That the Israelis could successfully attack the urban area convinced the PLO leadership that destruction of their forces was inevitable. For this reason they negotiated a cease-fire and a withdrawal on Israeli terms.

Avoid the Attrition Approach

A-23. The Israelis carefully focused their attacks on objects that were decisive and would have the greatest impact on the PLO: the known PLO headquarters and refugee centers. Other areas of West Beirut were essentially ignored. For example, the significant Syrian forces in West Beirut were not the focus of Israeli attention even though they had significant combat power. Selectively ignoring portions of the urban area allowed the Israelis to focus their combat power on the PLO and limit both friendly casualties and collateral damage.

Control the Essential and Preserve Critical Infrastructure

A-24. The Israeli siege assured Israeli control of the essential infrastructure of Beirut. The initial Israeli actions secured East Beirut and the city's water, power, and food supplies. The Israelis also dominated Beirut's international airport, closed all the sea access, and controlled all routes into and out of the city. They controlled and preserved all that was critical to operating the city and this put them in a commanding position when negotiating with the PLO.

Minimize Collateral Damage

A-25. The Israeli army took extraordinary steps to limit collateral damage, preserve critical infrastructure, and put in place stringent rules of engagement (ROE). They avoided randomly using grenades in house clearing, limited the use of massed artillery fires, and maximized the use of precision weapons. With this effort, the Israelis extensively used Maverick missiles because of their precise laser guidance and small warheads.

A-26. The strict ROE, however, conflicted with operational guidance that mandated that Israeli commanders minimize their own casualties and adhere to a rapid timetable. The nature of the environment made fighting slow. The concern for civilian casualties and damage to infrastructure declined as IDF casualties rose. They began to bring more field artillery to bear on Palestinian strong points and increasingly employed close air support. This tension underscores the delicate balance that Army commanders will face between minimizing collateral damage and protecting infrastructure while accomplishing the military objective with the least expenditure of resources—particularly soldiers. ROE is but one tool among many that a commander may employ to adhere to this UO fundamental.

Understand the Human Dimension

A-27. The Israelis had a noteworthy (although imperfect and at times flawed) ability to understand the human dimension during their operations against the PLO in Beirut. This was the result of two circumstances. First, the PLO was a threat with which the Israeli forces were familiar after literally decades of conflict. Second, through a close alliance and cooperation with Lebanese militia, the Israelis understood a great deal regarding the attitudes and disposition of the civil population both within and outside Beirut.

Separate Noncombatants from Combatants

A-28. Separating combatants from noncombatants was a difficult but important aspect of the Beirut operation. The Israelis made every effort to positively identify the military nature of all targets. They also operated a free passage system that permitted the passage of all civilians out of the city through Israeli lines. The need to impose cease-fires and open lanes for civilians to escape the fighting slowed IDF operations considerably. Additionally, Israeli assumptions that civilians in urban combat zones would abandon areas where fighting was taking place were incorrect. In many cases, civilians would try to stay in their homes, leaving only after the battle had begun. In contrast, the PLO tied their military operations closely to the civilian community to make targeting difficult. They also abstained from donning uniforms to make individual targeting difficult.

A-29. Earlier in OPERATION PEACE FOR GALILEE when the IDF attacked PLO forces located in Tyre, Israeli psychological operations convinced 30,000 Lebanese noncombatants to abandon their homes and move to beach locations outside the city. However, the IDF was subsequently unable to provide food, water, clothing, shelter, and sanitation for these displaced civilians. IDF commanders compounded the situation by interfering with the efforts by outside relief agencies to aid the displaced population (for fear that the PLO would somehow benefit). Predictably, many civilians tried to return to the city complicating IDF maneuver and targeting—that which the separation was designed to avoid. IDF commanders learned that, while separation is important, they must also adequately plan and prepare for the subsequent control, health, and welfare of the noncombatants they displace.

Restore Essential Services

A-30. Since essential services were under Israeli command, and had been since the beginning of the siege, the Israelis had the ability to easily restore these resources to West Beirut as soon as they adopted the cease-fire.

Transition Control

A-31. In the rear areas of the Israeli siege positions, the Israeli army immediately handed over civic and police responsibility to civil authorities. This policy of rapid transition to civil control within Israeli lines elevated the requirement for the Israeli army to act as an army of occupation. The Israeli army believed the efficient administration of local government and police and the resulting good will of the population more than compensated for the slightly increased force protection issues and the increased risk of PLO infiltration.

A-32. Upon the cease-fire agreement, Israeli forces withdrew to predetermined positions. International forces under United Nations (UN) control supervised the evacuation of the PLO and Syrian forces from Beirut. These actions were executed according to a meticulous plan developed by the Israeli negotiators and agreed to by the PLO. Israeli forces did not take over and occupy Beirut as a result of the 1982 siege (an occupation did occur later but as a result of changing situations).

SUMMARY

A-33. The Israeli siege of West Beirut demonstrates many of the most demanding challenges of urban combat. Apart from Israel's poor understanding of the strategic influence of the media on operations (a lesson that they learned and subsequently applied to future operations), the IDF's successful siege of Beirut emerged from their clear understanding of national strategic objectives and close coordination of diplomatic efforts with urban military operations. A key part of that synchronization of capabilities was the understanding that the efforts of the IDF would be enhanced if they left an escape option open to the PLO. This option was the PLO's supervised evacuation that occurred after the siege. Although the PLO was not physically destroyed, the evacuation without arms and to different host countries effectively shattered the PLO's military capability. Had Israel insisted on the physical destruction of the PLO in Beirut, the overall operation might have failed because that goal may not have been politically obtainable in view of the costs in casualties, collateral damage, and international opinion.

This page intentionally left blank.

Appendix B

An Urban Focus to the Intelligence Preparation of the Battlefield

Maneuvers that are possible and dispositions that are essential are indelibly written on the ground. Badly off, indeed, is the leader who is unable to read this writing. His lot must inevitably be one of blunder, defeat, and disaster.

Infantry in Battle

The complexity of the urban environment and increased number of variables (and their infinite combinations) increases the difficulty of providing timely, relevant, and effective intelligence support to urban operations (UO). Conducted effectively, however, the intelligence preparation of the battlefield (IPB) allows commanders to develop the situational understanding necessary to visualize, describe, and direct subordinates in successfully accomplishing the mission.

IPB is the systematic process of analyzing the threat and environment in a specific geographic area—the area of operations (AO) and its associated area of interest (see figure B-1). It provides the basis for intelligence support to current and future UO, drives the military decision-making process, and supports targeting and battle damage assessment. The procedure (as well as each of its four steps) is performed continuously throughout the planning, preparation, and execution of an urban operation.

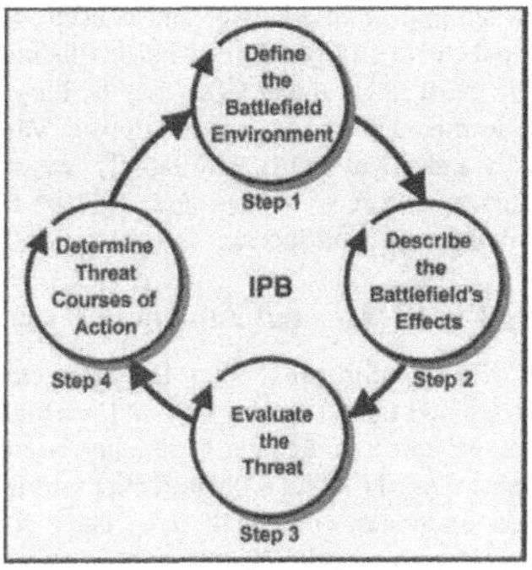

Figure B-1. The steps of IPB

UNAFFECTED PROCESS

B-1. The IPB process is useful at all echelons and remains constant regardless of the operation or environment. However, an urban focus to IPB stresses some aspects not normally emphasized for IPBs conducted for operations elsewhere. The complex urban mosaic is composed of the societal, cultural, or civil dimension of the urban environment; the overlapping and interdependent nature of the urban infrastructure; and the multidimensional terrain. This mosaic challenges the conduct of an urban-focused IPB. There is potential for full spectrum operations to be executed near-simultaneously as part of a single major operation occurring in one urban area with multiple transitions. Such multiplicity stresses the importance of a thorough, non-stop IPB cycle aggressively led by the commander and executed by the entire staff. Overall, the art of applying IPB to UO is in properly employing the steps to the specific operational environment. In UO, this translates to analyzing the significant characteristics of the environment, the role that its populace has in threat evaluation, and understanding how these affect the planning and execution of UO. FM 2-01.3 details how to conduct IPB; FM 34-3 has the processes and procedures for producing all-source intelligence. This appendix supplements the information found there; it does not replace it.

INCREASED COMPLEXITY

B-2. Uncovering intricate relationships takes time, careful analysis, and constant refinement to determine actual effects on friendly and threat courses of action (COAs). These relationships exist among—

- Urban population groups.
- The technical aspects of the infrastructure.
- The historical, cultural, political, or economic significance of the urban area in relation to surrounding urban and rural areas or the nation as a whole.
- The physical effects of the natural and man-made terrain.

B-3. A primary goal of any IPB is to accurately predict the threat's likely COA (step four—which may include political, social, religious, informational, economic, and military actions). Commanders can then develop their own COAs that maximize and apply combat power at decisive points. Understanding the decisive points in the urban operation allows commanders to select objectives that are clearly defined, decisive, and attainable.

Reducing Uncertainty and Its Effects

B-4. Commanders and their staffs may be unfamiliar with the intricacies of the urban environment and more adept at thinking and planning in other environments. Therefore, without detailed situational understanding, commanders may assign missions that their subordinate forces may not be able to achieve. As importantly, commanders and their staffs may miss critical opportunities because they appear overwhelming or impossible (and concede the initiative to the threat). They also may fail to anticipate potential threat COAs afforded by the distinctive urban environment. Commanders may fail to recognize that the least likely threat COA may be the one adopted precisely because it is least likely and, therefore, may be intended to maximize surprise. Misunderstanding the urban environment's effect on potential friendly and threat COAs may rapidly lead to mission failure and the unnecessary loss of Soldiers' lives and other resources. A thorough IPB of the urban environment can greatly reduce uncertainty and contribute to mission success.

Training, Experience, and Functional Area Expertise

B-5. Not all information about the urban environment is relevant to the situation and mission—hence the difficulty and the reason for IPB and intelligence analysis. Although it may appear daunting, institutional education, unit training, and experience at conducting intelligence support to UO will improve the ability to rapidly sort through all the potential information to separate the relevant from merely informative. (This applies to any new or difficult task.) The involvement and functional expertise of the entire staff will allow commanders to quickly identify the important elements of the environment affecting their operations. Fortunately, IPB as part of the entire intelligence process (plan, prepare, collect, process, and produce) is comprehensive enough to manage the seemingly overwhelming amounts of information coming from many sources. Accomplished properly, it allows commanders to recognize opportunities often without complete information.

B-6. As in any operational environment, tension exists between the desire to be methodical and the need to create the tempo necessary to seize, retain, and exploit the initiative necessary for decisive UO. Quickly defining the significant characteristics of the urban environment requiring in-depth evaluation (not only what we need to know but what is possible to know) allows rapid identification of intelligence gaps (what we know versus what we don't know). Such identification leads to information requirements and will drive the intelligence, surveillance, and reconnaissance (ISR) plan (how will we get the information we need). Commanders must carefully consider how to develop focused priority information requirements (PIR) to enable collectors to more easily weed relevant information from the plethora of information. Commanders can then make better decisions and implement them faster than urban threats can react.

Amplified Importance of Civil (Societal) Considerations

B-7. The Army focuses on warfighting. The experiences in urban operations gained at lower echelons often center on the tactics of urban offensive and defensive operations where the influences of terrain and enemy frequently dominate. At higher echelons, the terrain and enemy are still essential considerations, but the societal component of the urban environment is considered more closely and throughout the operational process. Moreover, the human or civil considerations gain importance in civil support or stability operations regardless of the echelon or level of command. In addition to the echelon and the type of operation, a similar relationship exists between the key elements of the urban environment and other situational factors. These factors can include where the operation lies within the range of operations or the level of war and the conventional or unconventional nature of the opposing threat. Figure B-2 graphically represents the varying significance of these elements to urban IPB. Overall, population effects are significant in how they impact the threat, Army forces, and overall accomplishment of strategic and operational goals.

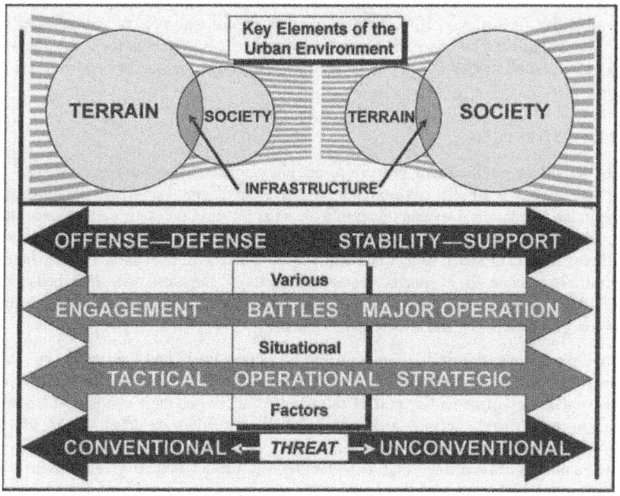

Figure B-2. Relativity of key urban environment elements

B-8. Describing the battlefield's effects—step two in IPB—ascribes meaning to the characteristics analyzed. It helps commanders understand how the environment enhances or degrades friendly and threat forces and capabilities. It also helps commanders understand how the environment supports the population. It also explains how changes in the "normal" urban environment (intentional or unintentional and because of threat or friendly activities) may affect the population. Included in this assessment are matters of perception. At each step of the IPB process, commanders must try to determine the urban society's perceptions of ongoing activities to ensure Army operations are viewed as intended. Throughout this process, commanders, staffs, and analysts cannot allow their biases—cultural, organizational, personal, or cognitive—to markedly influence or alter their assessment (see FM 34-3). This particularly applies when they analyze the societal aspect of the urban environment. With so many potential groups and varied interests in such a limited area, misperception is always a risk.

SIGNIFICANT CHARACTERISTICS

B-9. Urban intelligence analysis must include consideration of the urban environment's distinguishing attributes—man-made terrain, society, and infrastructure—as well as the underlying natural terrain (to include weather) and the threat. Because the urban environment is so complex, it is often useful to break it into categories. Then commanders can understand the intricacies of the environment that may affect their operations and assimilate this information into clear mental images. Commanders can then synthesize these images of the environment with the current status of friendly and threat forces and develop a desired end state. Then they can determine the most decisive sequence of activities that will move their forces from the current state to the end state. Identifying and understanding the environment's characteristics (from a friendly, threat, and noncombatant perspective) allows commanders to establish and maintain situational understanding. Then they can develop appropriate COAs and rules of engagement that will lead to decisive mission accomplishment.

B-10. Figures B-3, B-4, and B-5 are not intended to be all-encompassing lists of urban characteristics. Instead, they provide a starting point or outline useful for conducting an urban-focused IPB and analysis that can be modified to fit the specific operational environment and meet the commander's requirements. Commanders and staffs can compare the categories presented with those in the civil affairs area study and assessment format found in FM 41-10 and the IPB considerations for stability and civil support operations found in ST 2-91.1.

INTERCONNECTED SYSTEMS

B-11. Since the urban environment comprises an interconnected "system of systems," considerations among the key elements of the environment will overlap during an urban intelligence analysis. For example, boundaries, regions, or areas relate to a physical location on the ground. Hence, they have urban terrain implications. These boundaries, regions, or areas often stem from some historical, religious, political, administrative, or social aspect that could also be considered a characteristic of the urban society. Overlaps can also occur in a specific category, such as infrastructure. For instance, dams are a consideration for their potential effects on transportation and distribution (mobility), administration and human services (water supply), and energy (hydroelectric).

B-12. This overlap recognition is a critical concern for commanders and their staffs. In "taking apart" the urban environment and analyzing the pieces, commanders and staffs cannot lose perspective of how each piece interacts with any other and as part of the whole. Otherwise, their vision will be shortsighted, and they will fail to recognize the second- and third-order effects of their proposed COAs; the actual end state differing dramatically from the one envisioned by the commander. The increased density of combatants and noncombatants, infrastructure, and complex terrain means that a given action will likely have unintended consequences—positive or negative. Those consequences will be more widely felt and their impact will spread in less time than in other environments. These unintended results may have important strategic and operational consequences. The multiple ways these dynamic urban elements and characteristics combine make it necessary to approach each urban environment as a unique challenge for intelligence analysis.

URBAN TERRAIN AND WEATHER

Terrain

B-13. Earlier admonitions that civil considerations are more closely considered in UO do not necessarily mean that consideration for urban terrain is de-emphasized. In every urban operation, terrain and its effects on both threat and friendly forces must be assessed and understood. Then commanders can quickly choose and exploit the terrain (and weather conditions) that best supports their missions. Effective terrain analysis thoroughly assesses urban structures as well as the ground on which they stand (see figure B-3 and FM 5-33). An analysis of urban terrain first considers broader urban characteristics and effects and then progresses to a more detailed examination.

Natural Terrain

B-14. The natural terrain features beyond the urban area and beneath urban structures significantly influence unit operations. They dictate where buildings can be constructed, the slope and pattern of streets, and even the broad urban patterns that develop over longer periods of time, thereby influencing a unit's scheme of maneuver. The military aspects of terrain—observation and fields of fire, avenues of approach, key terrain, obstacles, and cover and concealment (OAKOC)—remain critical to the analysis of natural terrain in, under, and around urban areas. Fortunately, commanders and their staffs are normally accustomed to this type of analysis.

Forms and Functions

- Cores
- Industrial areas
  - Toxic industrial material production and storage facilities
  - Standard signs and markings for toxic chemicals
- Outlying high-rise areas
- Residential areas and shantytowns
- Commercial ribbon areas
- Forts and military bases

Broad Urban Patterns

- Types
  - Satellite
  - Network
  - Linear
  - Segment
- Dominant or central hub (if any)
- Area covered (square miles)

Street Patterns

- Basic Types
  - Radial
  - Grid
  - Irregular (planned and unplanned)
- Variations
  - Rayed
  - Radial-Ring
  - Contour-forming
  - Combined
- Widths

Construction and Placement

- Construction
  - Mass or framed
  - Light or heavy clad
  - Material (dirt, wood, stone, brick, cinder block, concrete, steel, and glass)
  - Density and thickness (roofs, floors, and interior and exterior walls)
  - Load bearing walls and columns
  - Height (floors)
  - Doors, windows, fire escapes, and other openings
  - Interior floor plan (including crawl spaces, elevators, and stairs)
- Placement
  - Random
  - Close-orderly block
  - Dispersed
- Ownership

Military Aspects of Terrain: OAKOC

- Observation and fields of fire
  - Smoke (fire), dust (explosions), and flying debris
  - Rubble
  - Engagement ranges (including minimum safe distances and backblast factors) and obliquity/angle (ricochets)
  - Elevation and depression considerations
  - Lasers and reflective concerns
- Avenues of approach (mobility corridors)
  - Airspace
  - Surface
  - Supersurface
  - Subsurface
- Key terrain
  - Landmarks
  - Buildings of significant cultural, social, political, historical, or economic significance
- Obstacles
  - Rubble and vehicles
  - Steep embankments
  - Medians
  - Inadequate bridges and overpasses (destroyed, weight-restricted, or narrow)
  - Tunnels and underpasses (destroyed or narrow)
  - Mines and roadside improvised explosive devices
  - Masking of fires
  - Burning buildings or other fire hazards
  - Rivers and lakes
- Cover and concealment
  - Building protection
  - Weapon penetration (single shot and multiple rounds) considerations
  - Rubble and vehicles

Figure B-3. Significant urban terrain characteristics

Analysis of an Urban Area's Underlying Terrain: Mitrovica, Kosovo – 1999

An urban area's underlying terrain provides many clues into its history, economy, society, and current situation. Mitrovica, Kosovo is an illustrative example. The Ibar River creates a natural line of communications through the middle of the city as well as an obstacle that bisects the urban area. This bisection naturally divides the two resident ethnic groups: Albanians and Serbs. The separation became significant at both the strategic and tactical levels during 1999 deployments to Kosovo. Army forces had to ensure that the Orthodox Church located south of the Ibar was accessible to Serbs residing in the north. North Atlantic Treaty Organization (NATO)

> peacekeepers built a footbridge across the river that allowed reliable, safe passage. The natural feature separating the two groups assisted NATO troops in maintaining stability in the region.

Man-Made Terrain

B-15. Building composition, frontages, placement, forms and functions, size, floor plans, and window and door locations affect maneuver, force positioning, and weapons deployment considerations. Angles, displacement, surface reflection, and antenna locations influence command and control. Structures also influence ISR operations. The increased density and volume created by man-made structures increases how much information commanders and their staffs collect and assess as well as the number of forces required. Building materials and construction will also influence force structures to include weapons and equipment required. Heavily constructed buildings combined with hot and cold extremes may affect target identification for thermal sights. Thick walls, for example, may make combat vehicle identification difficult by distorting hotspots. Additionally, the increased use of heaters and warming fires may clutter thermal sights with numerous hotspots. The ability to maneuver through the urban dimensions—airspace, supersurface, surface, and subsurface—and shoot through walls, ceilings, and floors also creates increased psychological stress. The physical characteristics of man-made terrain can also be analyzed using OAKOC.

Weather

B-16. Weather and its effects are often considered when examining the military aspects of terrain. Military aspects of weather include temperature (heat and cold), light conditions, precipitation (cloud cover, rain, snow, fog, and smog), and wind. Their military effects during UO are similar to any operational environment (see FM 34-81 and FM 34-81-1). Extremes of heat and cold affect weapon systems and the Soldiers that operate them. The extra luminescence provided by the ambient light of an urban area, unless controlled, may affect night vision capabilities and the ability of the Army to "own the night." Precipitation affects mobility and visibility. Smog inversion layers are common over cities. An inversion layer may trap smoke and chemicals in the air and increase ground and air temperatures to the detriment of Soldiers' health. (If the conditions are severe enough, it might require the use of protective masks.) Changes in temperature as a result of air inversions can also affect thermal sights during crossover periods of warm to cold and vice versa. Winds, which may increase as they are funneled through urban canyons, may—

- Increase other weather effects (for example, wind chill).
- Decrease visibility (blowing debris, sand, rain, and snow).
- Spread radiation, biological, and chemical hazards.
- Adversely affect low-altitude air mobility and airborne ISR platforms.

B-17. Commanders also analyze weather for its potential effect on civilians and civilian infrastructure as well as Soldiers and military equipment. Rain might create sewage overflow problems in urban areas with collapsed sewage infrastructure and in refugee camps, increasing disease and even creating panic. Rain and flooding may also make some subsurface areas impassable or extremely hazardous to civilian and military forces alike. Chemical agents and TIM may be washed into underground systems creating toxic hotspots. These effects become more pronounced as chemical agents or TIM are absorbed by brick and unsealed concrete sewers. Other weather effects on UO can include—

- Heavy snowfall in an urban area that may paralyze area transportation and distribution infrastructure, hindering the urban administration's ability to provide vital human services (police, fire fighting, medical, and rescue). Heavy rains and flooding may have similar effects especially on poorly designed and constructed roads or roads that have been damaged by tracked vehicles.
- Extreme hot and cold weather climates that increase the dependence (and military significance) of many elements of the infrastructure. For example, the energy infrastructure may be critical; without it, civilians may not be adequately cooled or heated or they may not be able to cook their food.

- In urban areas located in tropical regions, it can rain at the same time each day during the wet season. Threat forces may attack during these periods knowing aircraft will have difficulty responding. Bad weather also reduces the effectiveness of surveillance, direct and indirect fire, and logistic support.
- Inclement weather may preclude demonstrations or rallies by threats. Good weather may mean a maximum turnout of civilians for events such as festivals, sporting events, and other social, cultural, or religious activities.
- Severe weather may affect psychological, civil-military and humanitarian assistance operations. Heavy rains and severe dust storms may disrupt leaflet drops, construction projects, food and water distribution, and medical and veterinary assistance programs.

URBAN SOCIETY

B-18. This manual shows that societal considerations take on added importance in UO. Critical to operational success is knowing and understanding which groups live in an urban area, what relationships exist among them, and how each population group will respond to friendly and threat activities. Often determining any of this is very difficult. Cultural acuity is also essential in helping commanders and their staffs to view the urban area as the residents view it. The demographics presented depict what conditions exist, while the other categories help to explain the root causes or why conditions exist (see figure B-4 on page B-8). Other categories besides basic demographics that are important to gain this understanding include health, history, leadership, ethnicity and culture, religion, and government and politics.

From Cultural Awareness to Understanding and Competence

B-19. This societal understanding must go beyond a superficial awareness of gestures and taboos to developing a thorough understanding of the organizing principles that make the urban society hold together and function as a coherent entity—or not. In UO, particularly longer-term stability operations, commanders use this knowledge and insight to design and plan operations that work effectively within that society. For example in OPERATION IRAQI FREEDOM, Army units, as they shifted to long-term stability operations, were required to recruit, organize, equip, and train indigenous security forces in order to be able to transition security responsibility from U.S. forces to the burgeoning Iraqi government. Army commanders were successful—to a point. The first units that they built and trained attained a remarkable level of unity and tactical ability in a relatively short period of time. The newly formed units effectively fought alongside Army forces. However when prodded into conducting operations with minimal Army supervision, many Iraqi units lost unity and coherence. Army commanders initially failed to account for a basic Iraqi societal consideration: their society is patriarchal—each member or organization in Iraqi society always seeks a patron or sponsor to support them and provide for their needs. When Iraqi units trained and operated with Army forces, this relationship was satisfied. Operating on their own, however, Iraqi commanders and units lost this patronage without an equivalent Iraqi replacement. In hindsight, the establishment of an effective and more robust ministry of defense, either as a precursor or simultaneously with the establishment of tactical units, may have provided the Iraqi security forces the patronage that their culture required.

Resource the Societal Analysis

B-20. Aside from friendly and threat forces, the society is the only thinking component of the urban environment able to rapidly impact the urban operation. (Even people going about their daily routines can unwittingly hamper the mission.) Urban residents create conditions for restrictive rules of engagement, increase stress on Soldiers and logistic capabilities, and confuse threat identification (see Threat Considerations in this appendix). Demographic, health, safety, ethnic, and cultural concerns will be essential considerations in most UO. Other situational factors—the mission, enemy, and time available—dictate the balance between the level of detail and analysis to support the overall urban operation with the level of detail that commanders and their staffs can achieve. However, an analysis that fails to devote enough time and resources to understanding the urban society can find large elements of the population turned against the Army force. Detailed analysis of the urban society will help save time and ISR resources. While not necessarily a resource saver in the short term, a thorough understanding of the urban society and its culture ultimately contributes to faster achievement of strategic goals.

Population Demographics

- General population size
  - Village
  - Town
  - City
  - Metropolis
  - Megalopolis
- Group size based on race age, sex, political affiliation, economics, religion, tribe, clan, gang, criminal activities, or other significant grouping
  - Significant US or allied populations
  - Distribution, densities, and physical boundaries and overlaps
  - Majority, minority, and dominant groups
- Increasing or decreasing migration trends
  - Dislocated civilians
- Nongovernmental organizations (NGOs)
  - Local
  - National
  - International
- Languages (distribution, dialects, relationship to social structure)
- Educational levels and literacy rates
- Crime rates
- Birth and death rates
- Labor statistics and considerations
  - Skilled and unskilled
  - Imported and exported
  - Unemployment
  - Standard wages and per capita income
  - Workday and workweek norms

Health

- Diseases
- Nutritional deficiencies
- Local standards of care
- Pollution and environmental hazards (air, water, food, and soil)
- Health workers (types, numbers, and degree of skill)

History

- General and for a specific group
  - Internal or external
  - Recent conflicts
- Relationship with US, allies, and other participating multinational forces
- Applicable international treaties
- Status-of-forces agreements
- Antagonists/protagonists
- Heroes
- Events, facts, and dates considered important or celebrated
- Urban area's historical importance

Leadership and Prominent Personalities

- Identification, location, and prioritization of influential leaders (exploitation, evacuation, protection, etc.)
- Affiliation (ethnic, religion, military, government, industry, criminal, or entertainment)
- Education attained
- Organization and distribution of power
- Associations among different leaders and groups

Ethnicity and Culture

- Values, moral codes, taboos, and insults (verbal and non-verbal)
- Attitudes towards age, sex, and race (including same-sex interaction)
- Role of the clan, tribe, or family
- Biases between ethnic groups
- Privacy and individuality
- Recreation, entertainment, and humor
- Fatalism or self-determination
- Exchanges of gifts
- Displays of emotion
- Lines of authority
- Dating and marriage
- Greetings, leave-takings, and gestures
- Visiting practices
- Alcohol and drug use
- Important holidays, festivals, sporting, or entertainment events
- Eating and dietary practices
- Significance of animals and pets
- Urban-rural similarities and differences
- Driving habits
- Clothing

Religion

- Sects, divisions, and overlaps
- Religious biases and problems
- Relationship and influence on government, politics, economics, and education
- Impact on ethnic and cultural beliefs
- Key events or celebrations (daily, weekly, monthly, or annually)
- Funeral and burial practices

Government and Politics

- Present and past forms
- Organization and powers (executive, legislative, judicial, and administrative divisions)
- Scheduled elections and historical turnouts
- Degree of control over the population
  - Identification required
  - Border-crossing procedures
- Relationship with US or multinational governments, national governments, and criminal elements
- Political factions and boundaries
- Political traditions
- Grievances
- Censorship
- Nepotism and other clan, tribal, or social ties
- Civil defense and disaster preparedness (organization, plans, training, equipment, and resources)
  - Evacuation routes
- Legal system
  - System of laws
  - Applicable treaties
  - Courts and tribunals
  - Procedures
  - Records (birth and deeds)
- Property control
- Monetary system (formal and informal)
- Domestic and foreign trade
  - Taxation and tariffs
  - Customs requirements
  - Rationing and price controls
  - Economic performance and contribution to gross national product
- Economic aid
- Perception of relative deprivation
- Trade unions
- Competition with the black market and organized crime

Figure B-4. Significant urban societal characteristics

Transportation and Distribution

- Water
  - Shipyards and other port and harbor facilities
  - Inland waterways, canals, and locks
  - Offshore pipeline berths
  - Cargo storage and handling
  - Types and number of ships, boats, and ferries
  - Dams
- Streets and roads
  - Bridges and fords
  - Over- and underpasses
  - Raised embankments, tunnels, culverts, and other subterranean features (widths and clearances)
  - Parking areas (surface, subsurface, and supersurface)
  - Weight restrictions
  - Traffic light operations
  - Traffic patterns
  - Widths
  - Surface materials
- Rail
  - Lines
  - Terminals
  - Switchyards and junctions
  - Subways, bridges, elevated rail lines, and underpasses (clearances)
  - Track gauges
  - Types and number of rolling stock
  - Electrification
- Air
  - Airfields and runways (including capabilities)
  - Heliports and helipads (including rooftop)
  - Types and number of aircraft
  - Cargo storage and handling
- Trucking companies and delivery services
- Available material-handling equipment
- Rush hour and market time considerations
- Seasonal (weather) effects
- Rubble effects
- Impact of dislocated civilians and migration patterns
- Likely population congregation points
- Identifiable primary and alternate lines of communication

Economics and Commerce

- Industries
  - Types and locations
  - Important companies (including US or allied)
  - Military production facilities
- Sources of raw materials
- Use of toxic industrial materials and biological agents
  - Agriculture (insecticides, herbicides, and fertilizers)
  - Manufacturing
  - Cleaning
  - Research
- Food types, quantities, and sources
  - Requirements and availability
  - Storage and processing
  - Cleanliness standards
- Stores, shops, restaurants, hotels, and strip malls
- Recreation facilities
  - Outdoor and amusement parks
  - Stadiums and other sports facilities
- Machine shops
- Brick and lumber yards
- Banking and investment institutions

Administration and Human Services

- Police and fire protection
  - Headquarters, station, and key facilities locations
  - Organization and strengths
  - Equipment
  - Functions, authority, and jurisdictional boundaries
  - Contract guard services
- Welfare and public assistance
  - Monetary assistance
  - Orphanages
  - Elderly care facilities
- Water supply systems
  - Water sources and storage (lakes, reservoirs, cisterns, pools, and public baths)
  - Water treatment and quality
  - Pumping stations and other distribution methods (trucks, bottles)
  - Hydrant locations
- Snow removal capabilities
- Street light operations
- Health facilities
  - Hospitals
  - Emergency medical services
  - Mental institutions
  - Medical supplies and equipment
  - Research and pharmaceutical buildings
  - Blood banks
- Governmental buildings
  - Embassies
  - Capitol building
  - Legislative, judicial, and ministry buildings
  - Hall of records
- NGOs providing assistance
- Effects of military control measures on providing vital human services
- Waste and sanitation
  - Types (solid, sewage, and toxic)
  - Collection, processing, and disposal
  - Dumps or landfills
  - Drainage systems

Cultural

- Religious buildings and places of worship
- Shrines, monuments and other historical structures
- Schools
- Museums
- Theaters
- Libraries

Energy

- Types
  - Electric
  - Oil
  - Coal
  - Natural gas
  - Nuclear
  - Solar
  - Hydroelectric
  - Geothermal
- Facilities
  - Production and processing
  - Storage
- Distribution
  - Pipelines (above and below ground)
  - Power lines (overhead and underground)
  - Water, rail, and road
  - Potential hazards

Communication and Information

- Print media
  - Newspapers, periodicals, and pamphlets
  - Billboards and posters
  - Postal facilities
- Telephone facilities
  - Wire or wireless
  - Facsimile machines
- Telegraph facilities
- Radio facilities
- Police, fire, and rescue systems
- Security systems
- Television facilities
- Computers and the Internet
- Antennas, towers, relay stations, and lines (surface and subsurface)
- Integration of space-based capabilities
- Public forums and speech
- Low-technology media (car horns, drums, graffiti, and burning fires)
- Key media organizations and reporters
  - Local
  - International
  - US

Figure B-5. Significant urban infrastructure characteristics

URBAN INFRASTRUCTURE

B-21. Functional and analytical interdependence is readily apparent when examining urban infrastructures (see figure B-5 on page B-9). They are composed of physical structures or facilities and people. Hence, much of the analysis conducted for terrain and society can apply when assessing the urban infrastructure. For example, commanders, staffs, and analysts could not effectively assess the urban economic and commercial infrastructure without simultaneously considering labor. All aspects of the society relate and can be used to further analyze the urban work force since they are a sub-element of the urban society. Similarly, the OAKOC aspects used to evaluate terrain may also apply to the urban infrastructure, especially considerations of key terrain. Depending on the mission, elements of the urban infrastructure may become key terrain. While infrastructure analysis is not an engineer-specific task, Army engineers may provide the general and geospatial engineering expertise necessary to better understand the urban terrain and infrastructure.

THREAT CONSIDERATIONS

> *... [T]he adversaries of freedom ... send arms, agitators, aid, technicians and propaganda to every troubled area.... [S]ubversives and saboteurs and insurrectionists . ... [possess] the power to conscript talent and manpower for any purpose. and long experience in the techniques of violence and subversion....It is a contest of will and purpose as well as force and violence—a battle for minds and souls as well as lives and territory.*
>
> John F. Kennedy
> Message to Congress, 5 May 1961

B-22. Chapter 3 outlines the instability and uncertainty of the strategic environment. Commanders, staffs, and analysts identify and analyze the threat's composition, strength, disposition, leadership, training, morale, weapons and capabilities, vulnerabilities, internal logistics and external support, doctrine (if any), strategy or modus operandi, and tactics. The threat can take a variety of forms:

- Conventional military forces.
- Paramilitary forces.
- Guerrillas and insurgents.
- Terrorists.
- Militia or special police organizations.

B-23. A general study of guerrilla and insurgent tactics, techniques, and procedures may prove beneficial to many types of operations regardless of the actual composition or type of threat forces. Insurgent strategies and tactics may work especially well in this complex environment and will likely be a part of any threat COA. Particularly, commanders should understand how a threat might restrict itself by the laws of land warfare and similar conventions, or exploit the use of these conventions to its own gain. Commanders can refer to FM 31-20-3 and FMI 3-07.22 for more information. For many of the above threats, no doctrinal templates may exist. Intelligence staffs will need to identify and track common threat patterns and tactics, techniques, and procedures as they develop and change during the course of the operation. Understanding culture will be critical to determining threat motives that drive decisions. Comman415081ders, staffs, and analysts must evaluate, update (or create), and manage threat databases early (and continuously) during the operational process.

URBAN ENVIRONMENTAL THREATS

B-24. While threats vary, they share a common characteristic: the capability and intent to conduct violence against Army forces to negatively influence mission accomplishment. These threats are often the most recognizable for forces trained for warfighting—these are often the enemy. In order for Army units to be able to conduct full spectrum urban operations, commanders broaden their concept of the threat when analyzing the urban environment's terrain, societal, and infrastructure characteristics. This analysis includes many environmental dangers (potentially affecting both sides of a conflict as well as noncombatants) such as—

- Natural disasters (earthquakes, tornados, hurricanes, volcanic eruptions, fires, floods, tidal waves or tsunamis, and heavy snows).
- Food shortages.
- Water shortages (both quantity and quality).
- Rampant disease.
- Pollution and toxic industrial materials.

B-25. A critical difference between the latter forms of threat and the former is the lack of intent to do harm. The latter may stand alone as threats, or these conditions may be created, initiated, or used by the enemy or a hostile civilian group as a weapon or tool. Threat analysis includes identifying and describing how each relevant characteristic of the area of operations can hinder mission accomplishment. This analysis, particularly during stability or civil support operations, will require extensive coordination and cooperation with urban civil authorities, law enforcement, and numerous governmental and nongovernmental organizations.

CIVILIANS

B-26. In a major combat operation or campaign where offensive and defensive operations are conducted against a conventional enemy, threat identification is more readily accomplished. However, adaptation may be necessary to even further broaden the threat concept to include specific elements of the urban society and, in some instances, nongovernmental organizations (NGOs) and other civilian agencies working in the urban area. In many stability operations, this modification can account for opposing armed forces that are not an enemy but are a threat to the mission. As discussed in Chapter 9, Army forces in many stability or civil support operations must avoid classifying or thinking of these threats as the enemy. Further, commanders ensure that Soldiers do not shift the animosity that they feel toward urban threats onto the civilian population. In OPERATION IRAQI FREEDOM, for example, some Soldiers began to refer to Iraqi civilians as "hajjis" in reference to the annual pilgrimage—the Hajj—that forms one of the tenets or pillars of Islamic faith. Those who have completed the pilgrimage are entitled to add the phrase al-Hajj or hajji (pilgrim) to their name. However, the Soldiers were not using the term in its proper context but rather to denigrate. In response, commanders prohibited the inappropriate use of this term as one of many measures to stress and develop cultural understanding and respect for the people that Army forces were there to assist.

Need for More Accurate Categories

B-27. Urban commanders recognize that threat evaluation is not solely a straightforward assessment of the capabilities of a known, armed enemy. Instead, they develop a system of categories that mitigates potential situational uncertainty and better indicates the level of threat (or utility) that civilian groups pose for Army forces conducting UO. Categories can range from a simple construct—threat, neutral, or friendly—to a more refined one (see figure B-6 on page B-12). In most UO, a more intricate approach is necessary to detect and monitor shifts in key relevant relationships. While this analysis is important in any urban operations, it is especially critical when Army forces are opposed by urban insurgent and guerrilla forces.

B-28. In any system, commanders consider that the classification of a group is relative to the perspective from which it is viewed. This is an especially important consideration in multinational and interagency UO. During operations in Somalia, for example, U.S. Army forces may have viewed a particular clan as a hostile element. The United Nations' Italian contingent, with their colonial background in the region, may have considered the same clan as neutral or even an ally.

Shifting and Overlapping Groups

B-29. Commanders recognize that no system of categorization will precisely classify any given group; no system can reflect the overall nature and complexity of the urban society. A single group may fit in a particular category. It may also have components in two or more categories simultaneously. Often, it can shift among categories during an operation. A given group may have individuals in it who have interests identical to or different from that group and these individual interests may change over time.

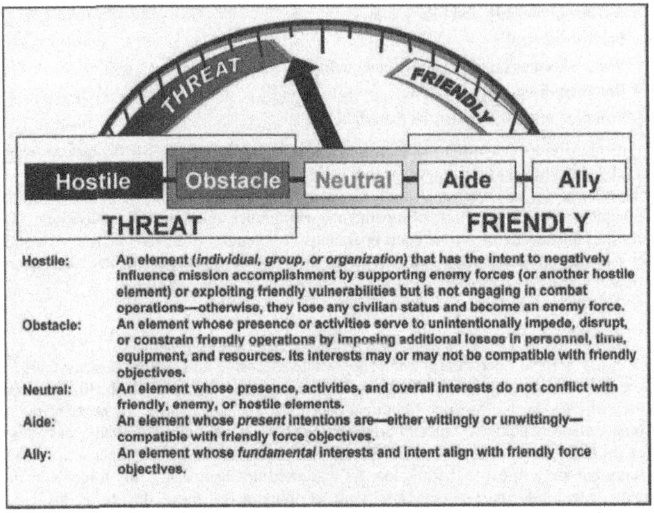

Figure B-6. A refined approach to civilian allegiance

B-30. A peace enforcement operation illustrates the varying nature of groups. An identified criminal group might be classified as an obstacle to the commander's mission because its illegal activities impede unit progress. Its compelling interest, however, is to make money rather than resist the actions of friendly forces. In the same operation, one of the armed belligerents may be intent on disrupting the peace process and would be, therefore, classified as a hostile. (Again, they are not an enemy unless the belligerents are engaged in planned combat operations against the peacekeeping force. Commanders should consider, however, that under international law, if any civilian takes a direct or active role in hostilities, that civilian can be subject to attack.) The belligerent force may finance the criminal organization to assist in further obstructing the peace mission. In this case, the criminal organization moves from being an obstacle to being a hostile.

B-31. This classification effort, therefore, is not a one-time undertaking; commanders must constantly review and update it. Groups or individuals can be influenced into assisting either the friendly or opposing force. People will also act opportunistically, shifting support and alliances as perceived advantages arise. Even seemingly passive and law abiding members of the urban society may conduct themselves in unexpected ways given the right conditions (mob violence, for example).

Shifting Civilian Interests and Intent

Among other applicable lessons (see also the vignette in Chapter 9), the Los Angeles riots of 1992 illustrate how urban population groups can shift their relative positions due to changing conditions in an urban AO. Several gangs exist in the Los Angeles area. Usually, these gangs are hostile to one another. During the riots, however, several rival gangs formed a "united front" against what was seen as a larger obstacle to their own interests: law enforcement. As a result, the hostile gangs became one another's aides during that time.

Similarities, Differences, Capabilities, and Vulnerabilities

B-32. Similarities and differences in interests and interdependencies between groups are often focal points for analysis (and the allocation of ISR assets). They may indicate how commanders may influence, coerce, or align civilian interests and intentions with mission objectives. Simultaneously, commanders should consider an analysis of the civilian element's (individual, group, or organization) capability to influence the accomplishment of friendly objectives. They should also consider civilians' vulnerabilities and dependencies. If a civilian group's fundamental interests align with friendly objectives and this group has the intent to assist friendly forces, it is clearly an ally. However, with limited or no capability, a specific group will not help move the commander any closer to achieving his desired end state and accomplishing the mission. Then the commander would normally limit the resources expended on this group to those necessary for maintaining their commitment to common goals and objectives.

B-33. In contrast, commanders may provide a group with resources to enhance or create the capability to assist in mission accomplishment. They may do this if they felt that the assistance gained (or reduction to threat support) exceeds the potential diminishment of their own force's capabilities from losing those same resources. Commanders would also consider the group's dependencies (such as food, infrastructure, and protection) and overall vulnerability to outside influence. If vulnerable to friendly influence or control (understanding urban societal considerations and matters of perception are critical in this regard), then forces are likely susceptible to enemy or hostile manipulation. Even if commanders can generate extra resources (and not significantly affect their own force's capabilities), they should still conduct this same cost-benefit analysis to determine which civilian group (if any) should receive resources.

Greatest Potential Nearer the Center

B-34. The most critical population sectors often are those nearest the middle of the spectrum shown in figure B-6, particularly if their capabilities (or potential) significantly degrade or enhance mission accomplishment. These are the obstacle, neutral, or aide categories. If their beliefs and interests can be adequately understood, commanders may have great chance to influence the population segment and significantly contribute to mission accomplishment. In the fight for information superiority, persuading a neutral population to become an aide may provide critical information required about the urban environment and threat forces operating within.

Political and Strategic Concerns

B-35. The aide category may be of significant political or strategic concern. An aide group may be invaluable for accomplishing intermediate objectives but become a vulnerability to accomplishing a larger stability operation. (Even an urban offensive or defensive operation is likely to transition to a stability operation.) For example, commanders may provide resources to a criminal organization to assist in defeating insurgent forces during urban combat operations. Once these forces have been subdued, the interests (monetary gain and defeat of the threat) linking friendly forces with this criminal element disappear. What may remain is a criminal organization with more power than a reconstituted or newly established law enforcement agency and a truly destabilizing force. This also illustrates the second- and third-order thinking that will be required of commanders and their staffs during UO.

URBAN INTELLIGENCE TOOLS AND PRODUCTS

B-36. Developing tools and products to assist with situational understanding is not the task of intelligence sections alone. Commanders and all elements of their staffs must develop and adapt products and tools suited to their particular requirements. Listed below are some of the analytic tools and products that may help meet those requirements in a complex environment (see figure B-7 on page B-14 and FMI 2-91.4). Standard tools and products include: modified combined obstacle overlays and doctrinal, situation, event, and decision support templates or matrices. In addition to these standards aids, staffs and analysts may develop or produce other innovative tools to assist commanders in their situational understanding of the complex urban environment. These tools may contain overlapping information as different types of information are compared to determine patterns among them. Staffs and analysts may also initiate requests for products (or information) from their higher headquarters or other agencies with the technical means or control over assets when the capability lies outside the Army force's means. The tools that may be developed or requested include—

- Imagery.
- Three-dimensional representations.
- Infrastructure blueprints.
- Hydrographic surveys.
- Psychological profiles.
- Matrices, diagrams, or charts.
- Various urban overlays.

B-37. There are many software applications available to the Army that can be used to conduct intelligence analysis and create many of the relevant products described above. These applications currently range from such programs as the Analyst Notebook and Crimelink which have link analysis, association matrix, and pattern analysis software tools to the Urban Tactical Planner developed by the Topographic Engineering Center as an operational planning tool. The focus of the following information, however, is the various types of tools that can be developed and used to understand the urban environment rather than the constantly evolving and changing software or hardware that may be used to create them.

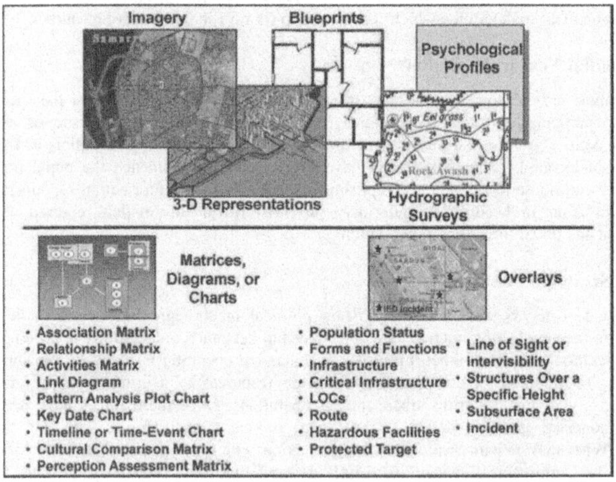

Figure B-7. Urban intelligence tools and products

IMAGERY

B-38. Recent satellite imagery or aerial photography will be required for most types of UO. Such images clarify vague and inaccurate maps and other graphic representations. Satellite assets provide responsive data input into the geographic information systems (GIS). (The National Geospatial Intelligence Agency [NGA] is one important source of imagery and map data.) GIS will often form the basis for creating the three-dimensional representations and the various overlays described below. Frequently updated (or continuous real-time) satellite or aerial imagery may be required for detailed pattern analysis and maintaining accurate situational understanding. For example, imagery taken during an area's rainy season may appear significantly altered during the summer months. Finally, digital, hand-held still and video cameras, particularly at the tactical level, will also be critical in developing and attaining situational understanding in a complex urban environment.

THREE-DIMENSIONAL REPRESENTATIONS

B-39. Often, physical or computer-generated (virtual) three-dimensional representations may be required to achieve situational understanding. These models or computer representations include specific sections of the urban area or specific buildings or structures. Such detail is particularly important for special operating forces and tactical-level units. These units require detail to achieve precision, increase the speed of the operation, and lessen friendly casualties and collateral damage.

INFRASTRUCTURE BLUEPRINTS

B-40. Urban police, fire, health, public utilities, city engineer, realty, and tourist agencies, and other urban organizations often maintain current blueprints and detailed maps. Such documents may prove useful to update or supplement military maps or to clarify the intricacies of a specific infrastructure. They may prove critical in operations that require detailed information to achieve the speed and precision required for success. Without such detail, analysts determine interior configurations based on a building's outward appearance. That task generally becomes more difficult as the building size increases.

HYDROGRAPHIC SURVEYS

B-41. Many urban areas are located along the world's littoral regions and major rivers. Therefore, commanders may need current hydrographic surveys to support amphibious, river crossing, riverine, and sustainment operations.

PSYCHOLOGICAL PROFILES

B-42. Psychological profiles analyze how key groups, leaders, or decision makers think or act—their attitudes, opinions, and views. They include an analysis of doctrine and strategy, culture, and historical patterns of behavior. The degree to which the attitudes, beliefs, and backgrounds of the military either reflect or conflict with the urban populace's (or civilian leadership's) core values is extremely important in this analysis. Psychological profiles help to assess the relative probability of a threat (or noncombatant group) adopting various COAs as well as evaluating a threat's vulnerability to deception. These profiles are derived from open-source intelligence as well as signals and human intelligence.

MATRICES, DIAGRAMS, AND CHARTS

B-43. Matrices, diagrams, and charts help to identify and understand key relationships among friendly and threat forces and other significant elements of the urban environment. While similar, each looks at information in a different way to uncover hidden patterns and connections useful in understanding the complex urban environment, particularly the threat and the urban society. These tools and products might include—

- **Association Matrix**. The association matrix helps identify the nature and relationship between individuals. Association matrices also help to identify those personalities and associations needing a more in-depth analysis necessary to establish the degree of relationship, contacts, or knowledge between individuals. Threat organizational structure and functions are uncovered as connections between personalities are made.

- **Relationship Matrix**. Relationship matrices are an extension of association matrices described above. They are used to examine the relationship between groups and organizations—threat forces, friendly forces, NGOs, the media, and the various elements of the urban population. Urban terrain and infrastructure may also be included as analytical factors to help expose relationships. These matrices graphically depict how human elements of the urban area interact with the physical elements to promote or degrade mission success. A significant relationship matrix may be a comparison of cultural perspectives—ideology, politics, religion, acceptable standards of living, and mores—between urban population groups and Army (and multinational) forces to help understand and accurately predict a civilian element's actions. The relationships discovered may suggest ways that commanders can shape the environment. For example, conflicting ideals or issues between two disparate threat forces may be emphasized to cause threats to focus their resources against each other rather than Army forces. Conversely, stressing common ideals identified between opposing noncombatant organizations may cause these groups to unite and work harmoniously toward a goal reflecting their commonalities.
- **Activities Matrix**. An activities matrix connects individuals to organizations, professions, events, activities, or addresses. Information from this matrix (combined with information from association and relationship matrices) also helps to link personalities, uncover the structure of an organization (threat or otherwise), and recognize differences, similarities, and dependencies for possible exploitation.
- **Link Diagram**. This tool graphically depicts many of the relationships and associations described above—people, groups, events, locations, or other factors deemed significant in any given situation. It helps commanders and analysts visualize how people and other factors are interrelated to determine key links.
- **Pattern Analysis Plot Chart**. This chart depicts the times and dates of a selected activity (such as ambushes, sniper and mortar attacks, bombings, and demonstrations) to search for patterns of activity for predictive purposes as well as to discern intent. This analysis can be conducted using a time-event wheel as well as by plotting events on maps using multiple historical overlays (analog or digital). (Compare with Incidents Overlay covered later in this chapter.) During pattern analysis, commanders and analysts consider not only what is occurring but also what is not occurring (or ceases to occur).
- **Key Event Chart**. In many urban operations, particularly stability operations, key holidays, historic events, and other significant cultural or political events can be extremely important for commanders to understand. More than just dates, these charts depict what can be expected to happen on each particular event. In Bosnia, for example, weddings are often held on Fridays and celebratory fire is a common occurrence. Understanding this cultural phenomenon could reduce collateral damage and accompanying civilian deaths.
- **Timeline or Time-Event Chart**. Timelines are a list of significant historical dates with relevant information and analysis that provide the commander with a record of past activities necessary to understand current operational conditions. A timeline may highlight a specific feature of the present situation such as population movements or political shifts or outline the general chronological record of the urban area, perhaps highlighting the activities of a certain population sector. Understanding past events also helps commanders predict future reactions to proposed COAs. Similarly to link diagrams, time-event charts display large amounts of information in a small space. These charts may help commanders discover larger-scale patterns of activity and relationships.
- **Cultural Comparison Matrix**. Commanders must avoid ethnocentricity—assuming that only one cultural perspective exists (often their own) and using that as the single lens with which to view the situation. To avoid this common obstacle to an accurate understanding of the urban society, commanders use this matrix to compare local ideology, attitudes, beliefs, and acceptable standards of living with associated U.S. and Army norms and values. This matrix not only highlights differences but also similarities.

- **Perception Assessment Matrix**. All of the above tools contribute to developing the perception assessment matrix; they allow commanders to see the urban environment from the perspective of its inhabitants. The perception assessment matrix is another tool that uses this refined perspective and the framework outlined in the matrix to predict the urban population's perception of a proposed COA. Although perceptions are not actions, they—more than reality—drive decision-making and resultant civilian activity. (As an example, legitimacy is a critical factor in developing many COAs that affect the civilian population, particularly during stability operations. This legitimacy must be viewed from the perspective of the inhabitants and not based on the Army commander's own culturally-shaped perceptions. Moreover, the legitimacy of an operation may need to be considered from other perspectives as well. These might include those of the American public, coalition partners and their publics, and those of neighboring states in the region.) This same matrix can later be used as a foundation to track and gauge the effectiveness of the chosen COA with favorable perception as one criterion of success. Commanders must be wary that perceptions are not fixed. They may change based on factors that the commander may not be able to control or influence. Therefore, commanders must continually monitor perceptions for deviation.

Various Urban Overlays

B-44. Staffs can produce various map overlays. These overlays depict physical locations of some aspect critical to the planning and conduct of the urban operation. Given adequate lead time, NGA can produce many overlays as an integrated map product (including satellite imagery). These overlays can include the—

- **Population Status Overlay**. This tool depicts the physical location of various groups identified by any significant social category such as religion, tribe, or language. During offensive and defensive operations, it may simply be where significant numbers of people are "huddled" or located throughout the battlefield. Population dispersal can vary significantly through the day, particularly at night, and must be considered as part of the overall analysis leading to the development of this tool.
- **Forms and Functions Overlay**. Based on the urban model, this overlay depicts the urban core or central business district, industrial areas, outlying high-rise areas, commercial ribbon areas, and residential areas, to include shantytowns.
- **Infrastructure Overlay**. This overlay is actually a series of overlays. It depicts identifiable subsystems in each form of urban infrastructure: communications and information, transportation and distribution, energy, economics and commerce, administration and human services, and cultural. Each subsystem can be broken down into more detail. Infrastructure data may be used to develop three other overlays—

  - **Critical Infrastructure Overlay**. This tool displays specific elements of the urban infrastructure that, if harmed, will adversely affect the living conditions of the urban society to the detriment of the mission. These elements may include power generation plants, water purification plants and pumping stations, and sewage treatment plants. This information could be coded as part of the overall infrastructure overlay.

  - **Lines of Communications (LOCs) Overlay**. The LOCs overlay highlights transportation systems and nodes, such as railways, road, trails, navigable waterways, airfields, and open areas for drop zones and landing zones. It also includes subsurface areas and routes such as sewage, drainage, and tunnels and considers movement between supersurface areas. The LOCs overlay and the route overlay (below) consider traffic conditions, times, and locations, to include potential points where significant portions of the urban population may congregate.

  - **Route Overlay**. This overlay emphasizes mobility information to assist commanders and planners in determining what forces and equipment can move along the urban area's mobility corridors. Pertinent data includes street names, patterns, and widths; bridge, underpass, and overpass locations; load capacities; potential sniper and ambush locations (which may be its own overlay); and key navigational landmarks such as major roads and highways, rivers and canals, cemeteries, bridges, stadiums, and churches. The structures over a specific height overlay and subsurface overlay may assist in its development. As with the LOCs overlay, commanders, planners, and analysts think in all dimensions.

- **Line of Sight or Intervisibility Overlay**. This product creates a profile view (optical or electronic) of the terrain from the observer's location to other locations or targets. It can show trajectory or flight-line masking as well as obstructed or unobstructed signal pathways.
- **Structures over a Specific Height Overlay**. This level of detail may also be critical to communications, fires, and Army airspace command and control (air mobility corridors especially low-level flight profiles). Incorporated as part of this overlay, it may include floors or elevations above limitations for particular weapon systems at various distances from the structure.
- **Subsurface Area Overlay**. As an alternate to the building or structure height overlay, this product provides the locations of basements, underground parking garages, sewers, tunnels, subways, naturally-occurring subterranean formations, catacombs, and other subsurface areas. Similar to elevation "dead spaces," this overlay may show areas that exceed depression capabilities of weapon systems and potential threat ambush locations—again, affecting maneuver options.
- **Urban Logistic Resources Overlay**. This product identifies the locations of urban logistic resources that may contribute to mission accomplishment. It may contain specific warehouse sites, hospitals and medical supply locations, viable food stores, building material locations, fuel storage areas, car or truck lots, maintenance garages, and appliance warehouses. (NGO locations, taken from an NGO relationship matrix, may be an essential, overlapping element of this overlay.)
- **Hazardous Facilities Overlay**. This overlay identifies urban structures with known or suspected chemical, biological, or radiological features, such as nuclear power plants, fertilizer plants, oil refineries, pharmaceutical plants, and covert locations for producing weapons of mass destruction. These locations are critical to maneuver and fire planning.
- **Protected Target Overlay**. This overlay depicts terrain that should not be destroyed or attacked based on restrictions due to international, host-nation, or U.S. law and subsequent rules of engagement. These may include schools, hospitals, historical or other culturally significant monuments, and religious sites. This overlay may incorporate no-fire areas, such as special operations forces locations, critical infrastructure, logistic sources, and hazardous sites that must be protected as part of the commander's concept of the operation.
- **Incident Overlay**. Similar to the pattern analysis plot chart, this product depicts the location of different threat actions and types of tactics employed to uncover recurring routines, schemes, methods, tactics, or techniques and overall threat interests, objectives, or the desired end state.

B-45. The above IPB tools and products constitute a small sampling of what staffs and analysts can produce. They are limited only by their imaginations and mission needs (not all tools presented above may be relevant or necessary to every operation). Many products can be combined into a single product or each can generate further products of increasing level of detail. This is similar to transparent overlays positioned one atop another on a map. As discussed earlier, technology—software and hardware—will allow for more urban data to be combined, compared, analyzed, displayed, and shared—ultimately allowing scaleable products that can expand and contract to meet the intelligence requirements of any echelon of command down to the individual Soldier. The challenge will remain, however, to provide timely, accurate, complete, and relevant information in an understandable, usable, and share-capable form—share-capable among different levels of command as well as among multinational partners and various governmental and (as required) nongovernmental organizations—without overloading the commander.

Appendix C

Operations in Somalia: Applying the Urban Operational Framework to Stability Operations

It's impossible for an American mother to believe that a Somali mother would raise children to avenge the clan.

Major General Thomas M. Montgomery

GENERAL SITUATION

C-1. Following decades of political unrest and the fall of Somali dictator Siad Barre, a civil war broke out as 14 clans vied for power. The resulting nation composed of hostile social factions was held together by weak political alliances—none strong enough to unite and lead the country to national reconciliation. An ongoing drought led to famine and compounded the ethnic tensions and political instability. This volatile situation rapidly led to a phased U.S. involvement (see figure C-1). Army forces combined, sequenced, and proportionally emphasized the different types of operations to accomplish changing political objectives. Throughout all operations in Somalia, urban areas were critical to achieving mission success.

SOMALI OPERATIONS

INITIAL UN RESPONSE

C-2. The United Nations (UN) initially responded to requests for assistance from international relief organizations by sending supplies and other forms of humanitarian aid to Somalia. However, widespread looting, fighting between gangs, and other lawlessness prevented supplies from reaching the hungry and sick. Only 20 percent of the food entering the country reached the people who needed it. An estimated 25 percent of Somalia's 6 million people died of starvation or disease. In April 1992, the UN issued Security Council Resolution (UNSCR) 751 which authorized 50 unarmed observers, but the action had little effect. Under continuing pressure for additional measures to ensure the delivery of supplies and relief, the UN authorized 500 armed peacekeepers (furnished by Pakistan and transported by U.S. sea- and airlift) to protect humanitarian workers. The battalion's limited mission, designated UN Operations in Somalia I (UNOSOM I), encompassed safeguarding the unloading of ships and providing convoy security.

PROVIDE RELIEF (UNOSOM I)

C-3. In July 1992, the UN requested an increased airlift of supplies and the United States quickly responded. U.S. Central Command (USCENTCOM) activated joint task force (JTF) OPERATION PROVIDE RELIEF. Based on careful mission analysis, USCENTCOM limited the JTF's actions to—

- Deploying a humanitarian assistance survey team to assessing relief requirements.
- Providing an emergency airlift of supplies.
- Using Air Force cargo aircraft for daily relief sorties into Somalia.

C-4. USCENTCOM restricted the sorties to flying during daylight hours and to locations that would provide a permissive and safe environment. In mid-September 1992, the United States prudently expanded its role by stationing the amphibious ready group, USS Tarawa, offshore to provide support to the Pakistani security battalion and to provide security for U.S. airlift operations. The 11th Marine expeditionary unit (MEU) was on board the USS Tarawa to rapidly respond to any change in mission (see Appendix D for a description and the capabilities of a MEU).

Operation	Dates	UN Security Council Resolution	Relative Proportionality Between Types of Operations
PROVIDE RELIEF (UNOSOM I)	Aug 92 – Dec 92	UNSCR# 751 24 Apr 92	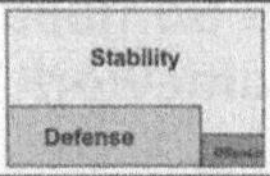
RESTORE HOPE (UNITAF)	Dec 92 – May 93	UNSCR# 794 3 Dec 92	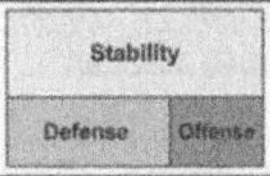
CONTINUED HOPE (UNOSOM II)	May 93 – Mar 94	UNSCR# 814 26 Mar 93	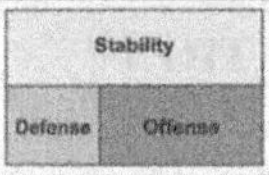

Figure C-1. Relative proportionality between types of operations

RESTORE HOPE (UNITAF)

C-5. By November 1992, the magnitude of the task, UN organizational deficiencies, and a continued lack of security precluded delivery of sufficient supplies to the needy. Notably, a ship laden with relief supplies was fired on in the harbor at Mogadishu, forcing its withdrawal before the supplies could be brought ashore, and a Pakistani peacekeeper was shot when his car was hijacked. Subsequently, the United States offered to provide forces and lead a UN-sponsored operation to reopen the flow of food to where it was needed most. In December 1992, the UN issued UNSCR 794, which authorized member states "to use all necessary means to establish a secure environment for humanitarian relief operations in Somalia" and demanded all "factions in Somalia immediately cease hostilities." To allay concerns of colonialism by a number of African countries, the UN Secretary-General was given oversight of the operation. The resolution also required soldiers to be withdrawn once order was restored; however, it provided no exit strategy. As clearly as possible, the USCENTCOM mission statement (below) for OPERATION RESTORE HOPE reflected the UN mandate:

> *When directed by the [President or the Secretary of Defense], [Commander, USCENTCOM] will conduct joint/combined military operations in Somalia to secure the major air and sea ports, key installations and food distribution points, to provide open and free passage of relief supplies, provide security for convoys and relief organization operations, and assist UN/[nongovernmental organizations] in providing humanitarian relief under UN auspices. Upon establishing a secure environment for uninterrupted relief operations, [Commander, USCENTCOM] terminates and transfers relief operations to UN peacekeeping forces.*

C-6. Mogadishu was the largest port in the country and the focal point of previous humanitarian relief activities of nongovernmental organizations (NGOs). It was also the headquarters of the coalition of 20 nations and over 30 active humanitarian relief organizations. As such, Mogadishu became the entry point for the operational buildup of the multinational force known as Unified Task Force (UNITAF) and the key logistic hub for all operations in Somalia. UNITAF immediately gained control over the flow of relief supplies into and through Mogadishu and stabilized the conflict among the clans. In less than a month, UNITAF forces expanded control over additional ports and interior airfields. They secured additional distribution sites in other key urban areas in the famine belt to include Baidoa, Baledogle, Gialalassi, Bardera, Belet Uen, Oddur, Marka, and the southern town of Kismayo (see figure C-2). With minimal force, the U.S.-led UNITAF established a secure environment that allowed relief to reach those in need, successfully fulfilling its limited—yet focused—mandate.

CONTINUED HOPE (UNOSOM II)

C-7. In March 1993, the UN issued UNSCR 814 establishing a permanent peacekeeping force, UNOSOM II. However, the orderly transition from UNITAF to UNOSOM II was repeatedly delayed until May 1993. (The UN Secretary-General urged the delay so that U.S. forces could effectively disarm bandits and rival clan factions in Somalia.) This resolution was significant in two critical aspects:

- It explicitly endorsed nation building with the specific objectives of rehabilitating the political institutions and economy of Somalia.
- It mandated the first ever UN-directed peace enforcement operation under the Chapter VII enforcement provisions of the Charter, including the requirement for UNOSOM II to disarm the Somali clans. The creation of a peaceful, secure environment included the northern region that had declared independence and had hereto been mostly ignored.

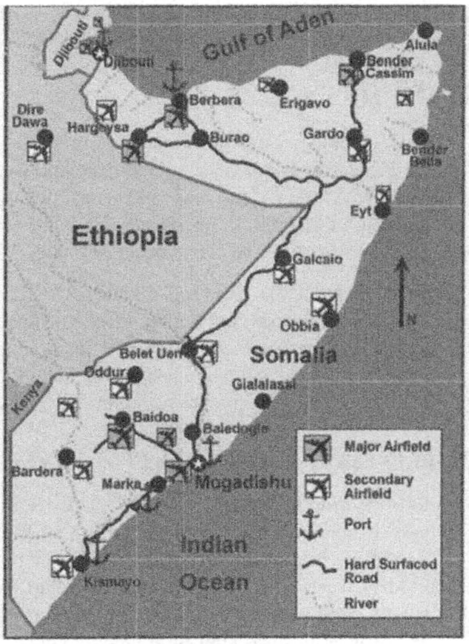

Figure C-2. Map of Somalia

C-8. These far-reaching objectives exceeded the limited mandate of UNITAF as well as those of any previous UN operation. Somali clan leaders rejected the shift from a peacekeeping operation to a peace enforcement operation. They perceived the UN as having lost its neutral position among rival factions. A more powerful clan leader, General Mohammed Farah Aideed (leader of the Habr Gidr clan), aggressively turned against the UN operation and began a radio and propaganda campaign. This campaign characterized UN soldiers as an occupation force trying to recolonize Somalia.

C-9. The mounting crisis erupted in June 1993. Aideed supporters killed 24 Pakistani soldiers and wounded 57 in an ambush while the soldiers were conducting a short-warning inspection of one of Aideed's weapons arsenals. UNSCR 837, passed the next day, called for immediately apprehending those responsible and quickly led to a manhunt for Aideed. The United States deployed 400 rangers and other special operations forces (SOF) personnel to aid in capturing Aideed, neutralizing his followers, and assisting the quick reaction force (QRF), composed of 10th Mountain Division units, in maintaining the peace around Mogadishu.

PHASED WITHDRAWAL

C-10. On 3 October 1993, elements of Task Force (TF) Ranger (a force of nearly 100 Rangers and SOF operators) executed a raid to capture some of Aideed's closest supporters. Although tactically successful, 2 helicopters were shot down, 75 soldiers were wounded, and 18 soldiers were killed accomplishing the mission. The U.S. deaths, as well as vivid scenes of mutilation to some of the soldiers, increased calls to Congress for withdrawing U.S. forces from Somalia. The President then ordered reinforcements to protect U.S. Forces, Somalia (USFORSOM) as they began a phased withdrawal with a 31 March deadline. The last contingent sailed from Mogadishu on 25 March, ending OPERATION CONTINUED HOPE and the overall U.S. mission in Somalia.

C-11. Although U.S. forces did not carry out the more ambitious UN goals of nation building, they executed their missions successfully, relieving untold suffering through humanitarian assistance with military skill and professionalism. Operations in Somalia occurred under unique circumstances, yet commanders may glean lessons applicable to future urban stability operations, as well as civil support operations. In any operations, commanders balance changing mission requirements and conditions.

UNDERSTAND

C-12. Although accomplished to varying degrees, U.S. forces failed to adequately assess and understand the urban environment, especially the society. Somali culture stresses the unity of the clan; alliances are made with other clans only when necessary to elicit some gain. Weapons, overt aggressiveness, and an unusual willingness to accept casualties are intrinsic parts of the Somali culture. Women and children are considered part of the clan's order of battle.

C-13. Early in the planning for OPERATION RESTORE HOPE, U.S. forces did recognize the limited transportation and distribution infrastructure in Mogadishu. The most notable was the limited or poor airport and harbor facilities and its impact on the ability of military forces and organizations to provide relief. Therefore, a naval construction battalion made major improvements in roads, warehouses, and other facilities that allowed more personnel, supplies, and equipment to join the relief effort faster.

Understanding the Clan (The Human Dimension)

C-14. During OPERATION RESTORE HOPE, the UNITAF worked with the various clan leaders as the only recognized leadership remaining in the country. The UNITAF was under the leadership of LTG Robert B. Johnston and U.S. Ambassador to Somalia, Robert Oakley. In addition, UNITAF forces also tried to reestablish elements of the Somali National Police—one of the last respected institutions in the country that was not clan-based. This reinstated police force manned checkpoints throughout Mogadishu and provided crowd control at feeding centers. Largely because of this engagement strategy, the UNITAF succeeded in its missions of stabilizing the security situation and facilitating humanitarian relief. Before its termination, the UNITAF also worked with the 14 major Somali factions to agree to a plan for a transitional or transnational government.

C-15. The UN Special Representative of the Secretary-General, retired U.S. Navy Admiral Jonathon Howe, worked with the UNOSOM II commander, Turkish General Cevik Bir. During OPERATION CONTINUED HOPE, Howe and General Bir adopted a philosophy and operational strategy dissimilar to their UNITAF predecessors. Instead of engaging the clan leaders, Howe attempted to marginalize and isolate them. Howe initially attempted to ignore Aideed and other clan leaders in an attempt to decrease the warlord's power. Disregarding the long-established Somali cultural order, the UN felt that, in the interest of creating a representative, democratic Somali government, they would be better served by excluding the clan leadership. This decision ultimately set the stage for strategic failure.

Threat Strategy and Tactics

C-16. During OPERATION RESTORE HOPE, U.S. forces also failed to properly analyze and understand their identified threat's intent and the impact that the urban environment would have on his strategy, operations, and tactics. The UN began to view eliminating Aideed's influence as a decisive point when creating an environment conducive to long-term conflict resolution. Aideed's objective, however, remained to consolidate control of the Somali nation under his leadership—his own brand of conflict resolution. He viewed the UN's operational center of gravity as the well-trained and technologically advanced American military forces, which he could not attack directly. He identified a potential American vulnerability—the inability to accept casualties for an operation not vital to national interests—since most Americans still viewed Somalia as a humanitarian effort. If he could convince the American public that the price for keeping troops in Somalia would be costly, or that their forces were hurting as many Somalis as they were helping, he believed they would withdraw their forces. If U.S. forces left, the powerless UN would leave soon after, allowing Aideed to consolidate Somalia under his leadership.

Vulnerability and Risk Assessment

C-17. U.S. forces failed to assess, understand, and anticipate that Aideed would adopt this asymmetric approach and attack the American public's desire to remain involved in Somalia. By drawing U.S. forces into an urban fight on his home turf in Mogadishu, he could employ guerrilla insurgency tactics and use the urban area's noncombatants and its confining nature. Such tactics made it difficult for the U.S. forces to employ their technological superiority. If U.S. forces were unwilling to risk harming civilians, his forces could inflict heavy casualties on them, thereby degrading U.S. public support for operations in Somalia. If, on the other hand, the U.S. forces were willing to risk increased civilian casualties to protect themselves, those casualties would likely have the same effect.

C-18. However, an assessment and understanding of the Somali culture and society should have recognized the potential for Aideed's forces to use women and children as cover and concealment. Accordingly, the plan should have avoided entering the densely populated Bakara market district with such restrictive rules of engagement. As legitimacy is critical to stability operations, TF Ranger should have been prepared and authorized to employ nonlethal weapons, to include riot control gas, as an alternative to killing civilians or dying themselves.

C-19. U.S. forces also failed to assess and understand the critical vulnerability of their helicopters in an urban environment and the potential impact on their operations. TF Ranger underestimated the threat's ability to shoot down its helicopters even though they knew Somalis had attempted to use massed rocket-propelled grenade (RPG) fires during earlier raids. (Aideed brought in fundamentalist Islamic soldiers from Sudan, experienced in downing Russian helicopters in Afghanistan, to train his men in RPG firing techniques). In fact, the Somalis had succeeded in shooting down a UH-60 flying at rooftop level at night just one week prior to the battle. Instead, TF Ranger kept their most vulnerable helicopters, the MH-60 Blackhawks, loitering for forty minutes over the target area in an orbit that was well within Somali RPG range. The more maneuverable AH-6s and MH-6s could have provided the necessary fire support. Planning should have included a ready ground reaction force, properly task organized, for a downed helicopter and personnel recovery contingency.

C-20. Information operations considerations apply throughout the entire urban operational framework; however, operations security (OPSEC) is critical to both assessment and shaping. OPSEC requires continuous assessment throughout the urban operation particularly as it transitions among the spectrum of operations. As offensive operations grew during OPERATION CONTINUED HOPE, U.S. forces did little to protect essential elements of friendly information. Combined with the vulnerability of U.S. helicopters, Aideed's followers used U.S. forces' inattention to OPSEC measures to their advantage. The U.S. base in Mogadishu was open to public view and Somali contractors often moved about freely. Somalis had a clear view both day and night of the soldiers' billets. Whenever TF Ranger would prepare for a mission, the word rapidly spread through the city. On 3 October 1993, Aideed's followers immediately knew that aircraft had taken off and, based on their pattern analysis of TF Ranger's previous raids, RPG teams rushed to the rooftops along the flight paths of the task force's Blackhawks.

SHAPE

C-21. One of the most critical urban shaping operations is isolation. During OPERATION CONTINUED HOPE, U.S. forces largely discounted other essential elements of friendly information and did not establish significant public affairs and psychological operations (PSYOP) initiatives. In fact, Army forces lacked a public affairs organization altogether. Consequently, Aideed was not isolated from the support of the Somali people. This failure to shape the perceptions of the civilian populace coupled with the increased use of lethal force (discussed below) allowed Aideed to retain or create a sense of legitimacy and popular support. Ironically, the focus on Aideed helped create a popular figure among the warlords through which many noncompliant, hostile, and even neutral Somalis could rally behind.

C-22. During OPERATION RESTORE HOPE, Aideed conducted propaganda efforts through "Radio Aideed"—his own radio station. UNITAF countered these efforts with radio broadcasts. This technique proved so effective that Aideed called MG Anthony C. Zinni, UNITAF's director of operations, over to his house on several occasions to complain about UNITAF radio broadcasts. General Zinni responded, "If he didn't like what we said on the radio station, he ought to think about his radio station and we could mutually agree to lower the rhetoric." This approach worked.

ENGAGE

C-23. The complexity of urban operations requires unity of command to identify and effectively strike the center of gravity with overwhelming combat power or capabilities. Complex command and control relationships will only add to the complexity and inhibit a commander's ability to dominate and apply available combat power to accomplish assigned objectives. Stability operations as seen in Somalia required commanders to dominate within a supporting role and, throughout, required careful, measured restraint.

UNITY OF COMMAND (EFFORT)

C-24. During OPERATION RESTORE HOPE, UNITAF successfully met unity of command challenges through three innovations. First, they created a civil-military operations center (CMOC) to facilitate unity of effort between NGOs and military forces. Second, UNITAF divided the country into nine humanitarian relief sectors centered on critical urban areas that facilitated both relief distribution and military areas of responsibility. Third, to establish a reasonable span of control, nations that provided less than platoon-sized contingents were placed under the control of the Army, Marine Corps, and Air Force components.

C-25. On the other hand, during OPERATION CONTINUED HOPE, UNOSOM II command and control relationships made unity of command (effort) nearly impossible. The logistic components of USFORSOM were under UN operational control, while the QRF remained under USCENTCOM's combatant command—as was TF Ranger. However, the USCENTCOM commander was not in theater. He was not actively involved in planning TF Ranger's missions or in coordinating and integrating them with his other subordinate commands. It was left to TF Ranger to coordinate with the QRF as needed. Even in TF Ranger, there were dual chains of command between SOF operators and the Rangers. This underscores the need for close coordination and careful integration of SOF and conventional forces (see Chapter 4). It also emphasizes overall unity of command (or effort when command is not possible) among all forces operating in a single urban environment.

C-26. Following TF Ranger's 3 October mission, the command structure during OPERATION CONTINUED HOPE was further complicated with the new JTF-Somalia. This force was designed to protect U.S. forces during the withdrawal from Somalia. JTF-Somalia came under the operational control of USCENTCOM, but fell under the tactical control of USFORSOM. Neither the JTF nor USFORSOM controlled the naval forces that remained under USCENTCOM's operational control. However unity of effort (force protection and a rapid, orderly withdrawal) galvanized the command and fostered close coordination and cooperation among the semiautonomous units.

MEASURED RESTRAINT

C-27. During OPERATIONS PROVIDE RELIEF and RESTORE HOPE, U.S. forces dominated within their supporting roles. Their perseverance, adaptability, impartiality, and restraint allowed them to provide a stable, secure environment. Hence, relief organizations could provide the food and medical care necessary to reduce disease, malnourishment, and the overall mortality rate. However, during OPERATION CONTINUED HOPE, U.S. operations became increasingly aggressive under the UN mandate. However, peace enforcement also requires restraint and impartiality to successfully dominate and achieve political objectives. The increased use of force resulted in increased civilian casualties, which in turn reduced the Somalis' perception of U.S. legitimacy. As a result, most moderate Somalis began to side with Aideed and his supporters. Many Somalis, without effective information operations to shape their perceptions, felt that it was fine for foreign military forces to intervene in their country to feed the starving and even help establish a peaceful government, but not to purposefully target specific Somali leaders as criminals.

CONSOLIDATE AND TRANSITION

C-28. Across the spectrum of conflict, Army forces must be able to execute full spectrum operations not only sequentially but, as in the case of operations in Somalia, simultaneously. OPERATION PROVIDE RELIEF began primarily as a stability operation in the form of foreign humanitarian assistance. Later this operation progressed to include peacekeeping (another stability operation), defensive operations to protect

UN forces and relief supplies, and minimum offensive operations. As operations transitioned to OPERATION RESTORE HOPE, it became apparent that while foreign humanitarian assistance was still the principal operation, other operations were necessary. More forceful peacekeeping, show of force, and arms control (all forms of stability operations), and offensive and defensive operations grew increasingly necessary to establish a secure environment for uninterrupted relief operations. In the final phase of U.S. involvement during OPERATION CONTINUED HOPE, major changes to political objectives caused a major transition in the form of the stability operation to peace enforcement with an even greater increase in the use of force, offensively and defensively, to create a peaceful environment and conduct nation building. Finally, the TF Ranger raid demonstrated the need to maintain a robust, combined arms force capable of rapidly transitioning from stability operations (primarily peace operations) to combat operations.

SUMMARY

C-29. OPERATIONS PROVIDE RELIEF and RESTORE HOPE were unquestionably successes. Conversely, during OPERATION CONTINUED HOPE, the 3-4 October battle of Mogadishu (also known as the "Battle of the Black Sea") was a tactical success leading to an operational failure. TF Ranger succeeded in capturing 24 suspected Aideed supporters to include two of his key lieutenants. Arguably, given the appropriate response at the strategic level, it had the potential to be an operational success. After accompanying Ambassador Oakley to a meeting with Aideed soon after the battle, MG Zinni described Aideed as visibly shaken by the encounter. MG Zinni believed Aideed and his subordinate leadership were tired of the fighting and prepared to negotiate. Unfortunately, the U.S. strategic leadership failed to conduct the shaping actions necessary to inform and convince the American public (and its elected members of Congress) of the necessity of employing American forces to capture Aideed. The President was left with little recourse after the battle of Mogadishu but to avoid further military confrontation.

C-30. Despite this strategic failing, the operational commanders might have avoided the casualties, and any subsequent public and Congressional backlash, had they better communicated among themselves and worked with unity of effort. Recognizing the separate U.S. and UN chains of command, the UN Special Representative, along with the USCENTCOM, USFORSOM, and TF Ranger commanders, should have established the command and control architecture needed. This architecture would have integrated planning and execution for each urban operation conducted. These commanders failed to "operationalize" their plan. They did not properly link U.S. strategic objectives and concerns to the tactical plan. The TF Ranger mission was a direct operational attempt to obtain a strategic objective in a single tactical action. Yet, they failed to understand the lack of strategic groundwork, the threat's intent and capabilities, and the overall impact of the urban environment, to include the terrain and society, on the operation. Such an understanding may not have led to such a high-risk course of action but instead to one that de-emphasized military operations and emphasized a political solution that adequately considered the clans' influence.

This page intentionally left blank.

Appendix D

Joint and Multinational Urban Operations

[Joint force commanders] synchronize the actions of air, land, sea, space, and special operations forces to achieve strategic and operational objectives through integrated, joint campaigns and major operations. The goal is to increase the total effectiveness of the joint force, not necessarily to involve all forces or to involve all forces equally.

JP 3-0

As pointed out earlier, Army forces, brigade size and larger, will likely be required to conduct operations in, around, and over large urban areas in support of a joint force commander (JFC). The complexity of many urban environments, particularly those accessible from the sea, requires unique leveraging and integration of all the capabilities of U.S. military forces to successfully conduct the operation. This appendix discusses many of these capabilities; JP 3-06 details joint urban operations.

PURPOSE

D-1. In some situations, a major urban operation is required in an inland area where only Army forces are operating. Army commanders determine if the unique requirements of the urban environment require forming a joint task force (JTF) or, if not, request support by joint capabilities from the higher joint headquarters. Sometimes the nature of the operation is straightforward enough or the urban operation is on a small enough scale that conventional intraservice support relationships are sufficient to meet the mission requirements.

D-2. Most major urban operations (UO), however, require the close cooperation and application of joint service capabilities. A JTF may be designated to closely synchronize the efforts of all services and functions in an urban area designated as a joint operations area (JOA). If a large urban area falls in the context of an even larger ground force area of operations, a JTF dedicated to the urban operation may not be appropriate. These situations still require joint capabilities. In such cases, the responsible JFC designates support relations between major land units and joint functional commands. The major land units can consist of Army forces, Marine Corps forces, or joint forces land component command. The joint functional commands can consist of the joint special operations task force, joint psychological operations task force, or joint civil-military operations task force.

D-3. This appendix describes the roles of other services and joint combatant commands in UO. It provides an understanding that enables Army commanders to recommend when to form a JTF or to request support from the JFC. It also provides information so commanders can better coordinate their efforts with those of the JFC and the commanders of other services or components conducting UO. Lastly, this appendix describes some considerations when conducting UO with multinational forces.

SISTER SERVICE URBAN CAPABILITIES

D-4. Army forces conducting UO rely on other services and functional joint commands for specialized support in the urban environment. These capabilities are requested from and provided through the commanding JFC. Army forces request the assets and capabilities described in this annex through their higher headquarters to the joint command. The JFC determines if the assets will be made available, the appropriate command relationship, and the duration of the support. Army forces prepare to coordinate planning and execution with other services and to exchange liaison officers. These capabilities can greatly increase the Army's ability to understand, shape, engage, consolidate, and transition within the context of UO.

AIR FORCE

D-5. Air Force support is an important aspect of the Army force concept for urban operations. Air Force elements have a role to play in UO across the spectrum of operations.

D-6. Air Force intelligence, surveillance, and reconnaissance (ISR) systems contribute significantly to understanding the urban environment. These ISR systems include the E-8 Joint Surveillance Target Attack Radar System (JSTARS) (see figure D-1), U2 Dragon Lady, RC-135 Rivet Joint, or RQ-4 Global Hawk and RQ/MQ-1 Predator unmanned aerial systems. Air Force ISR systems can provide vital data to help assess threat intentions, threat dispositions, and an understanding of the civilian population. These systems also can downlink raw information in real-time to Army intelligence processing and display systems, such as the common ground station or division tactical exploitation system.

Figure D-1. USAF E-8 JSTARS platform

D-7. Air interdiction (AI) can be a vital component of shaping the urban operational environment. Often, AI of the avenues of approach into the urban area isolates the threat by diverting, disrupting, delaying, or destroying portions of the threat force before they can be used effectively against Army forces. AI is especially effective in major combat operations where restrictions on airpower are limited and the threat is more likely to be a conventionally equipped enemy. In 1991 during OPERATION DESERT STORM, AI helped prevent the Iraqi 5th Mechanized Division from reaching Khafji.

D-8. Precision-guided munitions (PGMs) delivered from combat aircraft can positively and directly affect the conduct of Army close combat actions in the urban area. Special munitions designed to penetrate reinforced concrete can provide unique support to land forces executing UO. Although laser-guided bombs are extremely accurate with an effective laser lock-on, environmental factors such as smoke, haze, and weather can adversely affect successful laser designation. If the launching aircraft can achieve a successful laser designation and lock-on, these weapons have devastating effects, penetrating deep into reinforced concrete before exploding with great force. If launched without a lock-on, or if the laser spot is lost, these weapons are unpredictable and can travel long distances before they impact. However, continual advancements in tactics and weapons systems, such as joint direct attack munition (JDAM) weapons, help decrease the adverse effects of these environmental factors. (For example, in 2003 during OPERATION IRAQI FREEDOM, the Air Force supported the Army's attack on An Najaf with JDAM-equipped F 16s and B 1s during a blinding sandstorm. With near zero visibility, JDAM weapons provided significant fire support, allowing Army forces to complete their mission and continue the attack north.) When PGMs are unavailable, general-purpose bombs from 500 to 2,000 pounds can also prove effective. However, there may be risk factors associated with their use such as collateral damage due to decreased accuracy, threat exposure to the delivering aircraft due to lower delivery altitudes, or reduced weapons effectiveness due to shallower penetration angles. (See FM 3-06.1 for a more detailed description of various aircraft weapons and capabilities.)

D-9. In addition to shaping and engaging the UO through firepower, commanders can use Air Force capabilities to improve and augment the urban transportation and distribution infrastructure. Air Force units can repair or improve airfields, revitalize civil aviation maintenance facilities, manage air-delivered cargo, and control civil and military air traffic. These latter capabilities particularly enhance urban stability and civil support operations. These capabilities may even be decisive. Air-delivered cargo and air traffic management, for example, were the decisive factors in U.S. forces' successful resistance of the Soviet blockade of Berlin in 1948. In OPERATION RESTORE HOPE, from December 1992 to May 1993, Air Force operations in Mogadishu were critical to airlifting and staging supplies and forces. The Air Force determined the ultimate success of the humanitarian assistance operation (see Appendix C).

D-10. In unique situations, such as the Berlin Blockade and OPERATION RESTORE HOPE, exercising Air Force urban capabilities may be the decisive action of the operation. Most often, however, Air Force capabilities will play a shaping role, sustaining role, or both in joint urban operations because of the inherent requirement to occupy terrain and interface with the population.

Marine Corps

D-11. The Marine Corps can assault across water obstacles into a defended urban environment. This capability is an invaluable tactical and operational tool. The mere threat of this capability can divert many enemy forces from other avenues of approach and obscure the true nature of an attack. The impact of the threat of amphibious assault was vividly demonstrated during OPERATION DESERT STORM where embarked Marine Forces diverted several Iraqi divisions to defensive positions along the coast and near Kuwait City.

D-12. The presence of Marine amphibious equipment, apart from Marine infantry, provides Army forces unique capabilities. In UO, the amphibious operation is often not an assault from the sea, but rather an assault river crossing. In 1950, the 7th U.S. Infantry Division used amphibious tractor support from the 1st Marine Division to conduct an assault river crossing of the Han River into downtown Seoul. See FM 3-31.1.

D-13. The worldwide deployment of Marine air-ground task forces (MAGTFs) enables a short notice response into any urban areas accessible from the sea. Typically, a deployed MAGTF is a Marine expeditionary unit (special operations capable) (MEU[SOC]). The MEU(SOC) can perform forcible entry operations, seize lodgments, and may execute these tasks anticipating reinforcement by Army or joint forces. They are also well positioned and equipped to rapidly reinforce Army forces already deployed in theater. The special-operations-capable training that these units accomplish before deploying includes urban warfare training and contributes to their value in UO. The MEU(SOC) is relatively small (its core unit is a Marine infantry battalion), is forward deployed, and has a wide spectrum of organic capabilities. It is an important asset in crisis stability and civil support operations.

D-14. The Marine expeditionary brigade is the MAGTF between a MEU and a Marine expeditionary force (the largest MAGTF). All MAGTFs include an aviation combat element, which is particularly skilled at providing accurate and timely close air support. The responsiveness and accuracy of Marine close air support aptly suits it to UO where the lethality of combat and the close range of engagements demand accurate and responsive fires.

D-15. Marine forces that conduct UO work well in littoral urban areas because of their unique relationship with naval forces and thus their capability to closely integrate land and sea operations. A supporting arms liaison team may be attached to Army forces at battalion level to provide ship-to-shore communications and coordination for naval gunfire support.

Navy

D-16. Many major urban areas are accessible from the sea. Therefore, Army commanders should understand how sea power can influence and support UO. The Navy brings several major capabilities to UO. These include naval gunfire support, naval air support, and port and coastal security. In addition, the Navy can offer secure command post sites and meeting areas; provide limited emergency shelter and critical care for Soldiers and, as necessary, civilians; and store and transport critical supplies ashore providing flexibility in meeting support requirements. The ability to provide support from "over the horizon" not only enhances force protection but also the legitimacy of urban and host nation authorities during transition.

D-17. Naval surface fire support (which includes naval gunfire support) particularly applies to forcible entry operations in littoral urban areas. It provides an initial indirect fire support capability until Army forces land ashore. However, naval gunfire support lacks a precision munitions capability. In defensive operations, naval gunfire support can also add major fires, especially during retrograde operations through an urban area. Naval gunfire support of the 3rd Infantry Division was the primary fire support for the last two days of the X Corps retrograde operation out of the port of Hungnam, North Korea, in December 1950.

D-18. Destroyers and cruisers, which mount one and two 127mm MK45 lightweight gun systems respectively (see figure D-2), are capable of providing naval gunfire support. This gun system can provide a rate of fire of 20 rounds per minute per gun to a range of approximately 23 kilometers. Destroyers, cruisers, and attack submarines are capable of precision strikes with Tomahawk cruise missiles. Depending on the location of the urban area, naval air can provide responsive aviation support to UO alleviating the need for fixed-wing aviation bases.

Figure D-2. USN MK45 lightweight gun system

D-19. Due to its flat trajectory, terrain masking affects naval gunfire more than field artillery. Naval gunfire also results in large range probable errors (the dispersion pattern of the naval gun is roughly elliptical with the long axis in the direction of fire). Hence, coverage of targets such as roads and airfields is most effective when the gun-target line (GTL) coincides with the long axis of the target. Very close supporting fire can be delivered when the GTL is parallel to the front line of troops. Oppositely, a GTL perpendicular to the front trace can endanger friendly forces. Within the limits of hydrographic conditions, the ship can maneuver to achieve a better GTL, but ship movement also makes it difficult to adjust fire. Overall, naval and air threats, bad weather, and large range probable errors make naval gunfire difficult and can cause cancellation of supporting fires.

D-20. Water terminals located in urban areas are usually the debarkation points for the bulk of Army forces as well as a joint force theater logistics requirement. Army forces are responsible for water terminal operations while naval capabilities protect these strategic and operationally vital facilities.

D-21. Naval coastal warfare (NCW) is the responsibility of the JFC and is often exercised through the Navy component commander. He may assign a naval coastal warfare commander for an appropriate geographic area. NCW includes coastal sea control, port security, and harbor defense. While coastal sea control is conducted in the environment of the open seas, port security and harbor security include the urban environment. Port security is the safeguarding of vessels, harbors, ports, waterfront facilities, and cargo from internal threats. It includes destruction, loss, or injury from sabotage or other subversive acts; accidents; thefts; or other causes of similar nature. The Navy's role in protecting essential urban infrastructure is often key. This task is also important when executing stability operations if the threats against urban infrastructure will likely be unconventional.

D-22. Harbor defense protects harbor approaches, harbors, anchorages, and ports from external threats. Harbor defense focuses on the conventional defense of port infrastructure. It is a task appropriate in a major operation or campaign and often includes port security as a subtask. The JFC executes NCW (focused on harbor defense and port security) using combined Navy capabilities including surface warfare, aviation, and naval special operations. See also discussions in this appendix on transportation command and special operations command.

COAST GUARD

D-23. The Coast Guard, like the Navy, can significantly influence the conduct of UO when the urban area is accessible from the sea. The U.S. Coast Guard (USCG) is the federal authority for port security and harbor defense of domestic facilities. When directed by the President, the USCG can augment the Navy in operations overseas. Historically, the entire USCG was under Navy control during both World Wars I and II. USCG elements deployed overseas and operated under Navy control during the Vietnam War and during OPERATIONS DESERT SHIELD and DESERT STORM.

D-24. The USCG uses surface warfare systems and aerial reconnaissance systems to conduct its missions. Its air systems are unarmed, whereas its surface systems are armed for self-defense and law enforcement operations. The USCG is experienced and adept at supporting other agencies, local governments, and law enforcement. Its capabilities can best support Army UO in civil support operations or the civil support aspects of an operation. The USCG works effectively against an unconventional threat and threats with on-water capability.

URBAN FUNCTIONAL COMBATANT COMMAND CAPABILITIES

D-25. Three commands provide urban functional combatant command capabilities. Transportation Command manages the global transportation system. Strategic Command integrates plans for and employs space forces ad assets. Special Operations Command is responsible for providing special operations forces.

TRANSPORTATION COMMAND

D-26. U.S. Transportation Command (USTRANSCOM) provides strategic air, land, and sea transportation for the Department of Defense to deploy, employ, sustain, and redeploy U.S. military forces worldwide. USTRANSCOM provides global transportation management, using an integrated transportation system across the spectrum of operations through its transportation component commands (TCCs). The TCC consists of Air Mobility Command (AMC), Military Sealift Command (MSC), and Surface Deployment and Distribution Command (SDDC).

D-27. During urban operations, USTRANSCOM, through its TCC, can provide common-user terminal services in support of strategic transportation movements to a theater of operations. AMC provides common-user airlift, air refueling, and aeromedical evacuation services. It is the worldwide aerial port manager and, where designated, the operator of common-user aerial ports of embarkation and aerial ports of debarkation. MSC provides common-user sealift services between seaports of embarkation (SPOEs) and seaports of debarkation (SPODs). SDDC provides common-user ocean terminal service and, where designated, serves as the single port manager at SPOEs and SPODs. The urban transport system, in the form of railheads, ports, and airfields, is integral to many urban operations across the spectrum of Army operations and often the objective of Army UO. Army forces planning, preparing, and executing urban operations engage the supported geographic commander of a combatant command or his component commands in all aspects of conducting UO when the objective is transport related.

D-28. In offensive operations, commanders use transportation expertise to identify the urban transportation infrastructure (both the terrain and social [human] aspects) that is secured or that can affect current and future operations. This analysis also includes second- and third-order support systems. USTRANSCOM is consulted regarding the degree of acceptable damage that the system can sustain and still meet mission requirements. During execution, USTRANSCOM units may integrate into the operation so they can begin operating the transportation systems as early as possible. During the Inchon landing of September 1950, Army forces had begun rail operations on D+1 and port operations under way by D+3 of the forcible entry.

D-29. In defensive and stability or civil support operations, USTRANSCOM units safeguard and prevent disruption of the transport system by the conduct of defensive operations or stability tasks.

USTRANSCOM advises Army commanders of the impact of defensive tactics, techniques, and procedures (TTP) on USTRANSCOM operations as well as security requirements for USTRANSCOM facilities. The JFC provides guidance to deconflict any issues that may arise from these potentially divergent missions and tasks.

Strategic Command

D-30. The United States Strategic Command (USSTRATCOM) is the combatant command that integrates and synchronizes space operations, information operations, integrated missile defense, global command and control (C2); ISR, global strike, and strategic deterrence to provide. A unified source for greater understanding of specific threats around the world and the means to rapidly respond to those threats including weapons of mass destruction. USSTRATCOM operates assigned forces through its service component commands: U.S. Army Space Command/Army Forces Strategic Command; Air Force Space Command, and Naval Net Warfare Command.

D-31. USSTRATCOM provides critical space systems that offer global coverage and potential for real-time and near real-time support to urban operations. Space systems are unconstrained by political boundaries. Commanders can use space systems during peacetime or times of crisis to monitor an urban area before inserting friendly forces. Space systems enhance operations and assist commanders in overcoming some of the physical challenges of the urban environment. These enhancement operations include ISR; positioning and navigation; environmental monitoring; and communications. Space-based imagery and sensors are important ISR capabilities that contribute to situational understanding throughout the depth, breadth, and height of the urban area. ISR systems can provide route and target information for mission planning, locate presurveyed missile launch sites, detect camouflage, assess threat operations and movements, and warn of hostile acts and reconnaissance. Positioning and navigation systems also assist situational understanding through links to digital information systems (INFOSYS) while assisting tactical navigation in ambiguous terrain found in some urban areas. These enhancements can include precise location and position information for urban fires, ingress and egress routes, and rendezvous coordination. Environmental monitoring systems can provide weather and ionospheric information needed to assess weapon selection, air routes, ground and water trafficability, and communications. Communications systems provide secure, survivable links between elements of Army and joint forces to disseminate plans, orders, and warnings. These systems may form a critical link in the INFOSYS that transmit data to assessment centers and intelligence to key decision makers. However, some of the same environmental influences and degradation as ground-based systems may affect space-based systems. Commanders should refer to FM 3-14 for more information on space support to Army operations.

Special Operations Command

D-32. U.S. Special Operations Command (USSOCOM) exercises combatant command of all active and reserve special operations forces stationed in the United States. USSOCOM also provides trained and combat-ready special operations forces (SOF) to the geographical combatant commanders and, when directed by the President or the Secretary of Defense, command-designated special operations.

D-33. SOF provide commanders with capabilities critical to success in the urban environment. The density of this environment in both space and time requires the careful integration of SOF and conventional forces. Army forces conducting UO have a clear and unambiguous C2 relationship with the SOF in the urban area to ensure coordination, massing of effects, and unity of effort.

D-34. Each service has unique special operations (SO) capabilities. For example, SOF can identify and seize or destroy key terrain or infrastructure in denied areas; secure or capture key personnel; identify supportive civilians (aides and allies) who can facilitate future infiltrations and the cache of critical supplies; counter urban insurgencies; and conduct unconventional warfare in enemy-held urban areas. SOF can also emplace covert sensors and surveillance devices, provide clandestine intelligence collection, and provide target acquisition information or emplace beacons (for weapons guidance or navigation) in the highly restrictive terrain of the urban environment. In multinational UO, they can provide coalition support teams with trained, culturally aware, language proficient, military liaison personnel with organic communications connectivity. Army leaders understand the SO capabilities available to ensure that they request the right support for Army UO and to ensure unity of effort within the urban JOA.

Army SOF

D-35. The Army provides five types of SOF units to USSOCOM. Two of these, psychological operations (PSYOP) and civil affairs, are discussed in Chapter 4.

Ranger Forces

D-36. The U.S. Army 75th Ranger Regiment can rapidly deploy light infantry forces from company through regimental size. Rangers specialize in direct action (DA) SO missions and focus on airfield seizure and raids. Typically, Ranger units turn over their objectives to conventional units upon mission completion. The Ranger DA capability especially applies to UO because of the many critical infrastructures that often prompt Army forces to engage in UO. Ranger capabilities are ideal for seizing critical facilities to preempt their defense or destruction. The size and combat power of Ranger units permit execution of offensive and defensive operations against enemy conventional units for periods of limited duration; austere combat service support capabilities limit the Ranger regiment's ability to sustain combat action without extensive augmentation.

Special Forces

D-37. U.S. Army Special Forces Command (Airborne) trains and prepares Army Special Forces (SF) to deploy and execute operational requirements for geographical combatant commanders. SF units are small and capable of extended operations in remote and hostile locations. SF units execute seven basic missions: foreign internal defense (FID), unconventional warfare (UW), counterproliferation of weapons of mass destruction, special reconnaissance (SR), DA, counterterrorism, and information operations. Important collateral activities consist of coalition support, combat search and rescue, counterdrug activities, humanitarian demining activities, humanitarian assistance, security assistance, and special activities.

D-38. SR, DA, and coalition support are particularly important in UO. SF units may physically penetrate an urban area to conduct SR to determine threat strengths, dispositions, and intentions. In some situations, SF units, due to their language and cultural training, can gather accurate information regarding the disposition and attitudes of the population. SF special reconnaissance can also determine or verify the functional status of urban infrastructure as well as conduct target acquisition, area assessment, and poststrike reconnaissance.

D-39. SF teams can execute terminal guidance and with qualified joint tactical air controllers can control operations for Army and joint precision fires. Although these teams possess limited organic combat power for their short-duration, DA missions, they can bring significant effects to bear against high-payoff targets. Yet, SF units have no capability to conduct a sustained defense of such targets.

D-40. SF units can advise, train, and assist urban indigenous movements already in existence to conduct unconventional warfare and possibly accompany these groups into combat. The upsurge in urban insurgency and terrorism has caused worldwide concern since it is not confined to developing countries. Present day dissident groups are well aware of this situation. These groups realize that to be successful, they must center the insurgent activities on the major cities or political center of their countries.

Special Operations Aviation Forces

D-41. Army special operations aviation provided by the 160th Special Operations Aviation Regiment operates primarily to support SO missions. They execute insertion, extraction, and resupply missions to support SO. Similar to conventional aviation, this capability may be more vulnerable to concealed air defenses when operating over hostile or unsecured urban terrain than in many other environments.

Navy SOF

Sea-Air-Land Teams

D-42. Navy sea-air-land teams (SEALs) specialize in water approaches to targets. They operate in small, squad-size teams and have many of the same capabilities as Army Special Forces (see above). Navy SEALs do not typically have the cultural and language training of Army Special Forces. Their capability to insert from sea gives them a unique ability to penetrate into urban areas that are accessible from the sea.

Special Boat Units

D-43. Special boat units (SBUs) employ, operate, and maintain various surface combatant craft (see figure D-3) to conduct and support naval and joint special operations, riverine warfare, and coastal patrol and interdiction. The SBU can infiltrate and exfiltrate forces; provide small-caliber gunfire support; and conduct coastal patrol, surveillance, harassment, and interdiction of maritime lines of communications. These units are ideal in UO that include ports and rivers. They can assist in port security, conduct river patrols, and participate in harbor defense. They are well suited for preventing sea infiltration by unconventional threats.

Figure D-3. USN MK V special

Operations CraftAir Force SOF

D-44. The Air Force has three primary SOF elements: special operations air units, battlefield Airmen, and combat aviation advisory units.

Special Operations Wings and Groups

D-45. Air Force special operations aviation elements operate both fixed-wing and rotary-wing aircraft (which include variants of both C-130s and MH-53s). Air Force SOF conduct core missions which include specialized air mobility, precision aerospace fires, aerospace surface interface, personnel recovery operations, specialized refueling, combat aviation advisory, information operations, and PSYOP. As a result, Air Force SOF conduct infiltrations, exfiltrations, resupplies, close air support of SOF ground elements, electronic warfare, and aerial C2 support.

D-46. The AC-130 U and AC-130 H model aircraft (see figure D-4) are designed specifically to provide close air support to ground SOF. However, they will be an invaluable asset to any ground force operating in an urban environment. Hence, Army commanders seek opportunities to train with AC-130 gunships prior to the conduct of UO. AC-130s are armed with one 40mm autocannon and one 105mm howitzer. The AC-130 U is also armed with one 25mm autocannon. This aircraft has night capability and is extremely accurate. Its fires are responsive and can be decisive in close urban combat. Its cannon and howitzer are accurate enough to concentrate fire onto a single spot to create a rooftop breach that allows fire to be directed deeper into the building. The aircraft can provide excellent covert illumination with its infrared spotlight. The AC-130 is very vulnerable to air defense systems, which friendly forces must suppress or destroy to effectively use this system. (During OPERATION DESERT STORM in 1991, an AC-130 was shot down over Kuwait.)

Battlefield Airmen

D-47. Battlefield Airmen consist of special tactics teams (STTs) and special operations weather teams (SOWTs). Combat controllers and pararescue personnel constitute STTs. STTs are specially tailored to meet mission criteria and may vary from a small three-man team to a larger twenty-five-man element. STTs support the UO Army commanders by—

- Performing air-land-sea personnel recovery operations.
- Providing terminal attack control or guidance.
- Establishing and operating navigational aids and beacons.
- Providing liaison to ground commanders.
- Providing visual flight rules and limited instrument flight rules air traffic control.
- Providing positive control of the terminal objective area aviation environment during SOF operations.

***Note*:** SOWTs are normally attached to Army SOF to provide weather observation and limited tactical forecasting.

Combat Aviation Advisory (CAA) Operations

D-48. CAA operations are part of an interagency process aimed toward freeing and protecting the host nation's population from subversion, lawlessness, and insurgency. Specifically, CAA operations are tailored to assess, train, advise, and assist foreign aviation forces in air operations employment and sustainability. CAA operations support geographical combatant commanders throughout the spectrum of conflict primarily by facilitating the availability, reliability, safety and interoperability, and integration of friendly and allied aviation forces supporting joint, interagency, and multinational forces and organizations. CAA operations provide assistance in the interrelated areas of FID, coalition support, UW, and humanitarian and disaster relief. CAA operations also include a liaison role in coalition support.

MULTINATIONAL CONSIDERATIONS

D-49. Army UO in foreign urban areas will often be joint and likely have a multinational component. When properly executed, integrating multinational forces into UO greatly enhances the operation's military (as well as political) effectiveness. Properly integrating multinational forces into UO requires a thorough understanding of both the urban environment and the nature of individual national forces. This understanding includes the political, cultural, and historical characteristics of the other national forces. Such understanding also includes the national force's doctrine and military capabilities, strengths, and weaknesses. In UO, these considerations are critical because these factors will alter the urban population's attitude toward multinational forces and the behavior of such forces as they interact with the urban population. Combining this understanding with effective C2 and an equal understanding of the urban environment results in effective multinational UO (see FM 100-8).

CULTURAL COMPATIBILITY

D-50. When evaluating and assigning UO tasks, Army commanders also consider the degree of interaction with the civilian population. A national force from a Muslim-majority country may work better with a civilian population that is also Muslim than a force having a different religion. In such a situation, the national force with the same religion as the urban population may be assigned tasks that require close relations with civilians. Army forces, in this situation, may be assigned tasks that are more remote from the population. National forces that have a national history of animosity to the civil population (or that sympathize with antagonists within the civilian population) are not used in tasks requiring diplomacy and close cooperation with the civilians or government.

Military Capabilities and Restrictions

D-51. Army forces are responsible to understand the military capabilities of national forces with which they work. Some national forces, as part of their normal capabilities, are adept at police functions that enable them to operate with little training in a law enforcement role. Other national forces specialize in small-unit, light infantry patrolling. These forces may be ideal in a stability operation. In contrast, a national force composed of conscripts and trained primarily in conventional warfare techniques may best work as a reinforcing force or may require extensive training before mission execution in an urban environment or a stability situation. Army commanders also consider the type of weapon systems with which participating multinational forces are equipped; they may be more or less effective in an urban environment. For example, some countries may still possess the Vulcan antiaircraft gun system (or similar weapon) that can be very useful in urban offensive and defensive operations.

D-52. Commanders in multinational UO must also understand nation-specific rules of engagement (ROE). Some nations will severely limit their participating military's use of lethal and nonlethal fires. On the other hand, other nations may not be as constrained as Army units. Ideally, commanders attempt to create ROE that are the same across the coalition but recognize that this may not be achievable. Hence, in an environment where minimization of collateral damage will be paramount, Army commanders will need to carefully consider national limitations on the employment of forces when planning and executing UO. Relatedly, commanders take into account each nation's ability to achieve precision effects with their available forces and systems.

Intelligence Sharing

D-53. Due to classification issues, sharing intelligence with coalition partners may be challenging; the United States may have close intelligence ties with some countries within the coalition and few or none with others. In many cases intelligence personnel from other countries have unique skills (and cultural perspectives) that can significantly contribute to the friendly intelligence effort. Reports from some coalition members may fill intelligence gaps for Army forces and the coalition as a whole. On the other hand, few countries have the sophisticated intelligence collection assets available to Army forces and information that Army units may provide could be critical to a coalition partner's force protection and overall mission success.

Information Management

D-54. National capabilities to collect, process, store, display, and disseminate information and establish a common operational picture vary considerably between coalition partners. As the capabilities of Army information systems increase, those of our future partners may not keep pace. Even when capabilities are similar, equipment may not be compatible. The complex nature of the urban environment and coalition language differences (and linguist shortages) will only serve to exacerbate incompatibilities and limitations. Therefore commanders must understand capabilities and limitations to ensure that orders, control measures, and information can be appropriately disseminated, displayed, and understood. Significantly, they must understand the coalition's ability to accurately locate and track friendly forces as well as identify friend from foe. Commanders must take the appropriate steps to ensure that essential intelligence and relevant information is provided to coalition partners in a form (and language) that they can accept and use. This will be critical to establishing the common operational picture necessary to facilitate rapid decision making and effective command and control across the coalition.

Logistic Requirements

D-55. Logistically, agreement among the multinational forces should include support consolidation whenever possible. Creating a multinational logistics office or coordination center works best when accomplished early. This office can coordinate local contracts as well as already agreed upon host-nation support. This coordination among participating nations will reduce the competition for local assets that could otherwise have detrimental effects on one or more participating nations. Additionally, many nations providing forces for a coalition effort may not be able to support them logistically and will rely almost entirely on the United States or another coalition army for logistic support. Finally, pooled resources such as transportation assets and maintenance personnel could potentially reduce deployment requirements for one or more coalition partners. As the coalition matures, the role of this office or center may be expanded to include command activities.

This page intentionally left blank.

Source Notes

These are sources quoted or paraphrased in this publication. They are listed by page number. Quotations are identified by the first few word of the quote. Where a quote is embedded within a paragraph, the paragraph number is listed. Boldface indicates the title of historical vignettes.

Chapter 1–The Urban Outlook

1-1 "Today's security environment demands more...": FM 1, *The Army*, 14 June 2005: 1-19.

1-4 **Rome: A Microcosm of Urban Warfare**: R. Ernest Dupuy and Trevor N. Dupuy, *The Encyclopedia of Military History: From 3500 B.C. to the Present* (San Francisco: Harper & Row, 1977).

1-6 **The Three Block War**: as quoted in Russell W. Glenn, Steven Hartman, and Scott Gerwehr, *Urban Combat Service Support Operations: The Shoulders of Atlas* (Santa Monica, CA: RAND, 2003): 26.

1-8 **Winning the Peace in Iraq: The Requirement for Full Spectrum Operations**: Peter W. Chiarelli and Patrick R. Michaelis, "Winning the Peace: The Requirement for Full-Spectrum Operations," *Military Review* (July/August 2005): 4.

Chapter 2–Understanding the Urban Environment

2-1 "From a planning perspective, commanders view cities….": FM 3-0, *Operations*, 14 June 2001: 6-19.

2-16 "Me and Somalia Against the World": cited in Andrew S. Natsios, "Food Through Force: Humanitarian Intervention and U.S. Policy," *The Washington Quarterly* (winter 1994): 136.

2-18 **Understanding the Effects of Unit and Soldier Actions in Iraq**: *Initial Impressions Report: Operation Iraqi Freedom—Stability Operations-Support Operations,* Center for Army Lessons Learned (CALL) Newsletter No. 04-13 (Fort Leavenworth, KS: Center for Army Lessons Learned, U.S. Training and Doctrine Command [TRADOC], 2004): ii.

2-21 "Whoever coined the phrase… ": General Sir Rupert Smith, "Wars in Our Time–A Survey of Recent and Continuing Conflicts," *World Defence Systems*, volume 3:2 (London: Royal United Services Institute for Defence Studies, 2001): 121.

2-25 **Understanding the Urban Environment: Paris – 1944**: Russell W. Glenn, "Urban Combat is Complex," *Proceedings* (February 2002): 65. Russell W. Glenn, Steven Hartman, and Scott Gerwehr, *Urban Combat Service Support Operations: The Shoulders of Atlas* (Santa Monica, CA: RAND, 2003): 10-11.

Chapter 3–Understanding the Urban Threat

3-1 "…[T]he United States could be forced to intervene...": *Quadrennial Defense Review Report*, 30 September 2001: 6.

3-3 "…Iraq made no direct effort...": Gregory Fontenot, E. J. Degen, and David Tohn, *On Point: US Army in Operation IRAQI FREEDOM* (Fort Leavenworth, KS: Combat Studies Institute Press, 2004): 388.

3-4 **Tempo**: 26th Infantry Regimental Association, *Aachen. Military Operations in Urban Terrain* (Lititz, PA: 26th Infantry Regimental Association, 1999); Irving Werstein, *The Battle of Aachen* (New York: Thomas Y. Crowell Company, 1962); Charles Whiting, *Bloody Aachen* (New York: Stein and Day, 1976).

3-6 "Chechen fighters sometimes disguised...": Olga Oliker, *Russia's Chechen Wars 1994–2000: Lessons from Urban Combat* (Santa Monica, CA: RAND, 2001): 21.

3-7 **Identifying Threats from Noncombatants**: Olga Oliker, *Russia's Chechen Wars 1994–2000: Lessons from Urban Combat* (Santa Monica, CA: RAND, 2001).

3-9 **Information and the Media**: Eric Hammel, *Fire in the Streets: The Battle for Hue, Tet 1968* (New York: Dell Publishing, 1991).

3-12 **Cultural and Religious Instability**: Sid Heal, "Crowds, Mobs and Nonlethal Weapons," *Military Review* (March/April 2000): 45–50.

3-13 **Food and Water Shortages**: Leif R. Rosenberger, "The Strategic Importance of the World Food Supply," *Parameters* (spring 1997): 84–105.

3-15 **Insurgencies and the Urban Society**: Roger Trinquier, *Modern Warfare: A French View of Counterinsurgency* (New York: Praeger Publisher, 1964): 16-17.

3-16 **Crime and Criminal Organizations**: Eugene Linden, "The Exploding Cities of the Developing World," *Foreign Affairs* (January/February 1996): 52–65; Andrew S. Natsios, "Commander's Guidance: A Challenge of Complex Humanitarian Emergencies," *Parameters* (summer 1996): 50–66.

Chapter 4–Understanding the Urban Environment's Effects on Warfighting Functions and Tactics

4-1 "War is, above all things, an art...": Francis V. Greene, "The Important Improvements in the Art of War During the Past Twenty Years and Their Probable Effect on Future Military Operations," *Journal of the Military Service Institution of the United States* 4, no. 13 (1883): 41.

4-13 "Fighting in a city is much...": Vasili I. Chuikov, *The Battle for Stalingrad* (New York: Holt, Rinehart, and Winston, 1964): as cited by JP 3-06, *Doctrine for Urban Operations*, 16 September 2002: 1-8.

4-14 "I heard small-arms fire and...": Mark A. B. Hollis, "Platoon Under Fire," *Infantry* (January/April 1998): 29–30.

4-17 **Example of Simple Communications Innovation: Israel's Six-Day War – 1967**: Trevor N. Dupuy and Paul Martell, *Flawed Victory: The Arab-Israeli Conflict and the 1982 War in Lebanon* (Fairfax, VA: Hero Books, 1986); Richard A. Gabriel, *Operation Peace for Galilee: The Israeli-PLO War in Lebanon* (New York: Hill and Wang, 1984).

Chapter 5–Contemplating Urban Operations

5-1 "We based all our further calculations...": S. M. Shtemenko, *The Soviet General Staff at War, 1941-1945: Book One* (Moscow: Progress Publishers, 1981): 317–318.

5-19 "Four hostile newspapers...": Napoleon Bonaparte, Justin Wintle (ed.), *The Dictionary of War Quotations* (New York: The Free Press, 1989): 73.

5-24 **Conventional and Special Forces Integration**: Gregory Fontenot, E. J. Degen, and David Tohn, *On Point: US Army in Operation IRAQI FREEDOM* (Fort Leavenworth, KS: Combat Studies Institute Press, 2004): 403.

Chapter 6–Foundations for Urban Operations

6-1 "Utilities such as electricity…": Richard Connaughton, John Pimlott, and Duncan Anderson, *The Battle for Manila: The Most Devastating Untold Story of World War II* (Novata, CA: Presidio Press, 1995): 103.

6-7 **Applying the Urban Operational Framework: Panama – 1989**: Ronald H. Cole, *Operation Just Cause* (Washington DC: Joint History Office, 1995); Thomas Donnelly, Margaret Roth, and Caleb Baker, *Operation Just Cause: The Storming of Panama* (New York: Lexington Books, 1991); John T. Fishel, *The Fog of Peace: Planning and Executing the Restoration of Panama* (Carlisle, PA: Strategic Studies Institute, 1992); Malcolm McConnell, *Just Cause: The Real Story of America's High-Tech Invasion of Panama* (New York: St. Martin's Press, 1991); John Embry Parkerson, Jr., "United States Compliance with Humanitarian Law Respecting Civilians During Operation Just Cause," *Military Law Review* (1991): 31–140; Jennifer Morrison Taw, *Operation Just Cause: Lessons for Operations Other Than War* (Santa Monica, CA: RAND, 1996).

6-12 "I'm talking about attacking...": William S. Wallace quoted in Michael R. Gordon, "Baghdad Targets Picked if Hussein Holes Up There," *New York Times* (7 March 2003): A11.

6-13 "Do not try to..." T. E. Lawrence, "Twenty-Seven Articles," *The Arab Bulletin* (20 August 1917): as cited in David H. Petraeus, "Learning Counterinsurgency: Observations from Soldiering in Iraq," *Military Review* (January/February 2006): 4–5.

Chapter 7–Urban Offensive Operations

7-1 "…Capture Suez City...": Abraham Adan, *On the Banks of the Suez* (Novato, CA: Presidio Press, 1980): 409.

7-2 **Operational Context of Urban Operations: Brittany Ports – 1944**: Kent H. Butts, "The Strategic Importance of Water," *Parameters* (spring 1997): 65–83; Paul Carell, *Stalingrad: The Defeat of the German 6th Army* (Atglen, PA: Schieffer Publishing, 1993); R. Ernest Dupuy and Trevor N. Dupuy, *The Encyclopedia of Military History: From 3500 B.C. to the Present* (San Francisco: Harper & Row, 1977); Ralph Peters, "Our Soldiers, Their Cities," *Parameters* (spring 1996): 43–50; John J. Peterson, *Into the Cauldron* (Clinton, MD: Clavier House, 1973).

7-10 **Forms of Attack in the Urban Offense: Metz – 1944**: John Colby, *War From the Ground Up: The 90th Division in WWII* (Austin, TX: Nortex Press, 1991); Anthony Kemp, *The Unknown Battle: Metz, 1944* (New York: Stein and Day, 1981).

7-17 **Isolating an Urban Area: Hue, Vietnam – 1968**: James R. Arnold, *Tet Offensive 1968: Turning Point in Vietnam* (London: Osprey, 1990); Eric Hammel, *Fire in the Streets: The Battle for Hue, Tet 1968* (New York: Dell Publishing, 1991).

7-21 "Often what seems to be...": Russell W. Glenn, *An Attack on Duffer's Downtown* (Santa Monica, CA: RAND, 2001): 28.

7-22 **Creative Task Organization: Using Artillery in the Direct Fire Role**: 26th Infantry Regimental Association, *Aachen. Military Operations in Urban Terrain* (Lititz, PA: 26th Infantry Regimental Association, 1999); Ronald H. Cole, *Operation Just Cause* (Washington DC: Joint History Office, 1995); Thomas Donnelly, Margaret Roth, and Caleb Baker, *Operation Just Cause: The Storming of Panama* (New York: Lexington Books, 1991); David Eshel, *Mid-East Wars: The Lebanon War 1982* (Hod Hasharon, Israel: Eshel-Dramit, 1983); Richard A. Gabriel, *Operation Peace for Galilee: The Israeli-PLO War in Lebanon* (New York: Hill and Wang, 1984); Malcolm McConnell, *Just Cause: The Real Story of America's High-Tech Invasion of Panama* (New York: St. Martin's Press, 1991); Irving Werstein, *The Battle of Aachen* (New York: Thomas Y. Crowell Company, 1962); Charles Whiting, *Bloody Aachen* (New York: Stein and Day, 1976); Bruce Allen Watson, *Sieges: A Comparative Study* (Westport, CT: Praeger, 1993).

7-26 **Bold Operational Maneuver to Seize an Urban Area: Inchon and Seoul, Korea – 1950**: Robert Debs Heinl, Jr., *Victory at High Tide: The Inchon-Seoul Campaign* (Annapolis, MD: Nautical & Aviation Publishing Company of America, 1979); Douglas MacArthur, *Reminiscences* (New York: McGraw-Hill, 1964); Shelby L. Stanton, *Ten Corps in Korea, 1950* (Novato, CA: Presidio Press, 1989).

Chapter 8–Urban Defensive Operations

8-1 "Generally, a modern city magnifies...": Christopher R. Gabel, "Military Operations on Urbanized Terrain: The 2d Battalion, 26th Infantry, at Aachen, October 1944," *Urban Combat Operations: tactics, Techniques and Procedure.* Center for Army Lessons Learned (CALL) Newsletter No. 99-16 (Fort Leavenworth, KS: Center for Army Lessons Learned, U.S. Training and Doctrine Command [TRADOC], November 1999): 1-9.

8-5 **Urban Defense in a Major Operation: Stalingrad –1942 to 1943**: Anthony Beevor, *Stalingrad* (New York: Penguin Books, 1998); Vasili I. Chuikov, *The Battle for Stalingrad* (New York: Holt, Rinehart, and Winston, 1964); William Craig, *Enemy at the Gates: The Battle for Stalingrad* (New York: Reader's Digest Press, 1973).

8-8 "What is the position...": Michael Dewar, *War in the Streets: The Story of Urban Combat from Calais to Khafji* (New York: Sterling Publishers, 1992): 8.

8-10 **Defensive Combat Power: Suez – 1973**: Abraham Adan, *On the Banks of the Suez* (Novato, CA: Presidio Press, 1980).

Chapter 9–Urban Stability and Civil Support Operations

9-1 "The Rangers were bound...": Mark Bowden, *Black Hawk Down: A Story of Modern War* (New York: Atlantic Monthly Press, 1999): 18.

9-5 "In wars of intervention...": Antoine-Henri Jomini, "The Art of War," in *Roots of Strategy: Book 2* (Harrisburg, PA: Stackpole Books, 1987): 441.

9-8 **Assessment of Security and Force Protection: Belfast, Northern Ireland**: Michael Dewar, *British Army in Northern Ireland* (New York: Sterling Publishers, 1985); Desmond Hamill, *Pig in the Middle: The Army in Northern Ireland, 1969-1984* (London: Methuen, 1985).

9-11 "If there is any lesson...": Frederick Funston, "How the Army Worked to Save San Francisco," *Cosmopolitan Magazine* (July 1906). Available at http: //www.sfmuseum.org/1906/cosmo.html.

9-12 **Civil Support and Coordination with Civilian Authorities: Los Angeles – 1992**: James D. Delk, *Fires & Furies: The LA Riots* (Palm Springs, CA: Etc. Publications, 1995); William W. Mendel, *Combat in Cities: The LA Riots and Operation Rio* (Fort Leavenworth, KS: Foreign Military Studies Office, 1996).

9-14 **Restraint: An Najaf, Iraq – 2003**: Gregory Fontenot, E. J. Degen, and David Tohn, *On Point: US Army in Operation IRAQI FREEDOM* (Fort Leavenworth, KS: Combat Studies Institute Press, 2004).

9-16 "I met immediately with the...": Jeffrey B. Van Sickle, "Stability Operations in Northern Iraq: Task Force Altun Kupri," *Infantry* (January/February 2005): 26.

Chapter 10–Urban Sustainment

10-1 "Even supply is different…" Ralph Peters, "Our Soldiers, Their Cities," *Parameters* (spring 1996): 48.

10-2 **Base Security: Tan Son Nhut, Vietnam – 1968**: James R. Arnold, *Tet Offensive 1968: Turning Point in Vietnam* (London: Osprey, 1990); Keith William Nolan, *The Battle for Saigon: Tet 1968* (New York: Pocket Books, 1996).

10-4 "[Sustainment] planning accounts for...": FM 3-0, *Operations*, 14 June 2001: 6-19.

10-8 "Greater friendly force density would appear...": Russell W. Glenn, Steven Hartman, and Scott Gerwehr, *Urban Combat Service Support Operations: The Shoulders of Atlas* (Santa Monica, CA: RAND, 2003): 4.

10-13 **Host Nation Repair Operations: OPERATION IRAQI FREEDOM**: Steven Leonard, personal experience written specifically for FM 3-06.

10-18 **Combat Stress: Chechnya – 1994 to 1996**: Timothy L. Thomas and Charles P. O'Hara, "Combat Stress in Chechnya: 'The Equal Opportunity Disorder'," *Army Medical Department Journal* (January/March 2000): 46–53.

10-22 "…[I]n an endeavor like that in Iraq…": David H. Petraeus, "Learning Counterinsurgency: Observations from Soldiering in Iraq," *Military Review* (January/February 2006): 4–5.

Appendix A–Siege of Beirut: An Illustration of the Fundamentals of Urban Operations

A-1 **Siege of Beirut: An Illustration of the Fundamentals of Urban Operations**: Trevor N. Dupuy and Paul Martell, *Flawed Victory: The Arab-Israeli Conflict and the 1982 War in Lebanon* (Fairfax, VA: Hero Books, 1986); David Eshel, *Mid-East Wars: The Lebanon War 1982* (Hod Hasharon, Israel: Eshel-Dramit, 1983); Richard A. Gabriel, *Operation Peace for Galilee: The Israeli-PLO War in Lebanon* (New York: Hill and Wang, 1984).

A-1 "The [Israeli Defense Forces] IDF had...": Richard A. Gabriel, *Operation Peace for Galilee: The Israeli-PLO War in Lebanon* (New York: Hill and Wang, 1984): 128.

Appendix B–An Urban Focus to the Intelligence Preparation of the Battlefield

B-1 "Maneuvers that are possible...": *Infantry in Battle, 2d ed.* (Richmond, VA: Garret & Massie, 1939; Reprint, Fort Leavenworth, KS: U.S. Army Command and General Staff College, 1981): 69.

B-5 **Analysis of an Urban Area's Underlying Terrain: Mitrovica, Kosovo – 1999**: Jamison Jo Medby and Russell W. Glenn, *Street Smart: Intelligence Preparation of the Battlefield for Urban Operations* (Santa Monica, CA: RAND, 2002).

B-10 "…[T]he adversaries of freedom...": John F. Kennedy, Message to Congress, 25 May 1961. Available at <http://www.jfklibrary.org/Historical+Resources/Archives>.

B-12 **Shifting Civilian Interests and Intent**: James D. Delk, *Fires & Furies: The LA Riots* (Palm Springs, CA: Etc. Publications, 1995); Jamison Jo Medby and Russell W. Glenn, *Street Smart: Intelligence Preparation of the Battlefield for Urban Operations* (Santa Monica, CA: RAND, 2002).

Appendix C–Operations in Somalia: Applying the Urban Operational Framework to Stability Operations

C-1 **Operations in Somalia: Applying the Urban Operational Framework to Stability and Reconstruction Operations**: Kenneth Allard, *Somalia Operations: Lessons Learned* (Washington, DC: National Defense University Press, 1995).

C-1 "It's impossible for an American mother...": Thomas M. Montgomery, interview, "Ambush in Mogasdishu," *Frontline*, WGBH Boston, 29 September 1998, <http://www.pbs.org/wgbh/pages/frontline/programs/1998.html>.

C-2 C-5 …" to use all necessary…": *United Nations Security Council Resolution 794* (1992) (New York: United Nations, 1992), <http://daccessdds.un.org/doc/UNDOC/GEN/N92/772/11/PDF>.

C-2 C-5 "When directed by the…": CENTCOM mission statement as cited in Kenneth Allard, *Somalia Operations: Lessons Learned* (Washington, DC: National Defense University Press, 1995): 13-14.

Appendix D–Joint and Multinational Urban Operations

D-1 "[Joint force commanders] synchronize the actions…" JP 3-0, *Doctrine for Joint Operations*, 10 September 2001: II-4.

This page intentionally left blank.

Glossary

SECTION I – ACRONYMS AND ABBREVIATIONS

AI	air interdiction
AIDS	acquired immune deficiency syndrome
AMC	Air Mobility Command
AO	area of operations
ARVN	Army of the Republic of Viet Nam
ASCOPE	areas, structures, capabilities, organizations, people, and events
BCT	brigade combat team
BDAR	battle damage assessment and repair
BENELUX	Belgium, Netherlands, and Luxembourg
C2	command and control
CA	civil affairs
CAA	combat aviation advisory
CASEVAC	casualty evacuation
CBRN	chemical, biological, radiological, nuclear
CBRNE	chemical, biological, radiological, nuclear, high-yield explosive
CCIR	commander's critical information requirement
CI	counterintelligence
CEA	captured enemy ammunition
CERP	commander's emergency response program
CLS	combat lifesaver
CMO	civil-military operations
CMOC	civil-military operations center
CNA	computer network attack
CND	computer network defense
CNE	computer network exploitation
CNO	computer network operations
COA	course of action
COG	center of gravity
CONUS	continental United States
COP	common operational picture
DA	direct action
DNBI	disease and nonbattle injury
DOTMLPF	doctrine, organization, training, materiel, leadership and education, personnel, and facilities
EAC	echelons above corps
EEFI	essential elements of friendly information

EOD	explosive ordnance disposal
EW	electronic warfare
FHP	force health protection
FID	foreign internal defense
FM	field manual; frequency modulation
FMI	field manual, interim
FSB	forward support battalion
G-9	assistant chief of staff, civil affairs
GIS	geographic information system
GTL	gun-target line
HRS	human resources support
HUMINT	human intelligence
IDF	Israeli Defense Forces
IED	improvised explosive device
IMINT	imagery intelligence
INFOSYS	information system
IO	information operations
IPB	intelligence preparation of the battlefield
ISR	intelligence, surveillance, and reconnaissance
JDAM	Joint Direct Attack Munition
JFC	joint force commander
JOA	joint operations area
JP	joint publication
JTF	joint task force
JUO	joint urban operations
LAN	local area network
LNO	liaison officer
LOC	line of communications
LOGCAP	logistics civilian augmentation program
LOS	line of sight
LPT	logistics preparation of the theater
MAGTF	Marine air-ground task force
MANPADS	man-portable air defense system
MEDEVAC	medical evacuation
METT-TC	mission, enemy, terrain and weather, troops and support available, time available, civil considerations
MEU	Marine expeditionary unit
MEU(SOC)	Marine expeditionary unit (special operations capable)
MHE	materials handling equipment
MLRS	Multiple Launch Rocket System
MP	military police

MSC	Military Sealift Command
MST	maintenance support team
NATO	North Atlantic Treaty Organization
NCW	naval coastal warfare
NGA	National Geospatial-Intelligence Agency
NGO	nongovernmental organization
NKPA	North Korean People's Army
NVA	North Vietnamese Army
OAKOC	observation and fields of fire, avenues of approach, key terrain, obstacles, and cover and concealment
ODA	operational detachment alpha
OIF	Operation Iraqi Freedom
OPSEC	operations security
PA	public affairs
PDF	Panamanian Defense Force
PGM	precision-guided munitions
PLO	Palestinian Liberation Organization
PMCS	preventive maintenance checks and services
POW	prisoner of war
PSYOP	psychological operations
QRF	quick reaction force
ROE	rules of engagement
RPG	rocket-propelled grenade
SBU	special boat unit
SDDC	Surface Deployment and Distribution Command
SEAL	sea-air-land (team)
SF	special forces
SIGINT	signals intelligence
SJA	Staff Judge Advocate
SO	special operations
SOF	special operations forces
SOFA	status-of-forces agreement
SOP	standing operating procedure
SOWT	special operations weather team
SPOD	seaport of debarkation
SPE	seaport of embarkation
SR	special reconnaissance
STT	special tactics team
SWET	sewer, water, electricity, trash
SWEAT	sewer, water, electricity, academics trash
SWEAT-MS	sewer, water, electricity, academics trash-medical, security

TACON	tactical control
TC	training circular
TCC	transportation component command
TF	task force
TIM	toxic industrial material
TTP	tactics, techniques, and procedures
UAS	unmanned aircraft system
UN	United Nations
UNITAF	Unified Task Force
UNOSOM	United Nations Operations in Somalia
UNSCR	UN Security Council resolution
UO	urban operations
USACE	United States Army Corps of Engineers
USAID	United States Agency for International Development
USCENTCOM	United States Central Command
USFORSOM	United States Forces, Somalia
USCG	United States Coast Guard
USTRANSCOM	United States Transportation Command
UXO	unexploded ordnance
UW	unconventional warfare
VC	Viet Cong
WMD	weapons of mass destruction

SECTION II – Terms and Definitions

agility – The ability to move and adjust quickly and easily. (FM 3-0)

air interdiction – Air operations conducted to destroy, neutralize, or delay the enemy's military potential before it can be brought to bear effectively against friendly forces at such distance from friendly forces that detailed integration of each air mission with the fire and movement of friendly forces is not required. (JP 1-02)

area defense – A type of defensive operation that concentrates on denying enemy forces access to designated terrain for a specific time rather than destroying the enemy outright. (FM 3-0)

area of interest – That area of concern to the commander, including the area of influence, areas adjacent thereto, and extending into enemy territory to the objectives of current or planned operations. This area also includes areas occupied by enemy forces who could jeopardize the accomplishment of the mission. (JP 1-02)

area of operations – An operational area defined by the joint force commander for land and naval forces. Areas of operations do not typically encompass the entire operational area of the joint force commander, but should be large enough for component commanders to accomplish their missions and protect their forces. (JP 1-02)

ARFOR – The senior Army headquarters and all Army forces assigned or attached to a combatant command, subordinate joint force command, joint functional command, or multinational command. (FM 3-0)

assessment – (Army) The continuous monitoring of the current situation and progress of an operation.(FMI 5-0.1)

asymmetry – Dissimilarities in organization, equipment, doctrine, and values between other armed forces (formally organized or not) and US forces. Engagements are symmetric if forces, technologies, and weapons are similar; they are asymmetric if forces, technologies, and weapons are different, or if a resort to terrorism and rejection of more conventional rules of engagement are the norm. (FM 3-0)

attack – An offensive operation that destroys or defeats enemy forces, seizes and secures terrain, or both. (FM 3-0)

attainability - One of the eight characteristics of combat service support: generating the minimum essential supplies and services necessary to begin operations. (FM 4-0)

attrition - The reduction of the effectiveness of a force caused by loss of personnel and materiel. (JP 1-02)

avenue of approach – An air or ground route of an attacking force of a given size leading to its objective or to key terrain in its path. (JP 1-02)

axis of advance – (Army) The general area through which the bulk of a unit's combat power must move. (FM 3-90)

battle – A set of related engagements that lasts longer and involves larger forces than an engagement. (FM 3-0)

battlefield organization – The allocation of forces in the area of operations by purpose. It consists of three all-encompassing categories of operations: decisive, shaping, and sustaining. (FM 3-0)

battle rhythm – The sequencing of command and control activities within a headquarters and throughout the force to facilitate effective command and control. (FMI 5-0.1)

branch – (Army) A contingency plan or course of action (an option built into the basic plan or course of action) for changing the mission, disposition, orientation, or direction of movement of the force to aid success of the current operation, based on anticipated events, opportunities, or disruptions caused by enemy actions. Army forces prepare branches to exploit success and opportunities, or to counter disruptions caused by enemy actions. (FM 3-0)

breakout – An operation conducted by an encircled force to regain freedom of movement or contact with friendly units. It differs from other attacks only in that a simultaneous defense in other areas of the perimeter must be maintained. (FM 3-90)

buffer zone – A defined area controlled by a peace operations force from which disputing or belligerent forces have been excluded. A buffer zone is formed to create an area of separation between disputing or belligerent forces and reduce the risk of renewed conflict. Also called **area of separation** in some United Nations operations. (JP 1-02)

bypass – A tactical mission task in which the commander directs his unit to maneuver around an obstacle, position, or enemy force to maintain the momentum of the operation while deliberately avoiding combat with an enemy force. (FM 3-90)

campaign – A series of related military operations aimed at accomplishing a strategic or operational objective within a given time and space. (JP 1-02)

canalize – (Army) A tactical mission task in which the commander restricts enemy movement to a narrow zone by exploiting terrain coupled with the use of obstacles, fires, or friendly maneuver. (FM 3-90)

capability - The ability to execute a specified course of action (a capability may or may not be accompanied by an intention) (JP 1-02)

casualty evacuation – (Army) A term used by nonmedical units to refer to the movement of casualties aboard nonmedical vehicles or aircraft. (FM 8-10-6)

centers of gravity – Those characteristics, capabilities, or sources of power from which a military force derives its freedom of action, physical strength, or will to fight. (JP 1-02)

civil affairs – Designated Active and Reserve component forces and units organized, trained, and equipped specifically to conduct civil affairs activities and to support civil-military operations. (JP 1-02)

civil considerations – The influence of manmade infrastructure, civilian institutions, and attitudes and activities of the civilian leaders, populations, and organizations within an area of operations on the conduct of military operations. (FM 6-0)

civil disturbance – Group acts of violence and disorder prejudicial to public law and order. (JP 1-02)

civil-military operations – The activities of a commander that establish, maintain, influence, or exploit relations between military forces, governmental and nongovernmental civilian organizations and authorities, and the civilian populace in a friendly, neutral, or hostile operational area in order to facilitate military operations, to consolidate and achieve US objectives. Civil-military operations may include performance by military forces of activities and functions normally the responsibility of the local, regional, or national government. These activities may occur prior to, during, or subsequent to other military actions. They may also occur, if directed, in the absence of other military operations. Civil-military operations may be performed by designated civil affairs, by other military forces, or by a combination of civil affairs and other forces. (JP 1-02)

civil-military operations center – An ad hoc organization, normally established by the geographic combatant commander or subordinate joint force commander, to assist in the coordination of activities of engaged military forces, and other United States Government agencies, nongovernmental organizations, and regional and international organizations. There is no established structure, and its size and composition are situation dependent. (JP 1-02)

civil support operations – Domestic operations that address the consequences of man-made or natural accidents and incidents beyond the capabilities of civilian authorities. (FM 1)

clear – (Army) 1. A tactical mission task that requires the commander to remove all enemy forces and eliminate organized resistance within an assigned area. (FM 3-90) 2. The total elimination or neutralization of an obstacle that is usually performed by follow-on engineers and is not done under fire. (FM 3-34.2)

close air support – Air action by fixed- and rotary-wing aircraft against hostile targets that are in close proximity to friendly forces and that require detailed integration of each air mission with the fire and movement of those forces. (JP 1-02)

close combat – Combat carried out with direct fire weapons, supported by indirect fire, air-delivered fires, and nonlethal engagement means. Close combat defeats or destroys enemy forces or seizes and retains ground. (FM 3-0)

coalition – An ad hoc arrangement between two or more nations for common action. (JP 1-02)

collateral damage – Unintentional or incidental injury or damage to persons or objects that would not be lawful military targets in the circumstances ruling at the time. Such damage is not unlawful so long as it is not excessive in light of the overall military advantage anticipated from the attack. (JP 1-02)

combatant command – A unified or specified command with a broad continuing mission under a single commander established and so designated by the President, through the Secretary of Defense and with the advice and assistance of the Chairman of the Joint Chiefs of Staff. Combatant commands typically have geographic or functional responsibilities. (JP 1-02)

combat power – The total means of destructive and/or disruptive force which a military unit/formation can apply against the opponent at a given time. (JP 1-02)

combined arms – The synchronized or simultaneous application of several arms—such as infantry, armor, field artillery, engineers, air defense, and aviation—to achieve an effect on the enemy that is greater than if each arm were used against the enemy separately or in sequence. (FM 3-0)

combined arms team – (Army) Two or more arms mutually supporting one another, usually consisting of infantry, armor, cavalry, aviation, field artillery, air defense artillery, and engineers. (FM 3-90).

command and control system – (Army) The arrangement of personnel, information management, procedures, and equipment and facilities essential for the commander to conduct operations. (FM 6-0)

commander's critical information requirements – (Army) – Elements of information required by commanders that directly affect decisionmaking and dictate the successful execution of military operations. (FM 3-0)

commander's intent – (Army) A clear, concise statement of what the force must do and the conditions the force must meet to succeed with respect to the enemy, terrain, and civil considerations that represent the operation's desired end state. (FMI 5-0.1)

commander's visualization – The mental process of developing situational understanding, determining a desired end state, and envisioning how the force will achieve that end state. (FMI 5-0.1)

common operational picture – (Army) An operational picture tailored to the user's requirements, based on common data and information shared by more than one command. (FM 3-0)

computer network attack – Actions taken through the use of computer networks to disrupt, deny, degrade, or destroy information resident in computers or computer networks, or the computers and networks themselves. (JP 1-02)

computer network defense – Actions taken through computer networks to protect, monitor, analyze, detect and respond to unauthorized activity within Department of Defense information systems and computer networks. (JP 1-02)

concept of operations – (Army) How commanders see the actions of subordinate units fitting together to accomplish the mission. As a minimum, the description includes the scheme of maneuver and concept of fires. The concept of operations expands the commander's selected course of action and expresses how each element of the force will cooperate to accomplish the mission. (FM 3-0)

constraint – A restriction placed on the command by a higher command. A constraint dictates an action or inaction, thus restricting the freedom of action a subordinate commander has for planning. (FM 5-0)

control – (Army) 1. The regulation of forces and warfighting functions to accomplish the mission in accordance with the commander's intent. (FMI 5-0.1) 2. A tactical mission task that requires the commander to maintain physical influence over a specified area to prevent its use by an enemy. (FM 3-90)

controlled exchange – The removal of serviceable parts, components, or assemblies from unserviceable, economically reparable equipment and their immediate reuse in restoring a like item of equipment to a combat operable or serviceable condition. (FM 4-30.3)

control measure – A means of regulating forces or warfighting functions. (FMI 5-0.1)

counterattack – (Army) A form of attack by part or all of a defending force against an enemy attacking force with the general objective of denying the enemy his goal in attacking. See FM 3-0.

counterdeception – Efforts to negate, neutralize, diminish the effects of, or gain advantage from a foreign deception operation. Counterdeception does not include the intelligence function of identifying foreign deception operations. (JP 1-02)

counterdrug – Those active measures taken to detect, monitor, and counter the production, trafficking, and use of illegal drugs. (JP 1-02)

counterinsurgency – Those military, paramilitary, political, economic, psychological, and civic actions taken by a government to defeat insurgency. (JP 1-02)

counterintelligence – Information gathered and activities conducted to protect against espionage, other intelligence activities, sabotage, or assassinations conducted by or on behalf of foreign governments or elements thereof, foreign organizations, or foreign persons, or international terrorist activities. (JP 1-02)

countermobility operations – The construction of obstacles and emplacement of minefields to delay, disrupt, and destroy the enemy by reinforcement of the terrain. The primary purpose of countermobility operations is to slow or divert the enemy, to increase time for target acquisition, and to increase weapon effectiveness. (JP 1-02)

counterpropaganda – Programs of products and actions designed to nullify propaganda or mitigate its effects. (FM 3-05.30)

counterterrorism – Operations that include the offensive measures taken to prevent, deter, preempt, and respond to terrorism. (JP 1-02)

course of action – 1. Any sequence of activities that an individual or a unit may follow. 2. A possible plan open to an individual or a commander that would accomplish or is related to the accomplishment of a mission. 3. The scheme adopted to accomplish a job or mission. 4. A line of conduct in an engagement. 5. A product of the Joint Operation Planning and Execution System concept development phase. (JP 1-02)

cover – (Army) 1. Protection from the effects of fires. (FM 6-0) 2. A form of security operation whose primary task is to protect the main body by fighting to gain time while also observing and reporting information and preventing enemy ground observation of and direct fire against the main body. Unlike a screening or guard force, the covering force is a self-contained force capable of operating independently of the main body. (FM 3-90)

dead space – 1. An area within the maximum range of a weapon, radar, or observer, which cannot be covered by fire or observation from a particular position because of intervening obstacles, the nature of the ground, the characteristics of the trajectory, or the limitations of the pointing capabilities of the weapon. 2. An area or zone which is within range of a radio transmitter, but in which a signal is not received. 3. The volume of space above and around a gun or guided missile system into which it cannot fire because of mechanical or electronic limitations. (JP 1-02)

debarkation – The unloading of troops, equipment, or supplies from a ship or aircraft. (JP 1-02)

decision point – (Army) An event, area, or point in the battlespace where and when the friendly commander will make a critical decision. (FM 5-0)

decision support matrix – An aid used by the commander and staff to make battlefield decisions. This matrix is a staff product of the wargaming process that lists the decision point, location of the decision point, the criteria to be evaluated at the point of decision, the action or operations to occur at the decision point, and the unit or element that is to act and has responsibility to observe and report the information affecting the criteria for the decision. (FM 5-0)

decision support template – (Army) A staff product initially used in the wargaming process that graphically represents the decision points and projected situations, and indicates when, where, and under what conditions a decision is most likely to be required to initiate a specific activity or event. (FM 5-0)

decisive engagement – In land and naval warfare, an engagement in which a unit is considered fully committed and cannot maneuver or extricate itself. In the absence of outside assistance, the action must be fought to a conclusion and either won or lost with the forces at hand. (JP 1-02)

decisive operation – The operation that directly accomplishes the task assigned by the higher headquarters. Decisive operations conclusively determine the outcome of major operations, battles, and engagements. (FM 3-0)

decisive point – A geographic place, specific key event, critical system or function that allows commanders to gain a marked advantage over an enemy and greatly influence the outcome of an attack. (JP 1-02)

defeat – A tactical mission task that occurs when an enemy force has temporarily or permanently lost the physical means or the will to fight. The defeated force's commander is unwilling or unable to pursue his adopted course of action, thereby yielding to the friendly commander's will, and can no longer interfere to a significant degree with the actions of friendly forces. Defeat can result from the use of force or the threat of its use. (FM 3-90)

defeat in detail – Concentrating overwhelming combat power against separate parts of a force in sequence rather than defeating the entire force at once. (FM 3-90)

defense in depth – The siting of mutually supporting defense positions designed to absorb and progressively weaken attack, prevent initial observations of the whole position by the enemy, and to allow the commander to maneuver the reserve. (JP 1-02)

defensive information operations – The integration and coordination of policies and procedures, operations, personnel, and technology to protect and defend information and information systems. Defensive information operations ensure timely, accurate, and relevant information access while denying adversaries the opportunity to exploit friendly information and information systems for their own purposes. (FM 3-0)

defensive operations – Operations that defeat an enemy attack, buy time, economize forces, or develop conditions favorable for offensive operations. Defensive operations alone normally cannot achieve a decision. Their purpose is to create conditions for a counteroffensive that allows Army forces to regain the initiative. (FM 3-0)

delay – A form of retrograde in which a force under pressure trades space for time by slowing the enemy's momentum and inflicting maximum damage on the enemy without, in principle, becoming decisively engaged. (FM 3-90)

depth – The extension of operations in time, space, and resources. (FM 3-0)

destroy – 1. A tactical mission task that physically renders an enemy force combat-ineffective until it is reconstituted. 2. To damage a combat system so badly that it cannot perform any function or be restored to a usable condition without being entirely rebuilt. (FM 3-90)

direct action – Short-duration strikes and other small-scale offensive actions conducted as a special operation in hostile, denied, or other politically sensitive environments and which employ specialized military capabilities to seize, destroy, capture, exploit, recover, or damage designated targets. Direct action differs from conventional offensive actions in the level of physical and political risk, operational techniques, and the degree of discriminate and precise use of force to achieve specific objectives. (JP 1-02)

direct approach – The application of combat power directly against the enemy center of gravity or the enemy's principal strength. (FM 3-90)

direct fire – Fire delivered on a target using the target itself as a point of aim for either the weapon or the director. (JP 1-02)

disinformation – Information disseminated primarily by intelligence organizations or other covert agencies designed to distort information or deceive or influence United States decisionmakers, United States forces, coalition allies, key actors, or individuals via indirect or unconventional means. (FM 3-13)

displaced person – (Army) A civilian who is involuntarily outside the national boundaries of his or her country or as an internally displaced person is a civilian involuntarily outside his area or region within his country. (FM 2-0)

display – (Army) An information management activity: to represent relevant information in a usable, easily understood audio or visual form tailored to the needs of the user that conveys the common operational picture for decisionmaking and exercising command and control functions (FM 6-0)

disposition – 1. Distribution of the elements of a command within an area, usually the exact location of each unit headquarters and the deployment of forces subordinate to it. (JP 1-02)

disrupt – 1. A tactical mission task in which a commander integrates direct and indirect fires, terrain, and obstacles to upset an enemy's formation or tempo, interrupt his timetable, or cause his forces to commit prematurely or attack in piecemeal fashion. (FM 3-90) 2. An engineer obstacle effect that focuses fire planning and obstacle effort to cause the enemy to break up his formation and tempo, interrupt his timetable, commit breaching assets prematurely, and attack in a piecemeal effort. (FM 90-7)

distribution system – That complex of facilities, installations, methods, and procedures designed to receive, store, maintain, distribute, and control the flow of military materiel between the point of receipt into the military system and the point of issue to using activities and units. (JP 1-02)

doctrinal template – A model based on known or postulated adversary doctrine. Doctrinal templates illustrate the disposition and activity of adversary forces and assets conducting a particular operation unconstrained by the effects of the battlespace. They represent the application of adversary doctrine under ideal conditions. Ideally, doctrinal templates depict the threat's normal organization for combat, frontages, depths, boundaries and other control measures, assets available from other commands, objective depths, engagement areas, battle positions, and so forth. Doctrinal templates are usually scaled to allow ready use with geospatial products. (JP 1-02)

doctrine – Fundamental principles by which the military forces or elements thereof guide their actions in support of national objectives. It is authoritative but requires judgment in application. (JP 1-02)

economy – One of the eight characteristics of combat service support: providing the most efficient support at the least cost to accomplish the mission. (FM 4-0)

economy of force – One of the nine principles of war: allocate minimum essential combat power to secondary efforts. (FM 3-0)

effect – (Army) A result, outcome, or consequence of an action. (FMI 5-0.1)

electromagnetic deception – The deliberate radiation, re-radiation, alteration, suppression, absorption, denial, enhancement, or reflection of electromagnetic energy in a manner intended to convey misleading information to an enemy or enemy electromagnetic-dependent weapons, thereby degrading or neutralizing the enemy's combat capability. Among the types of electromagnetic deception are: a. **manipulative electromagnetic deception**—Actions to eliminate revealing, or convey misleading, electromagnetic telltale indicators that may be used by hostile forces; b. **simulative electromagnetic deception**—Actions to simulate friendly, notional, or actual capabilities to mislead hostile forces; c. **imitative electromagnetic deception**—The introduction of electromagnetic energy into enemy systems that imitates enemy emissions. (JP 1-02)

electromagnetic spectrum – The range of frequencies of electromagnetic radiation from zero to infinity. It is divided into 26 alphabetically designated bands. (JP 1-02)

electronic attack—That division of electronic warfare involving the use of electromagnetic energy, directed energy, or antiradiation weapons to attack personnel, facilities, or equipment with the intent of degrading, neutralizing, or destroying enemy combat capability and is considered a form of fires. Electronic attack includes: 1. actions taken to prevent or reduce an enemy's effective use of the electromagnetic spectrum, such as jamming and electromagnetic deception, and 2. employment of weapons that use either electromagnetic or directed energy as their primary destructive mechanism (lasers, radio frequency weapons, particle beams), or antiradiation weapons. (JP 1-02)

electronic warfare – Any military action involving the use of electromagnetic and directed energy to control the electromagnetic spectrum or to attack the enemy. The three major subdivisions within electronic warfare are: electronic attack, electronic protection, and electronic warfare support. (JP 1-02)

electronic warfare support—That division of electronic warfare involving actions tasked by, or under direct control of, an operational commander to search for, intercept, identify, and locate or localize sources of intentional and unintentional radiated electromagnetic energy for the purpose of immediate threat recognition, targeting, planning, and conduct of future operations. Thus, electronic warfare support provides information required for immediate decisions involving electronic warfare operations and other tactical actions such as threat avoidance, targeting, and homing. Electronic warfare support data can be used to produce signals intelligence, provide targeting for electronic or destructive attack, and produce measurement and signature intelligence. (JP 1-02)

embarkation – The process of putting personnel and/or vehicles and their associated stores and equipment into ships and/or aircraft. (JP 1-02)

encirclement – An operation where one force loses its freedom of maneuver because an opposing force is able to isolate it by controlling all ground lines of communications. (FM 3-0)

end state – (Army) At the operational and tactical levels, the conditions that, when achieved, accomplish the mission. At the operational level, these conditions attain the aims set for the campaign or major operation. (FM 3-0)

engagement – A tactical conflict, usually between opposing lower echelon maneuver forces. (JP 1-02)

envelopment – (Army) – A form of maneuver in which an attacking force seeks to avoid the principal enemy defenses by seizing objectives to the enemy rear to destroy the enemy in his current positions. At the tactical level, envelopments focus on seizing terrain, destroying specific enemy forces, and interdicting enemy withdrawal routes. (FM 3-0).

essential elements of friendly information – (Army) The critical aspects of a friendly operation that, if known by the enemy, would subsequently compromise, lead to failure, or limit success of the operation, and, therefore, must be protected from enemy detection. (FM 3-13)

essential task – A specified or implied task that must be executed to accomplish the mission. Essential tasks are always included in the unit's mission statement. (FM 5-0)

exfiltration – The removal of personnel or units from areas under enemy control by stealth, deception, surprise, or clandestine means. (JP 1-02)

exploitation – 1. Taking full advantage of success in military operations, following up initial gains, and making permanent the temporary effects already achieved. 2. An offensive operation that usually follows a successful attack and is designed to disorganize the enemy in depth. (JP 1-02)

explosive ordnance disposal – The detection, identification, on-site evaluation, rendering safe, recovery, and final disposal of unexploded explosive ordnance. It may also include explosive ordnance which has become hazardous by damage or deterioration. (JP 1-02)

firepower – 1. The amount of fire which may be delivered by a position, unit, or weapon system. 2. Ability to deliver fire. (JP 1-02)

fires – The effects of lethal or nonlethal weapons. (JP 1-02)

fire support coordinating measure – A measure employed by land or amphibious commanders to facilitate the rapid engagement of targets and simultaneously provide safeguards for friendly forces. (JP 1-02)

fix – (Army) 1. A tactical mission task where a commander prevents the enemy from moving any part of his force from a specific location for a specific period of time. 2. An engineer obstacle effect that focuses fire planning and obstacle effort to slow an attacker's movement within a specified area, normally an engagement area. (FM 3-90)

flexibility – One of the eight characteristics of combat service support: being able to adapt combat service support structures and procedures to changing situations, missions, and concepts of operations. (FM 4-0)

force projection – The ability to project the military element of national power from the continental United States (CONUS) or another theater in response to requirements for military operations. Force projection operations extend from mobilization and deployment of forces to redeployment to CONUS or home theater. (JP 1-02)

force protection – Actions taken to prevent or mitigate hostile actions against Department of Defense personnel (to include family members), resources, facilities, and critical information. These actions conserve the force's fighting potential so it can be applied at a decisive time and place and incorporates the coordinated and synchronized offensive and defensive measures to enable the effective employment of the joint force while degrading opportunities for the enemy. Force protection does not include actions to defeat the enemy or protect against accidents, weather, or disease. (JP 1-02)

force tailoring – The process of determining the right mix and sequence of units for a mission. (FM 3-0)

foreign humanitarian assistance – Programs conducted to relieve or reduce the results of natural or manmade disasters or other endemic conditions such as human pain, disease, hunger, or privation that might present a serious threat to life or that can result in great damage to or loss of property. Foreign humanitarian assistance (FHA) provided by US forces is limited in scope and duration. The foreign assistance provided is designed to supplement or complement the efforts of the host nation civil authorities or agencies that may have primary responsibility for providing FHA. FHA operations are those conducted outside the United States, its territories, and possessions. (JP 1-02)

foreign internal defense – Participation by civilian and military agencies of a government in any of the action programs taken by another government to free and protect its society from subversion, lawlessness, and insurgency. (JP 1-02)

fratricide – The unintentional killing or wounding of friendly personnel by friendly firepower. (FM 3-0)

frontal attack – (Army) A form of maneuver in which the attacking force seeks to destroy a weaker enemy force or fix a larger enemy force in place over a broad front. (FM 3-0)

full spectrum operations – The range of operations Army forces conduct in war and military operations other than war. (FM 3-0)

graphic control measure – A symbol used on maps and displays to regulate forces and warfighting functions. (FMI 5-0.1)

gun-target line – An imaginary straight line from gun to target. (JP 1-02)

hazardous material – Any substance which has a human health hazard associated with it. Special storage, use, handling, and shipment safety procedures and protocols must be followed to help protect against accidental exposure. Hazardous materials are specifically identified under federal law. (FM 3-100.4)

high angle fire – (Army) Fire delivered to clear an obstacle (such as a hill) that low-angle fire cannot, or fire delivered to attack targets on the reverse side of an obstacle (such as a hill) that cannot be attacked with low-angle or direct fire. (FM 6-30).

host nation – A nation that receives the forces and/or supplies of allied nations, coalition partners, and/or NATO organizations to be located on, to operate in, or to transit through its territory. (JP 1-02)

host-nation support – Civil and/or military assistance rendered by a nation to foreign forces within its territory during peacetime, crises or emergencies, or war based on agreements mutually concluded between nations. (JP 1-02)

human intelligence – A category of intelligence derived from information collected and provided by human sources. (JP 1-02)

imagery intelligence – Intelligence derived from the exploitation of collection by visual photography, infrared sensors, lasers, electro-optics, and radar sensors such as synthetic aperture radar wherein images of objects are reproduced optically or electronically on film, electronic display devices, or other media. (JP 1-02)

implied task – A task that must be performed to accomplish a specified task or mission, but is not stated in the higher headquarters order. (FM 5-0)

indirect fire – Fire delivered on a target that is not itself used as a point of aim for the weapons or the director. (JP 1-02)

infiltration – (Army) 1. A form of maneuver in which an attacking force conducts undetected movement through or into an area occupied by enemy forces to occupy a position of advantage in the enemy rear while exposing only small elements to enemy defensive fires. (FM 3-90)

information – (Army) 1. In the general sense, the meaning humans assign to data. 2. In the context of the cognitive hierarchy, data that have been processed to provide further meaning. (FM 6-0)

information fratricide – The results of employing information operations elements in a way that causes effects in the information environment that impede the conduct of friendly operations or adversely affect friendly forces. (FM 3-13)

information management – The provision of relevant information to the right person at the right time in a usable form to facilitate situational understanding and decisionmaking. It uses procedures and information systems to collect, process, store, display, and disseminate information. (FM 3-0)

information operations – (Army) The employment of the core capabilities of electronic warfare, computer network operations, psychological operations, military deception, and operations security, in concert with specified supporting and related capabilities, to affect and defend information and information systems and to influence decisionmaking. (FM 3-13)

information requirements – (Army) All information elements the commander and staff require to successfully conduct operations, that is, all elements necessary to address the factors of METT-TC. (FM 6-0)

information superiority – The operational advantage derived from the ability to collect, process, and disseminate an uninterrupted flow of information while exploiting or denying an adversary's ability to do the same. (JP 1-02)

information system – (Army) The equipment and facilities that collect, process, store, display, and disseminate information. This includes computers—hardware and software—and communications, as well as policies and procedures for their use. (FM 3-0)

insurgency – An organized movement aimed at the overthrow of a constituted government through the use of subversion and armed conflict. (JP 1-02)

integration – One of the eight characteristics of combat service support: the total inclusion of Army combat service support into the operations process (plan, prepare, execute, assess), as well as into other logistics comnponents of the unified force. (FM 4-0).

intelligence – 1. The product resulting from the collection, processing, integration, analysis, evaluation, and interpretation of available information concerning foreign countries or areas. 2. Information and knowledge about an adversary obtained through observation, investigation, analysis, or understanding. (JP 1-02)

intelligence preparation of the battlefield – The systematic, continuous process of analyzing the threat and environment in a specific geographic area. Intelligence preparation of the battlefield (IPB) is designed to support the staff estimate and military decisionmaking process. Most intelligence requirements are generated as a result of the IPB process and its interrelation with the decisionmaking process. (FM 34-130)

intelligence, surveillance, and reconnaissance – An activity that synchronizes and integrates the planning and operation of sensors, assets, and processing, exploitation, and dissemination systems in direct support of current and future operations. This is an integrated intelligence and operations function. (JP 1-02)

intelligence, surveillance, and reconnaissance plan – An integrated plan for collection of information from all available sources and analysis of that information to produce intelligence to meet requirements. Specifically, a logical plan for transforming priority intelligence requirements into orders or requests to reconnaissance and surveillance assets to collect pertinent information within a required time limit. (FM 34-3)

interagency – Activities or operations conducted by or through coordination with two or more agencies or an agency and one or more Services of the same nation. (FM 3-07)

interdict – A tactical mission task where the commander prevents, disrupts, or delays the enemy's use of an area or route. (FM 3-90)

interior lines – A force operates on interior lines when its operations diverge from a central point. (FM 3-0)

isolate – A tactical mission task that requires a unit to seal off—both physically and psychologically—an enemy from his sources of support, deny an enemy freedom of movement, and prevent an enemy unit from having contact with other enemy forces. (FM 3-90)

joint force – A general term applied to a force composed of significant elements, assigned or attached, of two or more Military Departments, operating under a single joint force commander. (JP 1-02)

joint force commander – A general term applied to a combatant commander, subunified commander, or joint task force commander authorized to exercise combatant command (command authority) or operational control over a joint force. (JP 1-02)

joint operations – A general term to describe military actions conducted by joint forces, or by Service forces in relationships (e.g., support, coordinating authority), which, of themselves, do not create joint forces. (JP 1-02)

joint task force – A joint force that is constituted and so designated by the Secretary of Defense, a combatant commander, a subunified commander, or an existing joint task force commander. (JP 1-02)

key terrain – Any locality, or area, the seizure or retention of which affords a marked advantage to either combatant. (JP 1-02)

landing zone – Any specified zone used for the landing of aircraft. JP 1-02)

law of war – That part of international law that regulates the conduct of armed hostilities. (JP 1-02)

liaison – That contact or intercommunication maintained between elements of military forces or other agencies to ensure mutual understanding and unity of purpose and action. (JP 1-02)

line of communications – A route, either land, water, and/or air, that connects an operating military force with a base of operations and along which supplies and military forces move. (JP 1-02)

lines of operations – (Army) A line that defines the orientation of the force in time and space, or purpose, in relation to an enemy or objective. (FMI 5-0.1).

logistics – The science of planning and carrying out the movement and maintenance of forces. In its most comprehensive sense, those aspects of military operations which deal with: a. design and development, acquisition, storage, movement, distribution, maintenance, evacuation, and disposition of materiel; b. movement, evacuation, and hospitalization of personnel; c. acquisition or construction, maintenance, operation, and disposition of facilities; and d. acquisition and furnishing of services. (JP 1-02)

logistics-over-the-shore operations – The loading and unloading of ships without the benefit of deep draft-capable, fixed port facilities; or as a means of moving forces close to tactical assembly areas dependent on threat force capabilities. (JP 1-02)

logistics preparation of the theater – Actions taken by combat service support personnel to optimize means—force structure, resources, and strategic lift—of supporting the joint force commander's plan. (FM 4-0)

major operation – A series of tactical actions (battles, engagements, strikes) conducted by various combat forces of a single or several Services, coordinated in time and place, to accomplish operational and, sometimes, strategic objectives in an operational area. These actions are conducted simultaneously or sequentially in accordance with a common plan and are controlled by a single commander. (JP 1-02)

maneuver – One of the nine principles of war: place the enemy in a disadvantageous position through the flexible application of combat power. (FM 3-0)

Marine air-ground task force – The Marine Corps principal organization for all missions across the range of military operations, composed of forces task-organized under a single commander capable of responding rapidly to a contingency anywhere in the world. The types of forces in the Marine air-ground task force (MAGTF) are functionally grouped into four core elements: a command element, an aviation combat element, a ground combat element, and a combat service support element. The four core elements are categories of forces, not formal commands. The basic structure of the MAGTF never varies, though the number, size, and type of Marine Corps units comprising each of its four elements will always be mission dependent. The flexibility of the organizational structure allows for one or more subordinate MAGTFs to be assigned. In a joint or multinational environment, other Service or multinational forces may be assigned or attached. (FM 1-02)

Marine expeditionary force – The largest Marine air-ground task force (MAGTF) and the Marine Corps principal warfighting organization, particularly for larger crises or contingencies. It is task-organized around a permanent command element and normally consists of one or more Marine divisions, Marine aircraft wings, and Marine force service support groups. The Marine expeditionary force is capable of missions across the range of military operations, including amphibious assault and sustained operations ashore in any environment. It can operate from a sea base, a land base, or both. In a joint or multinational environment, it may contain other Service or multinational forces assigned or attached to the MAGTF. (FM 1-02)

Marine expeditionary unit – A Marine air-ground task force (MAGTF) that is constructed around an infantry battalion reinforced, a helicopter squadron reinforced, and a task-organized combat service support element. It normally fulfills Marine Corps' forward sea-based deployment requirements. The Marine expeditionary unit provides an immediate reaction capability for crisis response and is capable of limited combat operations. In a joint or multinational environment, it may contain other Service or multinational forces assigned or attached to the MAGTF. (FM 1-02)

mass – One of the nine principles of war: the effects of combat power at the decisive place and time. (FM 3-0)

mass casualty – Any large number of casualties produced in a relatively short period of time, usually as the result of a single incident such as a military aircraft accident, hurricane, flood, earthquake, or armed attack that exceeds logistical support capabilities. (JP 1-02)

measure of effectiveness – (Army) A criterion used to assess changes in system behavior, capability, or operational environment that is tied to measuring the attainment of an end state, achievement of an objective, or creation of an effect. (FMI 5-0.1)

medical evacuation – The timely and efficient movement of the wounded, injured, or ill while providing en route medical care to and between medical treatment facilities. (FM 4-02)

meeting engagement – A combat action that occurs when a moving force engages an enemy at an unexpected time and place. (FM 3-0)

METT-TC – A memory aid used in two contexts: (1) In the context of information management, the major subject categories into which relevant information is grouped for military operations: mission, enemy, terrain and weather, troops and support available, time available, civil considerations. (2) In the context of tactics, the major factors considered during mission analysis. (FM 6-0)

military deception – Actions executed to deliberately mislead adversary military decisionmakers as to friendly military capabilities, intentions, and operations, thereby causing the adversary to take specific actions (or inactions) that will contribute to the accomplishment of the friendly mission. (JP 1-02)

military decisionmaking process – A process that integrates the activities of the commander, staff and subordinate commanders in developing and operation plan or order. It establishes procedures for analyzing a mission; developing, analyzing, and comparing courses of action; selecting the best course of action; and producing an operation plan or order (FMI 5-0.1).

misinformation – Incorrect information from any source that is released for unknown reasons or to solicit a response or interest from a nonpolitical or nonmilitary target. (FM 3-13)

mission – 1. The task, together with the purpose, that clearly indicates the action to be taken and the reason therefor. 2. In common usage, especially when applied to lower military units, a duty assigned to an individual or unit; a task. (JP 1-02)

mission command – The conduct of military operations through decentralized execution based upon mission orders for effective mission accomplishment. Successful mission command results from subordinate leaders at all echelons exercising disciplined initiative within the commander's intent to accomplish missions. It requires an environment of trust and mutual understanding. (FM 6-0)

mission orders – A technique for completing combat orders that allows subordinates maximum freedom of planning and action in accomplishing missions and leaves the "how" of mission accomplishment to the subordinate. (FM 6-0)

mobile defense – (Army) A type of defensive operation that concentrates on the destruction or defeat of the enemy through a decisive attack by a striking force. (FM 3-0)

mobility corridor – Areas where a force will be canalized due to terrain restrictions. They allow military forces to capitalize on the principles of mass and speed and are therefore relatively free of obstacles. (JP 1-02)

mobility operations – Obstacle reduction by maneuver and engineer units to reduce or negate the effects of existing or reinforcing obstacles. The objective is to maintain freedom of movement for maneuver units, weapon systems, and critical supplies. (FM 3-34)

modified combined obstacle overlay – A joint intelligence preparation of the battlespace product used to portray the effects of each battlespace dimension on military operations. It normally depicts militarily significant aspects of the battlespace environment, such as obstacles restricting military movement, key geography, and military objectives. (JP 1-02)

monitoring – (Army) Continuous observation of the current situation to identify opportunities for the force, threats to the force, gaps in information, and progress according to the plan or order. (FMI 5-0.1)

movement to contact – A form of the offensive designed to develop the situation and to establish or regain contact. (JP 1-02)

naval coastal warfare – Coastal sea control, harbor defense, and port security, executed both in coastal areas outside the United States in support of national policy and in the United States as part of this Nation's defense. (JP 1-02)

naval gunfire support – Fire provided by Navy surface gun systems in support of a unit or units tasked with achieving the commander's objectives. A subset of naval surface fire support. (JP 1-02)

naval surface fire support – Fire provided by Navy surface gun and missile systems in support of a unit or units. (JP 1-02)

neutral – (Army) 1. An individual, group of individuals, organization, or nation that is not hostile to or in any way supportive of only one belligerent force in a hostile environment. (FM 3-07)

no-fire area – A land area, designated by the appropriate commander, into which fires or their effects are prohibited. (JP 1-02)

noncombatant – 1. An individual in an area of combat operations who is not armed and is not participating in any activity in support of any of the factions or forces involved in combat. 2. An individual, such as chaplain or medical personnel, whose duties do not involve combat. (FM 3-07)

noncombatant evacuation operations – Operations directed by the Department of State, the Department of Defense, or other appropriate authority whereby noncombatants are evacuated from foreign countries when their lives are endangered by war, civil unrest, or natural disaster to safe havens or to the United States. (JP 1-02)

nongovernmental organization – A private, self-governing, not-for-profit organization dedicated to alleviating human suffering; and/or promoting education, health care, economic development, environmental protection, human rights, and conflict resolution; and/or encouraging the establishment of democratic institutions and civil society. (JP 1-02)

nonlethal fires – Any fires that do not directly seek the physical destruction of the intended target and are designed to impair, disrupt, or delay the performance of enemy operational forces, functions, and facilities. Psychological operations, electronic warfare (jamming), and other command and control countermeasures are all nonlethal fire options. (FM 6-20)

nonlethal weapons – Weapons that are explicitly designed and primarily employed so as to incapacitate personnel or materiel, while minimizing fatalities, permanent injury to personnel and undesired damage to property and the environment. a. Unlike conventional lethal weapons that destroy their targets through blast, penetration, and fragmentation, nonlethal weapons employ means other than gross physical destruction to prevent the target from functioning. b. Nonlethal weapons are intended to have one, or both, of the following characteristics: (1) They have relatively reversible effects on personnel or materiel. (2) They affect objects differently within their area of influence. (JP 1-02)

objective – (Army) 1. One of the nine principles of war: direct every military operation toward a clearly defined, decisive and attainable objective. (FM 3-0) 2. A location on the ground used to orient operations, phase operations, facilitate changes of direction, and provide for unity of effort. (FM 3-90)

offensive – One of the nine principles of war: seize, retain, and exploit the initiative. (FM 3-0)

offensive information operations – The integrated use of assigned and supporting capabilities and activities, mutually supported by intelligence, to affect enemy decisionmakers or to influence others to achieve or promote specific objectives. (FM 3-0)

offensive operations – Operations which aim at destroying or defeating an enemy. Their purpose is to impose US will on the enemy and achieve decisive victory. (FM 3-0)

operation – 1. A military action or the carrying out of a strategic, operational, tactical, service, training, or administrative military mission. 2. The process of carrying on combat, including movement, supply, attack, defense, and maneuvers needed to gain the objectives of any battle or campaign. (JP 1-02)

operational control – Command authority that may be exercised by commanders at any echelon at or below the level of combatant command. Operational control is inherent in combatant command (command authority) and may be delegated within the command. When forces are transferred between combatant commands, the command relationship the gaining commander will exercise (and the losing commander will relinquish) over these forces must be specified by the Secretary of Defense. Operational control is the authority to perform those functions of command over subordinate forces involving organizing and employing commands and forces, assigning tasks, designating objectives, and giving authoritative direction necessary to accomplish the mission. Operational control includes authoritative direction over all aspects of military operations and joint training necessary to accomplish the missions assigned to the command. Operational control should be exercised through the commanders of subordinate organizations. Normally this authority is exercised through subordinate joint force commanders and Service and/or functional component commanders. Operational control normally provides full authority to organize commands and forces and to employ those forces as the commander in operational control considers necessary to accomplish assigned missions; it does not, in and of itself, include authoritative direction for logistics or matters of administration, discipline, internal organization, or unit training. (JP 1-02)

operational framework – The arrangement of friendly forces and resources in time, space, and purpose with respect to each other and the enemy or situation. It consists of the area of operations, battlespace, and battlefield organization. (FM 3-0)

operations process – The major command and control activities performed during operations: planning, preparation, execution with continuous assessment. These activities occur continuously throughout an operation, overlapping and recurring as required. (FMI 5-0.1)

operations security – A process of identifying critical information and subsequently analyzing friendly actions attendant to military operations and other activities to: a. identify those actions that can be observed by adversary intelligence systems; b. determine indicators hostile intelligence systems might obtain that could be interpreted or pieced together to derive critical information in time to be useful to adversaries; and c. select and execute measures that eliminate or reduce to an acceptable level the vulnerabilities of friendly actions to adversary exploitation. (JP 1-02)

paramilitary forces – Forces or groups distinct from the regular armed forces of any country, but resembling them in organization, equipment, training, or mission. (JP 1-02)

patrol – A detachment of ground, sea, or air forces sent out for the purpose of gathering information or carrying out a destructive, harassing, mopping-up, or security mission. (JP 1-02)

peace enforcement – Application of military force, or the threat of its use, normally pursuant to international authorization, to compel compliance with resolutions or sanctions designed to maintain or restore peace and order. (JP 1-02)

peacekeeping – Military operations undertaken with the consent of all major parties to a dispute, designed to monitor and facilitate implementation of an agreement (ceasefire, truce, or other such agreement) and support diplomatic efforts to reach a long-term political settlement. (JP 1-02)

peace operations – A broad term that encompasses peacekeeping operations and peace enforcement operations conducted in support of diplomatic efforts to establish and maintain peace. (JP 1-02)

penetration – (Army) A form of maneuver in which an attacking force seeks to rupture enemy defenses on a narrow front to disrupt the defensive system. (FM 3-90).

perceptions – Mental images the commander wants the deception target to believe are real. (FM 3-13)

phase line – A line utilized for control and coordination of military operations, usually an easily identified feature in the operational area. (JP 1-02)

physical destruction – The application of combat power to destroy or degrade adversary forces, sources of information, command and control systems, and installations. It includes direct and indirect forces from ground, sea, and air forces. Also included are direct actions by special operations forces. (FM 3-13)

physical security – 1. That part of security concerned with physical measures designed to safeguard personnel; to prevent unauthorized access to equipment, installations, material, and documents; and to safeguard them against espionage, sabotage, damage, and theft. 2. In communications security, the component that results from all physical measures necessary to safeguard classified equipment, material, and documents from access thereto or observation thereof by unauthorized persons. (JP 1-02)

pickup zone – A geographic area used to pick up troops or equipment by helicopter. (FM 90-4)

plan – A design for a future or anticipated operation. (FM 5-0)

planning – The process by which commanders (and staffs, if available) translate the commander's visualization into a specific course of action for preparation and execution, focusing on the expected results. (FMI 5-0.1)

port of debarkation – (DOD) The geographic point at which cargo or personnel are discharged. This may be a seaport or aerial port of debarkation; for unit requirements, it may or may not coincide with the destination. (JP 1-02)

port of embarkation – (DOD) The geographic point in a routing scheme from which cargo and personnel depart. This may be a seaport or aerial port from which personnel and equipment flow to a port of debarkation; for unit and nonunit requirements, it may or may not coincide with the origin. (JP 1-02)

principles of war – Principles that provide general guidance for conducting war and military operations other than war at the strategic, operational, and tactical levels. The nine principles of war are: objective, offensive, mass, economy of force, maneuver, unity of command, security, surprise, and simplicity (FM 3-0)

propaganda – Any form of communication in support of national objectives designed to influence the opinions, emotions, attitudes, or behavior of any group in order to benefit the sponsor, either directly or indirectly. (JP 1-02)

psychological operations – Planned operations to convey selected information and indicators to foreign audiences to influence their emotions, motives, objective reasoning, and ultimately the behavior of foreign governments, organizations, groups, and individuals. The purpose of psychological operations is to induce or reinforce foreign attitudes and behavior favorable to the originator's objectives. (JP 1-02)

public affairs – Those public information, command information, and community relations activities directed toward both the external and internal publics with interest in the Department of Defense. (JP 1-02)

pursuit – An offensive operation designed to catch or cut off a hostile force attempting to escape, with the aim of destroying it. (JP 1-02)

push – In logistics, the delivery of a predetermined amount of supplies to a user on a scheduled basis without the user requesting them. (FM 4-0)

railhead – A point on a railway where loads are transferred between trains and other means of transport. (JP 1-02)

reachback – The process of obtaining products, services, and applications, or forces, or equipment, or material from organizations that are not forward deployed. (JP 1-02)

rear area – (Army) For any particular command, the area extending forward from its rear boundary to the rear of the area assigned to the next lower level of command. This area is provided primarily for the performance of support functions, and is where the majority of the echelon's sustaining functions occur. (FM 3-0)

refugee – A person who, by reason of real or imagined danger, has left his home country or country of nationality and is unwilling or unable to return. (JP 1-02)

relevant information – All information of importance to commanders and staffs in the exercise of command and control. (FM 3-0)

relief in place – (Army) An tactical enabling operation in which, by direction of higher authority, all or part of a unit is replaced in an area by the incoming unit. (FM 3-90)

reorganization – Action taken to shift internal resources within a degraded unit to increase its level of combat effectiveness. (FM 100-9)

reserve – Portion of a body of troops which is kept to the rear or withheld from action at the beginning of an engagement, in order to be available for a decisive movement. (JP 1-02)

responsiveness – One of the eight principles of combat service support: providing the right support at the right place at the right time. (FM 4-0)

retrograde – A type of defensive operation that involves organized movement away from the enemy. (FM 3-0)

riot control agent – (Army) A substance that produces temporary irritating or disabling physical effects that disappear within minutes of removal from exposure. There is no significant risk of permanent injury, and medical treatment is rarely required. (FM 3-11.11)

risk – (DOD) Probability and severity of loss linked to hazards. (JP 1-02)

risk management – The process of identifying, assessing, and controlling risk arising from operational factors, and making decisions that balance risk cost with mission benefits. (JP 1-02)

rules of engagement – Directives issued by competent military authority that delineate the circumstances and limitations under which United States forces will initiate and/or continue combat engagement with other forces encountered. (JP 1-02)

running estimate – A staff section's continuous assessment of current and future operations to determine if the current operation is proceeding according to the commander's intent and if future operations are supportable. (FMI 5-0.1)

search and attack – A technique of conducting a movement to contact that shares many of the characteristics of an area security mission. (FM 3-0)

secure – (Army) 1. A tactical mission task that involves preventing a unit, facility, or geographical location from being damaged or destroyed as a result of enemy action. (FM 3-90)

security – One of the nine principles of war: never permit the enemy to acquire an unexpected advantage (FM 3-0).

seize – (Army) A tactical mission task that involves taking possession of a designated area using overwhelming force. (FM 3-90)

sequel – (Army) An operation that follows the current operation. It is a future operation that anticipates possible outcomes – success, failure, or stalemate – of the current operation. (FM 3-0).

shaping operation – An operation at any echelon that creates and preserves conditions for the success of decisive operations. (FM 3-0)

show of force – An operation designed to demonstrate US resolve that involves increased visibility of US deployed forces in an attempt to defuse a specific situation that, if allowed to continue, may be detrimental to US interests or national objectives. (JP 1-02)

signals intelligence – 1. A category of intelligence comprising either individually or in combination all communications intelligence, electronic intelligence, and foreign instrumentation signals intelligence, however transmitted. 2. Intelligence derived from communications, electronics, and foreign instrumentation signals. (JP 1-02)

simplicity – One of the nine principles of war: prepare clear, uncomplicated plans, and clear, concise orders to ensure thorough understanding. (FM 3-0) [Note: "simplicity" is also one of the eight characteristics of combat service support.]

situational understanding – (Army) The product of applying analysis and judgment to the common operational picture to determine the relationship among the factors of METT-TC. (FM 3-0)

space operations – The employment of space system capabilities that provide the means to enhance command and control, facilitate the maneuver of forces, reduce the commander's uncertainty, and improve fire support, air defense, intelligence collection, and combat service support operations which will support strategic, operational, and tactical missions across the operational continuum in the near, mid, and far term. (FM 3-14)

special operations – Operations conducted in hostile, denied, or politically sensitive environments to achieve military, political, economic, and/or informational objectives employing military capabilities for which there is no broad conventional force requirement. These operations often require covert, clandestine, or low visibility capabilities. Special operations are applicable across the rnage of military operations. They can be conducted independently or in conjunction with operations of conventional forces or other government agencies and may include operations through, with, or by indigenous or surrogate forces. Special operations differ from conventional operations in degree of physical and political risk, operational techniques, mode of employment, independence from friendly support, and dependence on detailed operational intelligence and indigenous assets. (JP 1-02)

special operations forces – Those Active and Reserve Component forces of the Military Services designated by the Secretary of Defense and specifically organized, trained, and equipped to conduct and support special operations. (JP 1-02)

special reconnaissance – Reconnaissance and surveillance actions conducted by special operations forces in hostile, denied, or politically sensitive environments to collect or verify information of strategic or operational significance, employing military capabilities not normally found in conventional forces. These forces provide an additive capability for commanders and supplement other conventional reconnaissance and surveillance actions. (JP 1-02)

stability operations – Operations that promote and protect US national interests by influencing the threat, political, and information dimensions of the operational environment through a combination of peacetime developmental, cooperative activities and coercive actions in response to crisis. (FM 3-0)

status-of-forces agreement – An agreement that defines the legal position of a visiting military force deployed in the territory of a friendly state. Agreements delineating the status of visiting military forces may be bilateral or multilateral. Provisions pertaining to the status of visiting forces may be set forth in a separate agreement, or they may form a part of a more comprehensive agreement. These provisions describe how the authorities of a visiting force may control members of that force and the amenability of the force or its members to the local law or to the authority of local officials. To the extent that agreements delineate matters affecting the relations between a military force and civilian authorities and population, they may be considered as civil affairs agreements. (JP 1-02)

strategy – The art and science of developing and employing instruments of national power in a synchronized and integrated fashion to achieve theater, national and/or multinational objectives. (JP 1-02)

striking force – A committed force organized to conduct the decisive attack in a mobile defense. It normally comprises the maximum combat power available to the commander at the time of the attack. (FM 3-0)

strong point – (Army) A heavily fortified battle position tied to a natural or reinforcing obstacle to create an anchor for the defense or to deny the enemy decisive or key terrain. (FM 3-90)

surprise – One of the nine principles of war: strike the enemy at a time or place or in a manner for which he is unprepared. (FM 3-0)

survivability – (Army) 1. Concept which includes all aspects of protecting personnel, weapons, and supplies while simultaneously deceiving the enemy. Survivability tactics include building a good defense; employing frequent movement; using concealment, deception, and camouflage; and constructing fighting and protective positions for both individuals and equipment. Encompasses planning and locating position sites, designing adequate overhead cover, analyzing terrain conditions and construction materials, selecting excavation methods, and countering the effects of direct and indirect fire weapons. (FM 5-103) 2. One of the eight characteristics of combat service support: being able to shield support functions from destruction or degradation. (FM 4-0)

sustainability - One of the eight characteristics of combat service support: the ability to maintain continuous support, rear area and base security, movement control, terrain management, and infrastructure development. (FM 4-0)

sustaining operation – An operation at any echelon that enables shaping and decisive operations by providing combat service support, rear area and base security, movement control, terrain management, and infrastructure development. (FM 3-0)

synchronization – 1. The arrangement of military actions in time, space, and purpose to produce maximum relative combat power at a decisive place and time. See FM 3-0. 2. In the intelligence context, application of intelligence sources and methods in concert with the operation plan. (JP 1-02)

tactical combat force – A combat unit, with appropriate combat support and combat service support assets, that is assigned the mission of defeating Level III threats. (JP 1-02)

tactical control – Command authority over assigned or attached forces or commands, or military capability or forces made available for tasking, that is limited to the detailed direction and control of movements or maneuvers within the operational area necessary to accomplish missions or tasks assigned. Tactical control is inherent in operational control. Tactical control may be delegated to, and exercised at any level at or below the level of combatant command. When forces are transferred between combatant commands, the command relationship the gaining commander will exercise (and the losing commander will relinquish) over these forces must be specified by the Secretary of Defense. Tactical control provides sufficient authority for controlling and directing the application of force or tactical use of combat support assets within the assigned mission or task. (JP 1-02)

tactical level of war –The level of war at which battles and engagements are planned and executed to accomplish military objectives assigned to tactical units or task forces. Activities at this level focus on the ordered arrangement and maneuver of combat elements in relation to each other and to the enemy to achieve combat objectives. (JP 1-02)

target acquisition – The detection, identification, and location of a target in sufficient detail to permit the effective employment of weapons. (JP 1-02)

targeting – The process of selecting and prioritizing targets and matching the appropriate response to them, taking account of operational requirements and capabilities. (JP 1-02)

target reference point – An easily recognizable point on the ground (either natural or man-made) used to initiate, distribute, and control fires. Target reference points (TRPs) can also designate the center of an area where the commander plans to distribute or converge the fires of all his weapons rapidly. They are used by task force and below, and can further delineate sectors of fire within an engagement area. TRPs are designated using the standard target symbol and numbers issued by the fire support officer. Once designated, TRPs also constitute indirect fire targets. (FM 3-90)

task organization – A temporary grouping of forces designed to accomplish a particular mission. (FM 3-0)

task-organizing – (Army) The process of allocating available assets to subordinate commanders and establishing their command and support relationships. (FM 3-0)

tempo – The rate of military action. (FM 3-0)

terminal control – 1. The authority to direct the maneuver of aircraft which are delivering ordnance, pasengers, or cargo to a specific location or target. Terminal control is a type of air control. 2. Any electronic, mechanical, or visual control given to aircraft to facilitate target acquisition and resolution. (JP 1-02)

terrorism – The calculated use of unlawful violence or threat of unlawful violence to inculcate fear; intended to coerce or to intimidate governments or societies in the pursuit of goals that are generally political, religious, or ideological. (JP 1-02)

terrorist – An individual who uses violence, terror, and intimidation to achieve a result. (JP 1-02)

throughput distribution – The bypassing of one or more intermediate supply echelons in the supply system to avoid multiple handling. (FM 4-0)

time-phased force and deployment data – The Joint Operation Planning and Execution System database portion of an operation plan; it contains time-phased force data, non-unit-related cargo and personnel data, and movement data for the operation plan, including the following: a. In-place units; b. Units to be deployed to support the operation plan with a priority indicating the desired sequence for their arrival at the port of debarkation; c. Routing of forces to be deployed; d. Movement data associated with deploying forces; e. Estimates of non-unit related cargo and personnel movements to be conducted concurrently with the deployment of forces; and f. Estimate of transportation requirements that must be fulfilled by common-user lift resources as well as those requirements that can be fulfilled by assigned or attached transportation resources. (JP 1-02)

toxic industrial material – A generic term for toxic radioactive compounds in solid, liquid, aerosolized or gaseous form. These may be used, or stored for use, for industrial, commercial, medical, military, or domestic purposes. Toxic industrial materials may be chemical, biological, or radioactive and described as toxic industrial chemical, toxic industrial biological, or toxic industrial radiological. (FM 3-11.4)

track – 1. To display or record the successive positions of a moving object. 2. To keep a gun properly aimed, or to point continuously a target-locating instrument at a moving target. (JP 1-02)

troop leading procedures – A dynamic process used by small unit leaders to analyze a mission, develop a plan, and prepare for an operation. (FM 5-0)

turning movement – (Army) A form of maneuver in which the attacking force seeks to avoid the enemy's principal defensive positions by seizing objectives to the enemy rear and causing the enemy to move out of his current positions or divert major forces to meet the threat. (FM 3-0)

unconventional warfare – A broad spectrum of military and paramilitary operations, normally of long duration, predominantly conducted through, with, or by indigenous or surrogate forces who are organized, trained, equipped, supported, and directed in varying degrees by an external source. It includes, but is not limited to, guerrilla warfare, subversion, sabotage, intelligence activities, and unconventional assisted recovery. (JP 1-02)

unexploded explosive ordnance – Explosive ordnance which has been primed, fused, armed, or otherwise prepared for action, and which has been fired, dropped, launched, projected, or placed in such a manner as to constitute a hazard to operations, installations, personnel, or material, and remains unexploded either by malfunction or design or for any other cause. (JP 1-02)

unity of command – One of the nine principles of war: for every objective, ensure unity of effort under one responsible commander.

unity of effort – Coordination and cooperation among all military forces and other organizations toward a commonly recognized objective, even if the forces and nonmilitary organizations are not necessarily part of the same command structure. (FM 6-0)

unmanned aircraft system – A powered aerial vehicle that does not carry a human operator, uses aerodynamic forces to provide lift, can fly automomously or is remotely operated, can be expendable or recoverable, and can carry a lethal or nonlethal payload. Ballistic or semi-ballistic vehicles, cruise missiles, and artillery projectiles are not considered unmanned aircraft systems.

urban area – A topographical complex where manmade construction or high population density is the dominant feature. See also **urban environment**; **urban operations**. (FM 3-06)

urban environment – The physical urban area, as well as the complex and dynamic interaction among its key components—the terrain (natural and manmade), the population, and the supporting infrastructure—as an overlapping and interdependent system of systems. (FM 3-06)

urban operations – Offense, defense, stability, and support operations conducted in a topographical complex and adjacent natural terrain where manmade construction and high population density are the dominant features. (FM 3-0)

versatility – The ability of Army forces to meet the global, diverse mission requirements of full spectrum operations. (FM 3-0)

visualize – To create and think in mental images. (FM 6-0)

vulnerability – 1. The susceptibility of a nation or military force to any action by any means through which its war potential or combat effectiveness may be reduced or its will to fight diminished. 2. The characteristics of a system that cause it to suffer a definite degradation (incapability to perform the designated mission) as a result of having been subjected to a certain level of effects in an unnatural (manmade) hostile environment. 3. In information operations, a weakness in information system security design, procedures, implementation, or internal controls that could be exploited to gain unauthorized access to information or an information system. (JP 1-02)

warfighting function – A group of tasks and systems (people, organizations, information, and processes), united by a common purpose, that commanders use to accomplish missions and training objectives. The six warfighting functions are: a. **command and control** – the related tasks and systems that support the commander in exercising authority and direction; b. **fire support** – the related tasks and systems that provide collective and coordinated use of Army indirect fires, joint fires, and offensive information operations; c. **intelligence** – the related tasks and systems that facilitate understanding of the enemy, terrain, weather, and civil considerations; d. **movement and maneuver** – the related tasks and systems that move forces to achieve a position of advantage in relation to the enemy; e. **protection** – the related tasks and systems that preserve the force so the commander can apply maximum combat power; f. **sustainment** – the related tasks and systems that provide support and services to ensure freedom of action, extend operational reach, and prolon endurance. (FMI 5-0.1)

weapons of mass destruction – Weapons that are capable of a high order of destruction and/or of being used in such a manner as to destroy large numbers of people. Weapons of mass destruction can be high explosives or nuclear, biological, chemical, and radiological weapons, but exclude the means of transporting or propelling the weapon where such means is a separable and divisible part of the weapon. (JP 1-02)

This page intentionally left blank.

References

Military publications are listed by title. When a field manual has been published under a new number for the first time, the old number is provided in parenthesis after the new number.

REQUIRED PUBLICATIONS

These documents must be available to intended users of this publication.

JOINT PUBLICATIONS

Joint publications are available at <http://www.dtic.mil/doctrine/jel/>.

JP 1-02. *Department of Defense Dictionary of Military and Associated Terms*, 12 April 2001.

JP 3-06. *Doctrine for Joint Urban Operations*. 16 September 2002.

ARMY PUBLICATIONS

Army doctrinal publications are available at <http://www.apd.army.mil/>.

FM 1-02 (101-5-1). *Operational Terms and Graphics*. 21 September 2004.

FM 34-130. *Intelligence Preparation of the Battlefield*. 8 July 1994. FM 34-130 will be republished as FM 2-01.3

FM 3-0 (100-5). *Operations*. 14 June 2001.

FM 3-06.11 (90-10-1). *Combined Arms Operations in Urban Terrain*. 28 February 2002.

FM 3-07 (100-20). *Stability Operations and Support Operations*. 20 February 2003. When updated, FM 3-07 will be retitled *Stability Operations*.

FM 3-90. *Tactics*. 04 July 2001.

FM 5-19 (100-14). *Composite Risk Management*. 21 August 2006.

FM 7-0. *Training the Force*. 22 October 2002.

FM 7-1. *Battle Focused Training*. 15 September 2003.

FM 5-0 (101-5). *Army Planning and Orders Production*. 20 January 2005.

RELATED PUBLICATIONS

These sources contain relevant supplemental information.

JOINT PUBLICATIONS

Joint publications are available at <http://www.dtic.mil/doctrine/jel/>.

JP 3-0. *Doctrine for Joint Operations*. 10 September 2001.

JP 3-05. *Doctrine for Joint Special Operations*. 17 December 2003.

JP 3-08. *Interagency Coordination During Joint Operations* (2 Volumes). 9 October 1996.

JP 3-16. *Joint Doctrine for Multinational Operations*. 05 April 2000.

JP 4-0. *Doctrine for Logistic Support of Joint Operations*. 06 April 2000.

JP 5-00.2. *Joint Task Force Planning Guidance and Procedures*. 13 January 1999.

Joint Military Operations Historical Collection. 15 July 1997.

ARMY PUBLICATIONS

Army doctrinal publications are available at <http://www.apd.army.mil/>.

Army Regulations

FM 1. *The Army*. 14 June 2005.

FM 2-0 (34-1). *Intelligence*. 17 May 2004.

FM 2-22.3 (34-52). *Human Intelligence Collector Operations*. 6 September 2006.

FM 3-05 (100-25). *Army Special Operations Forces*. 20 September 2006.

FM 3-05.20 (31-20). *Special Forces Operations*. 1 April 2004.

FM 3-05.30 (33-1). *Psychological Operations*. 15 April 2005.

FM 3-05.105. *Multiservice Tactics, Techniques, and Procedures for Special Operations Forces in Nuclear, Biological, and Chemical Environments*. 28 September 2001.

FM 3-05.212 (TC 31-25). *Special Forces Waterborne Operations*. 31 August 2004.

FM 3-05.301 (33-1-1). *Psychological Operations Tactics, Techniques, and Procedures*. 31 December 2003.

FM 3-05.401 (41-10-1). *Civil Affairs Tactics, Techniques, and Procedures*. 23 September 2003.

FM 3-06 (90-10). *Urban Operations*. 01 June 2003.

FM 3-06.1. *Aviation Urban Operations: Multi-service Tactics, Techniques, and Procedures for Aviation Urban Operations*. 09 July 2005.

FM 3-07.2. *Antiterrorism*. (Future Publication).

FM 3-11.3 ( ). *Multiservice Tactics, Techniques, and Procedures for Chemical, Biological, Radiological, and Nuclear Contamination Avoidance*. 2 February 2006.

FM 3-11.9 (3-9). *Potential Military Chemical/Biological Agents and Compounds*. 10 January 2005.

FM 3-11.14 (3-14). *Multiservice Tactics, Techniques, and Procedures for Nuclear, Biological, and Chemical Vulnerability Assessment*. 28 November 2004.

FM 3-11.19 (3-19). *Multiservice Tactics, Techniques, and Procedures for Nuclear, Biological, and Chemical Reconnaissance*. 30 July 2004.

FM 3-11.21 (3-21). *Multiservice Tactics, Techniques, and Procedures for Nuclear, Biological, and Chemical Aspects of Consequence Management*. 12 December 2001.

FM 3-11.22. *Weapons of Mass Destruction Civil Support Team Tactics, Techniques, and Procedures*. 06 June 2003.

FM 3-13 (100-6). *Information Operations: Doctrine, Tactics, Techniques, and Procedures*. 28 November 2003.

FM 3-14 (100-18). *Space Support to Army Operations*. 18 May 2005.

FM 3-19.4 (19-4). *Military Police Leaders' Handbook*. 04 March 2002.

FM 3-19.15 (19-15). *Civil Disturbance Operations*. 18 April 2005.

FM 3-19.17. *Military Working Dogs*. 22 September 2005.

FM 3-19.40 (19-40). *Military Police Internment/Resettlement Operations*. 01 August 2001.

FM 3-19.50. *Police Intelligence Operations*. 21 July 2006.

FM 3-22.40 (90-40). *NLW: Tactical Employment of Nonlethal Weapons*. 15 January 2003.

FM 3-28. *Civil Support Operations*. (Future Publication).

FM 3-31. *Joint Force Land Component Commander Handbook (JFLCC)*. 13 December 2001.

FM 3-31.1 (90-31). *AMCI: Army and Marine Corps Integration in Joint Operations*. 21 November 2001.

FM 3-34 (5-100, 5-114). *Engineer Operations*. 02 January 2004.

FM 3-34.2. *Combined Arms Breaching Operations*. 31 August 2000. FM 3-34.2 will be republished as FM 3-90.11

FM 3-34.230 (5-105). *Topographic Operations*. 03 August 2000.

FM 3-61.1. *Public Affairs Tactics, Techniques and Procedures*. 01 October 2000.

FM 3-90.12 (90-13). *Combined Arms Gap Crossing Operations*. TBP.

FM 3-97.6 (90-6). *Mountain Operations*. 28 November 2000.

FM 3-100.12. *Risk Management: Multiservice Tactics, Techniques, and Procedures for Risk Management*. 15 February 2001.

FM 3-100.38. *UXO: Multi-service Tactics, Techniques and Procedures for Unexploded Ordnance Operations*. 18 August 2005.

FM 4-0 (100-10). *Combat Service Support*. 29 August 2003.

FM 4-01.45. *Tactical Convoy Operations: Multiservice Tactics, Techniques, and Procedures for Tactical Convoy Operations*. 24 March 2005.

FM 4-02.51 (8-51). *Combat and Operational Stress Control*. 6 June 2006.

FM 4-30.16. *EOD: Multiservice Procedures for Explosive Ordnance Disposal in a Joint Environment*. 27 October 2005.

FM 4-30.51. *Unexploded Ordnance (UXO) Procedures*. 13 July 2006.

FM 5-01.12 (90-41). *JTF LNO Integration: Multiservice Tactics, Techniques, and Procedures for Joint Task Force (JTF) Liaison Officer Integration*. 27 January 2003.

FM 5-33. *Terrain Analysis*. 8 September 1992. FM 5-33 will be republished as FM 3 34.330.

FM 5-103. *Survivability*. 10 June 1985. FM 5-103 will be republished as FM 5-34.112.

FM 5-104. *General Engineering*. 12 November 1986. FM 5-104 will be republished as FM 3-34.250.

FM 5-170. *Engineer Reconnaissance*. 05 May 1988. FM 5-170 will be republished as FM 3 34.212.

FM 5-415. *Fire-Fighting Operations*. 09 February 1999. FM 5-415 will be republished as FM 3-34.281.

FM 6-0. *Mission Command: Command and Control of Army Forces*. 11 August 2003.

FM 6-22.5 (22-9). *Combat Stress*. 23 June 2000.

FM 7-15. *The Army Universal Task List (Incl. changes 2)*. 6 July 2006.

FM 20-32. *Mine/Countermine Operations*. 1 April 2005.

FM 22-51. *Leaders' Manual for Combat Stress Control*. 29 September 1994. FM 22-51 will be republished as FM 4-02.22.

FM 22-100. *Army Leadership: Be, Know, Do*. 31 August 1999. FM 22-100 will be republished as FM 6-22.

FM 23-10. *Sniper Training*. 17 August 1994. FM 23-10 will be republished as FM 3-22.10.

FM 27-10. *The Law of Land Warfare*. 18 July 1956. FM 27-10 will be republished as FM 1-04.10.

FM 27-100. *Legal Support to Operations*. 01 March 2000. FM 27-100 will be republished as FM 1-04.

FM 31-20-3. *Foreign Internal Defense Tactics, Techniques, and Procedures for Special Forces*. 20 September 1994. FM 31-20-3 will be republished as FM 3-05.202.

FM 31-50. *Combat in Fortified and Built-Up Areas*. March 1964.

FM 34-3. *Intelligence Analysis*. 15 March 1990. FM 34-3 will be republished as FM 2-33.4.

FM 34-60. *Counterintelligence*. 03 October 1995. FM 34-60 will be republished as FM 2-01.2.

FM 34-81. *Weather Support for Army Tactical Operations*. 31 August 1989. FM 34-81 will be republished as FM 2-33.2.

FM 34-81-1. *Battlefield Weather Effects*. 23 December 1992. FM 34-81-1 will be republished as FM 2-33.201.

FM 41-10. *Civil Affairs Operations*. 14 February 2000. FM 41-10 will be republished as FM 3-05.40.

FM 44-8. *Combined Arms for Air Defense*. 01 June 1999. FM 44-8 will be republished as FM 3-01.8.

FM 46-1. *Public Affairs Operations*. 30 May 1997. FM 46-1 will be republished as FM 3-61.

FM 90-8. *Counterguerrilla Operations*. 29 August 1986. FM 90-8 will be republished as FM 3-07.11.

FM 90-10. *Military Operations on Urbanized Terrain (MOUT) (How to Fight)*. 15 August 1979.

FM 100-5. *Operations*. 1944.

FM 100-8. *The Army in Multinational Operations.* 24 November 1997. FM 100-8 will be republished as FM 3-16.

FM 100-11. *Force Integration.* 15 January 1998. FM 100-11 will be republished as FM 3-100.11.

FMI 2-91.4. *Intelligence Support to Operations in the Urban Environment.* 30 June 2005. FMI 2-91.4 will be republished as FM 2-91.4.

FMI 3-07.22. *Counterinsurgency Operations.* 01 October 2004. FMI 3-07.22 will be republished as FM 3-07.22.

FMI 3-34.119. *Improvised Explosive Device Defeat.* 21 September 2005.

FMI 4-30.50. *Modular Explosive Ordnance Disposal Operations.* 31 July 2006.

FMI 5-0.1. *The Operations Process.* 31 March 2006.

TC 3-15. *Nuclear Accident and Incident Response and Assistance (NAIRA).* 27 December 1988.

TC 5-230. *Army Geospatial Guide for Commanders and Planners.* 28 November 2003.

TC 7-98-1. *Stability and Support Operations Training Support Package.* 05 June 1997.

TC 9-21-01. *Soldiers Improvised Explosive Device (IED) Awareness Guide: Iraq and Afghanistan Theaters of Operation.* 28 May 2004.

TC 90-1. *Training for Urban Operations.* 01 April 2002.

DEPARTMENT OF DEFENSE PUBLICATIONS

Defense Intelligence Agency Manual 58-11. *Defense Human Resources Intelligence Collection Management.* 20 August 1968.

Quadrennial Defense Review Report. 30 September 2001.

U.S. Army Training and Doctrine Command (TRADOC), Deputy Chief of Staff for Intelligence (DCSINT). *TRADOC DCSINT Handbook No. 1: A Military Guide to Terrorism in the Twenty-First Century* (Version2.0). Fort Leavenworth, KS: U.S. Government Printing Office, 2004.

NONMILITARY PUBLICATIONS

26th Infantry Regimental Association. *Aachen: Military Operations in Urban Terrain.* Lititz, PA: 26th Infantry Regimental Association, 1999.

Adan, Abraham. *On the Banks of the Suez.* Novato, CA: Presidio Press, 1980.

Akers, Frank H., Jr. and George B. Singleton. "Task Force Ranger: A Case Study Examining the Application of Advanced Technologies in Modern Urban Warfare." Oak Ridge, TN: National Security Program Office, November 2000.

Allard, Kenneth. *Somalia Operations: Lessons Learned.* Washington, DC: National Defense University Press, 1995.

Arnold, James R. *Tet Offensive 1968: Turning Point in Vietnam.* London: Osprey, 1990.

Bass, Bernard M. "Leading in the Army after Next." *Military Review* (March/April 1998): 46–58.

Beevor, Anthony. *Stalingrad.* New York: Penguin Books, 1998.

Bowden, Mark. *Black Hawk Down: A Story of Modern War.* New York: Atlantic Monthly Press, 1999.

Butts, Kent H. "The Strategic Importance of Water." *Parameters* (spring 1997): 65–83.

Carell, Paul. *Stalingrad: The Defeat of the German 6th Army.* Atglen, PA: Schieffer Publishing, 1993.

Celeski, Joseph D. "Joint Urban Operations: Special Forces in Urban Campaign Planning." *Special Warfare* (fall 2001): 2–13.

Chiarelli, Peter W. and Patrick R. Michaelis. "Winning the Peace: The Requirement for Full-Spectrum Operations." *Military Review* (July/August 2005): 4–17.

Chiarelli, Peter W., Patrick R. Michaelis, and Geoffrey A. Norman. "Armor in Urban Terrain: The Critical Enabler." *Armor* (March/April 2005): 7–12.

Chuikov, Vasili I. *The Battle for Stalingrad.* New York: Holt, Rinehart, and Winston, 1964.

Colby, John. *War From the Ground Up: The 90th Division in WWII.* Austin, TX: Nortex Press, 1991.

Cole, Ronald H. *Operation Just Cause.* Washington, DC: Joint History Office, 1995.

Connaughton, Richard, John Pimlott, and Duncan Anderson. *The Battle for Manila: The Most Devastating Untold Story of World War II.* Novata, CA: Presidio, 1995.

Craig, William. *Enemy at the Gates: The Battle for Stalingrad.* New York: Reader's Digest Press, 1973.

Delk, James D. *Fires & Furies: The LA Riots.* Palm Springs, CA: Etc. Publications, 1995.

Dewar, Michael. *British Army in Northern Ireland.* New York: Sterling Publishers, 1985.

______. *War in the Streets: The Story of Urban Combat from Calais to Khafji.* New York: Sterling Publishers, 1992.

Donnelly, Thomas, Margaret Roth, and Caleb Baker. *Operation Just Cause: The Storming of Panama.* New York: Lexington Books, 1991.

Dupuy, R. Ernest and Trevor N. Dupuy. *The Encyclopedia of Military History: From 3500 B.C. to the Present.* San Francisco: Harper & Row, 1977.

Dupuy, Trevor N. and Paul Martell. *Flawed Victory: The Arab-Israeli Conflict and the 1982 War in Lebanon.* Fairfax, VA: Hero Books, 1986.

Edwards, Sean J. A. *Freeing Mercury's Wings: Improving Tactical Communications in Cities.* Santa Monica, CA: RAND, 2001.

Eshel, David. *Mid-East Wars: The Lebanon War 1982.* Hod Hasharon, Israel: Eshel-Dramit, 1983.

Fishel, John T. *The Fog of Peace: Planning and Executing the Restoration of Panama.* Carlisle, PA: Strategic Studies Institute, 1992.

Fontenot, Gregory, E. J. Degen, and David Tohn. *On Point: US Army in Operation IRAQI FREEDOM.* Fort Leavenworth, KS: Combat Studies Institute Press, 2004.

Funston, Frederick. "How the Army Worked to Save San Francisco." *Cosmopolitan Magazine* (July 1906): <http://www.sfmuseum.org/1906/cosmo.html>.

Gabel, Christopher R. "Military Operations on Urbanized Terrain: The 2d Battalion, 26th Infantry, at Aachen, October 1944." *Urban Combat Operations: tactics, Techniques and Procedure.* Center for Army Lessons Learned (CALL) Newsletter No. 99-16. Fort Leavenworth, KS: Center for Army Lessons Learned, U.S. Training and Doctrine Command [TRADOC], November 1999: 1-9–1-13.

Gabriel, Richard A. *Operation Peace for Galilee: The Israeli-PLO War in Lebanon.* New York: Hill and Wang, 1984.

Glenn, Russell W. *An Attack on Duffer's Downtown.* Santa Monica, CA: RAND, 2001.

______. *Heavy Matter: Urban Operations' Density of Challenges.* Santa Monica, CA: RAND, 2000.

______. "Urban Combat is Complex." *Proceedings* (February 2002): 62–65.

______. "…We Band of Brothers": *The Call for Joint Urban Operations Doctrine.* Santa Monica, CA: RAND, 1999.

Glenn, Russell W., Steven Hartman, and Scott Gerwehr. *Urban Combat Service Support Operations: The Shoulders of Atlas.* Santa Monica, CA: RAND, 2003.

Gordon, Michael R. "Baghdad Targets Picked if Hussein Holes Up There." *New York Times* (7 March 2003): 1A2.

Grau, Lester W. "Something Old, Something New: Guerrillas, Terrorists, and Intelligence Analysis." *Military Review* (July/August 2004): 42–49.

Grau, Lester W. and Timothy Smith. *A 'Crushing' Victory: Fuel-Air Explosives and Grozny 2000.* Fort Leavenworth, KS: Foreign Military Studies Office, 2000.

Hamill, Desmond. *Pig in the Middle: The Army in Northern Ireland, 1969–1984.* London: Methuen, 1985.

Hammel, Eric. *Fire in the Streets: The Battle for Hue, Tet 1968*. New York: Dell Publishing, 1991.

Heal, Sid. "Crowds, Mobs and Nonlethal Weapons." *Military Review* (March/April 2000): 45–50.

Heinl, Robert Debs, Jr. *Victory at High Tide: The Inchon-Seoul Campaign*. Annapolis, MD: Nautical & Aviation Publishing Company of America, 1979.

Herzog, Cham. *The War of Atonement, October, 1973*. Boston: Little, Brown and Company, 1975.

Higham, Robin. *Bayonets in the Streets: The Use of Troops in Civil Disturbances*. Lawrence, KS: University Press of Kansas, 1969.

Hoar, Joseph P. "A CINC's Perspective." *JFQ Forum* (autumn 1993): 56–63.

Hollis, Mark A. B. "Platoon Under Fire." *Infantry* (January/April 1998): 27–34.

Infantry in Battle. 2d ed. Garret & Massie: Richmond, VA, 1939. Reprint, Fort Leavenworth, KS: U.S. Army Command and General Staff College, 1981.

Initial Impressions Report: Operation Iraqi Freedom—Stability Operations-Support Operations Center for Army Lessons Learned (CALL) Newsletter No. 04-13 (Fort Leavenworth, KS: Center for Army Lessons Learned, U.S. Army Training and Doctrine Command [TRADOC], 2004).

Initial Impressions Report: Operation Iraqi Freedom—Stability Operations-Support Operations. Center for Army Lessons Learned (CALL) Newsletter No. 03 (Fort Leavenworth, KS: Center for Army Lessons Learned, U.S. Army Training and Doctrine Command [TRADOC], 2003).

Jomini, Antoine-Henri. "The Art of War" in *Roots of Strategy: Book 2*. Harrisburg, PA: Stackpole Books, 1987.

Kemp, Anthony. *The Unknown Battle: Metz, 1944*. New York: Stein and Day, 1981.

Kennedy, John F. Message to Congress, 25 May 1961. Available at <http://www.jfklibrary.org/Historical+Resources/Archives>.

Linden, Eugene. "The Exploding Cities of the Developing World." *Foreign Affairs* (January/February 1996): 52–65.

Mabry, Robert L. et al. "United States Army Rangers in Somalia: An Analysis of Combat Casualties on an Urban Battlefield." *Journal of Special Operations Medicine* (fall 2001): 24–40.

MacArthur, Douglas. *Reminiscences*. New York: McGraw-Hill, 1964.

Manwaring, Max G. *Street Gangs: The New Urban Insurgency*. Carlisle, PA: Strategic Studies Institute, 2005.

Maxwell, Mark D. "The Law of War and Civilians on the Battlefield: Are We Undermining Civilian Protections?" *Military Review* (September/October 2004): 17–25.

McConnell, Malcolm. *Just Cause: The Real Story of America's High-Tech Invasion of Panama*. New York: St. Martin's Press, 1991.

Medby, Jamison Jo and Russell W. Glenn. *Street Smart: Intelligence Preparation of the Battlefield for Urban Operations*. Santa Monica, CA: RAND, 2002.

Mendel, William W. *Combat in Cities: The LA Riots and Operation Rio*. Fort Leavenworth, KS: Foreign Military Studies Office, 1996.

Monmonier, Mark. *How to Lie with Maps*. Chicago: University of Chicago Press, 1996.

Montgomery, Thomas M. Interview. "Ambush in Mogasdishu." *Frontline*, WGBH Boston. 29 September 1998. Available at <http://www.pbs.org/wgbh/pages/frontline/programs/1998.html>.

Moore, John C. "Sadr City: The Armor Pure Assault in Urban Terrain." *Armor* (November/December 2004): 31–37.

Natsios, Andrew S. "Commander's Guidance: A Challenge of Complex Humanitarian Emergencies." *Parameters* (summer 1996): 50–66.

______. "Food Through Force: Humanitarian Intervention and U.S. Policy." *The Washington Quarterly* (winter 1994): 136.

Nolan, Keith William. *The Battle for Saigon: Tet 1968*. New York: Pocket Books, 1996.

Oakley, Robert B. "An Envoy's Perspective." *JFQ Forum* (autumn 1993): 44–55.

Oliker, Olga. *Russia's Chechen Wars 1994–2000: Lessons from Urban Combat*. Santa Monica, CA: RAND, 2001.

Operation Restore Hope Lessons Learned Report: 3 December 1992 – 4 May 1993, Operations Other than War. Fort Leavenworth, KS: Center for Army Lessons Learned (CALL), U.S. Army Combined Arms Command (CAC), 1993. Reprint, Washington, DC: U.S. Government Printing Office, 1996.

Parkerson, John Embry, Jr. "United States Compliance with Humanitarian Law Respecting Civilians During Operation Just Cause." *Military Law Review* (1991): 31–140.

Peters, Ralph. "Our Soldiers, Their Cities." *Parameters* (spring 1996): 43–50.

Peterson, John J. *Into the Cauldron*. Clinton, MD: Clavier House, 1973.

Petraeus, David H. "Learning Counterinsurgency: Observations from Soldiering in Iraq." *Military Review* (January/February 2006): 2–12.

Ritchie, Elspeth C. and Robert Mott. "Caring for Civilians During Peace Keeping Missions: Priorities and Decisions." *Military Medicine* (Vol. 167, Supplement 3, 2002): 14–16.

Rosenberger, Leif R. "The Strategic Importance of the World Food Supply." *Parameters* (spring 1997): 84–105.

Schadlow, Nadia. "War and the Art of Governance." *Parameters* (autumn 2003): 85–94.

Shelton, H. Hugh and Timothy D. Vane. "Winning the Information War in Haiti." *Military Review* (November/December 1995): 3–9.

Shtemenko, S. M. *The Soviet General Staff at War, 1941-1945: Book One*. Moscow: Progress Publishers, 1981.

Smith, Rupert. "Wars in Our Time: A Survey of Recent and Continuing Conflicts." *World Defence Systems,* volume 3:2. London: Royal United Services Institute for Defence Studies, 2001.

Stanton, Shelby L. *Ten Corps in Korea, 1950*. Novato, CA: Presidio Press, 1989.

Taw, Jennifer Morrison. *Operation Just Cause: Lessons for Operations Other Than* War. Santa Monica, CA: RAND, 1996.

Thomas, Timothy L. "The Battle for Grozny: Deadly Classroom for Urban Combat." *Parameters* (summer 1999): 87–102.

Thomas, Timothy L. and Charles P. O'Hara. "Combat Stress in Chechnya: 'The Equal Opportunity Disorder'." *Army Medical Department Journal* (January/March 2000): 46–53.

Trinquier, Roger. *Modern Warfare: A French View of Counterinsurgency*. New York: Praeger Publisher, 1964.

United Nations Security Council Resolution 794 (1992). New York: United Nations, 1992. Available at <http://daccessdds.un.org/doc/UNDOC/GEN/N92/772/11/PDF>.

U.S. Agency for International Development, Bureau for Humanitarian Response, Office of Foreign Disaster Assistance. *Field Operations Guide for Disaster Assessment and Response, Version 3.0*. Washington, DC: U.S. Government Printing Office, 2003. Available at <http://www.usaid.gov/our_work/humanitarian_assistance/>.

U.S. Department of Transportation, Research and Special Programs Administration, Office of Hazardous Materials Initiative and Training. *2004 Emergency Response Guidebook*. Washington, DC: U.S. Government Printing Office, 2004. Available at <http://hazmat.dot.gov/pubs/erg/erg2004.pdf>.

U.S. Environmental Protection Agency, Office of Emergency and Remedial Response. *HAZMAT Team Planning Guidance*. Washington, DC: U.S. Government Printing Office, 1990.

Van Sickle, Jeffrey B. "Stability Operations in Northern Iraq: Task Force Altun Kupri." *Infantry* (January/February 2005): 25–29.

Watson, Bruce Allen. *Sieges: A Comparative Study*. Westport, CT: Praeger, 1993.

Werstein, Irving. *The Battle of Aachen*. New York: Thomas Y. Crowell Company, 1962.

Whiting, Charles. *Bloody Aachen*. New York: Stein and Day, 1976.

Wintle, Justin, ed. *The Dictionary of War Quotations*. New York: The Free Press, 1989.

Yates, Lawrence A. "Military Stability and Support Operations: Analogies, Patterns and Recurring Themes." *Military Review* (July/August 1997): 51–61.

RECOMMENDED READINGS

The following sources contain relevant supplemental information.

Alexander, John B. *Future War: Non-Lethal Weapons in Twenty-First-Century* Warfare. New York: St. Martin's Press, 1999.

Ashworth, G. J. *The City and War*. London: Routledge, 1991.

Barker, A. J. *Dunkirk: The Great Escape*. New York: David McKay, 1977.

Baumann, Robert F., Lawrence A. Yates, and Versalle F. Washington. "My Clan Against the World": *U.S. and Coalition Forces in Somalia, 1992-1994.* Fort Leavenworth, KS: Combat Studies Institute Press, 2004.

Birtle, Andrew J. *U.S. Army Counterinsurgency and Contingency Operations Doctrine: 1860-1941.* Washington, DC: Center of Military History, 2001.

Blaufarb, Douglas S. *The Counterinsurgency Era: U.S. Doctrine and Performance 1950 to Present.* New York: The Free Press, 1977.

Breuer, William B. *Hitler's Fortress Cherbourg: The Conquest of a Bastion*. New York: Stein and Day, 1984.

Brown, Stephen D. "PSYOP in Operation Uphold Democracy." *Military Review* (September/October 1996): 57–76.

Bulloch, Gavin. "Military Doctrine and Counterinsurgency: A British Perspective." *Parameters* (summer 1996): 4–16.

Urban Combat Operations: Tactics, Techniques and Procedures. Center for Army Lessons Learned (CALL) Newsletter No. 99-16 (Fort Leavenworth, KS: Center for Army Lessons Learned, U.S. Army Training and Doctrine Command [TRADOC], November 1999).

Cordesman, Anthony H. and Abraham R. Wagner. *The Lessons of Modern War—Volume II: The Iran-Iraq War*. Boulder, CO: Westview Press, 1990.

Corum, James S. and Wray R. Johnson. *Airpower in Small Wars: Fighting Insurgents and Terrorists.* Lawrence, KS: University Press of Kansas, 2003.

Desch, Michael C., ed. *Soldiers in Cities: Military Operations on Urban Terrain.* Carlisle Barracks, PA: Strategic Studies Institute, 2001.

Duffy, Christopher. *Fire & Stone: The Science of Fortress Warfare, 1660–1860.* Mechanicsburg, PA: Stackpole Books, 1996.

______. *The Fortress in the Age of Vauban and Frederick the Great, 1660–1789: Siege Warfare Volume I.* London: Routledge & Kegan Paul, 1985.

______. *Siege Warfare: The Fortress in the Early Modern World, 1494–1660.* London: Routledge & Kegan Paul, 1979.

Edwards, Sean J. A. *Mars Unmasked: The Changing Face of Urban Operations*. Santa Monica, CA: RAND, 2000.

Eikenberry, Karl W. "Take No Casualties." *Parameters* (summer 1996): 109–118.

Erickson, John. *The Road to Berlin: Continuing the History of Stalin's War with Germany.* Boulder, CO: Westview Press, 1983.

Gerwehr, Scott and Russell W. Glenn. *The Art of Darkness: Deception and Urban Operations*. Santa Monica, CA: RAND, 2000.

Glenn, Russell W. *Combat in Hell: A Consideration of Constrained Urban Warfare*. Santa Monica, CA: RAND, 1996.

______. *Marching Under Darkening Skies: The American Military and Impending Urban Operations Threat*. Santa Monica, CA: RAND, 1998.

Glenn, Russell W., ed. *Capital Preservation: Preparing for Urban Operations in the Twenty-First Century: Proceedings of the RAND Arroyo-TRADOC-MCWL-OSD Urban Operations Conference, March 22-23, 2000*. Santa Monica, CA: RAND, 2001.

Glenn, Russell W. et al. *Denying the Widow-Maker, Summary of Proceedings, RAND-DDBL Conference on Military Operations on Urbanized Terrain*. Santa Monica, CA: RAND, 1999.

______. *Getting the Musicians of Mars on the Same Sheet of Music: Army, Joint, Multinational, and Interagency C4ISR Interoperability*. Santa Monica, CA: RAND, 2000.

Glenn, Russell W. et al., eds. *The City's Many Faces: Proceedings of the RAND Arroyo-MCWL-J8 UWG Urban Operations Conference, April 13-14, 1999*. Santa Monica, CA: RAND, 2000.

Gouré, Leon. *The Siege of Leningrad*. Stanford, CA: Stanford University Press, 1962.

Grau, Lester W. and Jacob W. Kipp. "Urban Combat: Confronting the Specter." *Military Review* (July/August 1999): 9–17.

Groves, John R. "Operations in Urban Environments." *Military Review* (July/August 1998): 30–39.

Habeck, Mary R. Lecture. "Jihadist Strategies in the War on Terrorism." Washington, DC: The Heritage Foundation, 8 November 2004. Available at http://www.heritage.org/research/nationalsecurity/hl855.cfm.

Hahn, Robert F., II and Bonnie Jezior. "Urban Warfare and the Urban Warfighter of 2025." *Parameters* (summer 1999): 74–86.

Harman, Nicholas. *Dunkirk: The Patriotic Myth*. New York: Simon and Schuster, 1980.

Heckstall-Smith, Anthony. *Tobruk: The Story of a Siege*. New York: W. W. Norton, 1959.

Hibbert, Christopher. *The Battle of Arnhem*. London: Batsford, 1962.

House, John M. "The Enemy After Next." *Military Review* (March/April 1998): 22–28.

Jackson, W. G. F. *The Battle for Italy*. New York: Harper & Row, 1967.

Karslake, Basil. *1940 – The Last Act: The Story of the British Forces in France After Dunkirk*. Hamden, CT: Archon Books, 1979.

Kostof, Spiro. *The City Shaped: Urban Patterns and Meanings Through History*. London: Thames and Hudson, 1999.

______. *The City Assembled: The Elements of Urban Form Throughout History*. Boston: Little Brown and Company, 1999.

Kuehl, Daniel T. "Airpower vs. Electricity: Electric Power as a Target for Strategic Air Operations." *Journal of Strategic Studies* (March 1995): 237–266.

Larson, Eric. *Casualties and Consensus: The Historical Role of Casualties in Domestic Support of U.S. Military Operations*. Santa Monica, CA: RAND, 1996.

Lieven, Anatol. *Chechnya: Tombstone of Russian Power*. New Haven: Yale University Press, 1998.

______. "The World Turned Upside Down: Military Lessons of the Chechen War." *Armed Forces Journal International* (August 1998): 40–43.

McKee, Alexander. *Caen: Anvil of Victory*. New York: St. Martin's Press, 1964.

McLaurin, R. D. "Military Forces in Urban Antiterrorism." Aberdeen Proving Ground, MD: U.S. Army Human Engineering Laboratory, 1989.

McLaurin, R. D. and R. Miller. "Urban Counterinsurgency: Case Studies and Implications for U.S. Military Forces." Aberdeen Proving Ground, MD: U.S. Army Human Engineering Laboratory, 1989.

Mercier, Charles E. "Terrorists, WMD, and the U.S. Army Reserve." *Parameters* (autumn 1997): 98–118.

Milton, T. R. "Urban Operations: Future War." *Military Review* (February 1994): 37–46.

Moskin, J. Robert. *Among Lions: The Battle for Jerusalem, June 5–7, 1967*. New York: Ballantine Books, 1982.

Northam, Ray M. *Urban Geography*. New York: Wiley, 1975.

Peters, Ralph. "The Culture of Future Conflict." *Parameters* (winter 1995 96): 18–27.

______. "The Future of Armed Warfare." *Parameters* (autumn 1997): 50–59.

______. "The Human Terrain of Urban Operations." *Parameters* (spring 2000): 4–12.

Robertson, William G., ed. *Block by Block: The Challenges of Urban Operations*. Fort Leavenworth, KS: U.S. Army Command and Staff College Press, 2003.

Rosenau, William G. "'Every Room is a New Battle': The Lessons of Modern Urban Warfare." *Studies in Conflict and Terrorism* (October–December 1997): 371–394.

Scalard, Douglas. "People of Whom We Know Nothing: When Doctrine Isn't Enough." *Military Review* (July/August 1997): 4–11.

Scales, Robert H. "The Indirect Approach: How U.S. Military Forces Can Avoid the Pitfalls of Future Urban Warfare." *Armed Forces Journal International* (October 1998): 68–74.

Siegel, Adam B. *Eastern Exit: The Noncombatant Evacuation Operation (NEO) from Mogadishu, Somalia, in January 1991*. Alexandria, VA: Center for Naval Analyses, 1992.

Spiller, Roger J. *Sharp Corners: Urban Operations at Century's End*. Fort Leavenworth, KS: Combat Studies Institute, 2001.

Stech, Frank J. "Winning CNN Wars." *Parameters* (autumn 1994): 37–56.

Stevenson, Jonathan. *Losing Mogadishu: Testing U.S. Policy in Somalia*. Annapolis, MD: Naval Institute Press, 1995.

Taw, Jennifer Morrison and Bruce Hoffman. *The Urbanization of Insurgency: The Potential Challenge to U.S. Army Operations*. Santa Monica, CA: RAND, 1994.

Tully, Andrew. *Berlin: Story of a Battle*. Westport, CT: Greenwood Press, 1977.

United Nations Department of Economic and Social Affairs, Population Division. *World Urbanization Prospects: The 2003 Revision*. New York: United Nations, 2003.

Vick, Alan et al. *Aerospace Operations in Urban Environments: Exploring New Concepts*. Santa Monica, CA: RAND, 2000.

Wykes, Alan. *The Siege of Leningrad*. New York: Ballentine Books Inc., 1968.

Index

This page intentionally left blank.

FM 3-06
26 October 2006

By Order of the Secretary of the Army:

PETER J. SCHOOMAKER
General, United States Army
Chief of Staff

Official:

JOYCE E. MORROW
Administrative Assistant to the
Secretary of the Army
0627803

DISTRIBUTION:

Active Army, Army National Guard, and U. S. Army Reserve: To be distributed in accordance with initial distribution number 111233, requirements for FM 3-06.

*FM 23-10
17 August 1994

FIELD MANUAL No. 23-10

HEADQUARTERS
DEPARTMENT OF THE ARMY
Washington, DC, 17 August 1994

SNIPER TRAINING

CONTENTS

DISTRIBUTION RESTRICTION: Approved for public release; distribution is unlimited.

*This publication supersedes TC 23-14, 14 June 1989.

page

CHAPTER 4. FIELD TECHNIQUES

PREFACE

This field manual provides information needed to train and equip snipers and to aid them in their missions and operations. It is intended for use by commanders, staffs, trainers, snipers, and soldiers at training posts, Army schools, and units.

This manual is organized as a reference for snipers and leads the trainer through the material needed to conduct sniper training. Subjects include equipment, weapon capabilities, fundamentals of marksmanship and ballistics, field skills, mission planning, and skill sustainment. The left-handed firer can become a sniper, but all material in this book is referenced to the right-handed firer.

The proponent for this publication is Headquarters, United States Army Infantry School. Send comments and recommendations on DA Form 2028 (Recommended Changes to Publications and Blank Forms) directly to the Commandant, United States Army Infantry School, ATTN: ATSH-IN-S3, Fort Benning, GA 31905-5596.

This publication complies with the following international agreements:

STANAG 2020	Operational Situation Report
STANAG 2022	Intelligence Report
STANAG 2084	Handling and Reporting of Captured Enemy Equipment and Documents
STANAG 2103	Reporting Nuclear Detonations, Radioactive Fallout and Biological and Chemical Attacks, and Predicting Associated Hazards
STANAG 2934	Artillery Procedures—AARTY-1
STANAG 3204	Aeromedical Evacuation
STANAG 6004	Meaconing, Intrusion, Jamming, and Interference Report

Unless this publication states otherwise, masculine nouns and pronouns do not refer exclusively to men.

CHAPTER 1

INTRODUCTION

The sniper has special abilities, training and equipment. His job is to deliver discriminatory highly accurate rifle fire against enemy targets, which cannot be engaged successfully by the rifleman because of range, size, location, fleeting nature, or visibility. Sniping requires the development of basic infantry skills to a high degree of perfection. A sniper's training incorporates a wide variety of subjects designed to increase his value as a force multiplier and to ensure his survival on the battlefield. The art of sniping requires learning and repetitiously practicing these skills until mastered. A sniper must be highly trained in long-range rifle marksmanship and field craft skills to ensure maximum effective engagements with minimum risk.

1-1. MISSION

The primary mission of a sniper in combat is to support combat operations by delivering precise long-range fire on selected targets. By this, the sniper creates casualties among enemy troops, slows enemy movement, frightens enemy soldiers, lowers morale, and adds confusion to their operations. The secondary mission of the sniper is collecting and reporting battlefield information.

a. A well-trained sniper, combined with the inherent accuracy of his rifle and ammunition, is a versatile supporting arm available to an infantry commander. The importance of the sniper cannot be measured simply by the number of casualties he inflicts upon the enemy. Realization of the sniper's presence instills fear in enemy troop elements and influences their decisions and actions. A sniper enhances a unit's firepower and augments the varied means for destruction and harassment of the enemy. Whether a sniper is organic or attached, he will provide that unit with extra supporting fire. The sniper's role is unique in that it is the sole means by which a unit can engage point targets at distances beyond the effective range of the M16 rifle. This role becomes more significant when the target is entrenched or positioned among civilians, or during riot control missions. The fires of automatic weapons in such operations can result in the wounding or killing of noncombatants.

b. Snipers are employed in all levels of conflict. This includes conventional offensive and defensive combat in which precision fire is delivered at long ranges. It also includes combat patrols, ambushes, countersniper operations, forward observation elements, military operations in urbanized terrain, and retrograde operations in which snipers are part of forces left in contact or as stay-behind forces.

1-2. ORGANIZATION

In light infantry divisions, the sniper element is composed of six battalion personnel organized into three 2-man teams. The commander designates missions and priorities of targets for the team and may attach or place the team under the operational control of a company or platoon. They may perform dual missions, depending on the need. In the mechanized infantry battalions, the sniper element is composed of two riflemen (one team) located in a rifle squad. In some specialized units, snipers may be organized according to the needs of the tactical situation.

a. Sniper teams should be centrally controlled by the commander or the sniper employment officer. The SEO is responsible for the command and control of snipers assigned to the unit. In light infantry units, the SEO will be the reconnaissance platoon leader or the platoon sergeant. In heavy or mechanized units, the SEO may be the company commander or the executive officer. The duties and responsibilities of the SEO areas follows:

(1) To advise the unit commander on the employment of snipers.
(2) To issue orders to the team leader.
(3) To assign missions and types of employment.
(4) To coordinate between the sniper team and unit commander.
(5) To brief the unit commander and team leaders.
(6) To debrief the unit commander and team leaders.
(7) To train the teams.

b. Snipers work and train in 2-man teams. One sniper's primary duty is that of the sniper and team leader while the other sniper serves as the observer. The sniper team leader is responsible for the day-to-day activities of the sniper team. His responsibilities areas follows:

(1) To assume the responsibilities of the SEO that pertain to the team in the SEO'S absence.

(2) To train the team.
(3) To issue necessary orders to the team.
(4) To prepare for missions.
(5) To control the team during missions.

c. The sniper's weapon is the sniper weapon system. The observer has the M16 rifle and an M203, which gives the team greater suppressive fire and protection. Night capability is enhanced by using night observation devices.

1-3. PERSONNEL SELECTION CRITERIA

Candidates for sniper training require careful screening. Commanders must screen the individual's records for potential aptitude as a sniper. The rigorous training program and the increased personal risk in combat require high motivation and the ability to learn a variety of skills. Aspiring snipers must have an excellent personal record.

a. The basic guidelines used to screen sniper candidates are as follows:

(1) ***Marksmanship.*** The sniper trainee must be an expert marksman. Repeated annual qualification as expert is necessary. Successful participation in the annual competition-in-arms program and an extensive hunting background also indicate good sniper potential.

(2) ***Physical condition.*** The sniper, often employed in extended operations with little sleep, food, or water, must be in outstanding physical condition. Good health means better reflexes, better muscular control, and greater stamina. The self-confidence and control that come from athletics, especially team sports, are definite assets to a sniper trainee.

(3) ***Vision.*** Eyesight is the sniper's prime tool. Therefore, a sniper must have 20/20 vision or vision correctable to 20/20. However, wearing glasses could become a liability if glasses are lost or damaged. Color blindness is also considered a liability to the sniper, due to his inability to detect concealed targets that blend in with the natural surroundings.

(4) ***Smoking.*** The sniper should not be a smoker or use smokeless tobacco. Smoke or an unsuppressed smoker's cough can betray the sniper's position. Even though a sniper may not smoke or use smokeless tobacco on a mission, his refrainment may cause nervousness and irritation, which lowers his efficiency.

(5) ***Mental condition.*** When commanders screen sniper candidates, they should look for traits that indicate the candidate has the right qualities to be a sniper. The commander must determine if the candidate will pull the trigger at the right time and place. Some traits to look for

are reliability, initiative, loyalty, discipline, and emotional stability. A psychological evaluation of the candidate can aid the commander in the selection process.

(6) ***Intelligence.*** A sniper's duties require a wide variety of skills. He must learn the following:

- Ballistics.
- Ammunition types and capabilities.
- Adjustment of optical devices.
- Radio operation and procedures.
- Observation and adjustment of mortar and artillery fire.
- Land navigation skills.
- Military intelligence collecting and reporting.
- Identification of threat uniforms and equipment.

b. In sniper team operations involving prolonged independent employment, the sniper must be self-reliant, display good judgment and common sense. This requires two other important qualifications: emotional balance and field craft.

(1) ***Emotional balance.*** The sniper must be able to calmly and deliberately kill targets that may not pose an immediate threat to him. It is much easier to kill in self-defense or in the defense of others than it is to kill without apparent provocation. The sniper must not be susceptible to emotions such as anxiety or remorse. Candidates whose motivation toward sniper training rests mainly in the desire for prestige may not be capable of the cold rationality that the sniper's job requires.

(2) ***Field craft.*** The sniper must be familiar with and comfortable in a field environment. An extensive background in the outdoors and knowledge of natural occurrences in the outdoors will assist the sniper in many of his tasks. Individuals with such a background will often have great potential as a sniper.

c. Commander involvement in personnel selection is critical. To ensure his candidate's successful completion of sniper training and contribution of his talents to his unit's mission, the commander ensures that the sniper candidate meets the following prerequisites before attending the US Army Sniper School:

- Male.
- PFC to SFC (waiverable for MSG and above).
- Active duty or ARNG and USAR.
- Good performance record.

- No history of alcohol or drug abuse.
- A volunteer (with commander recommendation).
- Vision of 20/20 or correctable to 20/20.
- No record of disciplinary action.
- Expert marksman with M16A1 or M16A2 rifle.
- Minimum of one-year retrainability.
- Career management field 11.
- Pass APFT (70 percent, each event).

1-4. SNIPER AND OBSERVER RESPONSIBILITIES

Each member of the sniper team has specific responsibilities. Only through repeated practice can the team begin to function properly. Responsibilities of team members areas follows:

a. The sniper—

- Builds a steady, comfortable position.
- Locates and identifies the designated target.
- Estimates the range to the target.
- Dials in the proper elevation and windage to engage the target.
- Notifies the observer of readiness to fire.
- Takes aim at the designated target.
- Controls breathing at natural respiratory pause.
- Executes proper trigger control.
- Follows through.
- Makes an accurate and timely shot call.
- Prepares to fire subsequent shots, if necessary.

b. The observer—

- Properly positions himself.
- Selects an appropriate target.
- Assists in range estimation.
- Calculates the effect of existing weather conditions on ballistics.
- Reports sight adjustment data to the sniper.
- Uses the M49 observation telescope for shot observation.
- Critiques performance.

1-5. TEAM FIRING TECHNIQUES

A sniper team must be able to move and survive in a combat environment. The sniper team's mission is to deliver precision fire. This calls for a coordinated team effort. Together, the sniper and observer—

- Determine the effects of weather on ballistics.
- Calculate the range to the target.
- Make necessary sight changes.
- Observe bullet impact.
- Critique performance before any subsequent shots.

CHAPTER 2
EQUIPMENT

This chapter describes the equipment necessary for the sniper to effectively peform his mission. The sniper carries only what is essential to successfully complete his mission. He requires a durable rifle with the capability of long-range precision fire. The current US Army sniper weapon system is the M24. (See Appendix B for the M21 sniper weapon system.)

Section I
M24 SNIPER WEAPON SYSTEM

The M24 sniper weapon system is a 7.62-mm, bolt-action, six-shot repeating rifle (one round in the chamber and five rounds in the magazine). It is designed for use with either the M3A telescope (day optic sight) (usually called the M3A *scope)* or the metallic iron sights. The sniper must know the M24's components, and the procedures required to operate them (Figure 2-1, page 2-2). The deployment kit is a repair/maintenance kit with tools and repair parts for the operator to perform operator level maintenance (Figure 2-2, page 2-3.)

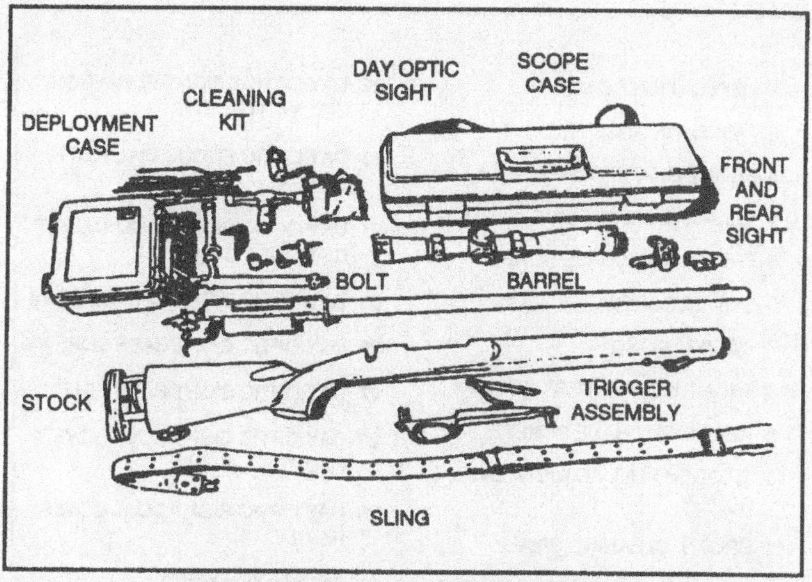

Figure 2-1. M24 sniper weapon system.

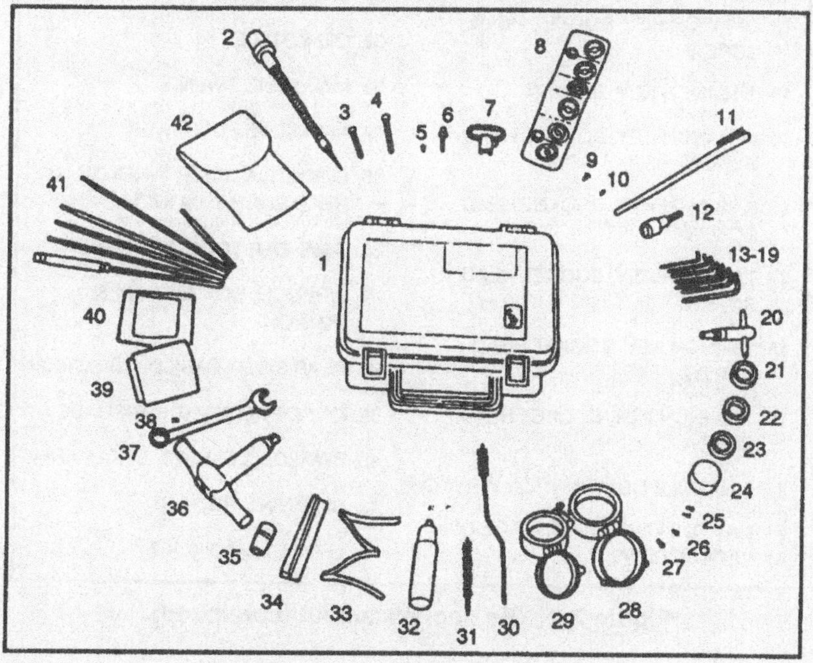

Figure 2-2. The deployment kit.

1 DEPLOYMENT CASE
2 FIRING PIN ASSEMBLY
3 FRONT GUARD SCREW
4 REAR GUARD SCREWS
5 FRONT SIGHT BASE SCREW
6 SWIVEL SCREW
7 SWIVEL, SLING
8 FRONT SIGHT INSERT KIT
9 REAR SIGHT BASE SCREW
10 TRIGGER PULL ADJUSTMENT SCREW
11 BRUSH, CLEANING SMALL
12 SOCKET WRENCH ATTACHMENT 3/8-INCH DRIVE HEX BIT 5/32-INCHES
13 050-INCH KEY, SOCKET HEAD SCREW
14 1/16-INCH KEY, SOCKET
15 5/64-INCH KEY, SOCKET HEAD SCREW
16 3/32-INCH KEY, SOCKET HEAD SCREW
17 7/64-INCH KEY, SOCKET HEAD SCREW
18 1/8-INCH KEY, SOCKET HEAD SCREW
19 5/32-INCH KEY, SOCKET HEAD SCREW
20 T-HANDLE COMBINATION WRENCH
21 DAY OPTIC SIGHT WINDAGE DIAL WITH SCREWS
22 DAY OPTICT SIGHT ELEVATION DIAL WITH SCEWS
23 DAY OPTIC FOCUS DIAL WITH SCREWS
24 DAY OPTIC SIGHT ADJUSTMENT DIAL DUST COVER
25 DAY OPTIC SIGHT RING SCREWS
26 DAY OPTIC SIGHT BASE SCREWS
27 DAY OPTIC SIGHT BASE REAR
28 DAY OPTIC SIGHT DUST COVER, FRONT
29 DAY OPTIC SIGHT DUST COVER, REAR
30 BRUSH, CHAMBER
31 BRUSH, BORE
32 OIL BOTTLE
33 MAGAZINE SPRING
34 MAGAZINE FOLLOWER
35 SOCKET, SOCKET WRENCH HEAD SCREW 1/2-INCH
36 T-HANDLE TORQUE WRENCH
37 WRENCH, BOX AND OPEN 1/2-INCH
38 REAR SIGHT BASE PLUG SCREW
39 DAY OPTIC SIGHT SUNSHADE
40 SWABS, CLEANING, SMALL ARMS
41 CLEANING ROD KIT
42 LENS CLEANING KIT

Figure 2-2. The deployment kit (continued).

2-1. OPERATIONS AND FUNCTIONS

To operate the M24 sniper weapon system, the sniper must know the information and instructions pertaining to the safety, bolt assembly, trigger assembly, and stock adjustment.

a. **Safety.** The safely is located on the right rear side of the receiver. When properly engaged, the safety provides protection against accidental discharge in normal usage.

(1) To engage the safety, place it in the "S" position (Figure 2-3).

(2) Always place the safety in the "S" position before handling, loading, or unloading the weapon.

(3) When the weapon is ready to be fired, place the safety in the "F" position (Figure 2-3).

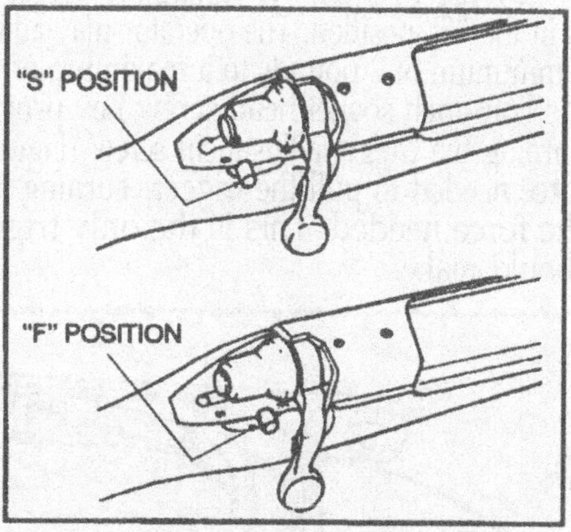

Figure 2-3. Safety.

b. **Bolt Assembly.** The bolt assembly locks the cartridge into the chamber and extracts the cartridge from the chamber.

(1) To remove the bolt from the receiver, release the internal magazine, place the safety in the "S" position, raise the bolt handle, and pull it back until it stops. Then push the bolt stop release (Figure 2-4) and pull the bolt from the receiver.

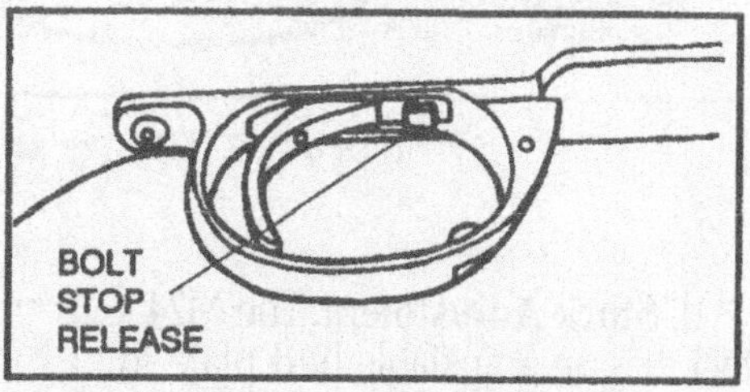

Figure 2-4. Bolt stop release.

(2) To replace the bolt, ensure the safety is in the "S" position, align the lugs on the bolt assembly with the receiver (Figure 2-5), slide the bolt all the way into the receiver, and then push the bolt handle down.

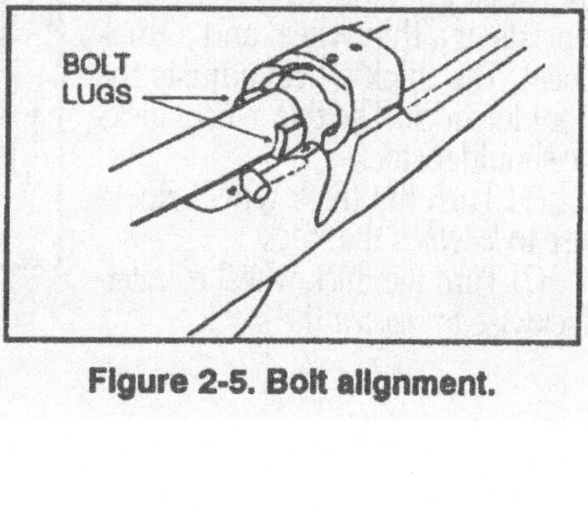

Figure 2-5. Bolt alignment.

WARNING
NEVER REMOVE THE TRIGGER MECHANISM, OR MAKE ADJUSTMENTS TO THE TRIGGER ASSEMBLY, EXCEPT FOR THE TRIGGER PULL FORCE ADJUSTMENT.

c. **Trigger Assembly.** Pulling the trigger fires the rifle when the safety is in the "F" position. The operator may adjust the trigger pull force from a minimum of 2 pounds to a maximum of 8 pounds. This is done using the 1/16-inch socket head screw key provided in the deployment kit. Turning the trigger adjustment screw (Figure 2-6) clockwise increases the force needed to pull the trigger. Turning it counterclockwise decreases the force needed. This is the only trigger adjustment the sniper should make.

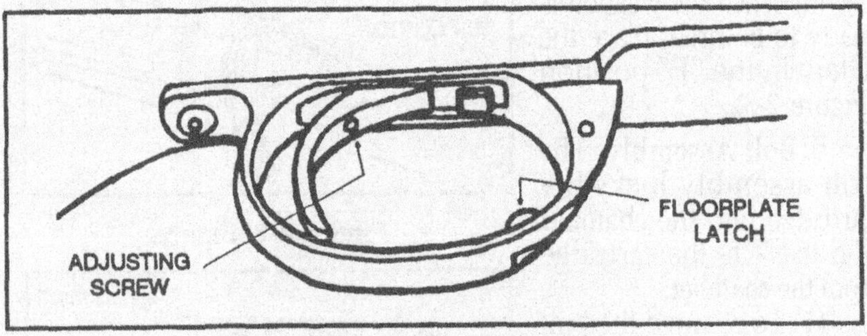

Figure 2-6. Trigger adjustment.

d. **Stock Adjustment.** The M24's stock has an adjustable butt plate to accommodate the length of pull. The stock adjustment (Figure 2-7) consists of a thin wheel and a thick wheel. The thick wheel adjusts the shoulder stock. The thin wheel locks the shoulder stock.

(1) Turn the thick wheel clockwise to *lengthen* the stock.

(2) Turn the thick wheel counterclockwise to *shorten* the stock.

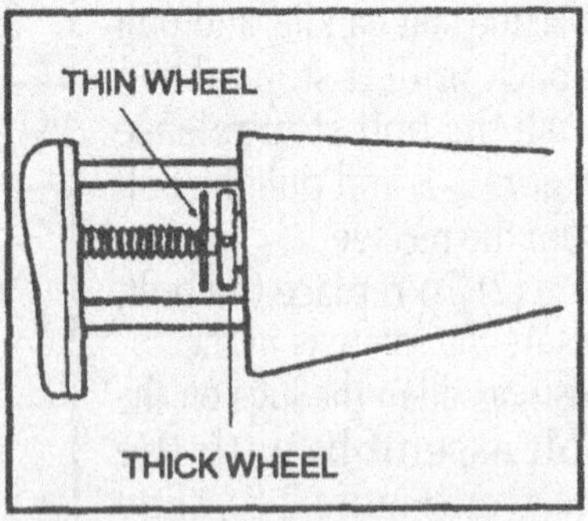

Figure 2-7. Stock adjustment.

(3) To lock the shoulder stock into position, turn the thin wheel clockwise against the thick wheel.

(4) To unlock the shoulder stock, turn the thin wheel counter-clockwise away from the thick wheel.

e. **Sling Adjustment** The sling helps hold the weapon steady without muscular effort. The more the muscles are used the harder it is to hold the weapon steady. The sling tends to bind the parts of the body used in aiming into a rigid bone brace, requiring less effort than would be necessary if no sling were used. When properly adjusted, the sling permits part of the recoil of the rifle to reabsorbed by the nonfiring arm and hand, removing recoil from the firing shoulder.

(1) The sling consists of two different lengths of leather straps joined together by a metal D ring (Figure 2-8). The longer strap is connected to the sling swivel on the rear stud on the forearm of the rifle. The shorter strap is attached to the sling swivel on the buttstock of the rifle. There are two leather loops on the long strap known as keepers. The keepers are used to adjust the tension on the sling. The frogs are hooks that are used to adjust the length of the sling.

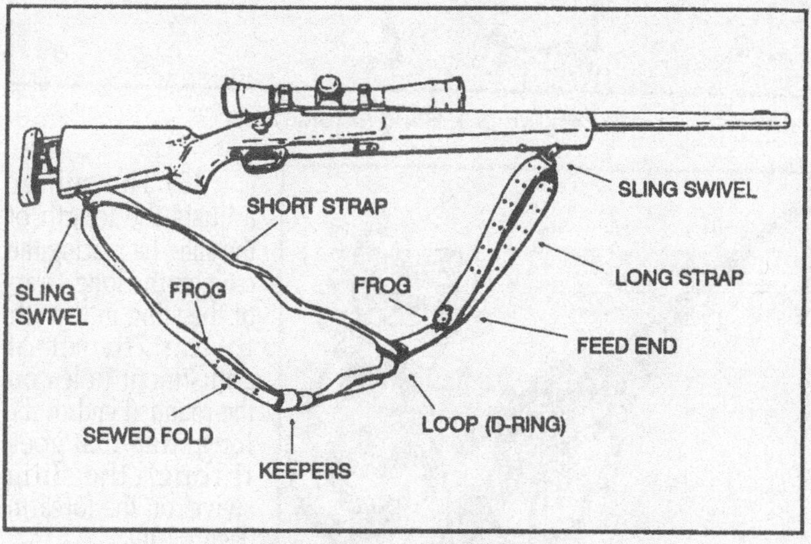

Figure 2-8. Leather Sling.

(2) To adjust the sling, the sniper disconnects the sling from the buttstock swivel. Then, he adjusts the length of the metal D ring that joins

the two halves of the sling. He then makes sure it is even with the comb of the stock when attaching the sling to the front swivel (Figure 2-9).

Figure 2-9. Sling adjustment.

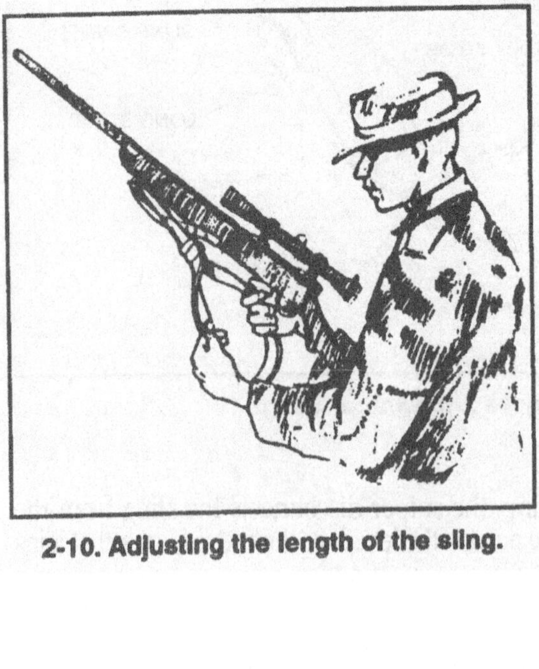

2-10. Adjusting the length of the sling.

(3) The sniper adjusts the length of the sling by placing the frog on the long strap of the sling in the 4th to the 7th set of adjustment holes on the rounded end of the long strap that goes through the sling swivel on the forearm (Figure 2-10).

(4) After adjusting the length, the sniper places the weapon on his firing hip and supports the

weapon with his firing arm. The sniper turns the sling away from him 90 degrees and inserts his nonfiring arm.

(5) The sniper slides the loop in the large section of the sling up the nonfiring arm until it is just below the armpit (Figure 2-11). He then slides both leather keepers down the sling until they bind the loop snugly round the nonfiring arm.

Figure 2-11. Placing the sling around the nonfiring arm.

Figure 2-12. Proper placement of the sling.

(6) The sniper moves his nonfiring hand from the outside of the sling to the inside of the sling between the rifle and the sling. The sniper then grasps the forearm of the weapon, just behind the sling swivel with his nonfiring hand. He forces it outward and away from his body with the nonfiring hand (Figure 2-12).

(7) The sniper pulls the butt of the

weapon into the pocket of his shoulder with the firing hand. He then grasps the weapon at the small of the stock and begins the aiming process.

2-2. INSPECTION

The sniper performs PMCS on the M24 SWS. Deficiencies that cannot be repaired by the sniper requires manufacturer repair. He must refer to TM 9-1005-306-10 that is furnished with each weapon system. The sniper must know this technical manual. He should cheek the following areas when inspecting the M24:

a. Check the appearance and completeness of all parts.

b. Check the bolt to ensure it locks, unlocks, and moves smoothly.

c. Check the safety to ensure it can be positively placed into the "S" and "F" positions easily without being too hard or moving too freely.

d. Check the trigger to ensure the weapon will not fire when the safety is in the "S" position, and that it has a smooth, crisp trigger pull when the safety is in the "F" position.

e. Check the trigger guard screws (rear of trigger guard and front of internal magazine) for proper torque (65 inch-pounds).

f. Check the scope mounting ring nuts for proper torque (65 inch-pounds).

g. Check the stock for any cracks, splits, or any contact it may have with the barrel.

h. Inspect the scope for obstructions such as dirt, dust, moisture, or loose or damaged lenses.

2-3. CARE AND MAINTENANCE

Maintenance is any measure taken to keep the M24 SWS in top operating condition. It includes inspection, repair, cleaning and lubrication-Inspection reveals the need for repair, cleaning, or lubrication. It also reveals any damages or defects. When sheltered in garrison and infrequently used, the M24 SWS must be inspected often to detect dirt, moisture, and signs of corrosion, and it must be cleaned accordingly. The M24 SWS that is in use and subject to the elements, however, requires no inspection for cleanliness, since the fact of its use and exposure is evidence that it requires repeated cleaning and lubrication.

a. **M24 SWS Maintenance.** The following materials are required for cleaning and maintaining the M24 SWS:

- One-piece plastic-coated .30 caliber cleaning rod with jag (36 inches).
- Bronze bristle bore brushes (.30 and .45 calibers).
- Cleaning patches (small and large sizes).

- Carbon cleaner.
- Copper cleaner.
- Rust prevention.
- Cleaner, lubricant, preservative.
- Rifle grease.
- Bore guide (long action).
- Swabs.
- Pipe cleaners.
- Medicine dropper.
- Shaving brush.
- Pistol cleaning rod.
- Rags.
- Camel's-hair brush.
- Lens tissue.
- Lens cleaning fluid (denatured or isopropyl alcohol).

b. **M24 SWS Disassembly.** The M24 SWS will be disassembled only when necessary, not for daily cleaning. For example, when removing an obstruction from the SWS that is stuck between the stock and the barrel. When disassembly is required, the recommended procedure is as follows:

- Place the weapon so that is it pointing in a safe direction.
- Ensure the safety is in the "S" position.
- Remove the bolt assembly.
- Loosen the mounting ring nuts on the telescope and remove the telescope.
- Remove the action screws.
- Lift the stock from the barrel assembly.
- For further disassembly, refer to TM 9-1005-306-10.

c. **M24 SWS Cleaning Procedures.** The M24 SWS must always be cleaned *before* and *after firing.*

(1) The SWS must always be cleaned *before firing*. Firing a weapon with a dirty bore or chamber will multiply and speed up any corrosive action. Oil in the bore and chamber of a SWS will cause pressures to vary and first-round accuracy will suffer. Clean and dry the bore and chamber before departure on a mission and use extreme care to keep the SWS clean and dry en route to the objective area. Firing a SWS with oil or moisture in the bore will cause smoke that can disclose the firing position.

(2) The SWS must be cleaned *after firing* since firing produces deposits of primer fouling, powder ashes, carbon, and metal fouling. Although ammunition has a noncorrosive primer that makes cleaning easier, the primer residue can still cause rust if not removed. Firing leaves two major types of fouling that require different solvents to remove *carbon* fouling and *copper* jacket fouling. The SWS must be cleaned within a reasonable time after firing. Use common sense when cleaning between rounds of firing. Repeated firing will not injure the weapon if it is properly cleaned before the first round is fired.

(3) Lay the SWS on a table or other flat surface with the muzzle away from the body and the sling down. Make sure not to strike the muzzle or telescopic sight on the table. The cleaning cradle is ideal for holding the SWS.

(4) Always clean the bore from the chamber toward the muzzle, attempting to keep the muzzle lower than the chamber to prevent the bore cleaner from running into the receiver or firing mechanism. Be careful not to get any type of fluid between the stock and receiver. If fluid does collect between the stock and receiver, the receiver will slide on the bedding every time the SWS recoils, thereby decreasing accuracy and increasing wear and tear on the receiver and bedding material.

(5) Always use a bore guide to keep the cleaning rod centered in the bore during the cleaning process.

(6) Push several patches saturated with carbon cleaner through the barrel to loosen the powder fouling and begin the solvent action on the copper jacket fouling.

(7) Saturate the bronze bristle brush (NEVER USE STAINLESS STEEL BORE BRUSHES-THEY WILL SCRATCH THE BARREL) with carbon cleaner (shake the bottle regularly to keep the ingredients mixed) using the medicine dropper to prevent contamination of the carbon cleaner. Run the bore brush through at least 20 times. Make sure the bore brush passes completely through the barrel before reversing its direction; otherwise, the bristles will break off.

(8) Use a pistol cleaning rod and a .45 caliber bronze bristle bore brush, clean the chamber by rotating the patch-wrapped brush 8 to 10 times. DO NOT scrub the brush in and out of the chamber.

(9) Push several patches saturated with carbon cleaner through the bore to push out the loosened powder fouling.

(10) Continue using the bore brush and patches with carbon cleaner until the patches have no traces of black/gray powder fouling and are green/blue. This indicates that the powder fouling has been removed and only copper fouling remains. Remove the carbon cleaner from the barrel

with several clean patches. This is important since solvents should never be mixed in the barrel.

(11) Push several patches saturated with copper cleaner through the bore, using a scrubbing motion to work the solvent into the copper. Let the solvent work for 10 to 15 minutes (NEVER LEAVE THE COPPER CLEANER IN THE BARREL FOR MORE THAN 30 MINUTES).

(12) While waiting, scrub the bolt with the toothbrush moistened with carbon cleaner and wipe down the remainder of the weapon with a cloth.

(13) Push several patches saturated with copper cleaner through the barrel. The patches will appear dark blue at first, indicating the amount of copper fouling removed. Continue this process until the saturated patches have no traces of blue/green. If the patches continue to come out dark blue after several treatments with copper cleaner, use the bronze brush saturated with copper cleaner to increase the scrubbing action. Be sure to clean the bronze brush thoroughly afterwards with hot running water (quick scrub cleaner/degreaser is preferred) as the copper cleaner acts upon its bristles as well.

(14) When the barrel is clean, dry it with several tight fitting patches. Also, dry the chamber using the .45 caliber bronze bristle bore brush with a patch wrapped around it.

(15) Run a patch saturated with rust prevention (*not* CLP) down the barrel and chamber if the weapon is to be stored for any length of time. Stainless steel barrels are not immune from corrosion. Be sure to remove the preservative by running dry patches through the bore and chamber before firing.

(16) Place a small amount of rifle grease on the rear surfaces of the bolt lugs. This will prevent galling of the metal surfaces.

(17) Wipe down the exterior of the weapon (if it is not covered with camouflage paint) with a CLP-saturated cloth to protect it during storage.

d. **Barrel Break-in Procedure.** To increase barrel life, accuracy, and reduce cleaning requirement the following barrel break-in procedure must be used. This procedure is best accomplished when the SWS is new or newly rebarreled. The break-in period is accomplished by polishing the barrel surface under heat and pressure. This procedure should only be done by qualified personnel. The barrel must be cleaned of all fouling, both powder and copper. The barrel is dried, and one round is fired. The barrel is then cleaned again using carbon cleaner and then copper cleaner. The barrel must be cleaned again, and another round is fired. The procedure must be repeated for a total of 10 rounds. After the 10th round the SWS is then tested for groups by firing three-round shot groups, with a complete barrel cleaning between shot groups for a total of five shot groups (15 rounds total).

The barrel is now broken in, and will provide superior accuracy and a longer usable barrel life. Additionally, the barrel will be easier to clean because the surface is smoother. Again the barrel should be cleaned at least every 50 rounds to increase the barrel life.

e. **Storage.** The M24 SWS should be stored (Figure 2-13) using the following procedures:

- Clear the SWS, close the bolt, and squeeze the trigger.
- Open the lens caps to prevent gathering of moisture.
- Hang the weapon upside down by the rear sling swivel.
- Place all other items in the system case.
- Transport the weapon in the system case during nontactical situations.
- Protect the weapon at all times during tactical movement.

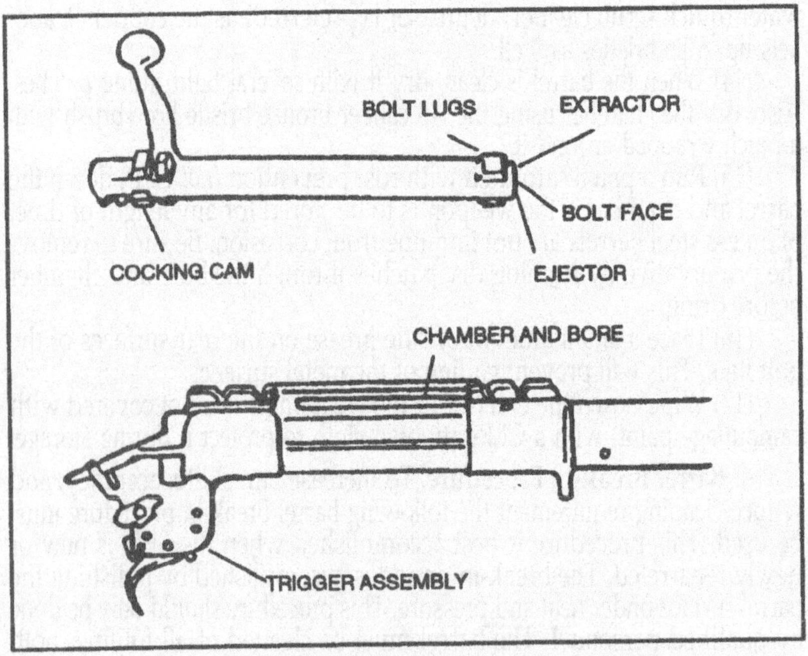

Figure 2-13. Maintenance for storing or using.

NOTE: Rod clean swabs through the bore before firing. This procedure ensures first-round accuracy and reduces the signature.

f. **Cold Climates.** In temperatures below freezing, the SWS must be kept free of moisture and heavy oil, both of which will freeze, causing the working parts to freeze or operate sluggishly. The SWS should be stored in a room with the temperature equal to the outside temperature. When the SWS is taken into a warm area, condensation occurs, thus requiring a thorough cleaning and drying before taking it into the cold. Otherwise, the condensation causes icing on exposed metal parts and optics. The firing pin should be disassembled and cleaned thoroughly with a decreasing agent. It should then be lubricated with CLP. Rifle grease hardens and causes the firing pin to fall sluggishly.

g. **Salt Water Exposure.** Saltwater and saltwater atmosphere have extreme and rapid corrosive effects on the metal parts of the SWS. During periods of exposure, the SWS must be checked and cleaned as often as possible, even if it means only lubricating the SWS. The SWS should always be well lubricated, including the bore, except when actually firing. Before firing, always run a dry patch through the bore, if possible.

h. **Jungle Operations** (High Humidity). In hot and humid temperatures, keep the SWS lubricated and cased when not in use. Protect the SWS from rain and moisture whenever possible. Keep ammunition clean and dry. Clean the SWS, the bore, and the chamber daily. Keep the caps on the telescope when not in use. If moisture or fungus develops on the inside of the telescope, replace it. Clean and dry the stock daily. Dry the carrying case and SWS in the sun whenever possible.

i. **Desert Operations.** Keep the SWS dry and free of CLP and grease except on the rear of the bolt lugs. Keep the SWS free of sand by using the carrying sleeve or carrying case when not in use. Protect the SWS by using a wrap. Slide the wrap between the stock and barrel, then cross over on top of the scope. Next, cross under the SWS (over the magazine) and secure it. The SWS can still be placed into immediate operation but all critical parts are covered. The sealed hard case is preferred in the desert if the situation permits. Keep the telescope protected from the direct rays of the sun. Keep ammunition clean and protected from the direct rays of the sun. Use a toothbrush to remove sand from the bolt and receiver. Clean the bore and chamber daily. Protect the muzzle and receiver from blowing sand by covering with a clean cloth. To protect the free-floating barrel of the SWS, take an 8- or 9-inch strip of cloth and tie a knot in each end. Before going on a mission, slide the cloth between the barrel and stock all the way to the receiver and leave it there. When in position, slide the cloth out, taking all restrictive debris and sand with it.

2-4. DISASSEMBLY

Occasionally, the weapon requires disassembly however, this should be done only when absolutely necessary, not for daily maintenance. An example of this would be to remove an obstruction that is stuck between the forestock and the barrel. When disassembly is required, the recommended procedure is as follows:

a. Point the rifle in a safe direction.

b. Put the safety in the "S" position.

c. Remove the bolt assembly.

d. Use the 1/2-inch combination wrench, loosen the front and rear mounting ring nuts (Figure 2-14) on the scope, and remove the scope.

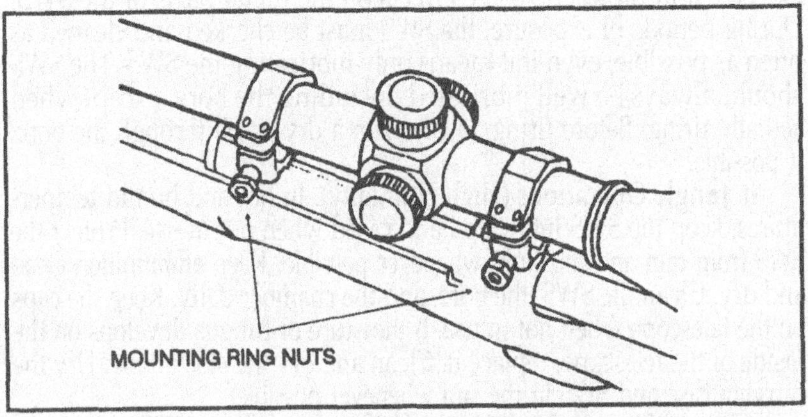

Figure 2-14. Mounting ring nuts.

e. Loosen the front and rear trigger guard screws (Figure 2-15).

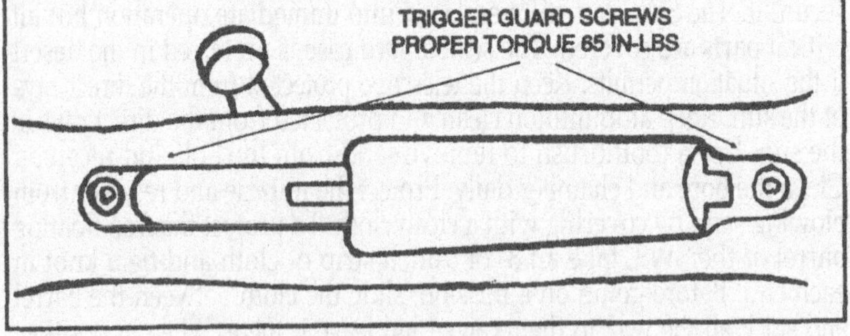

Figure 2-15. Trigger guard screws.

f. Lift the stock assembly from the barrel assembly (Figure 2-16).
g. Reassemble in reverse order.

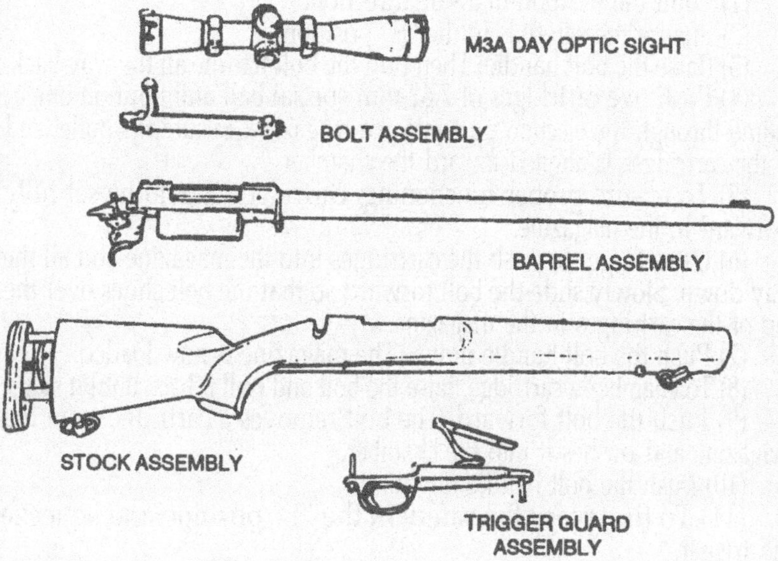

Figure 2-16. Disassembled weapon.

WARNING
ALWAYS KEEP FINGERS AWAY FROM THE TRIGGER UNTIL READY TO FIRE, MAKE SURE THE RIFLE IS NOT LOADED BY INSPECTING THE MAGAZINE AND CHAMBER, USE AUTHORIZED AMMUNITION AND CHECK THE CONDITION BEFORE LOADING THE RIFLE.

2-5. LOADING AND UNLOADING

Before loading, the sniper should ensure that the M24 SWS is on SAFE, and the bolt is in a forward position. Before unloading, he should ensure the M24 SWS is on SAFE, and the bolt is toward the rear.

a. **Loading.** The M24 has an internal, five-round capacity magazine. To load the rifle—

(1) Point the weapon in a safe direction.

(2) Ensure the safety is in the "S" position.

(3) Raise the bolt handle. Then pull the bolt handle all the way back.

(4) Push five cartridges of 7.62-mm special ball ammunition one at a time through the ejection port into the magazine. Ensure the bullet end of the cartridges is aligned toward the chamber.

(5) To ensure proper functioning, cartridges should be set fully rearward in the magazine.

(6) Use a finger to push the cartridges into the magazine and all the way down. Slowly slide the bolt forward so that the bolt slides over the top of the cartridges in the magazine.

(7) Push the bolt handle down. The magazine is now loaded.

(8) To chamber a cartridge, raise the bolt and pull it back until it stops.

(9) Push the bolt forward. The bolt removes a cartridge from the magazine and pushes it into the chamber.

(10) Push the bolt handle down.

(11) To fire, place the safety in the "F" position and squeeze the trigger.

WARNING

ENSURE THE CHAMBER AND MAGAZINE ARE CLEAR OF CARTRIDGES.

b. **Unloading.** To unload the M24 SWS—

(1) Point the muzzle in a safe direction.

(2) Ensure the safety is in the "S" position.

(3) Raise the bolt handle.

(4) Put one hand over the top ejection port. Slowly pull the bolt handle back with the other hand to remove the cartridge from the chamber.

(5) Remove the cartridge from the rifle.

(6) Put a hand under the floor plate.

(7) Push the floor plate latch to release the floor plate (Figure 2-17, page 2-18). The magazine spring and follower will be released from the magazine.

(8) Remove the released cartridges.

(9) Push in the magazine follower, then close the floor plate.

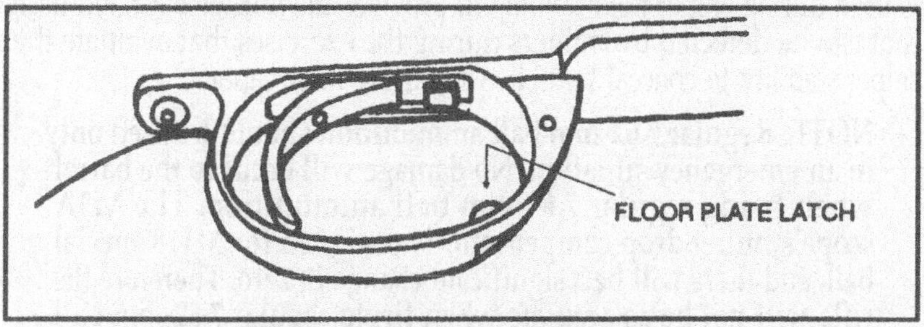

Figure 2-17. Floor plate latch.

2-6. STORAGE

The M24 SWS should be stored as follows:

a. Hang the weapon in an upside down position by the rear sling swivel.

b. Close the bolt and squeeze the trigger.

c. Open the lens caps to prevent gathering of moisture.

d. Place all other items in the system case.

e. Protect the weapon at all times during tactical movement. (See Chapter 4.)

Section II
AMMUNITION

The sniper uses the 7.62-mm special ball (M118) ammunition with the sniper weapon system. The sniper must rezero the weapon each time he fires a different type or lot of ammunition. This information should be maintained in the sniper data book.

2-7. TYPES AND CHARACTERISTICS

The types and characteristics of sniper ammunition are described in this paragraph.

a. **M118 Special Ball Bullet. The** 7.62-mm special ball (M118) bullet consists of a gilding metal jacket and a lead antimony slug. It is a boat-tailed bullet (rear of bullet is tapered) and weighs 173 grains. The tip of the bullet is not colored. The base of the cartridge is stamped with the year of manufacture and a circle that has vertical and horizontal lines, sectioning it into quarters. Its spread (accuracy standard) for a 10-shot group is no more than 12 inches at 550 meters (fired from an accuracy barrel in a test cradle).

b. **M82 Blank Ammunition.** The 7.62-mm M82 blank ammunition is used during sniper field training. It provides the muzzle blast and flash that can be detected by trainers during the exercises that evaluate the sniper's ability to conceal himself while firing his weapon.

NOTE: Regular 7.62-mm ball ammunition should be used only in an emergency situation. No damage will occur to the barrel when firing regular 7.62-mm ball ammunition. The M3A scope's bullet drop compensator is designed for M118 special ball, and there will be a significant change in zero. Therefore the rifle will not be as accurate when firing regular 7.62-mm ball ammunition. The 7.62-mm ball ammunition should be test fired and the ballistic data recorded in the data book.

2-8. ROUND-COUNT BOOK

The sniper maintains a log of the number of cartridge fired through the M24 SWS. It is imperative to accurately maintain the round-count book as the barrel should be replaced after 5,000 rounds of firing. The round-count book is issued and maintained in the arms room.

2-9. M24 MALFUNCTIONS AND CORRECTIONS

Table 2-1 does not reflect all malfunctions that can occur, or all causes and corrective actions. If a malfunction is not correctable, the complete weapon system must be turned in to the proper maintenance/supply channel for return to the contractor (see shipment, TM 9-1005-306-10).

MALFUNCTION	CAUSE	CORRECTION
Fail to fire	Safety in "S" position	1. Move safety to "F" position
	Defective ammunition	2. Eject cartridge
	Firing pin damaged	3. Change firing pin assembly
	Firing pin binds	4. Change firing pin assembly
	Firing pin protrudes	5. Change firing pin assembly
	Firing control out of adjustment	6. Turn complete system in to the maintenance/supply channel for return to contractor
	Trigger out of adjustment	7. Turn in as above
	Trigger does not retract	8. Turn in as above
	Trigger binds on trigger guard	9. Turn in as above
	Firing pin does not remain in the cocked position with bolt closed	10. Turn in as above
Bolt binds	Action screw protudes into bolt track	11. Turn in as above
	Scope base protrudes into bolt track	12. Turn in as above

Table 2-1. M24 malfunctions and corrections.

MALFUNCTION	CAUSE	CORRECTION
Fail to feed	Bolt override of cartridge	13. Seat cartridge fully rearward in magazine
	Cartridges stems chamber	14. Pull bolt fully rearward; remove stemmed cartridge from ejection port area; reposition cartridge fully in magazine
	Magazine in backward	15. Remove magazine spring, and reinstall with long leg follower
	Weak or broken magazine spring	16. Replace spring
Fail to eject	Broken ejector	17. Turn the complete weapon system in to the maintenance/supply channel for return to contractor
	Fouled ejector plunger	18. Inspect and clean bolt face; if malfunction continues, turn in as above
Fail to extract	Broken extractor	19. Turn in as above

Table 2-1. M24 malfunctions and corrections (continued).

Section III
SNIPER SIGHTING DEVICES

The sniper has two sighting devices: the M3A scope and iron sights. The M3A scope allows the sniper to see the cross hairs and the image of the target with identical sharpness. It can be easily removed and replaced with less than 1/2 minute of angle change in zero. However, the M3A scope should be left on the rifle. Iron sights are used only as a backup sighting system and can be quickly installed.

2-10. M3A SCOPE

The M3A scope is an optical instrument that the sniper uses to improve his ability to see his target clearly in most situations. Usually, the M3A scope presents the target at an increased size (as governed by scope magnification), relative to the same target at the same distance without a scope. The M3A scope helps the sniper to identify recognize the target. His increased sighting ability also helps him to successfully engage the target.

NOTE: The adjustment dials are under the adjustment dust cover.

a. **M3A Scope Adjustments.** The sniper must use the following adjustment procedures on the M3A scope:

(1) ***Focus adjustment dial.*** The focus adjustment dial (Figure 2-18) is on the left side of the scope barrel. This dial has limiting stops with the two extreme positions shown by the infinity mark and the largest dot. The focus adjustment dial keeps the target in focus. If the target is close, the dial is set at a position near the largest dot.

NOTE: Each minute of angle is an angular unit of measure.

(2) ***Elevation adjustment dial.*** The elevation adjustment dial (Figure 2-18) is on top of the scope barrel. This dial has calibrated index markings from 1 to 10. These markings represent the elevation setting adjustments needed at varying distances: 1 = 100 meters, 3 = 300 meters, 7 = 700 meters, and so on. Each click of the elevation dial equals 1 minute of angle.

(3) ***Windage adjustment dial.*** The windage adjustment dial (Figure 2-18) is on the right side of the scope barrel. This dial is used to make lateral adjustments to the scope. Turning the dial in the indicated direction moves the point of impact in that direction. Each click on the windage dial equals .5 minute of angle.

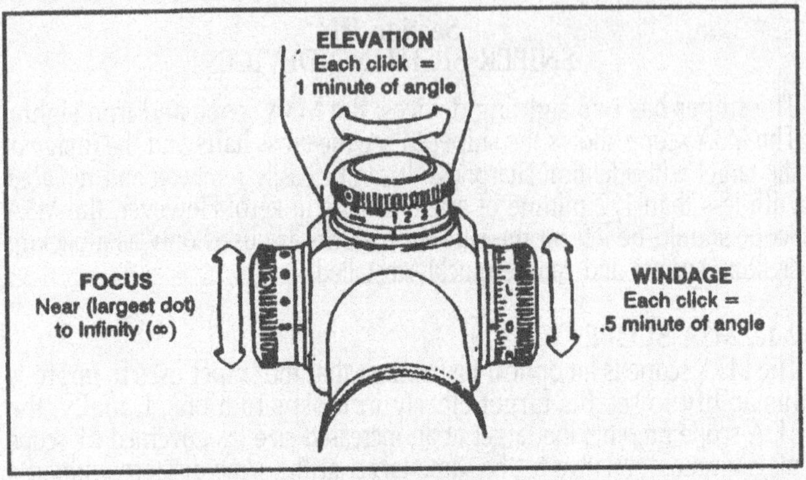

Figure 2-18. Focus, elevation, and windage adjustment dials.

(4) *Eyepiece adjustment.* The eyepiece (Figure 2-19) is adjusted by turning it in or out of the barrel until the reticle appears crisp and clear. Focusing the eyepiece should be done after mounting the scope. The sniper grasps the eyepiece and backs it away from the lock ring. He does not attempt to loosen the lock ring first; it loosens automatically when he backs away from the eyepiece (no tools needed). The eyepiece is turned several turns to move it at least 1/8 inch. It takes this much change to achieve any measurable effect on the focus. The sniper looks through the scope at the sky or a blank wall and checks to see if the reticle appears sharp and crisp. He locks the lock ring after achieving reticle clarity.

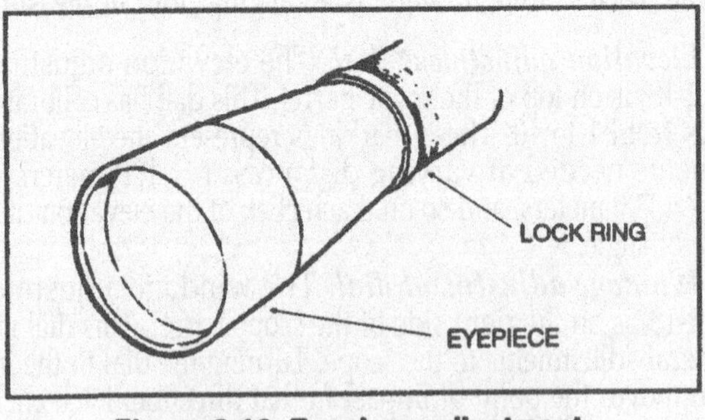

Figure 2-19. Eyepiece adjustment.

> **WARNINGS**
> 1. SECURELY FASTEN THE MOUNTING BASE TO THE RIFLE. LOOSE MOUNTING MAY CAUSE THE M3A SCOPE AND BASE MOUNT ASSEMBLY TO COME OFF THE RIFLE WHEN FIRING, POSSIBLY INJURING THE FIRER.
>
> 2. DURING RECOIL PREVENT THE M3A SCOPE FROM STRIKING THE FACE BY MAINTAINING AN AVERAGE DISTANCE OF 2 TO 3 INCHES BETWEEN THE EYE AND THE SCOPE.

b. **M3A Scope Mount.** The M3A scope mount has a baseplate with four screws; a pair of scope rings with eight ring screws, each with an upper and lower ring half with eight ring screws and two ring mounting bolts with nuts (Figure 2-20). The baseplate is mounted to the rifle by screwing the four baseplate screws through the plate and into the top of the receiver. The screws must not protrude into the receiver and interrupt the functioning of the bolt. After the baseplate is mounted, the scope rings are mounted.

NOTE: The M3A scope has two sets of mounting slots. The sniper selects the set of slots that provides proper eye relief (the distance that the eye is positioned behind the telescopic sight). The average distance is 2 to 3 inches. The sniper adjusts eye relief to obtain a full field of view.

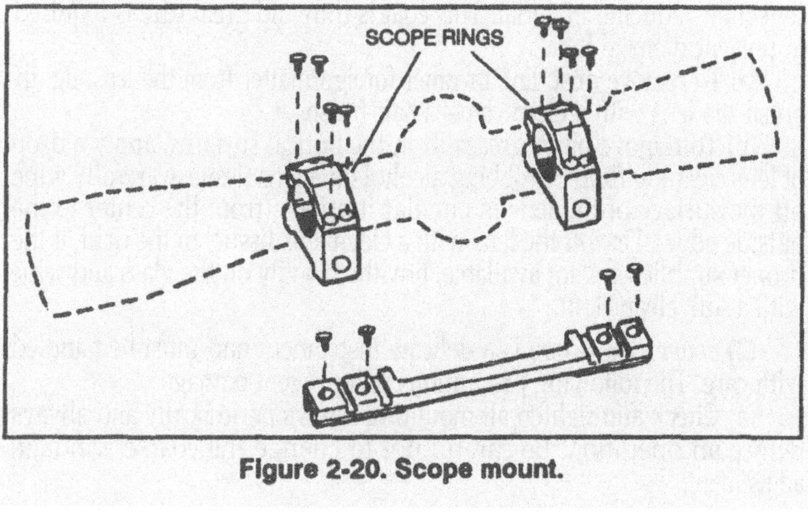

Figure 2-20. Scope mount.

(1) Before mounting the M3A scope, lubricate the threads of each mounting ring nut.

(2) Ensure smooth movement of each mounting ring nut and mount claw.

(3) Inspect for burrs and foreign matter between each mounting ring nut and mount claw. Remove burrs or foreign matter before mounting.

(4) Mount the sight and rings to the base.

NOTE: Once a set of slots is chosen, the same set should always be used in order for the SWS to retain zero.

(5) Ensure the mounting surface is free of dirt, oil, or grease.

(6) Set each ring bolt spline into the selected slot.

(7) Slide the rear mount claw against the base and finger-tighten the mounting ring nut.

(8) If the scope needs to be adjusted loosen the mounting ring nuts and align the ring bolts with the other set of slots on the base Repeat this process.

(9) Slide the front mount claw against the base, and finger-tighten the mounting ring nut.

(10) Use the T-handle torque wrench, which is preset to 65inch-pounds, to tighten the rear mounting ring nut.

c. **Care and Maintenance of the M3A Scope.** Dirt, rough handling, or abuse of optical equipment will result in inaccuracy and malfunction. When not in use, the rifle and scope should be cased, and the lens should be capped.

(1) ***Lens.*** The lens are coated with a special magnesium fluoride reflection-reducing material. This coat is thin and great care is required to prevent damage to it.

(a) To remove dust, lint, or other foreign matter from the lens, lightly brush the lens with a clean camel's-hair brush.

(b) To remove oil or grease from the optical surfaces, apply a drop of lens cleaning fluid or robbing alcohol on a lens tissue. Carefully wipe off the surface of the lens in circular motions (from the center to the outside edge). Dry off the lens with a clean lens tissue. In the field, if the proper supplies are not available, breathe heavily on the glass and wipe with a soft, clean cloth.

(2) ***Scope.*** The scope is a delicate instrument and must be handled with care. The following precautions will prevent damage

(a) Check and tighten all mounting screws periodically and always before an operation. Be careful not to change the coarse windage adjustment.

(b) Keep the lens free from oil and grease and never touch them with the fingers. Body grease and perspiration can injure them. Keep the cap on the lens.

(c) Do not force the elevation and windage screws or knobs.

(d) Do not allow the scope to remain in direct sunlight, and avoid letting the sun's rays shine through the lens. The lens magnify and concentrate sunlight into a pinpoint of intense heat, which is focused on the mil-scale reticle. This may melt the mil dots and damage the scope internally. Keep the lens covered and the entire scope covered when not in use.

(e) Avoid dropping the scope or striking it with another object. This could permanently damage the telescope as well as change the zero.

(f) To avoid damage to the scope or any other piece of sniper equipment, snipers or armorers should be the only personnel handling the equipment. Anyone who does not know how to use this equipment could cause damage.

(3) ***Climate conditions.*** Climate conditions play an important part in taking care of optical equipment.

(a) *Cold climates.* In extreme cold, care must be taken to avoid condensation and congealing of oil on the glass of the optical equipment. If the temperature is not excessive, condensation can be removed by placing the instrument in a warm place. Concentrated heat must not be applied because it causes expansion and damage can occur. Moisture may also be blotted from the optics with lens tissue or a soft, dry cloth. In cold temperatures, oil thickens and causes sluggish operation or failure. Focusing parts are sensitive to freezing oils. Breathing forms frost, so the optical surfaces must be cleaned with lens tissue, preferably dampened lightly with alcohol. DO NOT apply alcohol on the glass of the optics.

(b) *Jungle operations (high humidity).* In hot and humid temperatures, keep the caps on the scope when not in use. If moisture or fungus develops on the inside of the telescope, replace it.

(c) *Desert operations.* Keep the scope protected from the direct rays. of the sun.

(d) *Hot climate and salt water exposure.* The scope is vulnerable to hot, humid climates and salt water atmosphere. It MUST NOT be exposed to direct sunlight. In humid and salt air conditions, the scope must be inspected, cleaned, and lightly oiled to avoid rust and corrosion. Perspiration can also cause the equipment to rust; therefore, the instruments must be thoroughly dried and lightly oiled.

d. **M3A Scope Operation.** When using the M3A scope, the sniper looks at the target and determines the distance to it by using the mil dots on the reticle. The mil-dot reticle (Figure 2-21) is a duplex-style reticle that has thick outer sections and thin inner sections. Superimposed on the thin center section of the reticle is a series of dots. There are 4 dots on each side of the center and 4 dots above and below the center. These 4 dots are spaced 1 mil apart, and 1 mil from both the center and the start of the thick section of the reticle. This spacing allows the sniper to make close estimates of target range, assuming there is an object of known size (estimate) in the field of view. For example, a human target appears to be 6 feet tall, which equals 1.83 meters tall, and at 500 meters, 3.65 dots high (nominally, about 3.5 dots high). Another example is a l-meter target at a 1,000-meter range. This target is the height between 2 dots, or the width between 2 dots. If the sniper is given a good estimate of the object's size, then he may accurately determine target range using the mil-dot system.

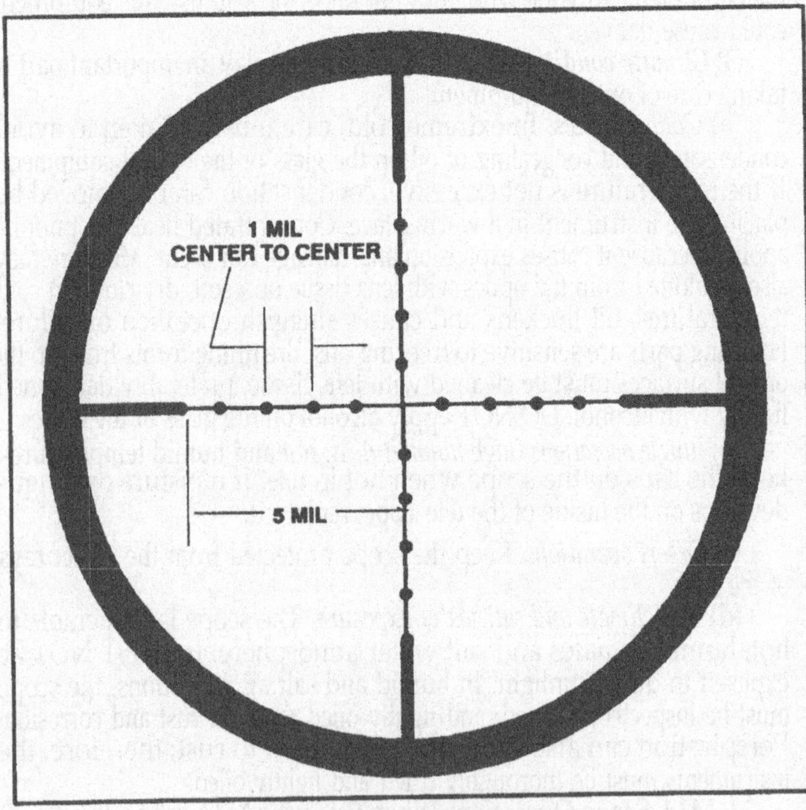

Figure 2-21. Mil-dot reticle.

e. **Zeroing.** Zeroing the M3A scope should be done on a known-distance range (preferably 900 meters long) with bull's-eye-type targets (200-yard targets, NSN SR1-6920-00-900-8204). When zeroing the scope, the sniper—

(1) Assumes a good prone-supported position 100 meters from the target.

(2) Ensures the "1" on the elevation dial is lined up with the elevation index line, and the "0" on the windage dial is lined up with the windage index line.

(3) Fires three rounds at the center of the target, keeping the same aiming point each time and triangulate.

(4) After the strike of the rounds has been noted, turns the elevation and windage dials to make the needed adjustments to the scope.

- Each click on the elevation dial equals one minute of angle.
- One minute of angle at 100 meters equals 1.145 inches or about 1 inch.
- Each click on the windage dial equals .5 minute of angle.
- .5 minute of angle at 100 meters equals about .5 inch.

(5) Repeats steps 3 and 4 until a three-round shot group is centered on the target.

(6) Once the shot group is centered, loosens the hex head screws on the elevation and windage dials. He turns the elevation dial to the index line marked "1" (if needed). He turns the windage dial to the index line marked "0" (if needed) and tighten the hex head screws.

(7) After zeroing at 100 meters and calibrating the dial, confirms this zero by firing and recording sight settings (see Chapter 3) at 100-meter increments through 900 meters.

f. **Field-Expedient Confirmation/Zeroing.** The sniper may need to confirm zero in a field environment. Examples are shortly after receiving a mission, a weapon was dropped, or excessive climatic changes as may be experienced by deploying to another part of the world. Two techniques of achieving a crude zero are the 25-yard/900-inch method and the observation of impact method.

(1) ***25-yard/900-inch method.*** Dial the scope to 300 meters for elevation and to "0" for windage. Aim and fire at a target that is at a 25-yard distance. Adjust the scope until rounds are impacting 5/8 of an inch above the point of aim. To confirm, set the elevation to 500 meters. The rounds should impact 2 1/4 inches above the point of aim.

(2) ***Observation of impact method.*** When a known distance range is unavailable, locate a target so that the observer can see the impact of

rounds clearly. Determine the exact range to the target, dial in the appropriate range, and fire. Watch the impact of the rounds; the observer gives the sight adjustments until a point of aim or point of impact is achieved.

2-11. IRON SIGHTS

Depending on the situation, a sniper may be required to deliver an effective shot at ranges up to 900 meters or more. This requires the sniper to zero his rifle with the iron sights and the M3A scope at most ranges that he can be expected to fire.

a. **Mounting.** To mount iron sights, the sniper must remove the M3A scope first.

(1) Attach the front sight to the barrel, align the front sight and the front sight base, and slide the sight over the base and tighten the screw (Figure 2-22).

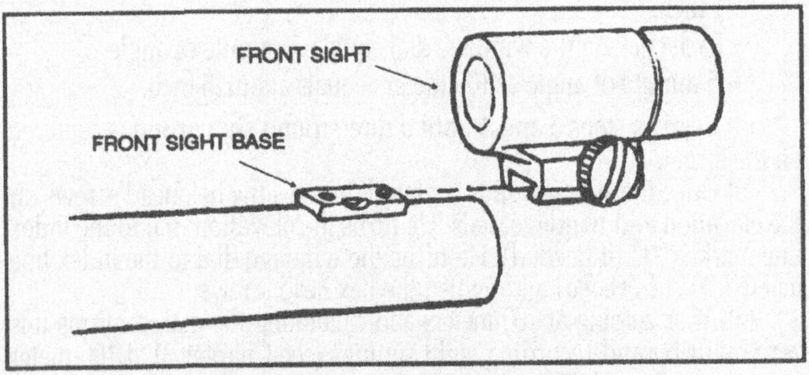

Figure 2-22. Front sight attachment.

(2) The aperture insert may be either skeleton or translucent plastic (Figure 2-23, page 2-30). The skeleton aperture is the most widely used. The translucent plastic aperture is preferred by some shooters and is available in clear plastic. Both apertures are available in various sizes. A common error is selecting an aperture that is too small. Select an aperture that appears to be at least twice the diameter of the bull's-eye. An aperture selected under one light condition may, under a different light, form a halo around the bull or make the bull appear indistinct or oblong. The aperture selected should reveal a wide line of white around the bull and allow the bull to standout in clear definition against this background.

Figure 2-23. Aperture Insert.

(3) Remove one of the three sets of screws from the rear sight base located on the left rear of the receiver. Align the rear sight with the rear sight base taking care to use the hole that provides the operator the desired eye relief. Then tighten the screw to secure the rear sight to the base.

NOTE: Operator-desired eye relief determines the set screw that must be removed.

b. **Adjustment Scales.** Adjustment scales are of the vernier type. Each graduation on the scale inscribed on the sight base equals 3 minutes of angle. (See the minutes of angle chart in Chapter 3.) Each graduation of the adjustable scale plates equals 1 minute of angle. To use the vernier-type adjustment scales—

(1) Note the point at which graduations on both the top and the bottom scales are aligned.

(2) Count the numbers of full 3 minutes of angle graduations from "0" on the fixed scale to "0" on the adjustable scale. Add this figure to the number of 1 minute of angle graduations from "0" on the adjustable scale to the point where the two graduations are aligned.

c. Zeroing. Zeroing iron sights should be done on the same type of range and targets as in paragraph 2-10a. To set a mechanical zero on the iron sights for windage, the sniper turns the windage dial all the way to the left or right, then he counts the number of clicks it takes to get from one side to the other. He divides this number by 2—for example, 120 divided by 2 equals 60. The sniper turns the windage dial 60 clicks back to the center. If the two zeros on the windage indicator plate do not align, he loosens the screw on the windage indicator plate and aligns the two zeros. The sniper uses the same procedure to set a mechanical zero for elevation. Once a mechanical zero has been set, he assumes a good prone-supported position, 100 meters from the target. He fires three rounds at the center of the target, observing the same aiming point each time. After noting the strike of the rounds, the sniper turns the *elevation* and *windage* dials to make needed adjustments to the iron sights as follows (Figure 2-24):

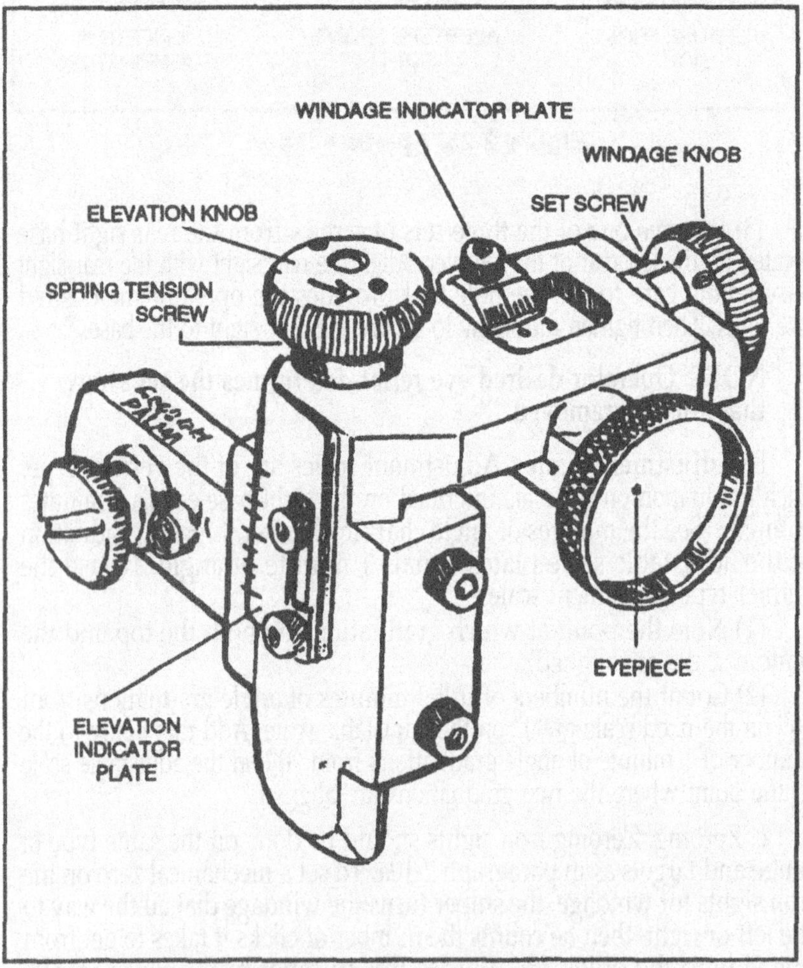

Figure 2-24. Zeroing adjustment dials.

(1) Each click of adjustment is 1/4 minute of angle (one minute of angle equals about 1 inch at 100 yards, 6 inches at 600 yards, and so forth). There are twelve 1/4 minutes of angle, equaling 3 minutes of angle adjustments in each dial revolution. The total elevation adjustment is 60 minutes of angle (600 inches at 1,000 yards) total windage adjustment is 36 minutes of angle (360 inches at 1,000 yards).

(2) Turn the elevation dial in the direction marked *UP* to raise the point of impact: turn the elevation dial in the opposite direction to lower the point of impact. Turn the windage dial in the direction marked *R* to move the point of impact to the right; then turn the windage dial in the opposite direction to move the point of impact to the left.

(3) Continue firing and adjusting shot groups until the point of aim or point of impact is achieved.

After zeroing the rifle sight to the preferred range, the sniper loosens the elevation and windage indicator plate screws with the socket head screw key provided. Now, he loosens the spring tension screw, aligns the "0" on the plate with the "0" on the sight body, and retightens the plate screws. Then the sniper loosens the spring tension screws and set screws in each dial, and aligns the "0" of the dial with the reference line on the sight. He presses the dial against the sight, tightens the set screws, and equally tightens the spring tension screws until a definite "click" can be felt when the dial is turned. This click can be sharpened or softened to preference by equally loosening or tightening the spring screws on each dial. The sniper makes windage and elevation corrections, and returns quickly to "zero" standard.

Section IV
OTHER EQUIPMENT

The sniper must use special equipment to reduce the possibility of detection. The types and characteristics are discussed in this section.

2-12. M16A1/A2 RIFLE WITH M203 GRENADE LAUNCHER

The observer carries the M16A1/A2 rifle with the M203 grenade launcher. The sniper, carrying the M24 SWS, lacks the firepower required to break contact with enemy forces-that is, ambush or chance contact. The rapid-fire ability of the M16A1/A2 rifle, combined with the destructive abilities of the M203 40-mm grenade launcher (Figure 2-25), gives the sniper team a lightweight, easily operated way to deliver the firepower required to break contact. (See FM 23-9 and FM 23-31, respectively, for the technical characteristics of these weapons.)

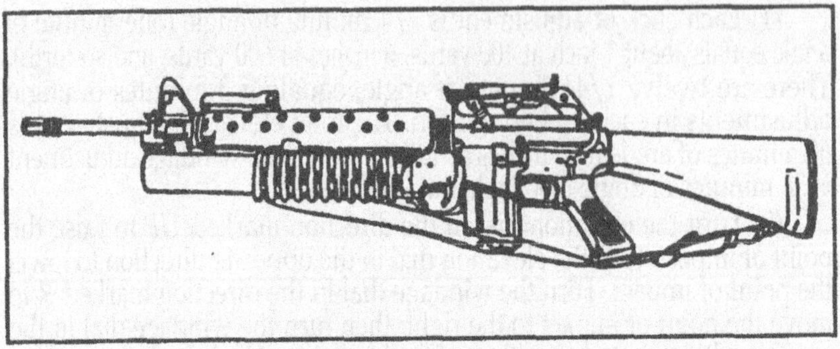

Figure 2-25. The M203 40-mm grenade launcher attached to M16A1 rifle.

2-13. IMAGE INTENSIFICATION AND INFRARED DEVICES

The sniper team employs night and limited visibility devices to conduct continuous operations.

a. **Night Vision Sight, AN/PVS-4.** The AN/PVS-4 is a portable, battery-operated, electro-optical instrument that can be hand-held for visual observation or weapon-mounted for precision fire at night (Figure 2-26). The observer can detect and resolve distant targets through the unique capability of the sight to amplify reflected ambient light (moon, stars, or sky glow). The sight is passive thus, it is free from enemy detection by visual or electronic means. This sight, with appropriate weapons adapter bracket, can be mounted on the M16 rifle.

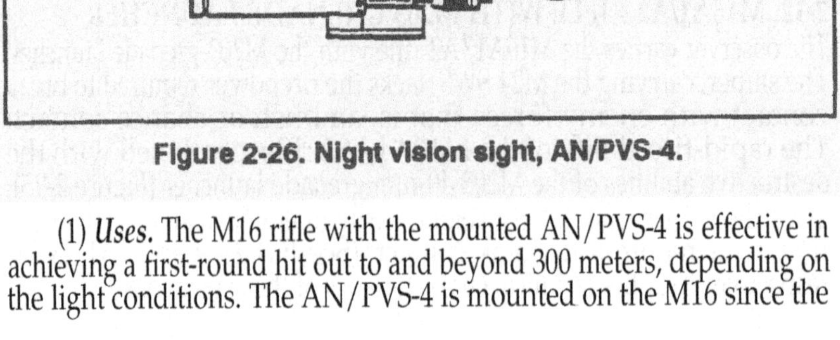

Figure 2-26. Night vision sight, AN/PVS-4.

(1) ***Uses.*** The M16 rifle with the mounted AN/PVS-4 is effective in achieving a first-round hit out to and beyond 300 meters, depending on the light conditions. The AN/PVS-4 is mounted on the M16 since the

nightsight's limited range does not make its use practical for the sniper weapon system. This avoids problems that may occur when removing and replacing the sniperscope. The nightsight provides an effective observation ability during night combat operations. The sight does not give the width, depth, or clarity of daylight vision; however, a well-trained operator can see enough to analyze the tactical situation, to detect enemy targets, and to place effective fire on them. The sniper team uses the AN/PVS-4 to accomplish the following:

(a) To enhance their night observation capability.

(b) To locate and suppress hostile fire at night.

(c) To deny enemy movement at night.

(d) To demoralize the enemy with effective first-round kills at night.

(2) ***Employment factors.*** Since the sight requires target illumination and does not project its own light source, it will not function in total darkness. The sight works best on a bright, moonlit night. When there is no light or the ambient light level is low (such as in heavy vegetation), the use of artificial or infrared light improves the sight's performance.

(a) Fog, smoke, dust, hail, or rain limit the range and decrease the resolution of the instrument.

(b) The sight does not allow seeing through objects in the field of view. For example, the operator will experience the same range restrictions when viewing dense wood lines as he would when using other optical sights.

(c) The observer may experience eye fatigue when viewing for prolonged periods. Viewing should be limited to 10 minutes, followed by a rest period of 10 minutes. After several periods of viewing, he can safely extend this time limit. To assist in maintaining a continuous viewing. capability and to reduce eye fatigue, the observer should use one eye then the other while viewing through the sight.

(3) **Zeroing.** The operator may zero the sight during daylight or darkness; however, he may have some difficulty in zeroing just 1 before darkness. The light level at dusk is too low to permit the operator to resolve his zero target with the lens cap cover in place, but it is still intense enough to cause the sight to automatically turnoff unless the lens cap cover is in position over the objective lens. The sniper normally zeros the sight for the maximum practical range that he can be expected to observe and fire, depending on the level of light.

b. **Night Vision Goggles, AN/PVS-5.** The AN/PVS-5 is a lightweight, passive night vision system that gives the sniper team another means of observing an area during darkness (Figure 2-27). The sniper normally carries the goggles, because the observer has the M16 mounted with the nightsight. The goggles make it easier to see due to their design. However, the same limitations that apply to the nightsight also apply to the goggles.

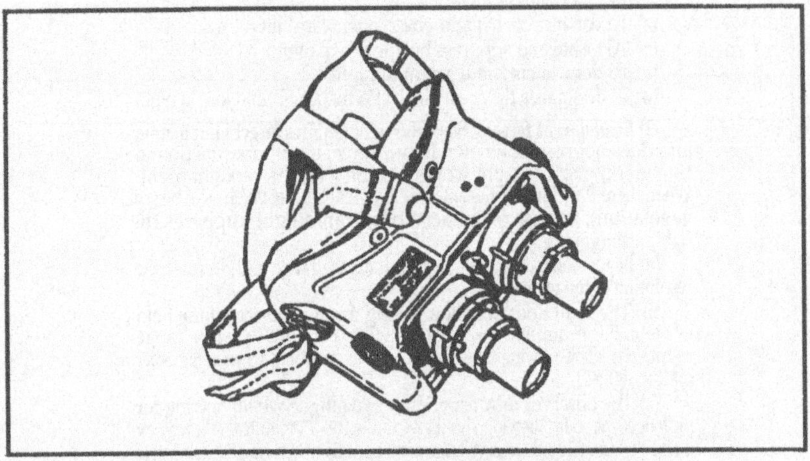

Figure 2-27. Night vision goggles, AN/PVS-5.

c. **Night Vision Goggles, AN/PVS-7 Series.** The night vision goggles, AN/PVS-7 series (Figure 2-28, page 2-36) has a better resolution and viewing ability than the AN/PVS-5 goggles. The AN/PVS-7 series goggles have a head-mount assembly that allows them to be mounted in front of the face so that both hands can be free. The goggles can be used without the mount assembly for hand-held viewing. (See TM 11-5855-262-10-1.)

d. **Laser Observation Set AN/GVS-5.** Depending on the mission, snipers can use the AN/GVS-5 to determine the range to the target. The AN/GVS-5 (LR) (Figure 2-29, page 2-36) is an individually operated, hand-held, distance-measuring device designed for distances from 200 to 9,990 meters (with an error of plus or minus 10 meters). It measures distances by firing an infrared beam at a target and by measuring the time the reflected beam takes to return to the operator. It then displays the target distance, in meters, inside the viewer. The reticle pattern in the viewer is graduated in 10-mil increments and has display lights to indicate low battery and multiple target hits. If the beam hits more than one target, the display gives a reading of the closest target hit. The beam that is fired from the set poses a safety hazard; therefore, snipers planning to use this equipment should be thoroughly trained in its safe operation. (See TM 11-5860-201-10.)

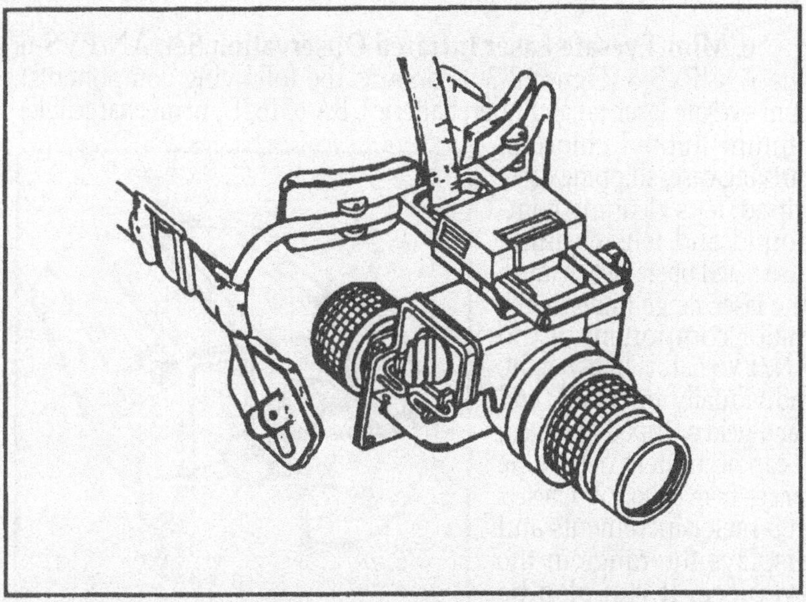

Figure 2-28. Night vision goggles, AN/PVS-7 series.

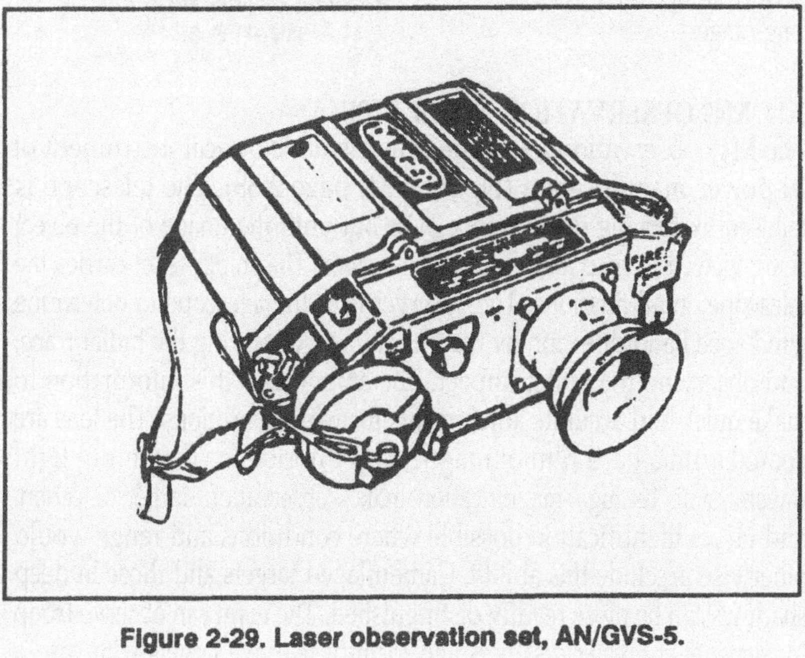

Figure 2-29. Laser observation set, AN/GVS-5.

e. **Mini-Eyesafe Laser Infrared Observation Set, AN/PVS-6.** The AN/PVS-6 (Figure 2-30) contains the following components: mini-eyesafe laser range finder; batteries, BA-6516/U, nonrechargeable, lithium thionyl chloride; carrying case; shipping case; tripod; lens cleaning compound and lens cleaning tissue; and operator's manual. The laser range finder is the major component of the AN/PVS-6. It is lightweight, individually operated, and hand-held or tripod mounted; it can accurately determine ranges from 50 to 9,995 meters in 5-meter increments and displays the range in the eyepiece. It can also be mounted with and boresighted to the night observation device, AN/TAS-6, long-range.

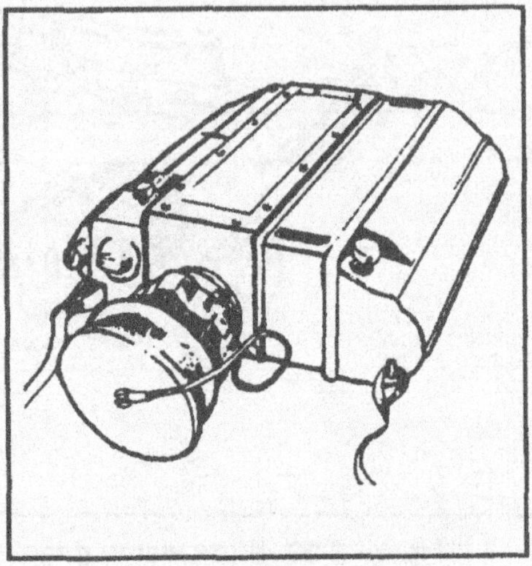

Figure 2-30. Mini-eyesafe laser infrared observation set, AV/PVS-6.

2-14. M49 OBSERVATION TELESCOPE

The M49 observation telescope is a prismatic optical instrument of 20-power magnification (Figure 2-31, page 2-38). The telescope is focused by turning the eyepiece in or out until the image of the object being viewed is crisp and clear to the viewer. The sniper team carries the telescope on all missions. The observer uses the telescope to determine wind speed and direction by reading mirage, observing the bullet trace, and observing the bullet impact. The sniper uses this information to make quick and accurate adjustments for wind conditions. The lens are coated with a hard film of magnesium fluoride for maximum light transmission. Its high magnification makes observation, target detection, and target identification possible where conditions and range would otherwise preclude this ability. Camouflaged targets and those in deep shadows can be more readily distinguished. The team can observe troop movements at greater distances and identify selective targets with ease.

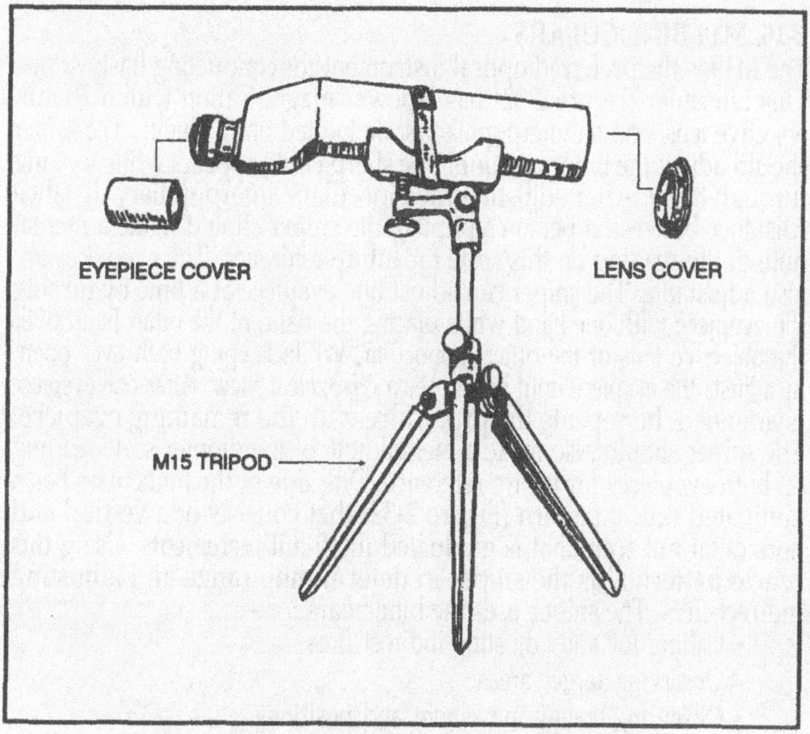

Figure 2-31. M49 observation telescope.

a. **Components.** Components of the telescope include a removable eyepiece and objective lens covers, an M15 tripod with canvas carrier, and a hard ease carrier for the telescope.

b. **Storage.** When storing the M49 observation telescope, the sniper must remove it from the hard case earner and remove the lens caps to prevent moisture from gathering on the inside of the scope. Maintenance consists of—

(1) Wiping dirt and foreign materials from the scope tube, hard case carrier, and M15 tripod with a damp rag.

(2) Cleaning the M49 lens with lens cleaning solution and lens tissue only.

(3) Brushing dirt and foreign agents from the M15 carrying case with a stiff-bristled brush; cleaning the threading of lens caps on the M49 and the tripod elevation adjustment screw on the M15 with a toothbrush, then applying a thin coat of grease and moving the lens caps and elevation adjustment screw back and forth to evenly coat threading.

2-15. M19 BINOCULARS

The M19 is the preferred optical instrument for conducting hasty scans. This binocular (Figure 2-32) has 7-power magnification with a 50-mm objective lens, and an interpupillary scale located on the hinge. The sniper should adjust the binocular until one sharp circle appears while looking through them. After adjusting the binoculars' interpupillary distance (distance between a person's pupils), the sniper should make a mental note of the reading on this scale for future reference. The eyepieces are also adjustable. The sniper can adjust one eyepiece at a time by turning the eyepiece with one hand while placing the palm of the other hand over the objective lens of the other monocular. While keeping both eyes open, he adjusts the eyepiece until he can see a crisp, clear view. After one eyepiece is adjusted, he repeats the procedure with the remaining eyepiece. The sniper should also make a mental note of the diopter scale reading on both eyepieces for future reference. One side of the binoculars has a laminated reticle pattern (Figure 2-32) that consists of a vertical and horizontal mil scale that is graduated in 10-mil increments. Using this reticle pattern aids the sniper in determining range and adjusting indirect-fires. The sniper uses the binoculars for—

- Calling for and adjusting indirect fires.
- Observing target areas.
- Observing enemy movement and positions.
- Identifying aircraft.
- Improving low-light level viewing.
- Estimating range.

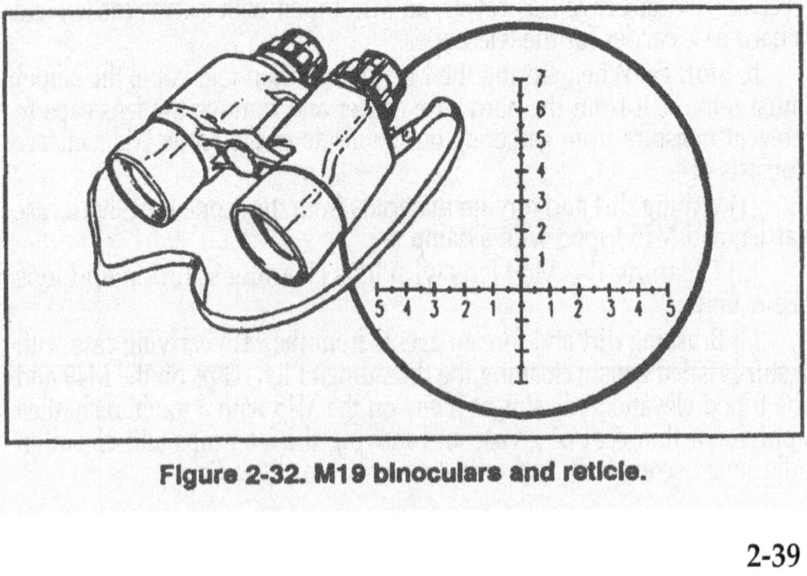

Figure 2-32. M19 binoculars and reticle.

2-16. M22 BINOCULARS

The M22 binoculars (Figure 2-33) can be used instead of the M19. These binoculars have the same features as the M19, plus fold-down eyepiece cups for personnel who wear glasses to reduce the distance between the eyes and the eyepiece. It also has protective covers for the objective and eyepiece lenses. The binoculars have laser protection filters on the inside of the objective lenses (direct sunlight can reflect off these lenses). The reticle pattern (Figure 2-33) is different than the M19 binocular reticle.

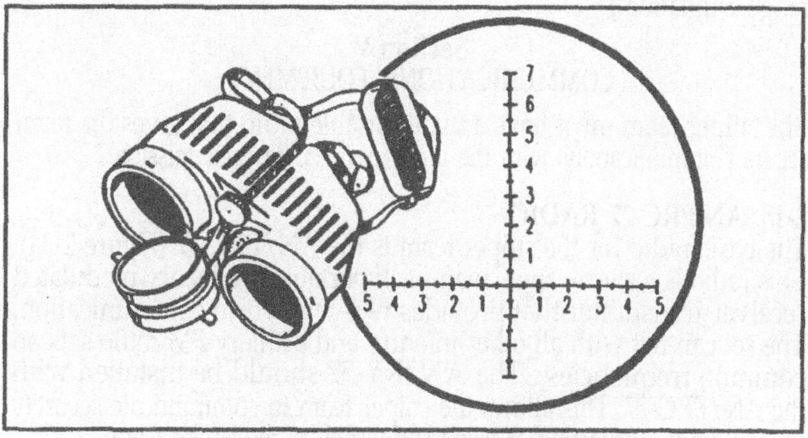

Figure 2-33. M22 binoculars and reticle.

2-17. OTHER SNIPER EQUIPMENT

Other equipment the sniper needs to complete a successful mission follows:

a. **Sidearms.** Each member of the team should have a sidearm, such as an M9, 9-mm Beretta, or a caliber .45 pistol. A sidearm gives a sniper the needed protection from a nearby threat while on the ground moving or while in the confines of a sniper position.

b. **Compass.** Each member of the sniper team must have a lensatic compass for land navigation.

c. **Maps.** The team must have military maps of their area of operations.

d. **Calculator.** The sniper team needs a pocket-size calculator to figure distances when using the mil-relation formula. Solar-powered calculators usually work well, but under low-light conditions, battery power may be preferred. If a battery-powered calculator is to be used in low-light conditions, it should have a lighted display.

e. **Rucksack.** The sniper's rucksack should contain at least a two-quart canteen, an entrenching tool, a first-aid kit, pruning shears, a sewing kit with canvas needles and nylon thread, spare netting and garnish, rations, and personal items as needed. The sniper also carries his ghillie suit (Chapter 4, paragraph 4-4) in his rucksack until the mission requires its use.

f. **Measuring Tape.** A standard 10-foot to 25-foot metal carpenter's tape allows the sniper to measure items in his operational area. This information is recorded in the sniper data book. (See Chapter 4 for range estimation.)

Section V
COMMUNICATIONS EQUIPMENT

The sniper team must have a man-portable radio that gives the team secure communications with the units involved in their mission.

2-18. AN/PRC-77 RADIO

The basic radio for the sniper team is the AN/PRC-77 (Figure 2-34). This radio is a short-range, man-pack portable, frequency modulated receiver-transmitter that provides two-way voice communication. The set can net with all other infantry and artillery FM radio sets on common frequencies. The AN/KY-57 should be installed with the AN/PRC-77. This allows the sniper team to communicate securely with all units supporting or being supported by the sniper team.

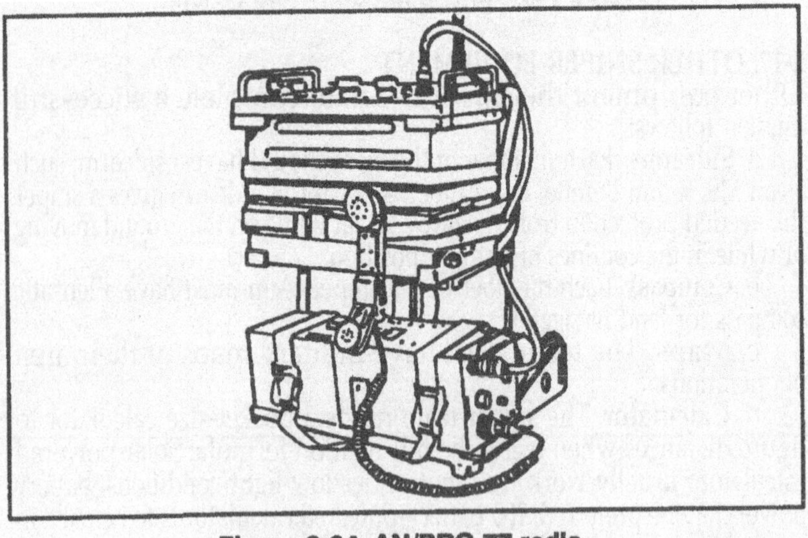

Figure 2-34. AN/PRC-77 radio.

2-19. AN/PRC-104A RADIO TRANSCEIVER

The AN/PRC-104A is a state-of-the-am lightweight radio transceiver that operates in the high frequency and in the upper part of the low frequency portions of the radio spectrum (Figure 2-35). The receiver/transmitter circuits can be tuned to any frequency between 2.0000 and 29.9999 MHz in 100 Hz increments, making it possible to tune up to 280,000 separate frequencies. The radio operates in the upper or lower side bank modes for voice communications, CW for Morse code, or FSK (frequency-shift keying) for transmission of teletype or other data.

a. In the man-pack configuration, the radio set is carried and operated by one man or, with the proper accessories, it can be configured for vehicle or fixed-station use. The radio set with antenna and handset weighs 15.7 pounds.

b. The control panel, human-engineered for ease of operation, makes it possible to adjust all controls even while wearing heavy gloves. Unlike older, similar radio sets, there are no front panel meters or indicator lights on the AN/PRC-104A. All functions that formerly required these types of indicators are monitored by the radio and communicated to the operator as special tones in the handset. This feature is highly useful during tactical blackout operations. The superior design and innovative features of the AN/PRC-104A radio set make it possible to maintain a reliable long-range communications link. The radio uses lightweight, portable equipment that can be operated by personnel who have minimum training.

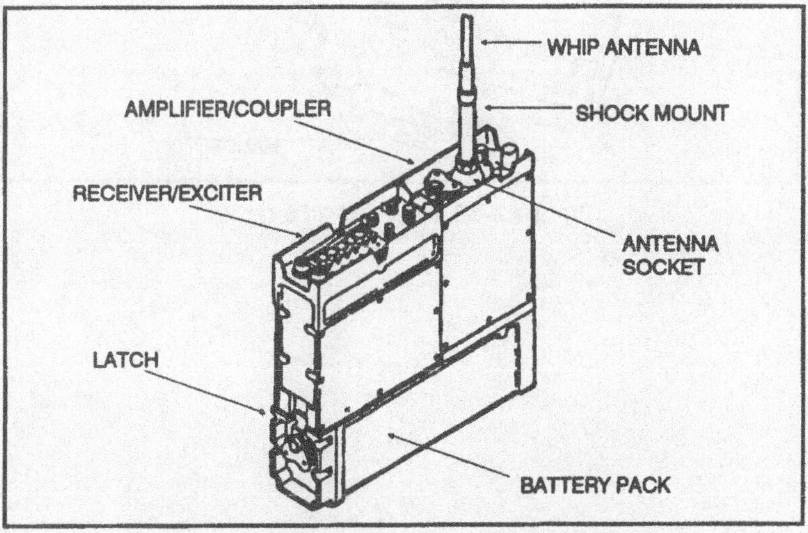

Figure 2-35. AN/PRC-104A radio transceiver.

2-20. AN/PRC-119 RADIO

The AN/PRC-119 (Figure 2-36) replaces the AN/PRC-77, although the AN/PRC-77 is still in use. The AN/PRC-l19 is a man-pack portable, VHF/FM radio that is designed for simple, quick operation using a 16-element keypad for push-button tuning. It can also be used for short-range and long-range operation for voice, FSK, or digital data communications. It can also be used for single-channel operation or in a jam-resistant, frequency-hopping mode, which can be changed as needed This radio has a built-in self-test with visual and audio readbacks. It is compatible with the AN/KY-57 for secure communications.

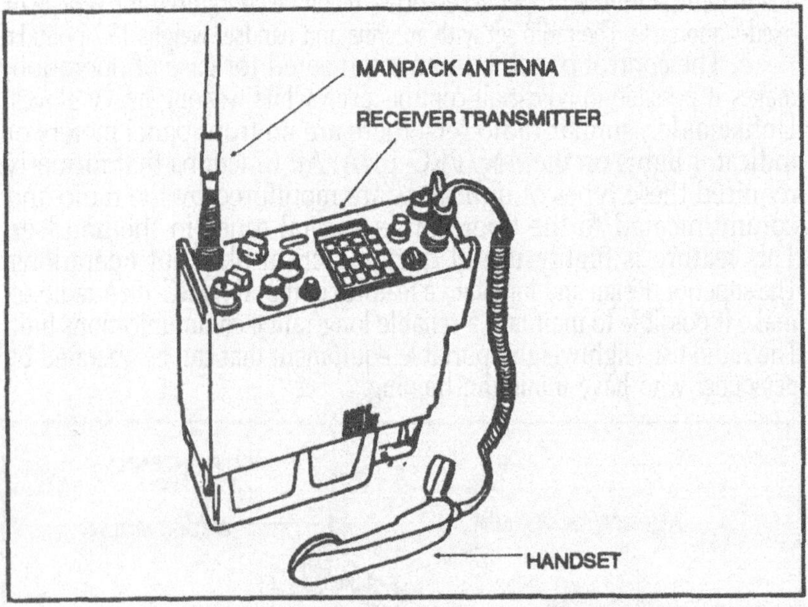

Figure 2-36. AN/PRC-119 radio.

CHAPTER 3

MARKSMANSHIP

Sniper marksmanship is an extension of basic rifle marksmanship and focuses on the techniques needed to engage targets at extended ranges. To successfully engage targets at increased distances, the sniper team must be proficient in marksmanship fundamentals and advanced marksmanship skills. Examples of these skills are determining the effects of weather conditions on ballistics, holding off for elevation and windage, engaging moving targets, using and adjusting scopes, and zeroing procedures. Markmanship skills should be practiced often.

Section I
FUNDAMENTALS

The sniper team must be thoroughly trained in the fundamentals of marksmanship. These include assuming a position, aiming, breath control, and trigger control. These fundamentals develop fixed and correct firing habits for instinctive application. Every sniper should periodically refamiliarize himself with these fundamentals regardless of his experience.

3-1. STEADY POSITION ELEMENTS

The sniper should assume a good firing position (Figure 3-1, page 3-2) in order to engage targets with any consistency. A good position enables the sniper to relax and concentrate when preparing to fire.

a. **Position Elements.** Establishing a mental checklist of steady position elements enhances the sniper's ability to achieve a first-round hit.

(1) ***Nonfiring hand.*** Use the nonfiring hand to support the butt of the weapon. Place the hand next to the cheat and rest the tip of the butt on it. Bail the hand into a fist to raise the weapon's butt or loosen the fist

to lower the weapon's butt. An effective method is to hold a sock full of sand in the nonfiring hand and to place the weapon butt on the sock. This reduces body contact with the weapon. To raise the butt, squeeze the sock and to lower it, loosen the grip on the sock.

(2) ***Butt of the stock.*** Place the butt of the stock firmly in the pocket of the shoulder. Insert a pad on the ghillie suit (see Chapter 4) where contact with the butt is made to reduce the effects of pulse beat and breathing, which can be transmitted to the weapon.

(3) ***Firing hand.*** With the firing hand, grip the small of the stock. Using the middle through little fingers, exert a slight rearward pull to keep the butt of the weapon firmly in the pocket of the shoulder. Place the thumb over the top of the small of the stock. Place the index finger on the trigger, ensuring it does not touch the stock of the weapon. This avoids disturbing the lay of the rifle when the trigger is squeezed.

(4) ***Elbows.*** Find a comfortable position that provides the greatest support.

(5) ***Stock weld.*** Place the cheek in the same place on the stock with each shot. A change in stock weld tends to cause poor sight alignment, reducing accuracy.

(6) ***Bone support.*** Bone support is the foundation of the firing position; they provide steady support of the weapon.

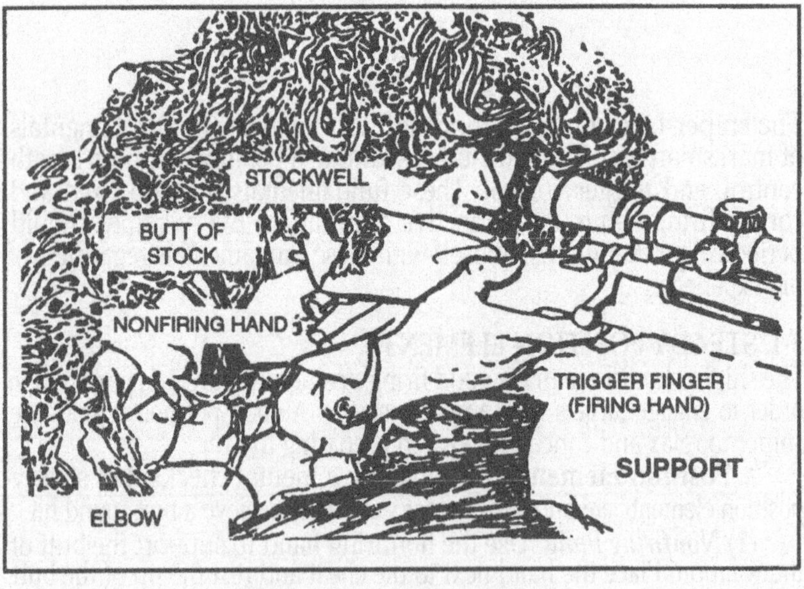

Figure 3-1. Firing position.

(7) ***Muscle relaxation.*** When using bone support, the sniper can relax muscles, reducing any movement that could be caused by tense or trembling muscles. Aside from tension in the trigger finger and firing hand, any use of the muscle generates movement of the sniper's cross hairs.

(8) ***Natural point of aim.*** The point at which the rifle naturally rest in relation to the aiming point is called natural point of aim.

(a) Once the sniper is in position and aimed in on his target, the method for checking for natural point of aim is for the sniper to close his eyes, take a couple of breaths, and relax as much as possible. Upon opening his eyes, the scope's cross hairs should be positioned at the sniper's preferred aiming point. Since the rifle becomes an extension of the sniper's body, it is necessary to adjust the position of the body until the rifle points naturally at the preferred aiming point on the target.

(b) Once the natural point of aim has been determined, the sniper must maintain his position to the target. To maintain his natural point of aim in all shooting positions, the natural point of aim can be readjusted and checked periodically.

(c) The sniper can change the elevation of the natural point of aim by leaving his elbows in place and by sliding his body forward or rearward. This raises or lowers the muzzle of the weapon, respectively. To maintain the natural point of aim after the weapon has been fired, proper bolt operation becomes critical. The sniper must practice reloading while in the prone position without removing the butt of the weapon from the firing shoulder. This may be difficult for the left-hand firer. The two techniques for accomplishing this task are as follows:

- After firing, move the bolt slowly to the rear while canting the weapon to the right. Execution of this task causes the spent cartridge to fall next to the weapon.
- After firing, move the bolt to the rear with the thumb of the firing hand. Using the index and middle fingers, reach into the receiver and catch the spent cartridge as it is being ejected. This technique does not require canting the weapon.

NOTE: The sniper conducts bolt operation under a veil or equivalent camouflage to improve concealment.

b. **Steady Firing Position.** On the battlefield, the sniper must assume a steady firing position with maximum use of cover and concealment. Considering the variables of terrain, vegetation, and tactical situations, the sniper can use many variations of the basic positions. When assuming a firing position, he must adhere to the following basic rules:

(1) Use any support available.

(2) Avoid touching the support with the barrel of the weapon since it interferes with barrel harmonics and reduces accuracy.

(3) Use a cushion between the weapon and the support to prevent slippage of the weapon.

(4) Use the prone supported position whenever possible.

c. **Types of Firing Positions.** Due to the importance of delivering precision fire, the sniper makes maximum use of artificial support and eliminates any variable that may prevent adhering to the basic rules. He uses the prone supported; prone unsupported; kneeling unsupported; kneeling, sling supported; standing supported; and the Hawkins firing positions.

(1) ***Prone supported position.*** The prone supported position is the steadiest position; it should be used whenever possible (Figure 3-2). To assume the prone supported position, the sniper should—

(a) Lie down and place the weapon on a support that allows pointing in the direction of the target. Keep the position as low as possible. (For field-expedient weapon supports, see paragraph 3-1d.)

(b) Remove the nonfiring hand from underneath the fore-end of the weapon by folding the arm underneath the receiver and trigger, grasping the rear sling swivel. This removes any chance of subconsciously trying to exert control over the weapon's natural point of aim. Keep the elbows in a comfortable position that provides the greatest support.

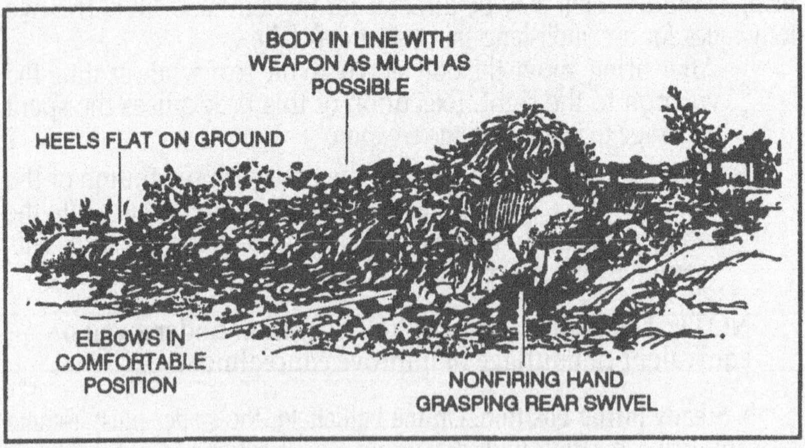

Figure 3-2. Prone supported position.

(c) Keep the body in line with the weapon as much as possible-not at an angle. This presents less of a target to the enemy and more body mass to absorb recoil.

(d) Spread legs a comfortable distance apart with the heels on the ground or as close as possible without causing strain.

(2) ***Prone unsupported position.*** The prone unsupported position (Figure 3-3) offers another stable firing platform for engaging targets. To assume this position, the sniper faces his target, spreads his feet a comfortable distance apart, and drops to his knees. Using the butt of the rifle as a pivot, the firer rolls onto his nonfiring side. He places the rifle butt in the pocket formed by the firing shoulder, grasps the pistol grip in his firing hand, and lowers the firing elbow to the ground. The rifle rests in the V formed by the thumb and fingers of the nonfiring hand The sniper adjusts the position of his firing elbow until his shoulders are about level, and pulls back firmly on the rifle with both hands. To complete the position, he obtains a stock weld and relaxes, keeping his heels close to the ground.

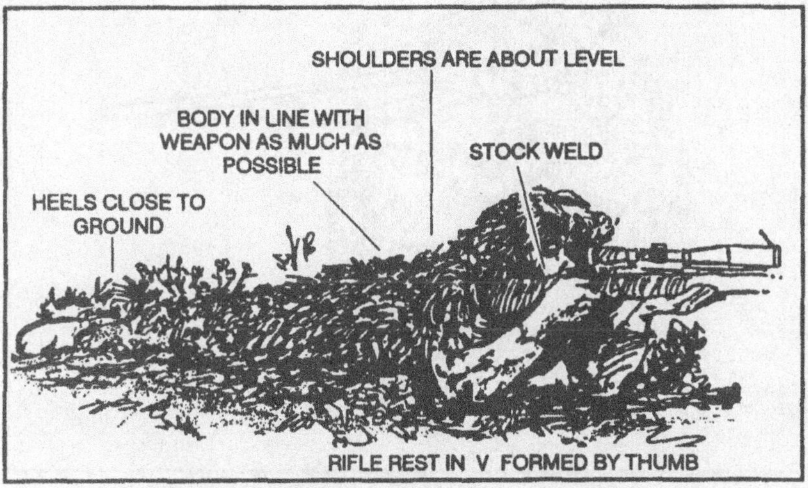

Figure 3-3. Prone unsupported position.

(3) ***Kneeling unsupported position.*** The kneeling unsupported position (Figure 3-4, page 3-6) is assumed quickly. It places the sniper high enough to see over small brush and provides for a stable position.

(a) Place the body at a 45-degree angle to the target.

(b) Kneel and place the right knee on the ground.

(c) Keep the left leg as perpendicular to the ground as possible; sit back on the right heel, placing it as directly under the spinal column as possible. A variation is to turn the toe inward and sit squarely on the right foot.

(d) Grasp the small of the stock of the weapon with the firing hand, and cradle the fore-end of the weapon in a crook formed with the left arm.

(e) Place the butt of the weapon in the pocket of the shoulder, then place the meaty underside of the left elbow on top of the left knee.

(f) Reach under the weapon with the left hand, and lightly grasp the firing arm.

(g) Relax forward and into the support position, using the left shoulder as a contact point. This reduces transmission of the pulsebeat into the sight picture.

(h) Lean against a tree, building, or vehicle for body support.

Figure 3-4. Kneeling unsupported position.

(4) ***Kneeling, sling supported position.*** If vegetation presents a problem, the sniper can raise his kneeling position by using the rifle sling. To assume the kneeling, sling supported position, he executes the first three steps for assuming a kneeling unsupported position. With the leather sling mounted to the weapon, the sniper turns the sling one-quarter turn to the left. The lower part of the sling will then form a loop.

(a) Place the left arm (nonfiring) through the loop; pull the sling up the arm and place it on the upper arm between the elbow and shoulder, but not directly over the biceps.

(b) Tighten the sling by sliding the sling keeper against the loop holding the arm.

(c) Rotate the left arm in a clockwise motion around the sling and under the rifle with the sling secured to the upper arm. Place the fore-end of the stock in the V formed by the thumb and forefinger of the left hand. Relax the left arm and hand, let the sling support the weight of the weapon.

(d) Place the butt of the rifle against the right shoulder and place the left elbow on top of the left knee (Figure 3-5). Pull the left hand back along the fore-end of the rifle toward the trigger guard to add to stability.

Figure 3-5. Kneeling, sling supported position.

(5) *Standing supported position.* The standing supported position is the least steady of the supported positions and should be used only as a last resort (Figure 3-6, page 3-8).

(a) To assume the standing supported position with horizontal support, such as a wall or ledge, the sniper proceeds as follows:

- Locate a solid object for support. Avoid branches as they tend to sway when wind is present.

- Form a V with the thumb and forefinger of the nonfiring hand.
- Place the nonfiring hand against the support with the fore-end of the weapon resting in the V of the hand. This steadies the weapon and allows quick recovery from recoil.
- Then place the butt of the weapon in the pocket of the shoulder.

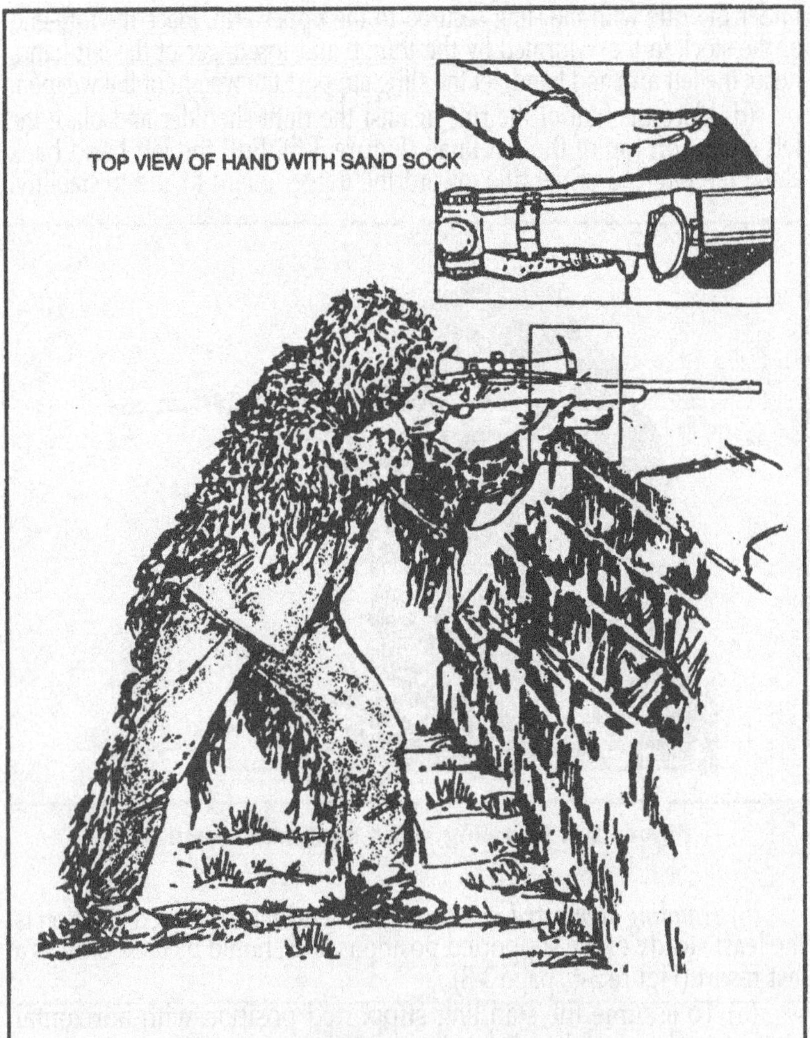

Figure 3-6. Standing supported position (horizontal support).

(b) To use vertical support (Figure 3-7), such as a tree, telephone pole, comer of building, or vehicle, the sniper proceeds as follows:

- Locate stable support. Face the target, then turn 45 degrees to the right of the target, and place the palm of the nonfiring hand at arm's length against the support.
- Lock the left arm straight, let the left leg buckle, and place body weight against the nonfiring hand. Keep the trail leg straight.
- Place the fore-end of the weapon in the V formed by extending the thumb of the nonfiring hand.
- Exert more pressure to the rear with the firing hand.

Figure 3-7. Standing supported position (vertical support).

(6) ***Hawkins position.*** The Hawkins position (Figure 3-8) is a variation of the prone unsupported position. The sniper uses it when firing from a low bank or a depression in the ground, over a roof, or so forth. It cannot be used on level ground since the muzzle cannot be raised high enough to aim at the target. It is a low-profile position with excellent stability and aids concealment. To assume this position, the sniper uses the weapon's sling and proceeds as follows:

CAUTION
LOCK THE NONFIRING ARM STRAIGHT OR THE FACE WILL ABSORB THE WEAPON'S RECOIL.

(a) After assuming a prone position, grasp the upper sling swivel and sling with the nonfiring hand, forming a fist to support the front of the weapon.

(b) Ensure the nonfiring arm is locked straight since it will absorb the weapon's recoil. Wearing a glove is advisable.

(c) Rest the butt of the weapon on the ground and place it *under* the firing shoulder.

The sniper can make minor adjustments in muzzle elevation by tightening or relaxing the fist of the nonfiring hand. If more elevation is required, he can place a support under the nonfiring fist.

Figure 3-8. Hawkins position.

d. **Field-Expedient Weapon Support.** Support of the weapon is critical to the sniper's success in engaging targets. Unlike a well-equipped firing range with sandbags for weapon support, the sniper can encounter situations where weapon support relies on common sense and imagination. The sniper should practice using these supports at every opportunity and select the one that best suits his needs. He must train as if in combat to avoid confusion and self-doubt. The following items are commonly used as field-expedient weapon supports

(1) ***Sand sock.*** The sniper needs the sand sock when delivering precision fire at long ranges. He uses a standard issue, olive-drab wool sock filled one-half to three-quarters full of sand and knotted off. He places it under the rear sling swivel when in the prone supported position for added stability (Figure 3-9). By limiting minor movement and reducing pulse beat, the sniper can concentrate on trigger control and aiming. He uses the nonfiring hand to grip the sand sock, rather than the rear sling swivel. The sniper makes minor changes in muzzle elevation by squeezing or relaxing his grip on the sock. He uses the sand sock as padding between the weapon and a rigid support also.

Figure 3-9. Sand sock.

(2) ***Rucksack.*** If the sniper is in terrain without any natural support, he may use his rucksack (Figure 3-10). He must consider the height and presence of rigid objects within the rucksack. The rucksack must conform to weapon contours to add stability.

Figure 3-10. Rucksack.

(3) *Sandbag.* The sniper can fill an empty sandbag (Figure 3-11) on site.

Figure 3-11. Sandbag.

(4) *Tripod.* The sniper can build a field-expedient tripod (Figure 3-12) by tying together three 12-inch long sticks (one thicker than the others) with 550 cord or the equivalent. When tying the sticks, he wraps the cord at the center point and leaves enough slack to fold the legs out into a triangular base. Then, he places the fore-end of the weapon between the three uprights.

(5) *Bipod.* The sniper can build a field-expedient bipod (Figure 3-12) by tying together two 12-inch sticks, thick enough to support the weight of the weapon. Using 550 cord or the equivalent, he ties the sticks at the center point, leaving enough slack to fold them out in a scissor-like manner. He then places the weapon between the two uprights. The bipod is not as stable as other field-expedient items, and it should be used only in the absence of other techniques.

(6) *Forked stake.* The tactical situation determines the use of the forked stake. Unless the sniper can drive a forked stake into the ground, this is the least desirable of the techniques; that is, he must use his nonfiring hand to hold the stake in an upright position (Figure 3-12). Delivering long-range precision fire is a near-impossibility due to the unsteadiness of the position.

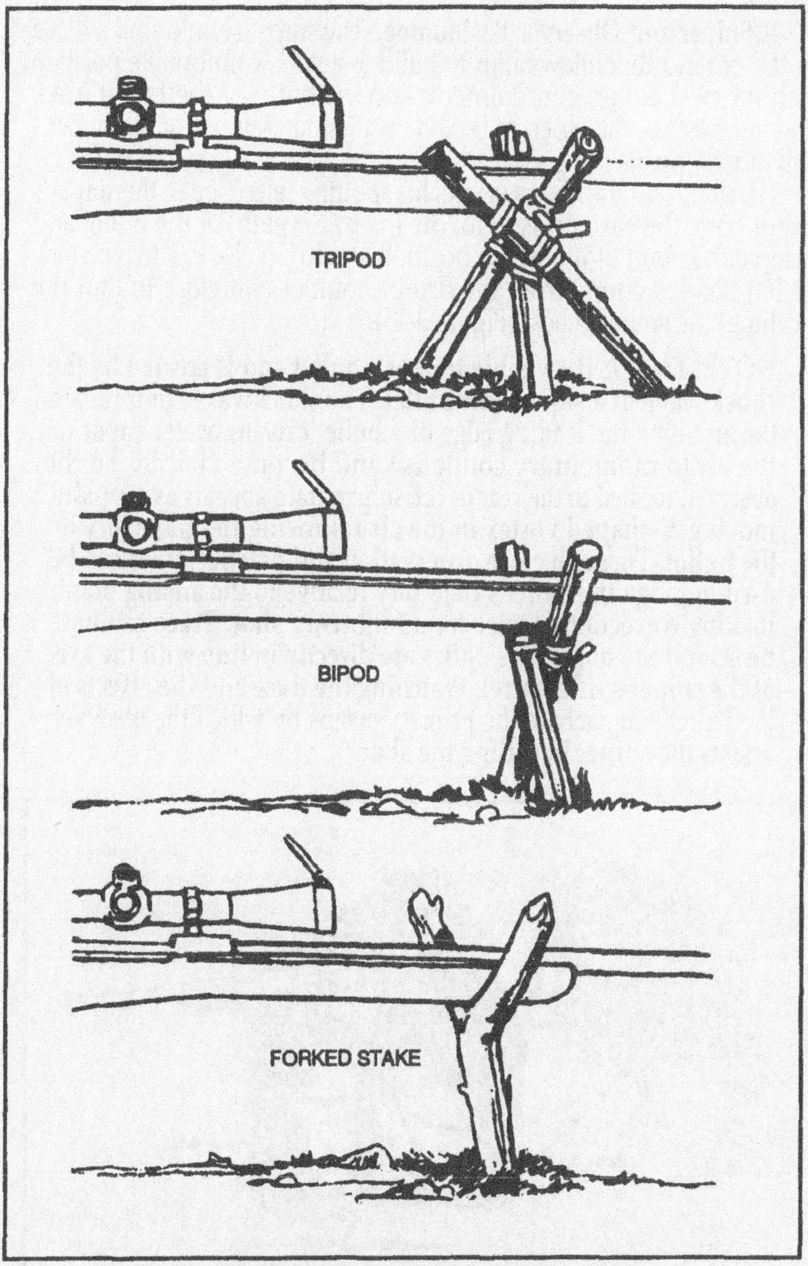

Figure 3-12. Field-expedient tripod, bipod, and forked stake.

e. **Sniper and Observer Positioning.** The sniper should find a place on the ground that allows him to build a steady, comfortable position with the best cover, concealment, and visibility of the target area. Once established, the observer should position himself out of the sniper's field of view on his firing side.

(1) The closer the observer gets his spotting telescope to the sniper's line of bore, the easier it is to follow the trace (path) of the bullet and observe the point of impact. A position at 4 to 5 o'clock (7 to 8 o'clock for left-handed firers) from the firing shoulder and close to (but not touching) the sniper is best (Figure 3-13).

NOTE: Trace is the visible trail of a bullet and is created by the shock wave of a supersonic bullet. The shockwave compresses the air along the leading edge of a bullet causing water vapor in the air to momentary condense and become visible. To the observer, located to the rear of the sniper, trace appears as a rapidly moving V-shaped vortex in the air following the trajectory of the bullet. Through close observation and practice, trace can be used to judge the bullet's trajectory relative to the aiming point, making corrections easier for a follow-up shot. Trace can best be seen if the observer's optics are directly in line with the axis of the sniper's rifle barrel. Watching the trace and the effects of the bullet's impact are the primary means by which the observer assists the sniper in calling the shot.

Figure 3-13. Sniper team positioning.

(2) If the sniper is without weapon support in his position, he uses the observer's body as a support (Figure 3-14). This support is not recommended since the sniper must contend with his own movement and the observer's body movement. The sniper should practice and prepare to use an observer supported position. A variety of positions can be used; however, the two most stable are when the observer is in a prone or sitting position.

(a) *Prone.* To assume the prone position, the observer lies at a 45-to 75-degree angle to the target and observes the area through his spotting telescope. The sniper assumes a a prone supported position, using the back of the observer's thigh for support. Due to the offset angle, the observer may only see the bullet impact.

Figure 3-14. Prone observer supported position.

(b) *Sitting.* If vegetation prevents the sniper from assuming a prone position, the sniper has the observer face the target area and assume a cross-legged sitting position. The observer places his elbows on his knees to stabilize his position. For observation, the observer uses binoculars held in his hands. The spotting telescope is not recommended due to its higher magnification and the unsteadiness of this position. The sniper is behind the observer in an open-legged, cross-legged, or kneeling position, depending on the target's elevation (Figure 3-15, page 3-16). The sniper places the fore-end of the weapon across the observer's left shoulder, stabilizing the weapon with the forefinger of the nonfiring hand. When using these positions, the sniper's effective engagement of targets at extended ranges is difficult and used only as a last resort. When practicing these positions, the sniper and observer must enter respiratory pause together to eliminate movement from breathing.

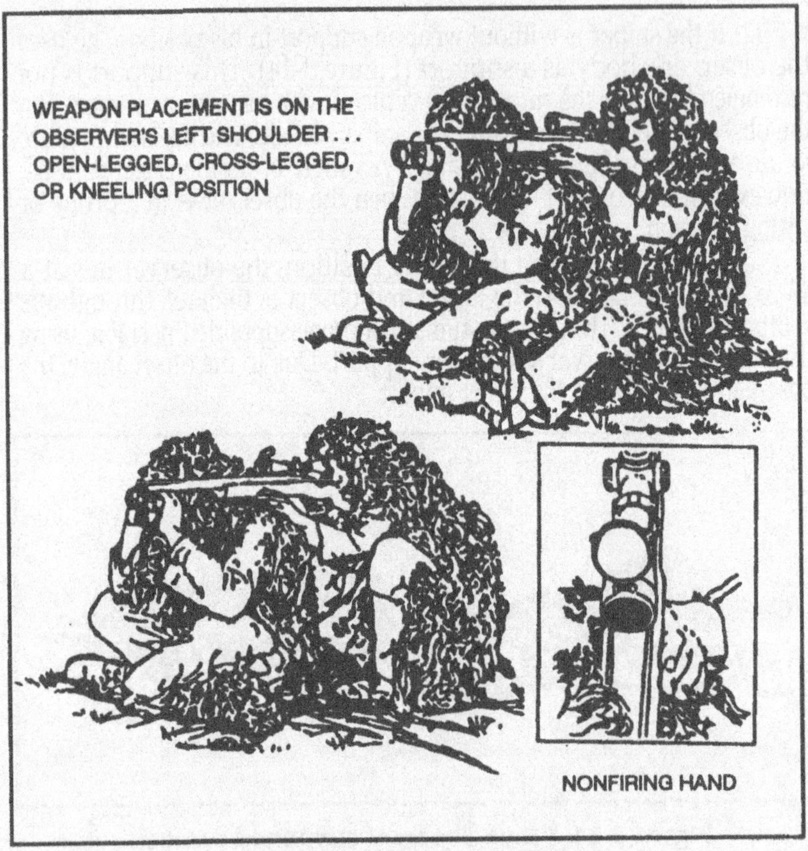

Figure 3-15. Sitting position.

3-2. AIMING

The sniper begins the aiming process by aligning the rifle with the target when assuming a firing position. He should point the rifle naturally at the desired point of aim. If his muscles are used to adjust the weapon onto the point of aim, they automatically relax as the rifle fires, and the rifle begins to move toward its natural point of aim. Because this movement begins just before the weapon discharge, the rifle is moving as the bullet leaves the muzzle. This causes inaccurate shots with no apparent cause (recoil disguises the movement). By adjusting the weapon and body as a single unit, rechecking, and readjusting as needed, the sniper achieves a true natural point of aim. Once the position is established, the sniper then aims the weapon at the exact point on the target. Aiming involves: eye relief, sight alignment, and sight picture.

a. **Eye Relief.** This is the distance from the sniper's firing eye to the rear sight or the rear of the scope tube. When using iron sights, the sniper ensures the distance remains consistent from shot to shot to preclude changing what he views through the rear sight. However, relief will vary from firing position to firing position and from sniper to sniper, according to the sniper's neck length, his angle of head approach to the stock, the depth of his shoulder pocket, and his firing position. This distance (Figure 3-16) is more rigidly controlled with telescopic sights than with iron sights. The sniper must take care to prevent eye injury caused by the scope tube striking his brow during recoil. Regardless of the sighting system he uses, he must place his head as upright as possible with his firing eye located directly behind the rear portion of the sighting system. This head placement also allows the muscles surrounding his eye to relax. Incorrect head placement causes the sniper to look out of the top or corner of his eye, resulting in muscular strain. Such strain leads to blurred vision and can also cause eye strain. The sniper can avoid eye strain by not staring through the telescopic or iron sights for extended periods. The best aid to consistent eye relief is maintaining the same stock weld from shot to shot.

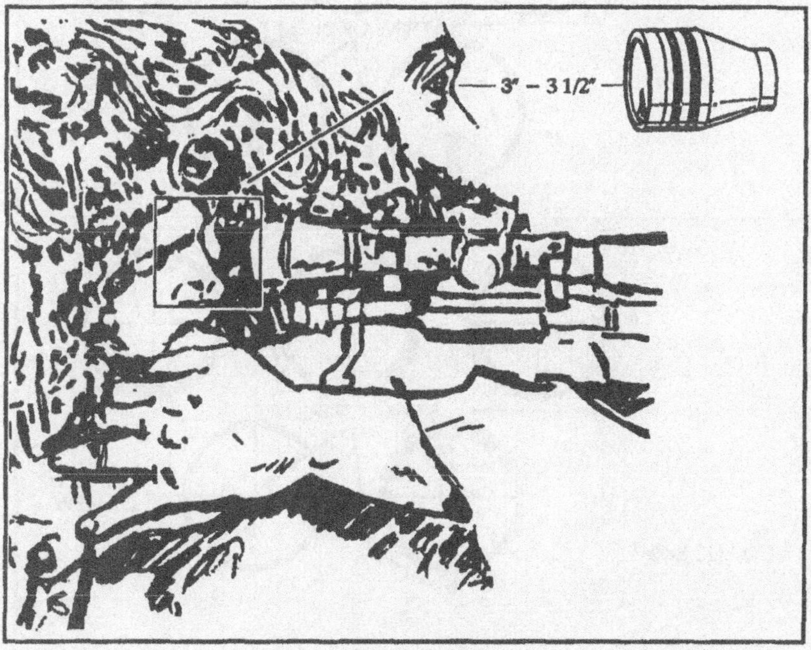

Figure 3-16. Eye relief.

b. **Sight Alignment.** With telescopic sights, sight alignment is the relationship between the cross hairs (reticle) and a full field of view as seen by the sniper. The sniper must place his head so that a full field of view fills the tube, with no dark shadows or crescents to cause inaccurate shots. He centers the reticle in a full field of view, ensuring the vertical cross hair is straight up and down so the rifle is not canted. Again, the center is easiest for the sniper to locate and allows for consistent reticle placement. With iron sights, sight alignment is the relationship between the front and rear sights as seen by the sniper (Figure 3-17). The sniper centers the top edge of the front sight blade horizontally and vertically within the rear aperture. (The center of aperture is easiest for the eye to locate and allows the sniper to be consistent in blade location.)

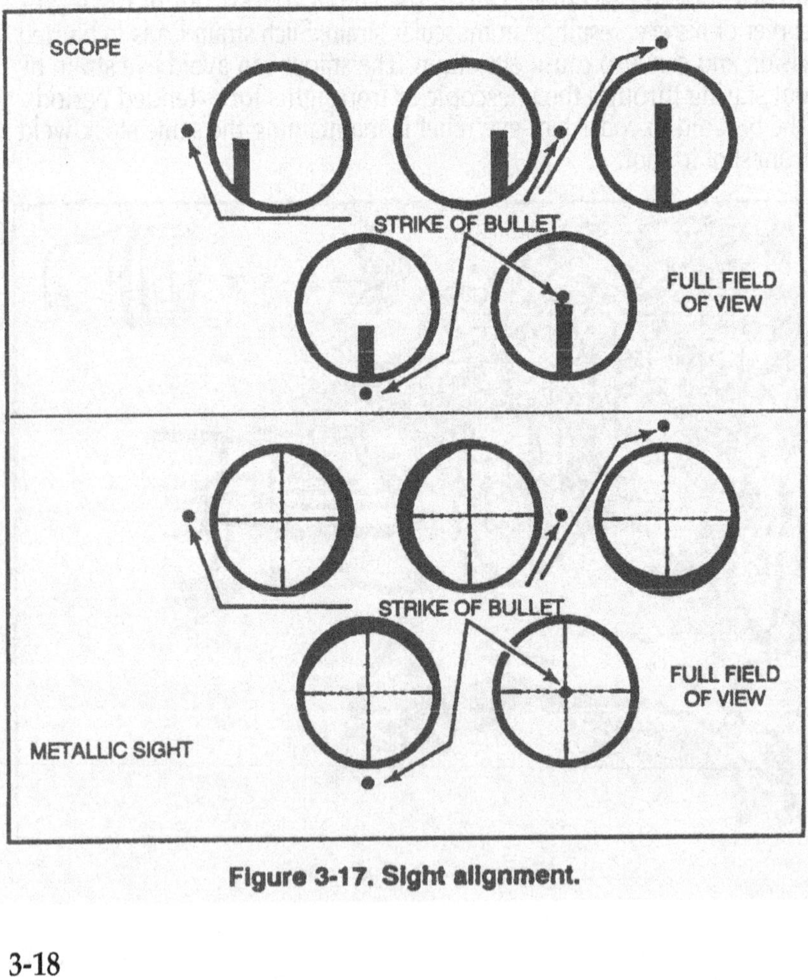

Figure 3-17. Sight alignment.

e. **Sight Picture.** With telescopic sights, the sight picture is the relationship between the reticle and full field of view and the target as seen by the sniper. The sniper centers the reticle in a full field of view. He then places the reticle center of the largest visible mass of the target (as in iron sights). The center of mass of the target is easiest for the sniper to locate, and it surrounds the intended point of impact with a maximum amount of target area. With iron sights, sight picture is the relationship between the rear aperture, the front sight blade, and the target as seen by the sniper (Figure 3-18). The sniper centers the top edge of the blade in the rear aperture. He then places the top edge of the blade in the center of the largest visible mass of the target (disregard the head and use the center of the torso).

Figure 3-18. Sight picture.

d. **Sight Alignment Error.** When sight alignment and picture are perfect (regardless of sighting system) and all else is done correctly, the shot will hit center of mass on the target. However, with an error insight alignment, the bullet is displaced in the direction of the error. Such an error creates an angular displacement between the line of sight and the line of bore. This displacement increases as range increases; the amount of bullet displacement depends on the size of alignment error. Close targets show little or no visible error. Distant targets can show great displacement or can be missed altogether due to severe sight misalignment. An inexperienced sniper is prone to this kind of error, since he is unsure of what correctly aligned sights look like (especially telescopic sights); a sniper varies his head position (and eye relief) from shot to shot, and he is apt to make mistakes while firing.

e. **Sight Picture Error.** An error in sight picture is an error in the placement of the aiming point. This causes no displacement between the line of sight and the line of bore. The weapon is simply pointed at the wrong spot on the target. Because no displacement exists as range increases, close and far targets are hit or missed depending on where the front sight or the reticle is when the rifle fires. All snipers face this kind of error every time they shoot. This is because, regardless of firing position stability, the weapon will always be moving. A supported rifle moves much leas than an unsupported one, but both still move in what is known as a *wobble area.* The sniper must adjust his firing position so that his wobble area is as small as possible and centered on the target. With proper adjustments, the sniper should be able to fire the shot while the front sight blade or reticle is on the target at, or very near, the desired aiming point. How far the blade or reticle is from this point when the weapon fires is the amount of sight picture error all snipers face.

f. **Dominant Eye.** To determine which eye is dominant, the sniper extends one arm to the front and points the index finger skyward to select an aiming point. With both eyes open, he aligns the index finger with the aiming point, then closes one eye at a time while looking at the aiming point. One eye will make the finger appear to move off the aiming point; the other eye will stay on the aiming point. The dominant eye is the eye that does not move the finger from the aiming point. Some individuals may have difficulty aiming because of interference from their dominant eye, if this is not the eye used in the aiming process. This may require the sniper to fire from the other side of the weapon (right-handed firer will fire left-handed). Such individuals must close the dominant eye while shooting.

3-3. BREATH CONTROL

Breath control is important with respect to the aiming process. If the sniper breathes while trying to aim, the rise and fall of his chest causes the rifle to move. He must, therefore, accomplish sight alignment during breathing. To do this, he first inhales then exhales normally and stops at the moment of natural respiratory pause.

a. A respiratory cycle lasts 4 to 5 seconds. Inhalation and exhalation require only about 2 seconds. Thus, between each respiratory cycle there is a pause of 2 to 3 seconds. This pause can be extended to 10 seconds without any special effort or unpleasant sensations. The sniper should shoot during this pause when his breathing muscles relax. This avoids strain on his diaphragm.

b. A sniper should assume his firing position and breathe naturally until his hold begins to settle. Many snipers then take a slightly deeper breath, exhale, and pause, expecting to fire the shot during the pause. If the hold does not settle enough to allow the shot to be fired, the sniper resumes normal breathing and repeats the process.

c. The respiratory pause should never feel unnatural. If it is too long, the body suffers from oxygen deficiency and sends out signals to resume breathing. These signals produce involuntary movements in the diaphragm and interfere with the sniper's ability to concentrate. About 8 to 10 seconds is the maximum safe period for the respiratory pause. During multiple, rapid engagements, the breathing cycle should be forced through a rapid, shallow cycle between shots instead of trying to hold the breath or breathing. Firing should be accomplished at the forced respiratory pause.

3-4. TRIGGER CONTROL

Trigger control is the most important of the sniper marksmanship fundamentals. It is defined as causing the rifle to fire when the sight picture is at its best, without causing the rifle to move. Trigger squeeze is uniformly increasing pressure straight to the rear until the rifle fires.

a. Proper trigger control occurs when the sniper places his firing finger as low on the trigger as possible and still clears the trigger guard, thereby achieving maximum mechanical advantage and movement of the finger to the entire rifle.

b. The sniper maintains trigger control beat by assuming a stable position, adjusting on the target, and beginning a breathing cycle. As the sniper exhales the final breath toward a natural respiratory pause, he secures his finger on the trigger. As the front blade or reticle settles at the desired point of aim, and the natural respiratory pause is entered, the sniper applies initial pressure. He increases the tension on the trigger during the respiratory pause as long as the front blade or reticle remains in the area of the target that ensures a well-placed shot. If the front blade or reticle moves away from the desired point of aim on the target, and the pause is free of strain or tension, the sniper stops increasing the tension on the trigger, waits for the front blade or reticle to return to the desired point, and then continues to squeeze the trigger. If movement is too large for recovery or if the pause has become uncomfortable (extended too long), the sniper should carefully release the pressure on the trigger and begin the respiratory cycle again.

c. As the stability of a firing position decreases, the wobble area increases. The larger the wobble area, the harder it is to fire the shot without reacting to it. This reaction occurs when the sniper—

(1) ***Anticipates recoil.*** The firing shoulder begins to move forward just before the round fires.

(2) ***Jerks the trigger.*** The trigger finger moves the trigger in a quick, choppy, spasmodic attempt to fire the shot before the front blade or reticle can move away from the desired point of aim.

(3) ***Flinches.*** The sniper's entire upper body (or parts thereof) overreacts to anticipated noise or recoil. This is usually due to unfamiliarity with the weapon.

(4) ***Avoids recoil.*** The sniper tries to avoid recoil or noise by moving away from the weapon or by closing the firing eye just before the round fires. This, again, is caused by a lack of knowledge of the weapon's actions upon firing.

3-5. FOLLOW-THROUGH

Applying the fundamentals increases the odds of a well-aimed shot being fired. When mastered, additional skills can make that first-round kill even more of a certainty. One of these skills is the follow-through.

a. Follow-through is the act of continuing to apply all the sniper marksmanship fundamentals as the weapon fires as well as immediately after it fires. It consists of—

(1) Keeping the head infirm contact with the stock (stock weld).
(2) Keeping the finger on the trigger all the way to the rear.
(3) Continuing to look through the rear aperture or scope tube.
(4) Keeping muscles relaxed.
(5) Avoiding reaction to recoil and or noise.
(6) Releasing the trigger only after the recoil has stopped.

b. A good follow-through ensures the weapon is allowed to fire and recoil naturally. The sniper/rifle combination reacts as a single unit to such actions.

3-6. CALLING THE SHOT

Calling the shot is being able to tell where the round should impact on the target. Because live targets invariably move when hit, the sniper will find it almost impossible to use his scope to locate the target after the round is fired. Using iron sights, the sniper will find that searching for a downrange hit is beyond his abilities. He must be able to accurately call his shots. Proper follow-through will aid in calling the shot. The dominant factor in shot calling is knowing where the reticle or blade is located when the weapon discharges. This location is called the *final focus point.*

a. With iron sights, the final focus point should be on the top edge of the front sight blade. The blade is the only part of the sight picture that is moving (in the wobble area). Focusing on it aids in calling the shot and detecting any errors insight alignment or sight picture. Of course, lining up the sights and the target initially requires the sniper to shift his focus from the target to the blade and back until he is satisfied that he is properly aligned with the target. This shifting exposes two more facts about eye focus. The eye can instantly shift focus from near objects (the blade) to far objects (the target).

b. The final focus is easily placed with telescopic sights because of the sight's optical qualities. Properly focused, a scope should present both the field of view and the reticle in sharp detail. Final focus should then be on the target. While focusing on the target, the sniper moves his head slightly from side to side. The reticle may seem to move across the target face, even though the rifle and scope are motionless. This movement is *parallax.* Parallax is present when the target image is not correctly focused on the reticle's focal plane. Therefore, the target image and the reticle appear to be in two separate positions inside the scope, causing the effect of reticle movement across the target. The M3A scope on the M24 has a focus adjustment that eliminates parallax in the scope. The sniper should adjust the focus knob until the target's image is on the same focal plane as the reticle. To determine if the target's image appears at the ideal location, the sniper should move his head slightly left and right to see if the reticle appears to move. If it does not move, the focus is properly adjusted and no parallax will be present.

3-7. INTEGRATED ACT OF FIRING

Once the sniper has been taught the fundamentals of marksmanship, his primary concern is his ability to apply it in the performance of his mission. An effective method of applying fundamentals is through the use of the integrated act of firing one round. The integrated act is a logical, step-by-step development of fundamentals whereby the sniper can develop habits that enable him to fire each shot the same way. The integrated act of firing can be divided into four distinct phases:

a. Preparation Phase. Before departing the preparation area, the sniper ensures that—

(1) The team is mentally conditioned and knows what mission they are to accomplish.

(2) A systematic check is made of equipment for completeness and serviceability including, but not limited to—

(a) Properly cleaned and lubricated rifles.

(b) Properly mounted and torqued scopes.

(c) Zero-sighted systems and recorded data in the sniper data book.

(d) Study of the weather conditions to determine their possible effects on the team's performance of the mission.

b. **Before-Firing Phase.** On arrival at the mission site, the team exercises care in selecting positions. The sniper ensures the selected positions support the mission. During this phase, the sniper—

(1) Maintains strict adherence to the fundamentals of position. He ensures that the firing position is as relaxed as possible, making the most of available external support. He also makes sure the support is stable, conforms to the position, and allows a correct, natural point of aim for each designated area or target.

(2) Once in position, removes the scope covers and checks the field(s) of fire, making any needed corrections to ensure clear, unobstructed firing lanes.

(3) Makes dry firing and natural point of aim checks.

(4) Double-checks ammunition for serviceability and completes final magazine loading.

(5) Notifies the observer he is ready to engage targets. The observer must be constantly aware of weather conditions that may affect the accuracy of the shots. He must also stay ahead of the tactical situation.

c. **Firing Phase.** Upon detection, or if directed to a suitable target, the sniper makes appropriate sight changes, aims, and tells the observer he is ready to fire. The observer then gives the needed windage and observes the target. To fire the rifle, the sniper should remember the key word, "BRASS." Each letter is explained as follows:

(1) ***Breathe.*** The sniper inhales and exhales to the natural respiratory pause. He checks for consistent head placement and stock weld. He ensures eye relief is correct (full field of view through the scope; no shadows present). At the same time, he begins aligning the cross hairs or front blade with the target at the desired point of aim.

(2) ***Relax.*** As the sniper exhales, he relaxes as many muscles as possible, while maintaining control of the weapon and position.

(3) ***Aim.*** If the sniper has a good, natural point of aim, the rifle points at the desired target during the respiratory pause. If the aim is off, the sniper should make a slight adjustment to acquire the desired point of aim. He avoids "muscling" the weapon toward the aiming point.

(4) **Squeeze.** As long as the sight picture is satisfactory, the sniper squeezes the trigger. The pressure applied to the trigger must be straight to the rear without disturbing the lay of the rifle or the desired point of aim.

d. After-Firing Phase. The sniper must analyze his performance If the shot impacted at the desired spot (a target hit), it may be assumed the integrated act of firing one round was correctly followed. If however, the shot was off call, the sniper and observer must check for Possible errors.

(1) Failure to follow the keyword, BRASS (partial field of view, breath held incorrectly, trigger jerked, rifle muscled into position, and so on).

(2) Target improperly ranged with scope (causing high or low shots).

(3) Incorrectly compensated for wind (causing right or left shots).

(4) Possible weapon/ammunition malfunction (used only as a last resort when no other errors are detected).

Once the probable reasons for an off-call shot is determined the sniper must make note of the errors. He pays close attention to the problem areas to increase the accuracy of future shots.

Section II
BALLISTICS

As applied to sniper marksmanship, types of ballistics may be defined as the study of the firing, flight, and effect of ammunition. Proper execution of marksmanship fundamentals and a thorough knowledge of ballistics ensure the successful completion of the mission. Tables and formulas in this section should be used only as guidelines since every rifle performs differently. Maximum ballistics data eventually result in a well-kept sniper data book and knowledge gained through experience.

3-8. TYPES OF BALLISTICS

Ballistics are divided into three distinct types: internal external, and terminal.

a. Internal-the interior workings of a weapon and the functioning of its ammunition.

b. External-the flight of the bullet from the muzzle to the target.

c. Termninal-what happens to the bullet after it hits the target. (See paragraph 3-16.)

3-9. TERMINOLOGY

To fully understand ballistics, the sniper should be familiar with the following terms:

a. Muzzle Velocity-the speed of the bullet as it leaves the rifle barrel, measured in feet per second. It varies according to various factors, such as ammunition type and lot number, temperature, and humidity.

b. Line of Sight- straight line from the eye through the aiming device to the point of aim.

c. Line of Departure-the line defined by the bore of the rifle or the path the bullet would take without gravity.

d. Trajectory-the path of the bullet as it travels to the target.

e. Midrange Trajectory/Maximum Ordinate-the highest point the bullet reaches on its way to the target. This point must be known to engage a target that requires firing underneath an overhead obstacle, such as a bridge or a tree. In attention to midrange trajectory may cause the sniper to hit the obstacle instead of the target.

f. Bullet Drop—how far the bullet drops from the line of departure to the point of impact.

g. Time of Flight-the amount of time it takes for the bullet to reach the target from the time the round exits the rifle.

h. Retained Velocity-the speed of the bullet when it reaches the target. Due to drag, the velocity will be reduced.

3-10. EFFECTS ON TRAJECTORY

To be effective, the sniper must know marksmanship fundamentals and what effect gravity and drag will have on those fundamentals.

a. **Gravity.** As soon as the bullet exits the muzzle of the weapon, gravity begins to pull it down, requiring the sniper to use his elevation adjustment. At extended ranges, the sniper actually aims the muzzle of his rifle above his line of sight and lets gravity pull the bullet down into the target. Gravity is always present, and the sniper must compensate for this through elevation adjustments or hold-off techniques.

b. **Drag.** Drag is the slowing effect the atmosphere has on the bullet. This effect decreases the speed of the bullet according to the air—that is, the less dense the air, the leas drag and vice versa. Factors affecting drag/density are temperature, altitude/barometric pressure, humidity, efficiency of the bullet, and wind.

(1) ***Temperature.*** The higher the temperature, the less dense the air. (See Section III.) If the sniper zeros at 60 degrees F and he fires at 80 degrees, the air is leas dense, thereby causing an increase in muzzle velocity and higher point of impact. A 20-degree change equals a one-minute elevation change in the strike of the bullet.

(2) ***Altitude/barometric pressure.*** Since the air pressure is less at higher altitudes, the air is less dense. Thus, the bullet is more efficient and impacts higher due to less drag. (Table 3-1 shows the approximate

effect of change of the point of impact from sea level to 10,000 feet if the rifle is zeroed at sea level.) Impact will be the point of aim at sea level. For example, a rifle zeroed at sea level and fired at a range of 700 meters at an altitude of 5,000 feet will hit 1.6 minutes high.

RANGE (METERS)	2,500 FEET *(ASL)	5,000 FEET (ASL)	10,000 FEET (ASL)
100	.05	.08	.13
200	.1	.2	.34
300	.2	.4	.6
400	.4	.5	.9
500	.5	.9	1.4
600	.6	1.0	1.8
700	1.0	1.6	2.4
800	1.3	1.9	3.3
900	1.6	2.8	4.8
1,000	1.8	3.7	6.0
*ABOVE SEA LEVEL			

Table 3-1. Point of impact rises as altitude increases (data are in MOA).

(3) ***Humidity.*** Humidity varies along with the altitude and temperature. Figure 3-19 considers the changes in altitudes. Problems can occur if extreme humidity changes exist in the area of operations. That is, when humidity goes up, impact goes down; when humidity goes down, impact goes up. Since impact is affected by humidity, a 20 percent change in humidity equals about one minute as a rule of thumb. Keeping a good sniper data book during training and acquiring experience are the best teachers.

(4) ***Efficiency of the bullet.*** This is called a *bullet's ballistic coefficient.* The imaginary perfect bullet is rated as being 1.00. Match bullets range

from .500 to about .600. The 7.62-mm special ball (M118) is rated at .530 (Table 3-2).

(5) ***Wind.*** Wind is discussed in Section III.

RANGE (METERS)	(A)	(B)	(C)	(D)
100	2,407	.7	NA	.1
200	2,233	3.0	1.5	.2
300	2,066	7.3	3.0	.4
400	1,904	14.0	3.5	.5
500	1,750	24.0	4.0	.7
600	1,603	37.6	4.5	.9
700	1,466	56.2	5.0	1.0
800	1,339	80.6	5.0	1.3
900	1,222	112.5	6.0	1.5
1,000	1,118	153.5	7.0	1.8

(A) RETAINED VELOCITY (FEET PER SECOND).
(B) MIDRANGE TRAJECTORY (INCHES).
(C) BULLET DROP IN 100-METER INCREMENTS (MINUTES).
(D) TIME OF FLIGHT (SECONDS).

Table 3-2. Muzzle velocity data for 7.62-mm special ball (M118).

3-11. ANGLE FIRING

Most practice firing conducted by the sniper team involves the use of military range facilities, which are relatively flat. However, as a sniper being deployed to other regions of the world, the chance exists for operating in a mountainous or urban environment. This requires target engagements at higher and lower elevations. Unless the sniper takes corrective action, bullet impact will be above the point of aim. How high the bullet hits is determined by the range and angle to the target (Table 3-3). The amount of elevation change applied to the telescope of the rifle for angle firing is known as *slope dope.*

RANGE (METERS)	SLANT DEGREES 5	10	15	20	25	30	35	40	45	50	55	60
100	.01	.04	.09	.16	.25	.36	.49	.63	.79	.97	1.2	1.4
200	.03	.09	.2	.34	.53	.76	1.	1.3	1.7	2.	2.4	2.9
300	.03	.1	.3	.5	.9	1.2	1.6	2.1	2.7	3.2	3.9	4.5
400	.05	.19	.43	.76	1.2	1.7	2.3	2.9	3.7	4.5	5.4	6.3
500	.06	.26	.57	1.	1.6	2.3	3.	3.9	4.9	6.	7.2	8.4
600	.08	.31	.73	1.3	2.	2.9	3.9	5.	6.3	7.7	9.2	10.7
700	.1	.4	.9	1.6	2.5	3.6	4.9	6.3	7.9	9.6	11.5	13.4
800	.13	.5	1.	2.	3.	4.4	5.9	7.7	9.6	11.7	14.	16.4
900	.15	.6	1.3	2.4	3.7	5.3	7.2	9.3	11.6	14.1	16.9	19.8
1,000	.2	.7	1.6	2.8	4.5	6.4	8.6	11.	13.9	16.9	20.2	23.7

*RANGE GIVEN IS SLANT RANGE (METERS), NOT MAP DISTANCE.

Table 3-3. Bullet rise at given angle and range in minutes.

Section III
EFFECTS OF WEATHER

For the highly trained sniper, the effects of weather are the main causes of error in the strike of the bullet. Wind, mirage, light, temperature, and humidity affect the bullet, the sniper, or both. Some effects are minor; however, sniping is often done in extremes of weather and all effects must be considered.

3-12. WIND CLASSIFICATION

Wind poses the biggest problem for the sniper. The effect that wind has on the bullet increases with range. This is due mainly to the slowing of the bullet's velocity combined with a longer flight time. This allows the wind to have a greater effect on the round as distances increase. The result is a loss of stability.

a. Wind also has a considerable effect on the sniper. The stronger the wind, the more difficult it is for him to hold the rifle steady. This can be partly offset by training, conditioning and the use of supported positions.

b. Since the sniper must know how much effect the wind will have on the bullet, he must be able to classify the wind. The best method is to use the clock system (Figure 3-19). With the sniper at the center of the clock and the target at 12 o'clock, the wind is assigned three values: full, half, and no value. Full value means that the force of the wind will have a full effect on the flight of the bullet. These winds come from 3 and 9 o'clock. Half value means that a wind at the same speed, but from 1,2,4,5,7,8, 10, and 11 o'clock, will move the bullet only half as much as a full-value wind. No value means that a wind from 6 or 12 o'clock will have little or no effect on the flight of the bullet.

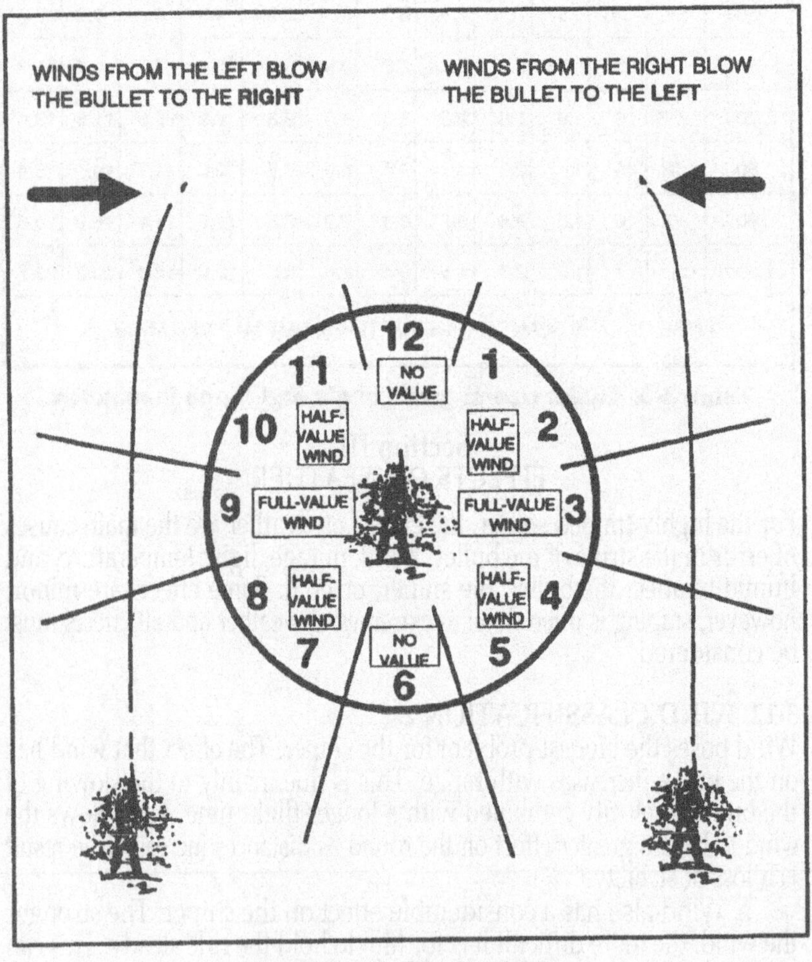

Figure 3-19. Clock system.

3-13. WIND VELOCITY

Before adjusting the sight to compensate for wind, the sniper must determine wind direction and velocity. He may use certain indicators to accomplish this. These are range flags, smoke, trees, grass, rain, and the sense of feel. However, the preferred method of determining wind direction and velocity is reading mirage (see paragraph d below). In most cases, wind direction can be determined simply by observing the indicators.

a. A common method of estimating the velocity of the wind during training is to watch the range flag (Figure 3-20). The sniper determines the angle between the flag and pole, in degrees, then divides by the constant number 4. The result gives the approximate velocity in miles per hour.

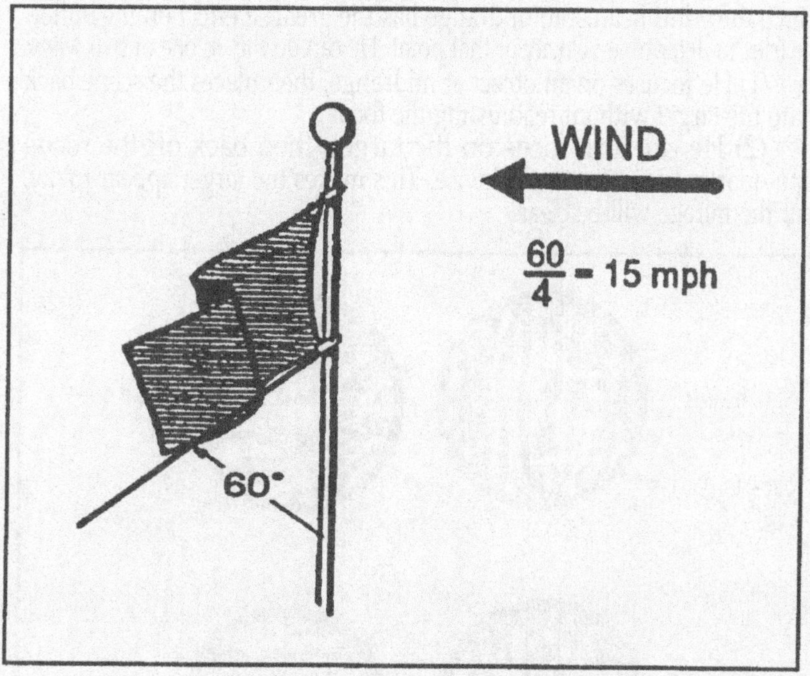

Figure 3-20. The Flag method.

b. If no flag is visible, the sniper holds a piece of paper, grass, cotton, or some other light material at shoulder level, then drops it. He then points directly at the spot where it lands and divides the angle between his body and arm by the constant number 4. This gives him the approximate wind velocity in miles per hour.

c. If these methods cannot be used, the following information is helpful in determining velocity. Winds under 3 miles per hour can barely be felt, although smoke will drift. A3- to 5-mile-per-hourwind can barely be felt on the face. With a 5- to 8-mile-per-hour wind, the leaves in the trees are in constant motion, and with a 12- to 15-mile-per-hour wind, small trees begin to sway.

d. A mirage is a reflection of the heat through layers of air at different temperatures and density as seen on a warm day (Figure 3-21). With the telescope, the sniper can see a mirage as long as there is a difference in ground and air temperatures. Proper reading of the mirage enables the sniper to estimate wind speed and direction with a high degree of accuracy. The sniper uses the M49 observation telescope to read the mirage. Since the wind nearest to midrange has the greatest effect on the bullet, he tries to determine velocity at that point. He can do this in one of two ways:

(1) He focuses on an object at midrange, then places the scope back onto the target without readjusting the focus.

(2) He can also focus on the target, then back off the focus one-quarter turn counterclockwise. This makes the target appear fuzzy, but the mirage will be clear.

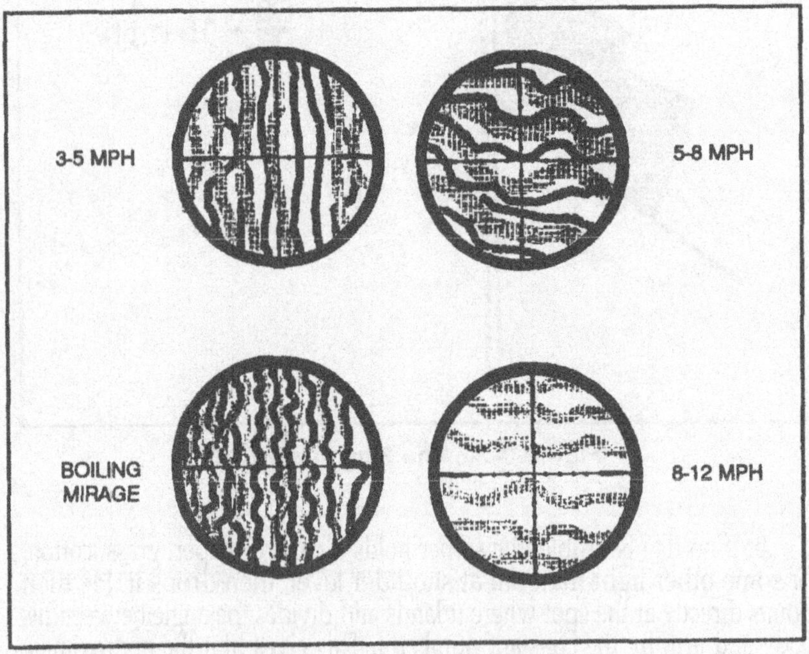

Figure 3-21. Types of mirages.

e. As observed through the telescope, the mirage appears to move with the same velocity as the wind, except when blowing straight into or away from the scope. Then, the mirage gives the appearance of moving straight upward with no lateral movement. This is called a *boiling mirage*. A boiling mirage may also be seen when the wind is constantly changing direction. For example, a full-value wind blowing from 9 o'clock to 3 o'clock suddenly changes direction. The mirage will appear to stop moving from left to right and present a boiling appearance. When this occurs, the inexperienced observer directs the sniper to fire with the "0" wind. As the sniper fires, the wind begins blowing from 3 o'clock to 9 o'clock, causing the bullet to miss the target therefore, firing in a "boil" can hamper shot placement. Unless there is a no-value wind, the sniper must wait until the boil disappears. In general, changes in the velocity of the wind, up to about 12 miles per hour, can be readily determined by observing the mirage. Beyond that speed, the movement of the mirage is too fast for detection of minor changes.

3-14. CONVERSION OF WIND VELOCITY TO MINUTES OF ANGLE

All telescopic sights have windage adjustments that are graduated in minutes of angle or fractions thereof. A minute of angle is 1/60th of a degree (Figure 3-22, page 3-34). This equals about 1 inch (1.145 inches) for every 100 meters.

EXAMPLE

1 MOA = 2 inches at 200 meters
1 MOA = 5 inches at 500 meters

a. Snipers use minutes of angle (Figure 3-22, page 3-34) to determine and adjust the elevation and windage needed on the weapon's scope. After finding the wind direction and velocity in miles per hour, the sniper must then convert it into minutes of angle, using the wind formula as a rule of thumb only. The wind formula is—

$$\frac{\text{RANGE (hundreds) divided by 100 VELOCITY (mph)}}{\text{CONSTANT}} = \text{Minutes full-value wind}$$

The constant depends on the target's range.

100 to 500	"C" =15
600	"C" =14
700 to 800	"C" =13
900	"C" =12
1,000	"C" =11

If the target is 700 meters away and the wind velocity is 10 mph, the formula is—

$$\frac{7 \times 10}{13} = 5.38 \text{ minutes or } 5\ 1/2 \text{ minutes}$$

This determines the number of minutes for a full-value wind. For a half-value wind, the 5.38 would be divided in half.

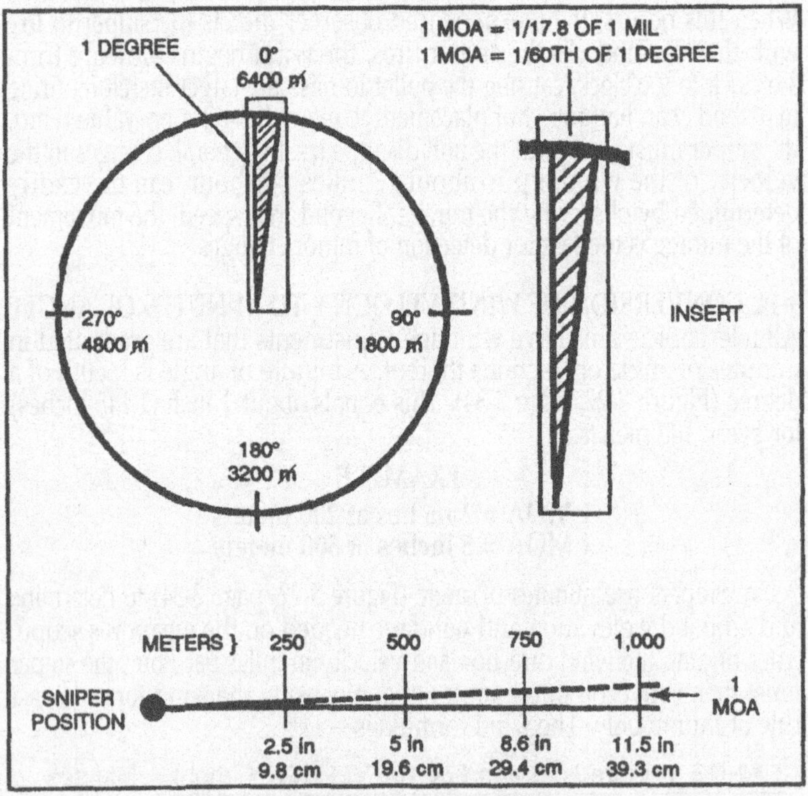

Figure 3-22. Minutes of angle.

b. The observer makes his own adjustment estimations, then compares them to the wind conversion table, which can be a valuable training tool. He must not rely on this table; if it is lost, his ability to perform the mission could be severely hampered. Until the observer gains skill in estimating wind speed and computing sight changes, he may refer to Table 3-4.

RANGE (METERS)	WIND VALUE	3 MPH MIN	3 MPH IN	5 MPH MIN	5 MPH IN	7 MPH MIN	7 MPH IN	10 MPH MIN	10 MPH IN
200	HALF	0.0	0.4	0.5	0.6	0.5	0.8	0.5	1.2
	FULL	0.5	0.8	0.5	1.2	1.0	1.7	1.0	2.4
300	HALF	0.5	0.9	0.5	1.3	0.5	1.9	1.0	2.7
	FULL	0.5	1.7	1.0	2.7	1.0	3.8	1.5	5.4
400	HALF	0.5	1.4	0.5	2.4	1.0	3.3	1.0	4.8
	FULL	0.5	2.9	1.0	4.8	1.5	6.7	2.0	9.6
500	HALF	0.5	2.3	0.5	3.8	1.0	5.3	1.5	7.5
	FULL	1.0	4.5	1.5	7.5	2.0	10.5	2.5	15.0
600	HALF	0.5	3.0	1.0	5.0	1.0	8.0	1.5	11.0
	FULL	1.0	7.0	1.5	11.0	2.5	15.0	3.5	21.0
700	HALF	0.5	4.0	'1.0	7.0	1.5	10.0	2.0	15.0
	FULL	1.0	9.0	2.0	15.0	2.5	21.0	4.0	29.0
800	HALF	0.5	6.0	1.0	10.0	1.5	13.0	2.0	19.0
	FULL	1.5	11.0	2.0	19.0	3.0	27.0	4.5	38.0
900	HALF	0.5	7.0	1.0	12.0	1.5	17.0	2.5	24.0
	FULL	3.5	15.0	2.5	24.0	3.5	34.0	5.0	49.0
1000	HALF	1.0	9.0	1.5	15.0	2.0	21.0	2.5	3.00
	FULL	1.5	18.0	2.5	30.0	4.0	42.0	5.5	60.0

RANGE (METERS)	WIND VALUE	12 MPH MIN	12 MPH IN	15 MPH MIN	15 MPH IN	18 MPH MIN	18 MPH IN	20 MPH MIN	20 MPH IN
200	HALF	0.5	1.3	1.0	1.8	1.0	2.2	1.0	2.4
	FULL	1.5	2.9	1.5	3.6	2.0	4.3	2.0	4.8
300	HALF	1.0	3.3	1.0	4.0	1.5	4.9	1.5	5.4
	FULL	2.0	6.5	2.5	8.1	3.0	9.8	3.5	10.9
400	HALF	1.5	5.8	1.5	7.2	2.0	8.6	2.0	9.6
	FULL	2.5	11.5	3.5	14.4	4.0	17.3	4.5	19.2
500	HALF	1.5	9.0	2.0	11.3	2.5	13.5	2.5	15.0
	FULL	3.5	18.0	4.0	22.6	5.0	27.0	5.5	30.0
600	HALF	1.5	13.0	2.5	16.0	3.0	19.0	3.5	22.0
	FULL	4.0	26.0	5.0	32.0	6.0	39.0	6.5	43.0
700	HALF	2.5	18.0	3.0	22.0	3.5	26.0	4.0	29.0
	FULL	4.5	35.0	6.0	44.0	7.0	53.0	7.5	59.0
800	HALF	2.5	23.0	3.5	29.0	4.0	35.0	4.5	38.0
	FULL	5.5	46.0	6.5	57.0	8.0	69.0	9.0	77.0
900	HALF	3.0	29.0	3.5	36.0	4.5	44.0	5.0	49.0
	FULL	6.0	58.0	7.5	73.0	9.0	97.0	10.0	97.0
1000	HALF	3.5	36.0	4.0	45.0	5.0	54.0	5.5	60.0
	FULL	6.5	72.0	8.0	90.0	10.0	103.0	11.5	120.0

Table 3-4. Wind conversion table.

3-15. EFFECTS OF LIGHT

Light does not affect the trajectory of the bullet; however, it does affect the way the sniper sees the target through the scope. This effect can be compared to the refraction (bending) of light through a medium, such as a prism or a fish bowl. The same effect, although not as drastic, can be observed on a day with high humidity and with sunlight from high angles. The only way the sniper can adjust for this effect is to refer to past firing recorded in the sniper data book. He can then compare different light and humidity conditions and their effect on marksmanship. Light may also affect firing on unknown distance ranges since it affects range determination capabilities.

3-16. EFFECTS OF TEMPERATURE

Temperature affects the firer, ammunition, and density of the air. When ammunition sits in direct sunlight, the bum rate of powder is increased, resulting in greater muzzle velocity and higher impact. The greatest effect is on the density of the air. As the temperature rises, the air density is lowered. Since there is leas resistance, velocity increases and once again the point of impact rises. This is in relation to the temperature at which the rifle was zeroed, If the sniper zeros at 50 degrees and he is now firing at 90 degrees, the point of impact rises considerably. How high it rises is best determined once again by past firing recorded in the sniper data book. The general role, however, is that when the rifle is zeroed, a 20-degree increase in temperature will raise the point of impact by one minute; conversely, a 20-degree decrease will drop the point of impact by one minute.

3-17. EFFECTS OF HUMIDITY

Humidity varies along with the altitude and temperature. The sniper can encounter problems if drastic humidity changes occur in his area of operation. Remember, if humidity goes up, impact goes down; if humidity goes down, impact goes up. As a rule of thumb, a 20-percent change will equal about one minute, affecting the point of impact. The sniper should keep a good sniper data book during training and refer to his own record.

Section IV
SNIPER DATA BOOK

The sniper data book contains a collection of data cards. The sniper uses the data cards to record firing results and all elements that had an effect on firing the weapon. This can vary from information about weather conditions to the attitude of the firer on that particular day. The sniper can refer to this information later to understand his weapon, the weather effects, and his shooting ability on a given day. One of the most important items of information he will record is the cold barrel zero of his weapon. A cold barrel zero refers to the first round fired from the weapon at a given range. It is critical that the sniper shoots the first round daily at different ranges. For example, Monday, 400 meters; Tuesday, 500 meters; Wednesday, 600 meters. When the barrel warms up, later shots begin to group one or two minutes higher or lower, depending on specific rifle characteristics. Information is recorded on DA Form 5785-R (Sniper's Data Card) (Figure 3-23). (A blank copy of this form is located in the back of this publication for local reproduction.)

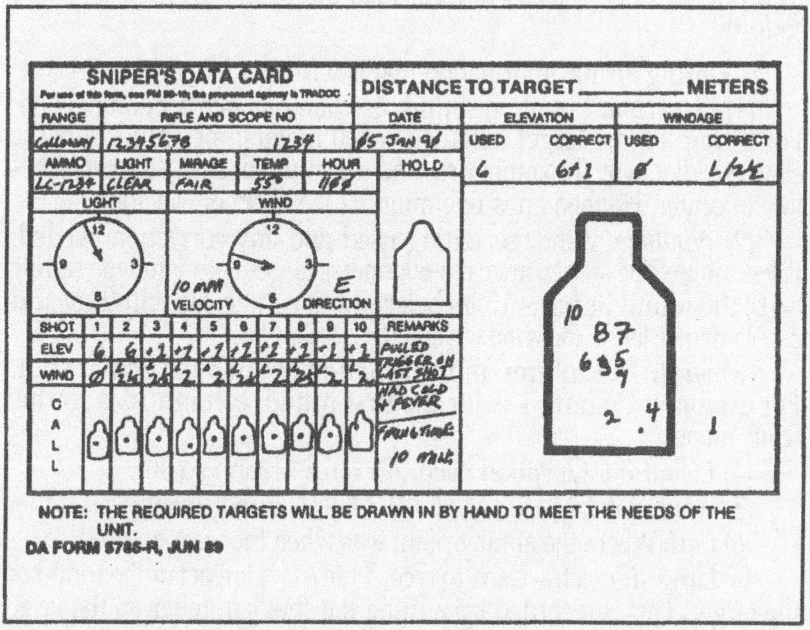

Figure 3-23. Example of completed DA Form 5785-R.

3-18. ENTRIES

Three phases in writing information on the data card (Figure 3-23) are *before firing*, *during firing*, and *after firing*.

a. **Before Firing.** Information that is written before firing is—

(1) ***Range.*** The distance to the target.

(2) ***Rifle and scope number.*** The serial numbers of the rifle and scope.

(3) ***Date.*** Date of firing.
(4) ***Ammunition.*** Type and lot number of ammunition.
(5) ***Light.*** Amount of light (overcast, clear, and so forth).
(6) ***Mirage.*** Whether a mirage can be seem or not (good, bad, fair, and so forth).
(7) ***Temperature.*** Temperature on the range.
(8) ***Hour.*** Time of firing.
(9) ***Light (diagram).*** Draw an arrow in the direction the light is shining.
(10) ***Wind.*** Draw an arrow in the direction the wind is blowing, and record its average velocity and cardinal direction (N, NE, S, SW, and so forth).

b. **During Firing.** Information that is written while firing is—
(1) ***Elevation.*** Elevation setting used and any correction needed. For example: The target distance is 600 meters; the sniper sets the elevation dial to 6. The sniper fires and the round hits the target 6 inches low of center. He then adds one minute (one click) of elevation (+1).
(2) ***Windage.*** Windage setting used and any correction needed. For example The sniper fires at a 600-meter target with windage setting on 0; the round impacts 15 inches right of center. He will then add 2 1/2 minutes left to the windage dial (L/2 1/2).
(3) ***Shot.*** The column of information about a particular shot. For example: Column 1 is for the first round; column 10 is for the tenth round.
(4) ***Elevation.*** Elevation used (6 +1, 6,6 –1, and so on).
(5) ***Wind.*** Windage used (L/2 1/2, O, R/1/2, and so on).
(6) ***Call.*** Where the aiming point was when the weapon fired.
(7) ***Large silhouette.*** Used to record the exact impact of the round on the target. This is recorded by writing the shot's number on the large silhouette in the same place it hit the target.

c. **After Firing.** After firing, the sniper records any comments about firing in the remarks section. This can be comments about the weapon, firing conditions (time allowed for fire), or his condition (nervous, felt bad, felt good, and so forth).

3-19. ANALYSIS

When the sniper leaves the firing line, he compares weather conditions to the information needed to hit the point of aim/point of impact. Since he fires in all types of weather conditions, he must be aware of temperature, light, mirage, and wind. The sniper must consider other major points or tasks to complete

a. Compare sight settings with previous firing sessions. If the sniper always has to fine-tune for windage or elevation, there is a chance he needs a sight change (slip a scale).

b. Compare ammunition by lot number for best rifle and ammunition combination.

c. Compare all groups fired under each condition. Check the low and high shots as well as those to the left and the right of the main group—the less dispersion, the better. If groups are tight, they are easily moved to the center of the target; if loose, there is a problem. Check the scope focus and make sure the rifle is cleaned correctly. Remarks in the sniper data book will also help.

d. Make corrections. Record corrections in the sniper data book, such as position and sight adjustment information, to ensure retention.

e. Analyze a group on a target. This is important for marksmanship training. The firer may not notice errors during firing, but errors become apparent when analyzing a group. This can only be done if the sniper data book has been used correctly. A checklist that will aid in shot group/performance analysis follows:

(1) Group tends to be low and right.
- Left hand not positioned properly.
- Right elbow slipping.
- Improper trigger control.

(2) Group scattered about the target.
- Incorrect eye relief or sight picture.
- Concentration on the target (iron sights).
- Stock weld changed.
- Unstable firing position.

(3) Good group but with several erratic shots.
- *Flinching.* Shots may be anywhere.
- *Bucking.* Shots from 7 to 10 o'clock.
- *Jerking.* Shots may be anywhere.

(4) Group strung up and down through the target.
- Breathing while firing.
- Improper vertical alignment of cross hairs.
- Stock weld changed.

(5) Compact group out of the target.
- Incorrect zero.
- Failure to compensate for wind.
- Bad natural point of aim.
- Scope shadow.

(6) Group center of the target out the bottom.
- Scope shadow.
- Position of the rifle changed in the shoulder.

(7) Horizontal group across the target.
- Scope shadow.
- Canted weapon.
- Bad natural point of aim.

Section V
HOLDOFF

Holdoff is shifting the point of aim to achieve a desired point of impact. Certain situations, such as multiple targets at varying ranges and rapidly changing winds, do not allow proper windage and elevation adjustments. Therefore, familiarization and practice of elevation and windage holdoff techniques prepare the sniper to meet these situations.

3-20. ELEVATION

This technique is used only when the sniper does not have time to change his sight setting. The sniper rarely achieves pinpoint accuracy when holding off, since a minor error in range determination or a lack of a precise aiming point might cause the bullet to miss the desired point. He uses holdoff with the sniperscope only if several targets appear at various ranges, and time does not permit adjusting the scope for each target.

a. The sniper uses holdoff to hit a target at ranges other than the range for which the rifle is presently adjusted. When the sniper aims directly at a target at ranges greater than the set range, his bullet will hit below the point of aim. At lesser ranges, his bullet will hit higher than the point of aim. If the sniper understands this and knows about trajectory and bullet drop, he will be able to hit the target at ranges other than that for which the rifle was adjusted. For example, the sniper adjusts the rifle for a target located 500 meters downrange and another target appears at a range of 600 meters. The holdoff would be 25 inches, that is, the sniper should hold off 25 inches above the center of visible mass in order to hit the center of mass of that particular target (Figure 3-24). If another target were to appear at 400 meters, the sniper would aim 14 inches below the ureter of visible mass in order to hit the center of mass (Figure 3-25).

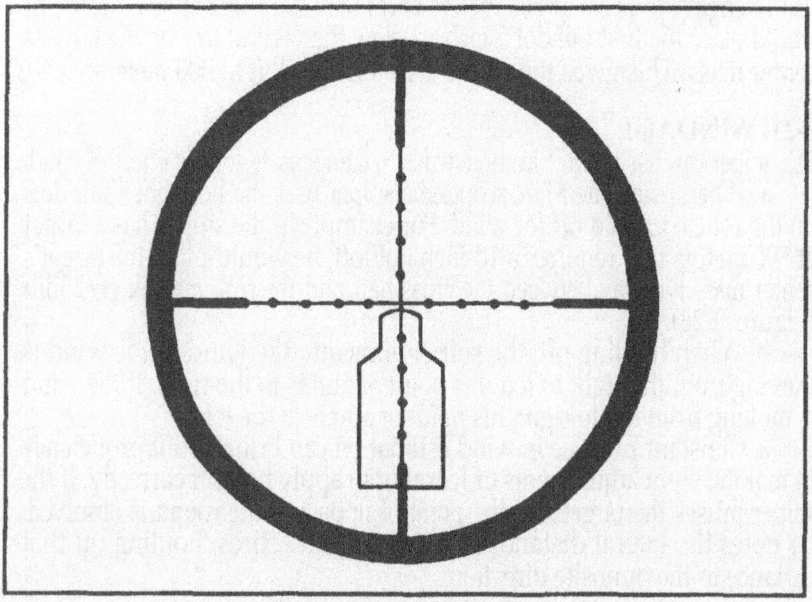

Figure 3-24. Elevation.

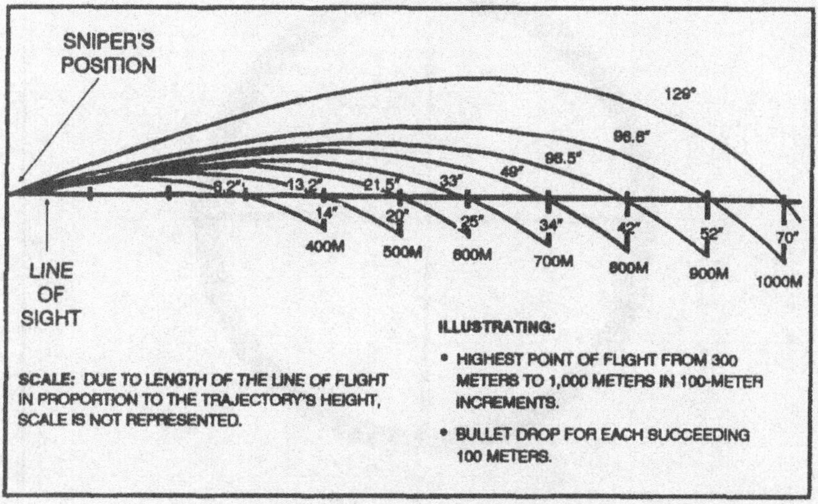

Figure 3-25. Trajectory chart.

b. The vertical mil dots on the M3A scope's reticle can be used as aiming points when using elevation holdoffs. For example, if the sniper has to engage a target at 500 meters and the scope is set at 400 meters, he would place the first mil dot 5 inches below the vertical line on the target's center mass. This gives the sniper a 15-inch holdoff at 500 meters.

3-21. WINDAGE

The sniper can use holdoff in three ways to compensate for the effect of wind.

a. When using the M3A scope, the sniper uses the horizontal mil dots on the reticle to hold off for wind. For example, if the sniper has a target at 500 meters that requires a 10-inch holdoff, he would place the target's center mass halfway between the cross hair and the first mil dot (1/2 mil) (Figure 3-26).

b. When holding off, the sniper aims into the wind. If the wind is moving from the right to left, his point of aim is to the right. If the wind is moving from left to right, his point of aim is to the left.

c. Constant practice in wind estimation can bring about proficiency in making sight adjustments or learning to apply holdoff correctly. If the sniper misses the target and the point of impact of the round is observed, he notes the lateral distance of his error and refires, holding off that distance in the opposite direction.

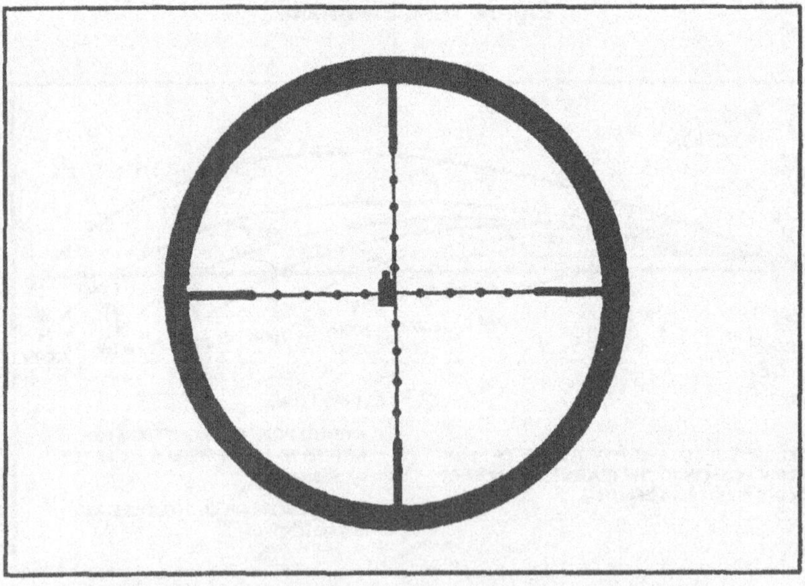

Figure 3-26. Holdoff for 7.62-mm special ball (M118).

Section VI
ENGAGEMENT OF MOVING TARGETS

Engaging moving targets not only requires the sniper to determine the target distance and wind effects on the round, but he must also consider the lateral and speed angle of the target, the round's time of flight, and the placement of a proper lead to compensate for both. These added variables increase the chance of a miss. Therefore, the sniper should engage moving targets when it is the only option.

3-22. TECHNIQUES

To engage moving targets, the sniper employs the following techniques:

- Leading.
- Tracking.
- Trapping or ambushing.
- Tracking and holding.
- Firing a snap shot.

a. **Leading.** Engaging moving targets requires the sniper to place the cross hairs ahead of the target's movement. The distance the cross hairs are placed in front of the target's movement is called a *lead.* There are four factors in determining leads:

(1) ***Speed of the target.*** As a target moves faster, it will move a greater distance during the bullet's flight. Therefore, the lead increases as the target's speed increases.

(2) ***Angle of movement.*** A target moving perpendicular to the bullet's flight path moves a greater lateral distance than a target moving at an angle away from or toward the bullet's path. Therefore, a target moving at a 45-degree angle covers less ground than a target moving at a 90-degree angle.

(3) ***Range to the target.*** The farther away a target is, the longer it takes for the bullet to reach it. Therefore, the lead must be increased as the distance to the target increases.

(4) ***Wind effects.*** The sniper must consider how the wind will affect the trajectory of the round. A wind blowing against the target's direction of movement requires less of a lead than a wind blowing in the same direction as the target's movement.

b. Tracking. hacking requires the sniper to establish an aiming point ahead of the target's movement and to maintain it as the weapon is fired. This requires the weapon and body position to be moved while following the target and firing.

c. **Trapping or Ambushing.** Trapping or ambushing is the sniper's preferred method of engaging moving targets. The sniper must establish an aiming point ahead of the target and pull the trigger when the target reaches it. This method allows the sniper's weapon and body position to remain motionless. With practice, a sniper can determine exact leads and aiming points using the horizontal stadia lines in the mil dots in the M3A.

d. **Tracking and Holding.** The sniper uses this technique to engage an erratically moving target. That is, while the target is moving, the sniper keeps his cross hairs centered as much as possible and adjusts his position with the target. When the target stops, the sniper quickly perfects his hold and fires. This technique requires concentration and discipline to keep from firing before the target comes to a complete halt.

e. **Firing a Snap Shot.** A sniper may often attempt to engage a target that only presents itself briefly, then resumes cover. Once he establishes a pattern, he can aim in the vicinity of the target's expected appearance and fire a snap shot at the moment of exposure.

3-23. COMMON ERRORS

When engaging moving targets, the sniper makes common errors because he is under greater stress than with a stationary target. There are more considerations, such as retaining a steady position and the correct aiming point, how fast the target is moving, and how far away it is. The more practice a sniper has shooting moving targets, the better he will become. Some common mistakes are as follows:

a. The sniper has a tendency to watch his target instead of his aiming point. He must force himself to watch his lead point.

b. The sniper may jerk or flinch at the moment his weapon fires because he thinks he must fire NOW. This can be overcome through practice on a live-fire range.

c. The sniper may hurry and thus forget to apply wind as needed. Windage must be calculated for moving targets just as for stationary targets. Failure to do this when squiring a lead will result in a miss.

3-24. CALCULATION OF LEADS

Once the required lead has been determined, the sniper should use the mil scale in the scope for precise holdoff. The mil scale can be mentally sectioned into 1/4-mil increments for leads. The chosen point on the mil scale becomes the sniper's point of concentration just as the cross hairs are for stationary targets. The sniper concentrates on the lead point and fires the weapon when the target is at this point. The following formulas are used to determine moving target leads:

TIME OF FLIGHT X TARGET SPEED = LEAD.

Time of flight= flight time of the round in seconds.

Target speed = speed the target is moving in fps.

Lead = distance aiming point must be placed ahead of movement in feet.

Average speed of a man during—

Slow patrol = 1 fps/0.8 mph

Fast patrol = 2 fps/1.3 mph

Slow walk = 4 fps/2.5 mph

Fast walk = 6 fps/3.7 mph

To convert leads in feet to meters:

LEAD IN FEET X 0,3048 = METERS

To convert leads in meters to mils:

$$\frac{\text{LEAD IN METERS} \times 1{,}000}{\text{RANGE TO TARGET}} = \text{MIL LEAD}$$

Section VII
NUCLEAR, BIOLOGICAL CHEMICAL

Performance of long-range precision fire is difficult at best. Enemy NBC warfare creates new problems for the sniper. Not only must the sniper properly execute the fundamentals of marksmanship and contend with the forces of nature, he must overcome obstacles presented by protective equipment. Testing conducted by the US Army Sniper School, Fort Benning, GA during 1989 to 1990 uncovered several problem areas. Evaluation of this testing discovered ways to help the sniper overcome these problems while firing in an NBC environment.

3-25. PROTECTIVE MASK

The greatest problem while firing the M24 with the M17-series protective mask was that of recoil breaking the seal of the mask. Also, due to filter elements and hard eye lenses, the sniper could not gain and maintain proper stock weld and eye relief. Additionally, the observer could not gain the required eye relief for observation through his M49 observation telescope. However, testing of the M25-series protective mask provided the following results:

a. Because of its separate filtering canister, the stock weld was gained and maintained with minimal effort.

b. Its flexible face shield allowed for excellent observation. This also allowed the sniper and observer to achieve proper eye relief, which was needed for observation with their respective telescopes.

3-26. MISSION-ORIENTED PROTECTION POSTURE

Firing while in MOPP has a significant effect on the ability to deliver precision fire. The following problems and solutions have been identified

a. **Eye Relief.** Special emphasis must be made in maintaining proper eye relief and the absence of scope shadow. Maintaining consistent stock weld is a must.

b. **Trigger Control.** Problems encountered with trigger control consist of the sense of touch and stock drag.

(1) ***Sense of touch.*** When gloves are worn, the sniper cannot determine the amount of pressure he is applying to the trigger. This is of particular importance if the sniper has the trigger adjusted for a light pull. 'Raining with a glove will be beneficial; however, the trigger should be adjusted to allow the sniper to feel the trigger without accidental discharge.

(2) ***Stock drag.*** While training, the sniper should have his observer watch his trigger finger to ensure that the finger and glove are not touching any part of the rifle but the trigger. The glove or finger resting on the trigger guard moves the rifle as the trigger is pulled to the rear. The sniper must wear a well-fitted glove.

c. **Vertical Sight Picture.** The sniper naturally cants the rifle into the cheek of the face while firing with a protective mask.

d. **Sniper/Observer Communications.** The absence of a voice emitter on the M2S-series protective mask creates an obstacle in relaying information. The team either speaks louder or uses written messages. A system of foot taps, finger taps, or hand signals may be devised. Communication is a must; training should include the development and practice of communications at different MOPP levels.

CHAPTER 4

FIELD TECHNIQUES

The primary mission of the sniper team is to eliminate selected enemy targets with long-range precision fire. How well the sniper accomplishes his mission depends on knowledge, understanding and application of various field techniques that allow him to move, hide, observe, and detect targets. This chapter discusses the field techniques and skills that the sniper must learn before employment in support of combat operations. The sniper's application of these skills will affect his survival on the battlefield.

Section I
CAMOUFLAGE

Camouflage is one of the basic weapons of war. It can mean the difference between a successful or unsuccessful mission. To the sniper team, it can mean the difference between life and death. Camouflage measures are important since the team cannot afford to be detected at any time while moving alone, as part of another element, or while operating from a firing position. Marksmanship training teaches the sniper to hit a target, and a knowledge of camouflage teaches him how to avoid becoming a target. Paying attention to camouflage fundamentals is a mark of a well-trained sniper. (See FM 5-20 for more details.)

4-1. TARGET INDICATORS

To become proficient in camouflage, the sniper team must first understand target indicators. Target indicators are anything a soldier does or fails to do that could result in detection. A sniper team must know and understand target indication not only to move undetected, but also to detect enemy movement. Target indicators are sound, movement, improper camouflage, disturbance of wildlife, and odors.

a. Sound.
- Most noticeable during hours of darkness.
- Caused by movement, equipment rattling, or talking.
- Small noises may be dismissed as natural, but talking will not.

b. Movement.
- Most noticeable during hours of daylight.
- The human eye is attracted to movement.
- Quick or jerky movement will be detected faster than slow movement.

c. Improper camouflage.
- Shine.
- Outline.
- Contrast with the background.

d. Disturbance of wildlife.
- Birds suddenly flying away.
- Sudden stop of animal noises.
- Animals being frightened.

e. Odors.
- Cooking.
- Smoking.
- Soap and lotions.
- Insect repellents.

4-2. BASIC METHODS

The sniper team can use three basic methods of camouflage. It may use one of these methods or a combination of all three to accomplish its objective. The three basic methods a sniper team can use are hiding, blending, and deceiving.

a. **Hiding.** Hiding is used to conceal the body from observation by lying behind an objector thick vegetation.

b. **Blending.** Blending is used to match personal camouflage with the surrounding area to a point where the sniper cannot be seen.

c. **Deceiving.** Deceiving is used to fool the enemy into false conclusions about the location of the sniper team.

4-3. TYPES OF CAMOUFLAGE

The two types of camouflage that the sniper team can use are *natural* and *artificial*.

a. **Natural.** Natural camouflage is vegetation or materials that are native to the given area. The sniper augments his appearance by using natural camouflage.

b. **Artificial.** Artificial camouflage is any material or substance that is produced for the purpose of coloring or covering something in order to conceal it. Camouflage sticks or face paints are used to cover all exposed areas of skin such as face, hands, and the back of the neck. The parts of the face that form shadows should be lightened, and the parts that shine should be darkened. The three types of camouflage patterns the sniper team uses are striping, blotching, and combination.

(1) ***Striping.*** Used when in heavily wooded areas and when leafy vegetation is scarce.

(2) ***Blotching.*** Used when an area is thick with leafy vegetation.

(3) ***Combination.*** Used when moving through changing terrain. It is normally the best all-round pattern.

4-4. GHILLIE SUIT

The ghillie suit is a specially made camouflage uniform that is covered with irregular patterns of garnish or netting (Figure 4-l).

a. Ghillie suits can be made from BDUs or one-piece aviator-type uniforms. Turning the uniform inside out places the pockets inside the suit. This protects items in the pockets from damage caused by crawling on the ground. The front of the ghillie suit should be covered with canvas or some type of heavy cloth to reinforce it. The knees and elbows should be covered with two layers of canvas, and the seam of the crotch should be reinforced with heavy nylon thread since these areas are prone to wear out quicker.

b. The garnish or netting should cover the shoulders and reach down to the elbows on the sleeves. The garnish applied to the back of the suit should be long enough to cover the sides of the sniper when he is in the prone position. A bush hat is also covered with garnish or netting. The garnish should belong enough to breakup the outline of the sniper's neck, but it should not be so long in front to obscure his vision or hinder movement.

e. A veil can be made from a net or piece of cloth covered with garnish or netting. It covers the weapon and sniper's head when in a firing position. The veil can be sewn into the ghillie suit or carried separately. A ghillie suit does not make one invisible and is only a camouflage base. Natural vegetation should be added to help blend with the surroundings.

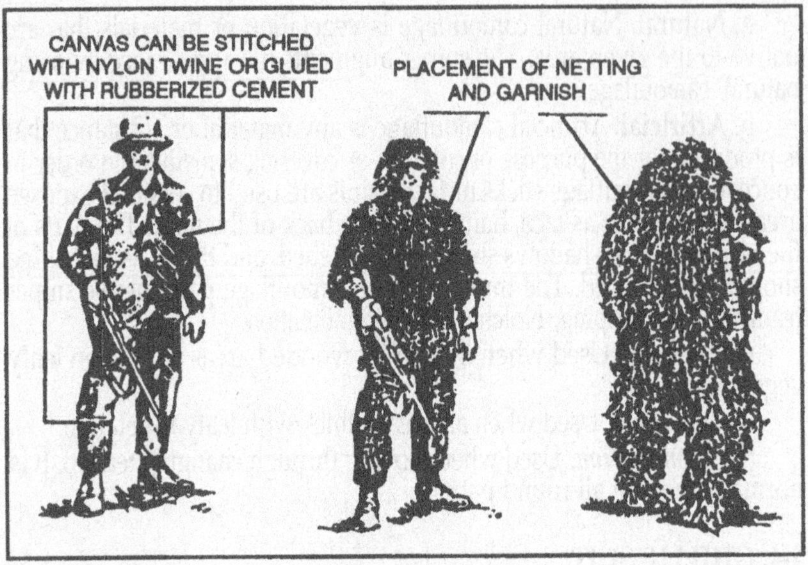

Figure 4-1. Ghillie suit.

4-5. FIELD-EXPEDIENT CAMOUFLAGE

The sniper team may have to use field-expedient camouflage if other means are not available. Instead of camouflage sticks or face paint, the team may use charcoal, walnut stain, mud, or whatever works. The team will not use oil or grease due to the strong odor. Natural vegetation can be attached to the body by boot bands or rubber bands or by cutting holes in the uniform.

a. The sniper team also camouflages its equipment. However, the camouflage must not interfere with or hinder the operation of the equipment.

(1) ***Rifles.*** The sniper weapon system and the M16/M203 should also. be camouflaged to break up their outlines. The sniper weapon system can be carried in a "drag bag" (Figure 4-2), which is a rifle case made of canvas and covered with garnish similar to the ghillie suit.

(2) ***Optics.*** Optics used by the sniper team must also be camouflaged to breakup the outline and to reduce the possibility of light reflecting off the lenses. Lenses can be covered with mesh-type webbing or nylon hose material.

(3) ***ALICE pack.*** If the sniper uses the ALICE pack while wearing the ghillie suit, he must camouflage the pack the same as the suit.

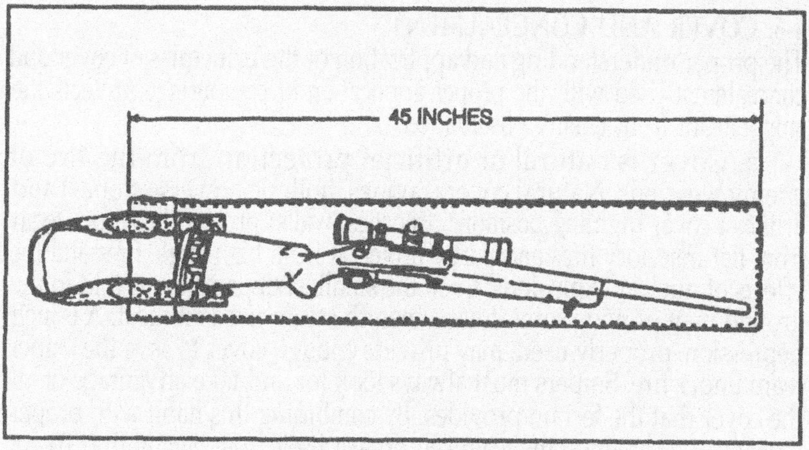

Figure 4-2. Drag bag.

b. The sniper team alters its camouflage to blend in with changes in vegetation and terrain in different geographic areas. Examples of such changes are as follows:

(1) ***Snow areas.*** Blending of colors is more effective than texture camouflage in snowy areas. In areas with heavy snow or in wooded areas with trees covered with snow, a full white camouflage suit should be worn. In areas with snow on the ground but not on the trees, white trousers with green and brown tops should be worn.

(2) ***Desert areas.*** In sandy desert areas that have little vegetation, the blending of tan and brown colors is important. In these areas, the sniper team must make full use of the terrain and the vegetation that is available to remain unnoticed.

(3) ***Jungle areas.*** In jungle areas, textured camouflage, contrasting colors, and natural vegetation must be used.

(4) ***Urban areas.*** In urban areas, the sniper team's camouflage should be a blended color (shades of gray usually work best). Texutred camouflage is not as important in these environments.

c. The sniper team must be camouflage conscious from the time it departs on a mission until it returns. It must constantly use the terrain, vegetation, and shadows to remain undetected. At no other time during the mission will the sniper team have a greater tendency to be careless than during its return to a friendly area. Fatigue and undue haste may override caution and planning. Therefore, the team needs to pay close attention to its camouflage discipline on return from missions.

4-6. COVER AND CONCEALMENT

The proper understanding and application of the principles of cover and concealment used with the proper application of camouflage protects the sniper team from enemy observation.

a. Cover is natural or artificial protection from the fire of enemy weapons. Natural cover (ravines, hollows, reverse slopes) and artificial cover (fighting positions, trenches, walls) protect the sniper team from flat trajectory fires and partly protect it from high-angle fires and the effects of nuclear explosions. Even the smallest depression or fold in the ground may provide some cover when the team needs it most. A 6-inch depression, properly used, may provide enough cover to save the sniper team under fire. Snipers must always look for and take advantage of all the cover that the terrain provides. By combining this habit with proper movement techniques, the team can protect itself from enemy fire. To get protection from enemy fire when moving, the team uses routes that put cover between itself and the enemy.

b. Concealment is natural or artificial protection from enemy observation. The surroundings may provide natural concealment that needs no change before use (bushes, grass, and shadows). The sniper team creates artificial concealment from materials such as burlap and camouflage nets, or it can move natural materials (bushes, leaves, and grass) from their original location. The sniper team must consider the effects of the change of seasons on the concealment provided by both natural and artificial materials. 'he principles of concealment include the following

(1) ***Avoid unnecessary movement.*** Remain still—movement attracts attention. The position of the sniper team is concealed when the team remains still, but the sniper's position is easily detected when the team moves. Movement against a stationary background makes the team stand out clearly. When the team must change positions, it moves carefully over a concealed route to a new position, preferably during limited visibility. Snipers move inches at a time, slowly and cautiously, always scanning ahead for the next position.

(2) ***Use all available concealment.*** Available concealment includes the following

(a) *Background.* Background is important the sniper team must blend with it to prevent detection. The trees, bushes, grass, earth, and man-made structures that form the background vary in color and appearance. This makes it possible for the team to blend with them. The team selects trees or bushes to blend with the uniform and to absorb the figure outline. Snipers must always assume they are under observation.

(b) *Shadows.* The sniper team in the open stands out clearly, but the sniper team in the shadows is difficult to see. Shadows exist under most conditions, day and night. A sniper team should never fire from the edge of a wood line; it should fire from a position inside the wood line (in the shade or shadows provided by the tree tops).

(3) ***Stay low to observe.*** A low silhouette makes it difficult for the enemy to see a sniper team. Therefore, the team observes from a crouch, a squat, or a prone position.

(4) ***Avoid shiny reflections.*** Reflection of light on a shiny surface instantly attracts attention and can be seen from great distances. The sniper uncovers his rifle scope only when indexing and aiming at a target. He uses optics cautiously in bright sunshine because of the reflections they cause.

(5) ***Avoid skylining.*** Figures on the skyline can be seen from a great distance, even at night, because a dark outline stands out against the lighter sky. The silhouette formed by the body makes a good target.

(6) ***Alter familiar outlines.*** Military equipment and the human body are familiar outlines to the enemy. The sniper team alters or disguises these revealing shapes by using the ghillie suit or outer smock that is covered with irregular patterns of garnish. The team must alter its outline from the head to the soles of the boots.

(7) ***Observe noise discipline.*** Noise, such as talking, can be picked up by enemy patrols or observation posts. The sniper team silences gear before a mission so that it makes no sound when the team walks or runs.

Section II
MOVEMENT

A sniper team's mission and method of employment differ in many ways from those of the infantry squad. One of the most noticeable differences is the movement technique used by the sniper team. Movement by teams must not be detected or even suspected by the enemy. Because of this, a sniper team must master individual sniper movement techniques.

4-7. RULES OF MOVEMENT

When moving, the sniper team should always remember the following rules

a. Always assume the area is under enemy observation.

b. Move slowly. A sniper counts his movement progress by feet and inches.

c. Do not cause overhead movement of trees, bushes, or tall grasses by rubbing against them.

d. Plan every movement and move in segments of the route at a time.

e. Stop, look, and listen often.

f. Move during disturbances such as gunfire, explosions, aircraft noise, wind, or anything that will distract the enemy's attention or conceal the team's movement.

4-8. INDIVIDUAL MOVEMENT TECHNIQUES

The individual movement techniques used by the sniper team are designed to allow movement without being detected. These movement techniques are sniper low crawl, medium crawl, high crawl, hand-and-knees crawl, and walking.

a. **Sniper Low Crawl.** The sniper low crawl (Figure 4-3) is used when concealment is extremely limited, when close to the enemy, or when occupying a firing position.

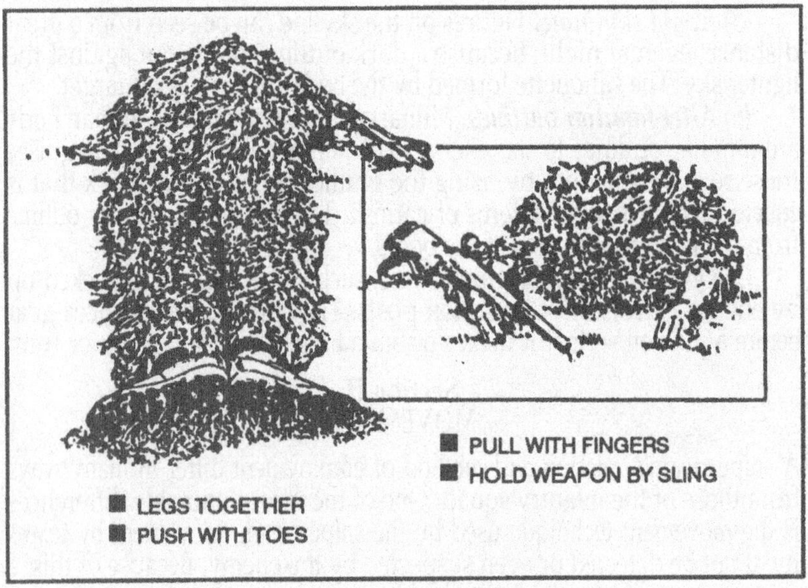

Figure 4-3. Sniper low crawl.

b. **Medium Crawl.** The medium crawl (Figure 4-4) is used when concealment is limited and the team needs to move faster-than the sniper low crawl allows. The medium crawl is similar to the infantryman's low crawl.

c. **High Crawl.** The high crawl (Figure 4-5) is used when concealment is limited but high enough to allow the sniper to raise his body off the ground. The high crawl is similar to the infantry high crawl.

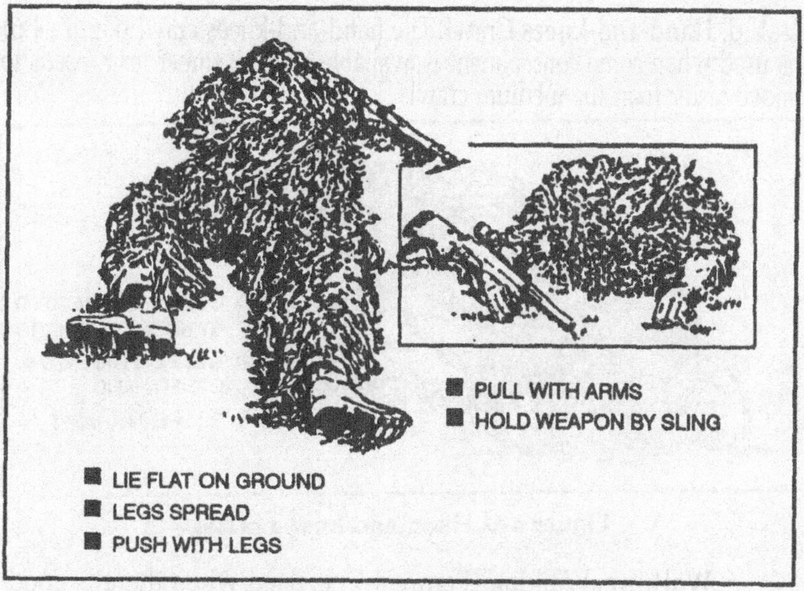

Figure 4-4. Medium crawl.

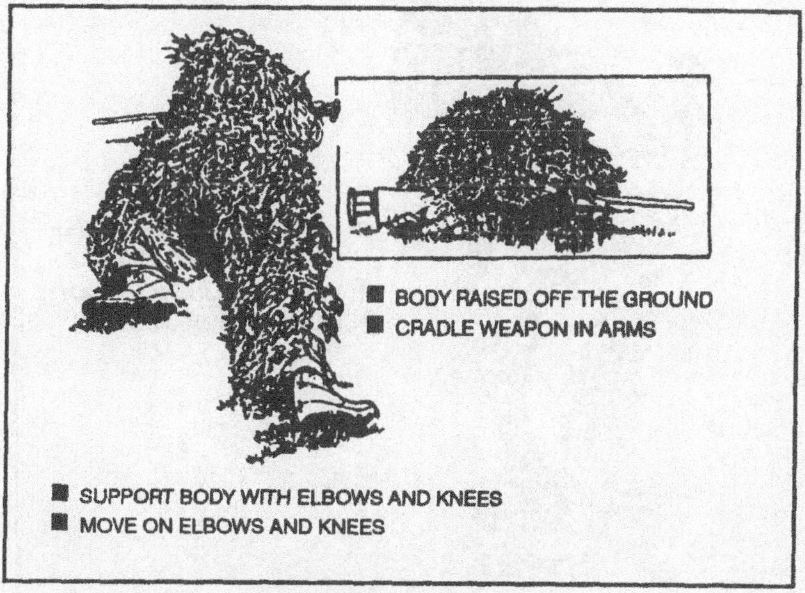

Figure 4-5. High crawl.

d. **Hand-and-knees Crawl.** The hand-and-knees crawl (Figure 4-6) is used when some concealment is available and the sniper team needs to move faster than the medium crawl.

Figure 4-6. Hand-and-knees crawl.

e. **Walking.** Walking (Figure 4-7) is used when there is good concealment, it is not likely the enemy is close, and speed is required.

Figure 4-7. Walking.

4-9. SNIPER TEAM MOVEMENT AND NAVIGATION

Due to lack of personnel and firepower, the sniper team cannot afford detection by the enemy nor can it successfully fight the enemy in sustained engagements.

a. When possible, the sniper team should be attached to a security element (squad/platoon). The security element allows the team to reach its area of operations quicker and safer than the team operating alone. Plus, the security element provides the team a reaction force should the team be detected. Snipers use the following guidelines when attached to a security element:

(1) The security element leader is in charge of the team while it is attached to the element.

(2) The sniper team always appears as an integral part of the element.

(3) The sniper team wears the same uniform as the element members.

(4) The sniper team maintains proper intends and positions in all formations.

(5) The sniper weapon system is carried in line and close to the body, hiding its outline and barrel length.

(6) All equipment that is unique to sniper teams is concealed from view (optics, ghillie suits, and so forth).

b. Once in the area of operation, the sniper team separates from the security element and operates alone. Two examples of a sniper team separating from security elements are as follows:

(1) The security element provides security while the team prepares for operation.

(a) The team dons the ghillie suits and camouflages itself and its equipment (if mission requires).

(b) The team ensures all equipment is secure and caches any nonessential equipment (if mission requires).

(c) Once the team is prepared, it assumes a concealed position, and the security element departs the area.

(d) Once the security element has departed, the team waits in position long enough to ensure neither itself nor the security element has been compromised. Then, the team moves to its tentative position.

(2) The security element conducts a short security halt at the separation point. The sniper team halts, ensuring they have good available concealment and know each other's location. The security element then proceeds, leaving the sniper team in place. The sniper team remains in position until the security element is clear of the area. The team then organizes itself as

required by the mission and moves on to its tentative position. This type of separation also works well in MOUT situations.

c. When selecting routes, the sniper team must remember its strengths and weaknesses. The following guidelines should be used when selecting routes:

(1) Avoid known enemy positions and obstacles.

(2) Seek terrain that offers the best cover and concealment.

(3) Take advantage of difficult terrain (swamps, dense woods, and so forth).

(4) Do not use trails, roads, or footpaths.

(5) Avoid built-up or populated areas.

(6) Avoid areas of heavy enemy guerrilla activity.

d. When the sniper team moves, it must always assume its area is under enemy observation. Because of this and the size of the team with the small amount of firepower it has, the team uses only one type of formation-the sniper movement formation. Characteristics of the formation are as follows:

(1) The observer is the point man; the sniper follows.

(2) The observer's sector of security is 3 o'clock to 9 o'clock; the sniper's sector of security is 9 o'clock to 3 o'clock (overlapping).

(3) Visual contact must be maintained even when lying on the ground.

(4) An interval of no more than 20 meters is maintained.

(5) The sniper reacts to the point man's actions.

(6) The team leader designates the movement techniques and routes used.

(7) The team leader designates rally points.

e. A sniper team must never become decisively engaged with the enemy. The team must rehearse immediate action drills to the extent that they become a natural and immediate reaction should it make unexpected contact with the enemy. Examples of such actions are as follows:

(1) *Visual contact.* If the sniper team sees the enemy and the enemy does not see the team, it freezes. If the team has time, it will do the following

(a) Assume the best covered and concealed position.

(b) Remain in position until the enemy has passed.

NOTE: The team will not initiate contact.

(2) ***Ambush.*** In an ambush, the sniper team's objective is to break contact immediately. One example of this involves performing the following

(a) The observer delivers rapid fire on the enemy.

(b) The sniper throws smoke grenades between the observer and the enemy.

(c) The sniper delivers well-aimed shots at the most threatening targets until smoke covers the area.

(d) The observer then throws fragmentation grenades and withdraws toward the sniper, ensuring he does not mask the sniper's fire.

(e) The team moves to a location where the enemy cannot observe or place direct fire on it.

(f) If contact cannot be broken, the sniper calls for indirect fires or a security element (if attached).

(g) If team members get separated, they should return to the next-to-last designated en route rally point.

(3) ***Indirect fire.*** When reacting to indirect fires, the team must move out of the area as quickly as possible. This sudden movement can result in the team's exact location and direction being pinpointed. Therefore, the team must not only react to indirect fire but also take actions to conceal its movement once it is out of the impact area.

(a) The team leader moves the team out of the impact area using the quickest route by giving the direction and distance (clock method).

(b) Team members move out of the impact area the designated distance and direction.

(c) The team leader then moves the team farther away from the impact area by using the most direct concealed route. They continue the mission using an alternate route.

(d) If team members get separated, they should return to the next-to-last designated en route rally point.

(4) ***Air attack.***

(a) Team members assume the best available covered and concealed positions.

(b) Between passes of aircraft, team members move to positions that offer better cover and concealment.

(c) The team does not engage the aircraft.

(d) Team members remain in positions until attacking aircraft depart.

(e) If team members get separated, they return to the next-to-last designated en route rally point.

f. To aid the sniper team in navigation, the team should memorize the route by studying maps, aerial photos, or sketches. The team notes distinctive features (hills, streams, roads) and its location in relation to the route. It plans an alternate route in case the primary route cannot be used. It plans offsets to circumvent known obstacles to movement. The team uses terrain countdown, which involves memorizing terrain features from the start point to the objective, to maintain the route. During the mission, the sniper team mentally counts each terrain feature, thus ensuring it maintains the proper route.

g. The sniper team maintains orientation at all times. As it moves, it observes the terrain carefully and mentally checks off the distinctive features noted in the planning and study of the route. Many aids are available to ensure orientation. The following are examples:

(1) The location and direction of flow of principal streams.

(2) Hills, valleys, roads, and other peculiar terrain features.

(3) Railroad tracks, power lines, and other man-made objects.

Section III
SELECTION, OCCUPATION, AND CONSTRUCTION OF SNIPER POSITIONS

Selecting the location for a position is one of the most important tasks a sniper team accomplishes during the mission planning phase of an operation. After selecting the location, the team also determines how it will move into the area to locate and occupy the final position.

4-10. SELECTION

Upon receiving a mission, the sniper team locates the target area and then determines the best location for a tentative position by using one or more of the following sources of information: topographic maps, aerial photographs, visual reconnaissance before the mission, and information gained from units operating in the area.

a. The sniper team ensures the position provides an optimum balance between the following considerations:

- Maximum fields of fire and observation of the target area.
- Concealment from enemy observation.
- Covered routes into and out of the position.
- Located no closer than 300 meters from the target area.
- A natural or man-made obstacle between the position and the target area.

b. A sniper team must remember that a position that appears to be in an ideal location may also appear that way to the enemy. Therefore, the team avoids choosing locations that are—

- On a point or crest of prominent terrain features.
- Close to isolated objects.
- At bends or ends of roads, trails, or streams.
- In populated areas, unless it is required.

c. The sniper team must use its imagination and ingenuity in choosing a good location for the given mission. The team chooses a location that not only allows the team to be effective but also must appear to the enemy to be the least likely place for a team position. The following are examples of such positions:

- Under logs in a deadfall area.
- Tunnels bored from one side of a knoll to the other.
- Swamps.
- Deep shadows.
- Inside rubble piles.

4-11. OCCUPATION

During the mission planning phase, the sniper also selects an objective rally point. From this point, the sniper team reconnoiters the tentative position to determine the exact location of its final position. The location of the ORP should provide cover and concealment from enemy fire and observation, be located as close to the selected area as possible, and have good routes into and out of the selected area.

a. From the ORP, the team moves foward to a location that allows the team to view the tentative position area (Figure 4-8 page 4-16). One member remains in this location to cover the other member who reconnoiters the area to locate a final position. Once a suitable location has been found, the covering team member moves to the position. While conducting the reconnaissance or moving to the position, the team—

- Moves slowly and deliberately, using the sniper low crawl.
- Avoids unnecessary movement of trees, bushes, and grass.
- Avoids making any noises.
- Stays in the shadows, if there are any.
- Stops, looks, and listens every few feet.

b. When the sniper team arrives at the firing position, it—

- Conducts a detailed search of the target area.
- Starts construction of the firing position, if required.

- Organizes equipment so that it is easily accessible.
- Establishes a system of observing eating resting, and latrine calls.

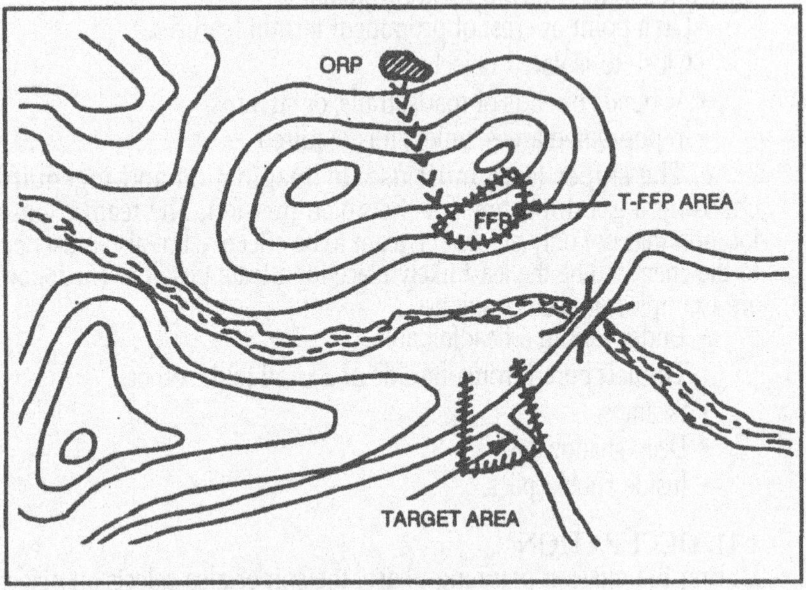

Figure 4-8. Tentative position areas.

4-12. CONSTRUCTION

A sniper mission always requires the team to occupy some type of position. These positions can range from a hasty position, which a team may use for a few hours, to a more permanent position, which the team could occupy. for a few days. The team should always plan to build its position during limited visibility.

a. **Sniper Position Considerations.** Whether a sniper team is in a position for a few minutes or a few days, the basic considerations in. choosing a type of position remain the same.

(1) ***Location:***

(a) *Type of terrain and soil.* Digging and boring of tunnels can be very difficult in hard soil or in fine, loose sand. The team takes advantage of what the terrain offers (gullies, holes, hollow tree stumps, and so forth).

(b) *Enemy location and capabilities.* Enemy patrols in the area may be close enough to the position to hear any noises that may accidentally be made during any construction. The team also considers the enemy's night vision and detection capabilities.

(2) ***Time:***

(a) *Amount of time to be occupied.* If the sniper team's mission requires it to be in position for a long time, the team constructs a position that provides more survivability. This allows the team to operate more effectively for a longer time.

(b) *Time required for construction.* The time required to build a position must be considered, especially during the mission planning phase.

(3) ***Personnel and equipment:***

(a) *Equipment needed for construction.* The team plans for the use of any extra equipment needed for construction (bow saws, picks, axes, and so forth).

(b) *Personnel needed for construction.* Coordination is made if the position requires more personnel to build it or a security element to secure the area during construction.

b. **Construction Techniques.** Belly and semipermanent hide positions can be constructed of stone, brick, wood, or turf. Regardless of material, every effort is made to bulletproof the front of the hide position. The team can use the following techniques:

- Pack protective jackets around the loophole areas.
- Emplace an angled armor plate with a loophole cut into it behind the hide loophole.
- Sandbag the loopholes from the inside.

(1) ***Pit.*** Hide construction begins with the pit since it protects the sniper team. All excavated dirt is removed (placed in sandbags, taken away on a poncho, and so forth) and hidden (plowed fields, under a log, or away from the hide site).

(2) ***Overhead cover.*** In a semipermanent hide position, logs should be used as the base of the roof. The sniper team places a dust cover over the base (such as a poncho, layers of empty sandbags, or canvas), a layer of dirt, and a layer of gravel, if available. The team spreads another layer of dirt, and then adds camouflage. Due to the various materials, the roof is difficult to conceal if not countersunk.

(3) ***Entrance.*** To prevent detection, the sniper team should construct an entrance door sturdy enough to bear a man's weight.

(4) ***Loopholes.*** The construction of loopholes (Figure 4-9, page 4-18) requires care and practice to ensure they afford adequate fields of fire. Loopholes must be camouflaged by foliage or other material that blends with or is natural to the surroundings.

(5) ***Approaches.*** It is vital that the natural appearance of the ground remains unaltered and camouflage blends with the surroundings.

Construction time is wasted if the enemy observes a team entering the hide; therefore, approached must be concealed. Teams try to enter the hide during darkness, keeping movement to a minimum and adhering to trail discipline. In built-up areas, a secure and quiet approach is needed. Teams must avoid drawing attention to the mission and carefully plan movement. A possible ploy is to use a house search with sniper gear hidden among other gear. Sewers may be used for movement also.

Figure 4-9. Loopholes in hide position.

c. **Hasty Position.** A hasty position is used when the sniper team is in a position for a short time and cannot construct a position due to the location of the enemy, or immediately assumes a position. The hasty position is characterized by the following

(1) ***Advantages:***

(a) *Requires no construction* The sniper team uses what is available for cover and concealment.

(b) *Can be occupied in a short time.* As soon as a suitable position is found, the team need only prepare loopholes by moving small amounts of vegetation or by simply backing a few feet away from the vegetation that is already thereto conceal the weapon's muzzle blast.

(2) ***Disadvantages:***

(a) *Affords no freedom of movement.* Any movement that is not slow and deliberate may result in the team being compromised.

(b) *Restricts observation of large areas.* This type of position is normally used to observe a specific target area (intersection, passage, or crossing).

(c) *Offers no protection from direct or indirect fires.*

(d) *Relies heavily on personal camouflage.* The team's only protection against detection is personal camouflage and the ability to use the available terrain.

(3) ***Occupation time.*** The team should not remain in this type of position longer than eight hours.

d. ***Expedient Position.*** When a sniper team is required to remain in position for a longer time than the hasty position can provide, an expedient position (Figure 4-10) should be constructed. The expedient position lowers the sniper's silhouette as low to the ground as possible, but it still allows him to fire and observe effectively. The expedient position is characterized by the following

(1) ***Advantages:***

(a) *Requires little construction.* This position is constructed by digging a hole in the ground just large enough for the team and its equipment. Soil dug from this position can be placed in sandbags and used for building firing platforms.

(b) *Conceals most of the body and equipment.* The optics, rifles, and heads of the sniper team are the only items that are above ground level in this position.

(c) Provides some protection from direct fires due to its lower silhouette.

Figure 4-10. Expedient position.

(2) ***Disadvantages:***

(a) *Affords little freedom of movement.* The team has more freedom of movement in this position than in the hasty position. Team members can lower their heads below ground level slowly to ensure a target indicator is not produced.

(b) *Allows little protection from indirect fires.* This position does not protect the team from shrapnel and debris falling into the position.

(c) *Exposes the head, weapons, and optics.* The team must rely heavily on the camouflaging of these exposed areas.

(3) ***Construction time:*** 1 to 3 hours (depending on the situation).

(4) ***Occupation time:*** 6 to 12 hours.

e. **Belly Hide.** The belly hide (Figure 4-11) is similar to the expedient position, but it has overhead cover that not only protects the team from the effects of indirect fires but also allows more freedom of movement. This position can be dugout under a tree, a rock, or any available object that provides overhead protection and a concealed entrance and exit. The belly hide is characterized by the following

(1) *Advantages:*

(a) *Allows some freedom of movement.* The darkened area inside this position allows the team to move freely. The team must remember to cover the entrance/exit door so outside light does not silhouette the team inside the position or give the position away.

(b) *Conceals all but the rifle barrel.* All equipment is inside the position except the rifle barrels. Depending on the room available to construct the position, the rifle barrels may also be inside.

(c) *Provides protection from direct and indirect fires.* The team should try to choose a position that has an object that will provide good overhead protection (rock tracked vehicle, rubble pile, and so forth), or prepare it in the same manner as overhead cover for other infantry positions.

(2) ***Disadvantages:***

(a) *Requires extra construction time.*

(b) *Requires extra materials and tools.* Construction of overhead cover requires saws or axes, waterproof material, and so forth.

(c) *Has limited space.* The sniper team will have to lay in the belly hide without a lot of variation in body position due to limited space and design of the position.

(3) ***Construction time:*** 4 to 6 hours.

(4) ***Occupation time:*** 12 to 48 hours.

Figure 4-11. Belly hide position.

f. **Semipermanent Hide.** The semipermanent hide (Figure 4-12, page 4-22) is used mostly in defensive situations. This position requires additional equipment and personnel to construct. However, it allows sniper teams to remain in place for extended periods or to be relieved in place by other sniper teams. Like the belly hide, this position can be constructed by tunneling through a knoll or under natural objects already in place. The semipermanent hide is characterized by the following

(1) **Advantages:**

(a) *Offers total freedom of movement inside the position.* The team members can move about freely. They can stand, sit, or even lie down.

(b) *Protects against direct and indirect fires.* The sniper team should look for the same items as mentioned in the belly hide.

(c) *Is completely concealed.* Loopholes are the only part of the position that can be detected. They allow for the smallest exposure possible; yet they still allow the sniper and observer to view the target area. These loopholes should have a large diameter (10 to 14 inches) in the interior of the position and taper down to a smaller diameter (4 to 8 inches) on the outside of the position. A position may have more than two sets of loopholes if needed to cover large areas. The entrance/exit to the position must be covered to prevent light from entering and highlighting the loopholes. Loopholes that are not in use should be covered from the inside with a piece of canvas or suitable material.

(d) *Is easily maintained for extended periods.* This position allows the team to operate effectively for a longer period.

(2) ***Disadvantages:***

(a) *Requires extra personnel and tools to construct.* This position requires extensive work and extra tools. It should not be constructed near the enemy. It should be constructed during darkness and be completed before dawn.

Figure 4-12. Semipermanent hide position.

(b) *Increases risk of detection.* Using a position for several days or having teams relieve each other in a position always increases the risk of detection.

(3) ***Construction time:*** 4 to 6 hours (4 personnel).

(4) ***Occupation time:*** 48 hours plus (relieved by other teams).

g. **Routines in Sniper Team positions.** Although the construction of positions may differ, the routines while in position are the same. The sniper and the observer should have a good firing platform. This gives the sniper a stable platform for the sniper weapon and the observer a platform for the optics. When rotating observation duties, the sniper weapon should remain in place, and the optics are handed from one member to the other. Sniper data book, observation logs, range cards, and the radio should be placed between the team where both members have easy access to them. A system of resting, eating, and latrine calls must be arranged between the team. All latrine calls should be done during darkness, if possible. A hole should be dug to conceal any traces of latrine calls.

4-13. POSITIONS IN URBAN TERRAIN

Positions in urban terrain are quite different than positions in the field. The sniper team normally has several places to choose. These can range from inside attics to street-level positions in basements. This type of terrain is ideal for a sniper, and a sniper team can stop an enemy's advance through its area of responsibility.

a. When constructing an urban position, the sniper team must be aware of the outside appearance of the structure. Shooting through loopholes in barricaded windows is preferred; the team must make sure all other windows are also barricaded. Building loopholes in other windows also provides more positions to engage targets. When building loopholes, the team should make them different shapes (not perfect squares or circles). Dummy loopholes also confuse the enemy. Positions in attics are also effective. The team removes the shingles and cuts out loopholes in the roof; however, they must make sure there are other shingles missing from the roof so the firing position loophole is not obvious.

(1) The sniper team should not locate the position against contrasting background or in prominent buildings that automatically draw attention. It must stay in the shadows while moving, observing, and engaging targets.

(2) The team must never fire close to a loophole. It should always back away from the hole as far as possible to hide the muzzle flash and to scatter the sound of the weapon when it fires. The snipers may be located in a different room than the loophole; however, they can make a hole through a wall to connect the rooms and fire from inside one room. The team must not fire continually from one position. (More than one position should be constructed if time and situation permit.) When constructing other positions, the team makes sure the target area can be observed. Sniper team positions should never be used by any personnel other than a sniper team.

b. Common sense and imagination are the sniper team's only limitation in the construction of urban hide positions. Urban hide positions that can be used are the room hide, crawl space hide, and rafter hide. The team constructs and occupies one of these positions or a variation thereof.

> **WARNING**
> **WHEN MOVING THROUGH SEWERS, TEAMS MUST BE ALERT FOR BOOBY TRAPS AND POISONOUS GASES.**

(1) ***Room hide position.*** In a room hide position, the sniper team uses an existing room and fires through a window or loophole (Figure 4-13). Weapon support may be achieved through the use of existing furniture-that is, desks or tables. When selecting a position, teams must notice both front and back window positions. To avoid. silhouetting, they may need to use a backdrop such as a dark-colored blanket, canvas, carpet, and a screen. Screens (common screening material) are important since they allow the sniper teams maximum observation and deny observation by the enemy. They must not. remove curtains; however, they can open windows or remove panes of glass. Remember, teams can randomly remove panes in other windows so the position is not obvious.

(2) ***Crawl space hide position.*** The sniper team builds a crawl space hide position in the space between floors in multistory buildings (Figure 4-14). Loopholes are difficult to construct, but a damaged building helps considerably. Escape routes can be holes knocked into the floor or ceiling. Carpet or furniture placed over escape holes or replaced ceiling tiles will conceal them until needed.

Figure 4-13. Room hide position.

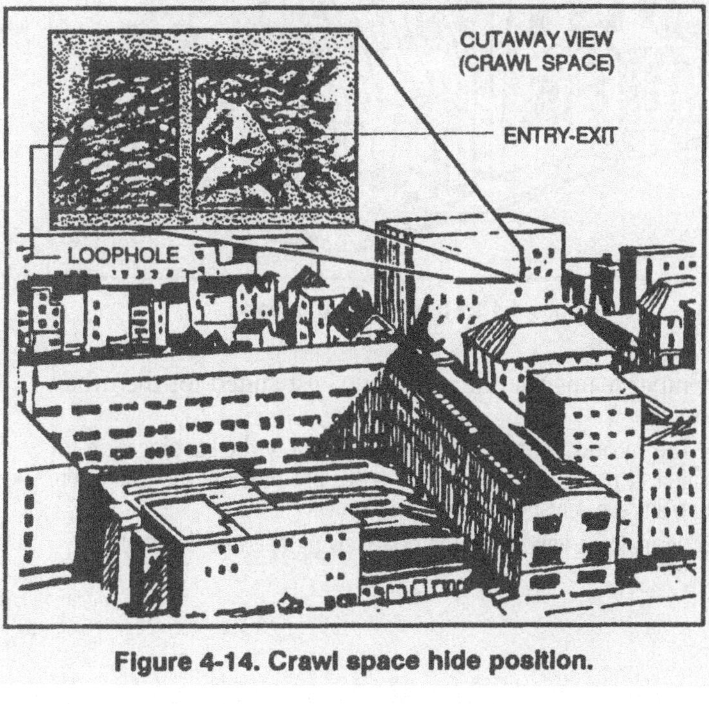

Figure 4-14. Crawl space hide position.

(3) ***Rafter hide position.*** The sniper team constructs a rafter hide position in the attic of an A-frame-type building. These buildings normally have shingled roofs (A and B, Figure 4-15). Firing from inside the attic around a chimney or other structure helps prevent enemy observation and fire.

Figure 4-15. Rafter hide positions.

c. Sniper teams use the technique best suited for the urban hide position.

(1) The second floor of a building is usually the best location for the position. It presents minimal dead space but provides the team more protection since passersby cannot easily spot it.

(2) Normally, a window is the best viewing aperture/loophole.

(a) If the window is dirty, do not clean it for better viewing.

(b) If curtains are prevalent in the area, do not remove those in the position. Lace or net-type curtains can be seen through from the inside, but they are difficult to see through from the outside.

(c) If strong winds blow the curtains open, staple, tack, or weight them.

(d) Firing a round through a curtain has little effect on accuracy however, ensure the muzzle is far enough away to avoid muzzle blast.

(e) When area routine indicates open curtains, follow suit. Set up well away from the loophole; however, ensure effective coverage of the assigned target area.

(3) Firing through glass should be avoided since more than one shot may be required. The team considers the following options:

(a) Break or open several windows throughout the position before occupation. This can be done during the reconnaissance phase of the operation; however, avoid drawing attention to the area.

(b) Remove or replace panes of glass with plastic.

(4) Other loopholes/viewing apertures are nearly unlimited.

- Battle damage.
- Drilled holes (hand drill).
- Brick removal.
- Loose boards/derelict houses.

(5) Positions can also beset up in attics or between the ceiling and roof. (See rafter hide positions.)

- Gable ends close to the eaves (shadow adding to concealment).
- Battle damage to gables and or roof.
- Loose or removed tiles, shingles, or slates.
- Skylights.

(6) The sniper makes sure the bullet clears the loophole. The muzzle must be far enough from the loophole to ensure the bullet's path is not in line with the bottom of the loophole.

(7) Front drops, usually netting, may have to be changed (if the situation permits) from dark to light colors at BMNT/EENT due to sunlight or lack of sunlight into the position.

(8) If the site is not multiroomed, partitions can be made by hanging blankets or nets to separate the operating area from the rest/administrative area.

(9) If sandbags are required, they can be filled and carried inside of rucksacks or can be filled in the basement, depending on the situation/location of the position site.

(10) Always plan an escape route that leads to the objective rally point. When forced to vacate the position, the team meets the security element at the ORP. Normally, the team will not be able to leave from the same point at which it gained access; therefore, a separate escape point may be required in emergency situations. The team must consider windows (other than the viewing apertures); anchored ropes to climb down buildings, or a small, preset explosive charge situated on a wall or floor for access into adjoining rooms, buildings, or the outside.

(11) The type of uniform or camouflage to be worn by the team will be dictated by the situation, how they are employed, and area of operation. The following applies:

(a) Most often, the BDU and required equipment are worn.

(b) Urban-camouflaged uniforms can be made or purchased. Urban areas vary in color (mostly gray [cinder block]; red [brick]; white [marble]; black [granite]; or stucco, clay, or wood). Regardless of area color, uniforms should include angular-line patterns.

(c) When necessary, most woodland-patterned BDUs can be worn inside out as they are a gray or green-gray color underneath.

(d) Soft-soled shoes or boots are the preferred footwear in the urban environment.

(e) Civilian clothing can be worn (native/host country populace).

(f) Tradesmen's or construction worker's uniforms and accessories can be used.

Section IV
OBSERVATION

Throughout history, battles have been won and nations conquered based on an accurate accounting and description of the opposing forces strength, equipment, and location. As the sniper team performs the secondary mission of collecting and reporting battlefield intelligence, the commander can act, rather than react. The purpose of observation is to gather facts and to provide information for a specific intent. Observation uses all of the sniper team's five senses but often depends on sight and hearing. For example, the sniper team is issued a PIR or OIR for a specific mission. Information gathered by the sniper team is reported, analyzed, and processed into intelligence reports. The sniper team's success depends upon its powers of observation. In addition to the sniperscope, the sniper team has an observation telescope, binoculars, night vision sight, and night vision goggles to enhance its ability to observe and engage targets. Team members must relieve each other when using this equipment since prolonged use can cause eye fatigue, greatly reducing the effectiveness of observation. Team members rotate periods of observation. During daylight, observation should be limited to 10 minutes followed by a 10-minute rest. When using night vision devices, the observer should limit his initial period of viewing to 10 minutes followed by a 10-minute rest. After several periods of viewing, he can extend the viewing period to 15 minutes and then a 15-minute rest.

4-14. HASTY AND DETAILED SEARCHES

While observing a target area, the sniper team alternately conducts two types of visual searches: hasty and detailed.

a. A hasty search is the first phase of observing a target area. The observer conducts a hasty search immediately after the team occupies the firing position. A hasty search consists of quick glances with binoculars at specific points, terrain features, or other areas that could conceal the enemy. The observer views the area closest to the team's position first since it could pose the most immediate threat. The observer then searches farther out until the entire target area has been searched. When the observer sees or suspects a target, he uses an M49 observation telescope for a detailed view of the target area. The telescope should not be used to search the area because its narrow field of view would take much longer to cover an area; plus, its stronger magnification can cause eye fatigue sooner than the binoculars.

b. After a hasty search has been completed, the observer then conducts a detailed search of the area. A detailed search is a closer, more thorough search of the target area, using 180-degree area or sweeps, 50 meters in depth, and overlapping each previous sweep at least 10 meters to ensure the entire area has been observed (Figure 4-16, page 4-30). Like the hasty search, the observer begins by searching the area closest to the sniper team position.

c. This cycle of a hasty search followed by a detailed search should be repeated three or four times. This allows the sniper team to become accustomed to the area; plus, the team will look closer at various points with each consecutive pass over the area. After the initial searches, the observer should view the area, using a combination of both hasty and detailed searches. While the observer conducts the initial searches of the area, the sniper should record prominent features, reference points, and distances on a range card. The team members should alternate the task of observing the area about every 30 minutes.

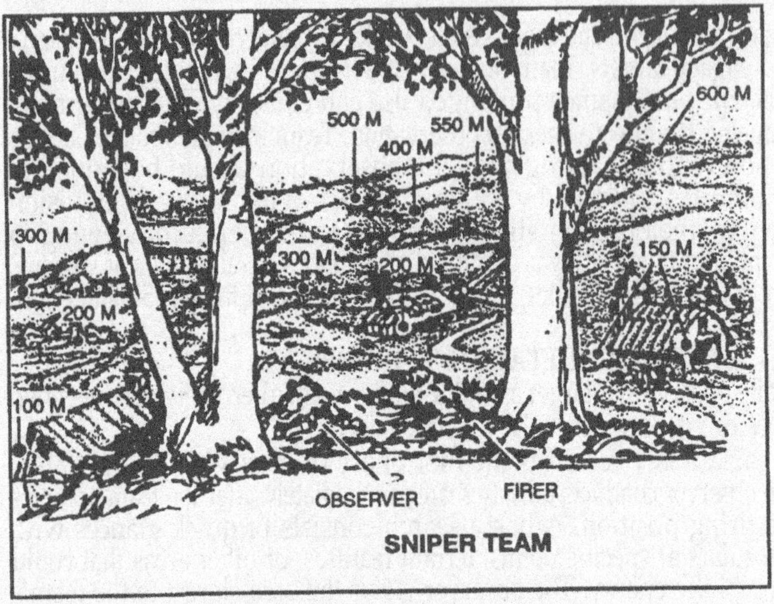

Figure 4-16. Detailed search.

4-15. ELEMENTS OF OBSERVATION

The four elements in the process of observation include awareness, understanding, recording, and response. Each of these elements may be accomplished as a separate processor accomplished at the same time.

a. **Awareness.** Awareness is being consciously attuned to a specific fact. A sniper team must always be aware of the surroundings and take nothing. for granted. The team also considers certain elements that influence and distort awareness.

(1) An object's size and shape can be misinterpreted if viewed incompletely or inaccurately.

(2) Distractions degrade the quality of observations.

(3) Active participation or degree of interest can diminish toward the event.

(4) Physical abilities (five senses) have limitations.

(5) Environmental changes affect accuracy.

(6) Imagination may cause possible exaggerations or inaccuracy.

b. **Understanding.** Understanding is derived from education, training, practice, and experience. It enhances the sniper team's knowledge about what should be observed, broadens its ability to view and consider all aspects, and aids in its evaluation of information.

c. **Recording.** Recording is the ability to save and recall what was observed. Usually, the sniper team has mechanical aids, such as writing utensils, sniper data book, sketch kits, tape recorders, and cameras, to support the recording of events; however, the most accessible method is memory. The ability to record, retain, and recall depends on the team's mental capacity (and alertness) and ability to recognize what is essential to record. Added factors that affect recording include

(1) The amount of training and practice in observation.

(2) Skill gained through experience.

(3) Similarity of previous incidents.

(4) Time interval between observing and recording.

(5) The ability to understand or convey messages through oral or other communications.

d. **Response.** Response is the sniper team's action toward information. It may be as simple as recording events in a sniper data book, making a communications call, or firing a well-aimed shot.

NOTE: See Chapter 9 for discussion on the keep-in-memory (KIM) game.

4-16. TWILIGHT TECHNIQUES

Twilight induces a false sense of security, and the sniper team must be extremely cautious. The enemy is also prone to carelessness and more likely to expose himself at twilight. During twilight, snipers should be alert to OP locations for future reference. The M3A telescope reticle is still visible and capable of accurate fire 30 minutes before BMNT and 30 minutes after EENT.

4-17. NIGHT TECHNIQUES

Without night vision devices, the sniper team must depend upon eyesight. Regardless of night brightness, the human eye cannot function at night with daylight precision. For maximum effectiveness, the sniper team must apply the following principles of night vision:

a. **Night Adaptation.** The sniper team should wear sunglasses or red-lensed goggles in lighted areas before departing on a mission. After departure, the team makes a darkness adaptation and listening halt for 30 minutes.

b. **Off-Center Vision.** In dim light, an object under direct focus blurs, appears to change, and sometimes fades out entirely. However, when the eyes are focused at different points, about 5 to 10 degrees away from an object, peripheral vision provides a true picture. This allows the light-sensitive portion of the eye, that not used during the day, to be used.

c. Factors Affecting Night Vision. The sniper team has control over the following night vision factors:

(1) Lack of vitamin A impairs night vision. However, an overdose of vitamin A will not improve night vision capability.

(2) Colds, fatigue, narcotics, headaches, smoking, and alcohol reduce night vision.

(3) Exposure to bright light degrades night vision and requires a readaption to darkness.

4-18. ILLUMINATION AIDS

The sniper team may occasionally have artificial illumination for observing and firing. Examples are artillery illumination fire, campfires, or lighted buildings.

a. Artillery Illumination Fire. The M301A2 illuminating cartridge provides 50,000 candlepower.

b. Campfires. Poorly disciplined enemy soldiers may use campfires, or fires may be created by battlefield damage. These opportunities give the sniper enough illumination for aiming.

c. Lighted Buildings. The sniper can use lighted buildings to eliminate occupants of the building or personnel in the immediate area of the light source.

Section V
TARGET DETECTION AND SELECTION

Recording the type and location of targets in the area helps the sniper team to determine engageable targets. The sniper team must select key targets that will do the greatest harm to the enemy in a given situation. It must also consider the use of indirect fire on targets. Some targets, due to their size or location, may be better engaged with indirect fire.

4-19. TARGET INDEXING

To index targets, the sniper team uses the prepared range card for a reference since it can greatly reduce the engagement time. When indexing a target to the sniper, the observer locates a prominent terrain feature near the target. He indicates this feature and any other information to the sniper to assist in finding the target. Information between team members varies with the situation. The observer may sound like an FO giving a call for fire to an FDC depending on the condition of the battlefield and the total number of possible targets from which to choose.

a. **Purpose.** The sniper team indexes targets for the following reasons:

(1) Sniper teams may occupy an FFP in advance of an attack to locate, index, and record target locations; and to decide on the priority of targets.

(2) Indiscriminate firing may alert more valuable and closer enemy targets.

(3) Engagement of a distant target may result in disclosure of the FFP to a closer enemy.

(4) A system is needed to remember location if several targets are sighted at the same time.

b. **Considerations.** The sniper team must consider the following factors when indexing targets:

(1) ***Exposure times.*** Moving targets may expose themselves for only a short time The sniper team must note the point of disappearance of each target, if possible, before engagement. By doing so, the team may be able to take several targets under fire in rapid succession.

(2) ***Number of targets.*** If several targets appear and disappear at the same time, the point of disappearance of each is hard to determine; therefore, sniper teams concentrate on the most important targets.

(3) ***Spacing/distance between targets.*** The greater the distance between targets, the harder it is to see their movement. In such cases, the team should locate and engage the nearest targets.

(4) ***Evacuation of aiming points.*** Targets that disappear behind good aiming points are easily recorded and remembered, targets with poor aiming points are easily lost. Assuming that two such targets are of equal value and danger, the team should engage the more dangerous aiming point target first.

c. **Determination of Location of Hidden Fires.** When using the *crack-thump method,* the team listens for the crack of the round and the thump of the weapon being fired. By using this method, the sniper can obtain both a direction and a distance.

(1) ***Distance to firer.*** *The* time difference between the crack and the thump can be converted into an approximate range. A one-second lapse between the two is about 600 yards with most calibers; a one-half-second lapse is about 300 yards.

(2) ***Location of firer.*** *By* observing in the direction of the thump and near the predetermined range, the sniper team has a good chance of seeing the enemy's muzzle flash or blast from subsequent shots.

(3) ***Limitations.***The crack-thump method has the following limitations

(a) Isolating the crack and thump is difficult when many shots are being fired.

(b) Mountainous areas, tall buildings, and so forth cause echoes and make this method ineffective.

d. **Shot-Hole Analysis.** Locating two or more shot holes in trees, walls, dummy heads, and so forth may make it possible to determine the direction of the shots. The team can use the dummy-head pencil method and triangulate on the enemy sniper's position. However, this method only works if all shots come from the same position.

4-20. TARGET SELECTION

Target selection may be forced upon the sniper team. A target moving rapidly may be lost while obtaining positive identification. The sniper team considers any enemy threatening its position as a high-value target. When selecting key targets, the team must consider the following factors:

a. **Threat to the Sniper Team.** The sniper team must consider the danger the target presents. This can be an immediate threat, such as an enemy element walking upon its position, or a future threat, such as enemy snipers or dog tracking teams.

b. **Probability of First-Round Hit.**The sniper team must determine the chances of hitting the target with the first shot by considering the following:

- Distance to the target.
- Direction and velocity of the wind.
- Visibility of the target area.
- Amount of the target that is exposed.
- Amount of time the target is exposed.
- Speed and direction of target movement.

c. **Certainty of Target's Identity.** The sniper team must be reasonably certain that the target it is considering is the key target.

d. **Target Effect on the Enemy.** The sniper team must consider what effect the elimination of the target will have on the enemy's fighting ability It must determine that the target is the one available target that will cause the greatest harm to the enemy.

e. **Enemy Reaction to Sniper Fire.** The sniper team must consider what the enemy will do once the shot has been fired. The team must be prepared for such actions as immediate suppression by indirect fires and enemy sweeps of the area.

f. **Effect on the Overall Mission.** The sniper team must consider how the engagement will affect the overall mission. The mission may be one of intelligence gathering for a certain period. Firing will not only alert the enemy to a team's presence, but it may also terminate the mission if the team has to move from its position as a result of the engagement.

4-21. KEY TARGETS

Key personnel targets can be identified by actions or mannerisms, by positions within formations, by rank or insignias, and or by equipment being worn or carried. Key targets can also include weapon systems and equipment. Examples of key targets areas follows:

a. **Snipers.** Snipers are the number one target of a sniper team. The enemy sniper not only poses a threat to friendly forces, but he is also the natural enemy of the sniper. The fleeting nature of a sniper is reason enough to engage him because he may never be seen again.

b. **Dog Tracking Teams.** Dog tracking teams pose a great threat to sniper teams and other special teams that may be working in the area. It is hard to fool a trained dog. When engaging a dog tracking team, the sniper should engage the dog's handler first. This confuses the dog, and other team members may not be able to control it.

c. **Scouts.** Scouts are keen observers and provide valuable information about friendly units. This plus their ability to control indirect fires make them dangerous on the battlefield. Scouts must be eliminated.

d. **Officers.** Officers are another key target of the sniper team. Losing key officers in some forces is such a major disruption to the operation that forces may not be able to coordinate for hours.

e. **Noncommissioned Officers.** Losing NCOs not only affects the operation of a unit but also affects the morale of lower ranking personnel,

f. **Vehicle Commanders and Drivers.** Many vehicles are rendered useless without a commander or driver.

g. **Communications Personnel.** In some forces, only highly trained personnel know how to operate various types of radios. Eliminating these personnel can be a serious blow to the enemy's communication network.

h. **Weapon Crews.** Eliminating weapon crews reduces the amount of fire on friendly troops.

i. **Optics on Vehicles.** Personnel who are in closed vehicles are limited to viewing through optics. The sniper can blind a vehicle by damaging these optic systems.

j. **Communication and Radar Equipment.** The right shot in the right place can completely ruin a tactically valuable radar or communication system. Also, only highly trained personnel may attempt to repair these systems in place. Eliminating these personnel may impair the enemy's ability to perform field repair.

k. **Weapon Systems.** Many high-technology weapons, especially computer-guided systems, can be rendered useless by one well-placed round in the guidance controller of the system.

Section VI
RANGE ESTIMATION

A sniper team is required to accurately determine distance, to properly adjust elevation on the sniper weapon system, and to prepare topographical sketches or range cards. Because of this, the team has to be skilled in various range estimation techniques.

4-22. FACTORS AFFECTING RANGE ESTIMATION

Three factors affect range estimation: nature of the target, nature of the terrain, and light conditions.

a. **Nature of the Target.**

(1) An object of regular outline, such as a house, appears closer than one of irregular outline, such as a clump of trees.

(2) A target that contrasts with its background appears to be closer than it actually is.

(3) A partly exposed target appears more distant than it actually is.

b. **Nature of the Terrain.**

(1) As the observer's eye follows the contour of the terrain, he tends to overestimate distant targets.

(2) Observing over smooth terrain, such as sand, water, or snow, causes the observer to underestimate distant targets.

(3) Looking downhill, the target appears farther away.

(4) Looking uphill, the target appears closer.

c. **Light Conditions.**

(1) The more clearly a target can be seen, the closer it appears.

(2) When the sun is behind the observer, the target appears to be closer.

(3) When the sun is behind the target, the target is more difficult to see and appears to be farther away.

4-23. RANGE ESTIMATION METHODS

Sniper teams use range estimation methods to determine distance between their position and the target.

a. **Paper-Strip Method.** The paper-strip method (Figure 4-17) is useful when determining longer distances (1,000 meters plus). When using this method, the sniper places the edge of a strip of paper on the map and ensures it is long enough to reach between the two points. Then he pencils in a tick mark on the paper at the team position and another at the distant location. He places the paper on the map's bar scale, located at the bottom center of the map, and aligns the left tick mark with the 0 on the scale. Then he reads to the right to the second mark and notes the corresponding distance represented between the two marks.

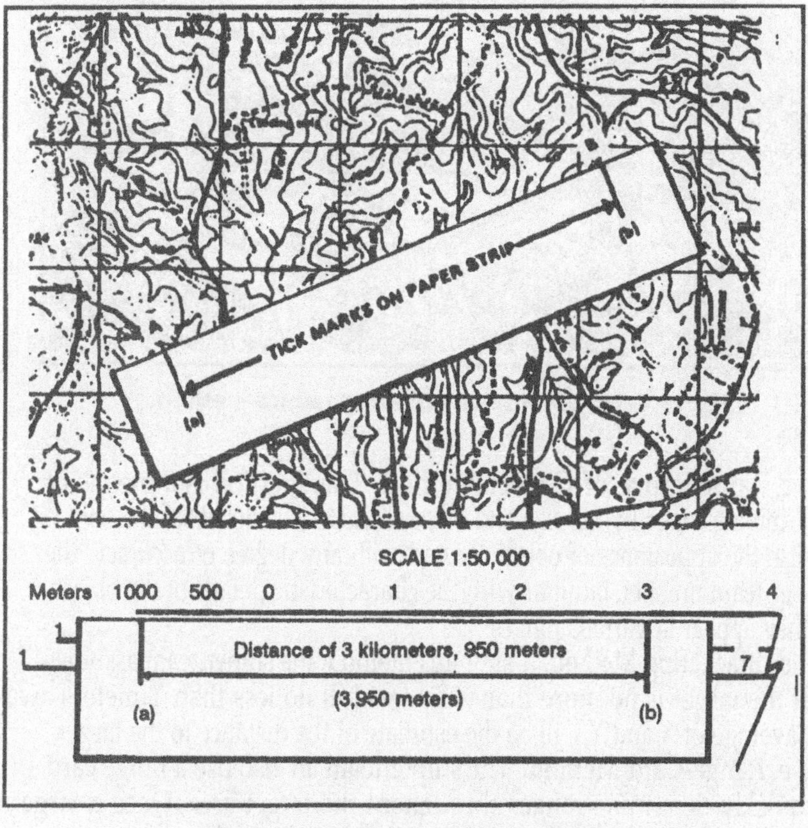

Figure 4-17. Paper-strip method.

b. **100-Meter-Unit-of-Measure Method.** To use this method (Figure 4-18, page 4-38), the sniper team must be able to visualize a distance of 100 meters on the ground. For ranges up to 500 meters, the team determines the number of 100-meter increments between the two objects it wishes to measure. Beyond 500 meters, it must select a point halfway to the object and determine the number of 100-meter increments to the halfway point, then double it to find the range to the object.

Figure 4-18. 100-meter-unit-of-measure method.

c. **Appearance-of-Object Method.** This method is a means of determining range by the size and other characteristic details of the object. To use the appearance-of-object method with any degree of accuracy, the sniper team must be familiar with the characteristic details of the objects as they appear at various ranges.

d. **Bracketing Method.** Using this method, the sniper team assumes that the target is no more than X meters but no less than Y meters away. An average of X and Y will be the estimate of the distance to the target.

e. **Range-Card Method.** The sniper team an also use a range card to quickly determine ranges throughout the target area. Once a target is seen, the team determines where it is located on the card and then reads the proper range to the target.

f. **Mil-Relation Formula.** The mil-relation formula is the preferred method of range estimation. This method uses a mil-scale reticle located in the M19 binoculars (Figure 4-19) or in the M3A sniperscope (Figure 4-20). The team must know the target size in inches or meters. Once the target size is known, the team then compares the target size to the mil-scale reticle and uses the following formula:

$$\frac{\text{Size of target in meters} \times 1{,}000}{\text{Size of object in mils}} = \text{Range to target in meters}$$

(To convert inches to meters, multiply the number of inches by .0254.)

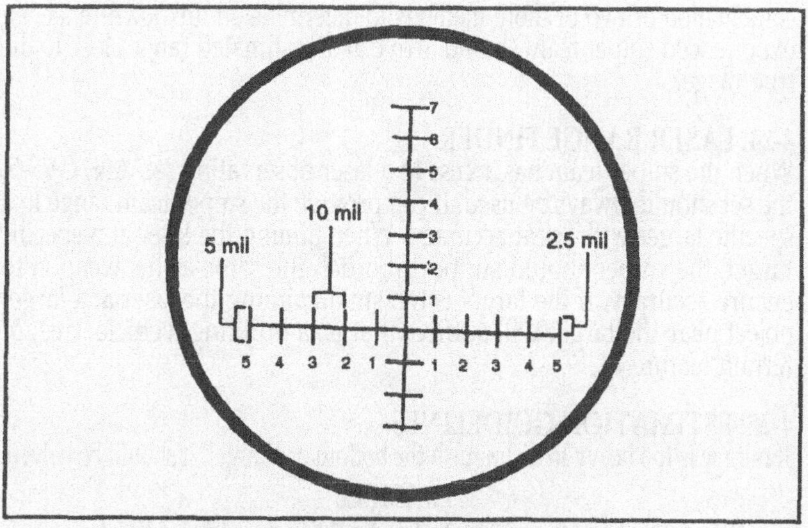

Figure 4-19. M19 mil-scale reticle.

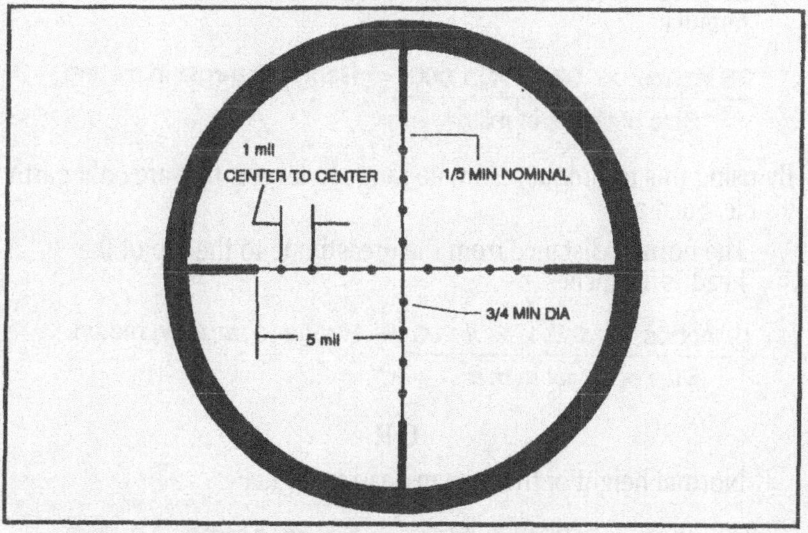

Figure 4-20. M3A mil-scale reticle.

g. **Combination Method.** In a combat environment, perfect conditions rarely exist. Therefore, only one method of range estimation may not be enough for the team's specific mission. Terrain with much dead space limits the accuracy of the 100-meter method. Poor visibility limits the use of the appearance-of-object method. However, by using a combination of two or more methods to determine an unknown range, an experienced sniper team should arrive at an estimated range close to the true range.

4-24. LASER RANGE FINDER

When the sniper team has access to a laser observation set, AN/GVS-5, the set should always be used. It can provide the sniper team range to a specific target with great accuracy. When aiming the laser at a specific target, the sniper should support it much the same as his weapon to ensure accuracy. If the target is too small, aiming the laser at a larger object near the target will suffice (that is, a building, vehicle, tree, or terrain feature.)

4-25. ESTIMATION GUIDELINES

If mirage is too heavy to distinguish the bottom of a target, it should be halved.

EXAMPLE

When the target is estimated to be 70 inches high, divide the height into one-half. Use the following mil-relation formula:

$$\frac{35 \text{ inches} \times .0254 \times 1{,}000}{\text{Size of target in mils}} = \text{Range to target in meters}$$

By using this technique, estimate range to targets that are only partly visible. Such as:

The normal distance from the breastbone to the top of the head is 19 inches.

$$\frac{19 \text{ inches} \times .0254 \times 1{,}000}{\text{Size of target in mils}} = \text{Range to target in meters}$$

OR

Normal height of the human head is 10 inches.

$$\frac{10 \text{ inches} \times .0254 \times 1{,}000}{\text{Size of target in mils}} = \text{Range to target in meters}$$

This example may prove to be of specific use when facing an enemy entrenched in bunkers or in dense vegetation.

a. The sniper team should keep a sniper data book complete with measurements.

(1) *Vehicles.*

- Height of road wheels.
- Vehicle dimensions.
- Length of main gun tubes on tanks.
- Lengths/sizes of different weapon systems.

(2) ***Average height of human targets in area of operation.***

(3) ***Urban environment.***

- Average size of doorways.
- Average size of windows.
- Average width of streets and lanes (average width of a paved road in the United States is 10 feet).
- Height of soda machines.

b. As the sniper team develops a sniper data book, all measurements are converted into constants and computed with different mil readings. An example of this is Table 4-1, which has already been computed for immediate use. This table should be incorporated into the sniper data book

TABLE FOR 6-FOOT MAN		
HEIGHT IN MILS	STANDING	SITTING/ KNEELING
1	2000	1000
1.5	1333	666
2	1000	500
2.5	800	400
3	666	333
3.5	571	286
4	500	250
4.5	444	222
5	400	200
5.5	364	182
6	333	167
6.5	308	154
7	286	143

TABLE FOR 5-FOOT 6-INCH MAN		
HEIGHT IN MILS	STANDING	SITTING/ KNEELING
1	1800	900
1.5	1200	600
2	900	450
2.5	750	375
3	600	300
3.5	514	257
4	450	225
4.5	400	200
5	360	180
5.5	327	164
6	300	150
6.5	277	139

Table 4-1. Range estimation table.

Section VII
INFORMATION RECORDS

The secondary mission of the sniper team is the collection and reporting of information. To accomplish this, the sniper team not only needs to be keen observers, but it also must accurately relay the information it has observed. To record this information, the team uses the sniper data book, which contains a range card, a military sketch, and an observation log.

4-26. RANGE CARD

The range card represents the target area drawn as seen from above with annotations indicating distances throughout the target area. Information is recorded on DA Form 5787-R (Sniper's Range Card) (Figure 4-21). (A blank copy of this form is located in the back of this publication for local reproduction.) The range card provides the sniper team with a quick-range reference and a means to record target locations, since it has preprinted range rings on it. These cards can be divided into sectors by using dashed lines. This provides the team members with a quick reference when locating targets-for example: "The intersection in sector A." A range card can be prepared on any paper the team has available. The sniper team position and distances to prominent objects and terrain features are drawn on the card. There is not a set maximum range on the range card, because the team may also label any indirect fire targets on its range card. Information contained on range cards includes:

a. Name, rank, SSN, and unit.

b. Method of obtaining range.

c. Left and right limits of engageable area.

d. Major terrain features, roads, and structures.

e. Ranges, elevation, and windage needed at various distances.

f. Distances throughout the area.

g. Temperature and wind. (Cross out previous entry whenever temperature, wind direction, or wind velocity changes.)

h. Target reference points (azimuth, distance, and description).

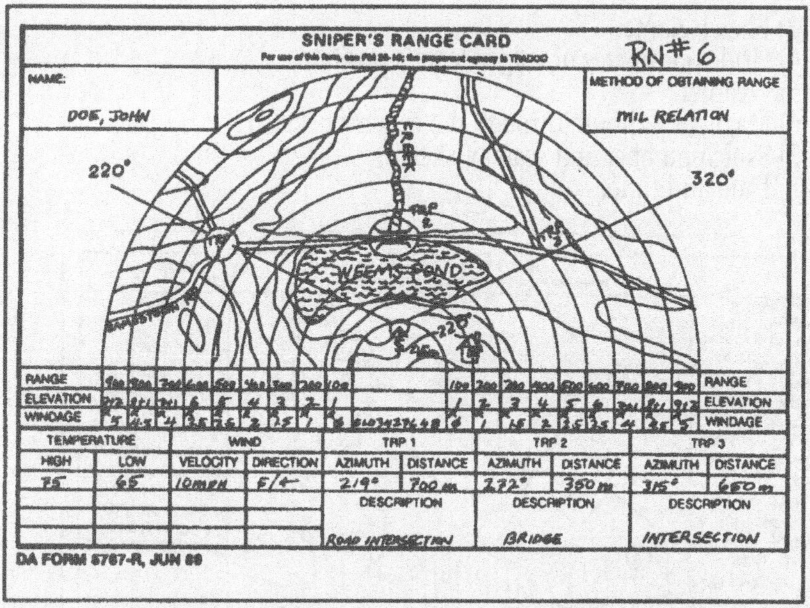

Figure 4-21. Example of completed DA Form 5787-R.

4-27. MILITARY SKETCH

DA Form 5788-R (Military Sketch) is used to record information about a general area, terrain features, or man-made structures that are not shown on a map. Military sketches provide intelligence sections a detailed, on-the-ground view of an area or object that is otherwise unobtainable. These sketches not only let the viewer see the area in different perspectives but also provide detail such as type of fences, number of telephone wires, present depth of streams, and so forth. There are two types of military sketches as stated in FM 21-26 panoramic sketches and topographic sketches. Information is recorded on DA Form 5788-R. (A blank copy of this form is located in the back of this publication for local reproduction.)

a. **Panoramic.** A panoramic sketch (Figure 4-22, page 4-44) is a representation of an area or object drawn to scale as seen from the sniper team's perspective. It shows details about a specific area or a man-made structure. Information considered in a panoramic sketch includes the following:

(1) Name, rank, SSN, and unit.

(2) Remarks section (two).

(3) Sketch name.
(4) Grid coordinates of sniper team's position.
(5) Weather.
(6) Magnetic azimuth through the center of sketch.
(7) Sketch number and scale of sketch.
(8) Date and time.

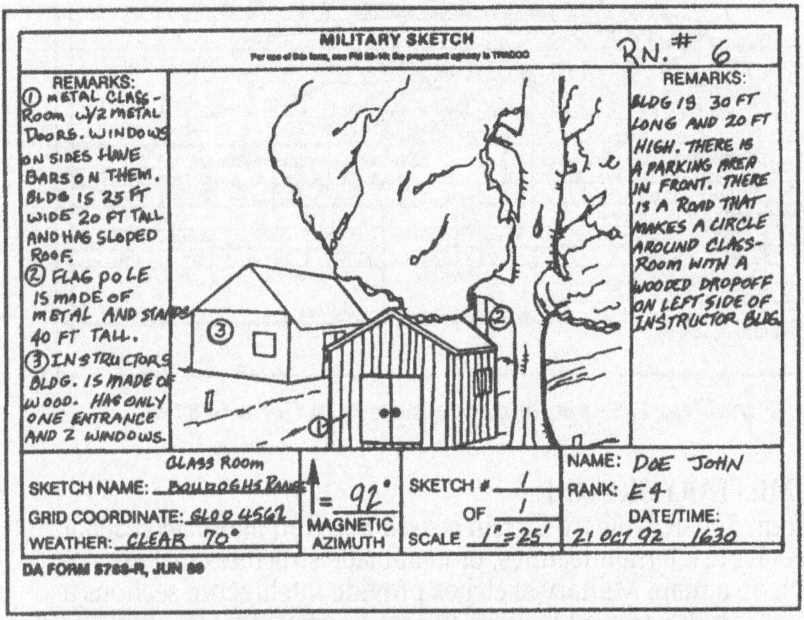

Figure 4-22. Example of completed DA Form 5788-R for panoramic sketch.

b. **Topographic Sketch.** A topographic sketch (Figure 4-23) is a topographic representation of an area drawn to scale as seen from above. It provides the sniper team with a method for describing large areas while showing reliable distance and azimuths between major features. This type of sketch is useful in describing road systems, flow of streams/rivers, or locations of natural and man-made obstacles. 'he field sketch can also be used as an overlay on the range card. Information contained in a field sketch includes the following

(1) Grid coordinates of the sniper team's position.
(2) Name, rank, SSN, and unit.
(3) Remarks.
(4) Sketch name.

(5) Grid coordinates.
(6) Weather.
(7) Magnetic azimuth.
(8) Sketch number and scale.
(9) Date and time.

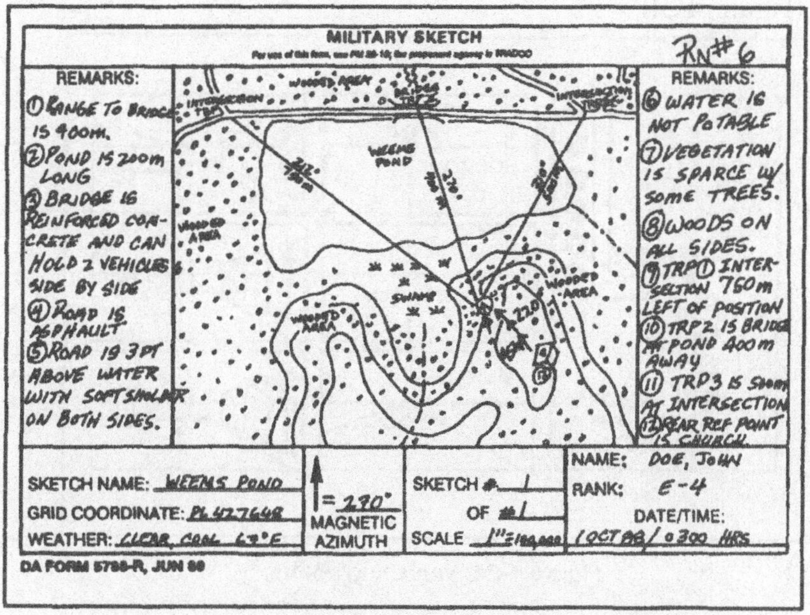

Figure 4-23. Example of completed DA Form 5788-R for topographic sketch.

c. **Guidelines for Drawing Sketches.** As with all drawings, artistic skill is an asset, but satisfactory sketches can be drawn by anyone with practice. The following are guidelines when drawing sketches:

(1) *Work from the whole to the part.* First determine the boundaries of the sketch. Then sketch the larger objects such as hills, mountains, or outlines of large buildings. After drawing the large objects in the sketch, start drawing the smaller details.

(2) *Use common shapes to show common objects.* Do not sketch each individual tree, hedgerow, or wood line exactly. Use common shapes to show these types of objects. Do not concentrate on the fine details unless they are of tactical importance.

(3) *Draw in perspective; use vanishing points.* Try to draw sketches in perspective. To do this, recognize the vanishing points of the area to

be sketched. Parallel lines on the ground that are horizontal vanish at a point on the horizon (Figure 4-24). Parallel lines on the ground that slope downward away from the observer vanish at a point below the horizon. Parallel lines on the ground that slope upward, away from the observer vanish at a point above the horizon. Parallel lines that recede to the right vanish on the right and those that recede to the left vanish on the left (Figure 4-24).

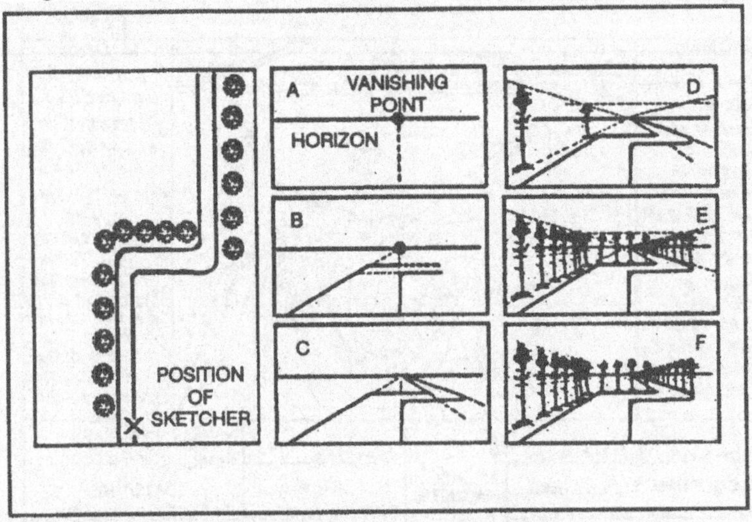

Figure 4-24. Vanishing points.

4-28. SNIPER DATA BOOK

The sniper data book is a written, chronological record of all activities and events that take place in a sniper team's area. It is used with military sketches and range cards; this combination not only gives commanders and intelligence personnel information about the appearance of the area, but it also provides an accurate record of the activity in the area. Information is recorded on DA Form 5786-R (Sniper's Observation Log) (Figure 4-25). (A blank copy of this form is in the back of this publication for local reproduction.) Information in the observation log includes: (Completion of this form is self-explanatory.)

a. Sheet number and number of total sheets.

b. Observer's name, rank, SSN, and unit.

c. Date and time of observation and visibility.

d. Grid coordinates of the sniper team's position.
e. Series number, time, and grid coordinates of each event.
f. The event that has taken place.
g. Action taken and remarks.

SNIPER'S OBSERVATION LOG
For use of this form, see FM 23-10; the proponent agency is TRADOC

SHEET____OF____SHEETS

ORIGINATOR:	DATE/TIME:	LOCATION:
DOE, JOHN R	1 OCT 92	GL03427648

SERIAL	TIME	GRID COORDINATE	EVENT	ACTIONS OR REMARKS
1	0300	GL03427648	OCCUPIED POSITION	OBSERVATION
2	0340	SAME	PFC JUDSON RESTED	NONE
3	0430	SAME	PFC JUDSON ASSUMED OBSERVATION	I RESTED
4	0530	SAME	BOTH OF US AWAKE	NONE
5	0630	SAME	PREPARED RANGE CARD AND TOPOGRAPHIC SKETCH	LIGHT ENOUGH TO SEE
6	0655	GL034276428	BRM CROSSED BRIDGE GOING SOUTH	OBSERVED
7	0700	GL034276428	PREPARED SKETCH OF BRIDGE GL03117631	COMPLETE
8	0900	GL034276428	MISSION COMPLETED – RETURN TO CP	END OF MISSION

DA FORM 5786-R, JUN 89

Figure 4-25. Example of completed DA Form 5786-R.

CHAPTER 5

MISSION PREPARATION

The sniper team uses planning factors to estimate the amount of time, coordinating and effort that must be expended to support the impending mission. Arms, ammunition, and equipment are METT-T dependent.

Section I
PLANNING AND COORDINATION

Planning and coordination are essential procedures that occur during the preparation phase of a mission.

5-1. MISSION ALERT

The sniper team may receive a mission briefing in either written or oral form (FRAGO). Usually, the team mission is stated specifically as to who, what, when, where, and why/how. On receipt of an order, the sniper analyzes his mission to ensure he understands it, then plans the use of available time.

5-2. WARNING ORDER

Normally, the sniper team receives the mission briefing. However, if the sniper receives the briefing, he prepares to issue a warning order immediately after the briefing or as soon as possible. He informs the observer of the situation and mission and gives him specific and general instructions. If the sniper team receives the mission briefing, the sniper should still present the warning order to the observer to clarify and emphasize the details of the mission briefing.

5-3. TENTATIVE PLAN

The sniper makes a tentative plan of how he intends to accomplish the mission. When the mission is complex and time is short, he makes a quick, mental estimate; when time is available, he makes a formal, mental estimate. The sniper learns as much as he can about the enemy and mission requirements and applies it to the terrain in the assigned area. Since an on-the-ground reconnaissance is not tactically feasible for most sniper operations, the sniper uses maps, pictomaps, or aerial photographs of the objective and surrounding area to help formulate his tentative plan. This plan is the basis for team preparation, coordination, movement, and reconnaissance.

5-4. COORDINATION CHECKLISTS

Coordination is continuous throughout the planning phase of the operation (see coordination checklists) (for example, aircraft, parachutes, or helicopters). Other items are left for the sniper to coordinate. He normally conducts coordination at the briefing location. To save time, he assigns tasks to the observer and has him report back with the results. However, the sniper is responsible for all coordination. He uses coordination checklists to verify mission-essential equipment for the mission. He coordinates directly with appropriate staff sections or the S3, or the SEO will provide the necessary information. The sniper may carry a copy of the coordination checklists to ensure he does not overlook an item that may be vital to the mission. Coordination with specific staff sections includes the following:

NOTE: Items may need coordination with more than one staff section; therefore, some items are listed under more than one heading.

a. **Intelligence.** The S2 informs the sniper of any changes in the situation as given in the OPORD or mission briefing. The sniper constantly updates the tentative plan with current information.

(1) Identification of the unit.
(2) Weather and light data.
(3) Terrain update.
- Aerial photos.
- Trails and obstacles not on map.

(4) Known or suspected enemy locations.
(5) weapons.
(6) Strength.
(7) Probable courses of action.
(8) Recent enemy activity.
(9) Reaction time of reaction forces.
(10) Civilian activity in area.
(11) Priority intelligence requirements and information requirements.
(12) Challenge and password.

b. **Operations.** The sniper coordinates with the operations section to receive the overall status of the mission.

(1) Identification of the unit.
(2) Changes in the friendly situation.
(3) Route selections and LZ and PZ selections.
(4) Linkup procedure.
(5) Transportation (other than air).
(6) Resupply (along with S4).
(7) Signal plan.
(8) Departure and reentry of forward units.
(9) Special equipment requirements.
(10) Adjacent units operating in the area of operations.
(11) Rehearsal areas.
(12) Method of insertion/extraction.
(13) Frequencies and call signs.

c. **Fire Support.** Usually, the sniper coordinates fire support with the fire support officer.

(1) Identification of the unit.
(2) Mission and objective.
(3) Routes to and from the objective (including alternate routes).
(4) Time of departure and expected time of return.
(5) Unit target list (fire plan).
(6) Fire support means available (artillery, mortar, naval gunfire, and aerial fire support to include Army, Navy, and Air Force).
(7) Ammunition available (to include different fuzes).
(8) Priority of fires.
(9) Control measures for fire support.

- Checkpoints.
- Boundaries.
- Phase lines.
- Fire support coordination measures.
- Priority targets (list TRPs).
- RFA.
- RFL.
- No-fire areas.
- Precoordinate authentication.

(10) Communications (include primary and alternate means, emergency signals, and code words and signals).

d. **Coordination with Forward Unit.** A sniper team that must move through a friendly forward unit must coordinate with the unit commander for a smooth, orderly passage. If there is no coordination time and place, the sniper sets the time and place with the S2 and S3. Then, he informs the forward unit and arranges assistance for the team's departure. Coordination is a two-way exchange of information.

(1) Identification (team leader, observer, and unit).

(2) Size of team.

(3) Time(s) and place(s) of departure and return, location(s) of departure point(s), IRPs, and detrucking points.

(4) General area of operation.

(5) Information on terrain and vegetation.

(6) Known or suspected enemy positions or obstacles.

(7) Possible enemy ambush sites.

(8) Latest enemy activity.

(9) Detailed information on friendly positions (for example, crew-served weapons or final protective fire).

(10) Fire and barrier plan.

(11) Support the forward unit can furnish. How long and what can they do?

- Fire support.
- Litter teams.
- Navigational signals and aids.
- Guides.
- Communications.
- Reaction units.
- Other.

(12) Call signs and frequencies and exchange of Vinson cryptographic variables.

- Pyrotechnic plans.
- Challenge and password.
- Emergency signals and codewords.
- Relieved unit (pass information to the relieving unit).

e. **Adjacent Unit Coordination.** Immediately after receiving the OPORD or mission briefing, the sniper coordinates with other units using the same area. If he is not aware of other units, he should check with the S3 to arrange coordination. The sniper exchanges the following information with other units or snipers operating in the same area:

(1) Identification of the unit.

(2) Mission and size of unit.

(3) Planned times and points of departure and reentry.
(4) Route.
(5) Fire support (planned) and control measures.
(6) Frequency, call signs, and exchange of Vinson cryptographic variables.
(7) Challenge and password and or number.
(8) Pyrotechnic plans.
(9) Any information that the unit may have about the enemy.

f. **Rehearsal Area Coordination.** The sniper coordinates with the S2 or S3.
(1) Identification of own unit.
(2) Mission.
(3) Terrain similar to objective site.
(4) Security of the area.
(5) Availability of aggressors.
(6) Use of blanks, pyrotechnics, live ammunition.
(7) Mockups available.
(8) Time the area is available (preferably when light conditions are close to the expected light conditions for the mission).
(9) Transportation.
(10) Coordination with other units using the area.

g. **Army Aviation Coordination.** The sniper coordinates with the supporting aviation unit commander through the S3 or S3 Air.
(1) ***Situation:***
(a) *Enemy forces:* Location, activity, probable course of action, and enemy air defense.
(b) *Weather:* Decision time/POC any delay for the mission.
(c) *Friendly forces:* Main mission, activity, boundaries, axis of movement.

(2) ***Mission:*** Task and purpose.

(3) ***Execution:***
(a) *Concept of the operation:* Overview of what requesting unit wants to accomplish with the air assault/air movement.
(b) *Coordinating instructions (PZ operation):*
- Direction of landing.
- Time of landing/flight direction.
- Location of PZ/alternate PZ.
- Loading procedures.
- Marking of PZ (panel, smoke, smoke munitions, lights).

- Flight route planned (start point, air control point, rally point).
- Formation of landing/flight/landing (LZ).
- Code words: PZ secure (before landing); PZ clear (lead plane, last plane); alternate PZ (at PZ en route, at landing zone); names of PZ/alternate PZ.
- TAC air/artillery.
- Number of passengers/planes for entire lift.
- Equipment carried by individuals.
- Secure PZ or not.
- Marking of key leaders (LZ operations).
- Direction of landing.
- Time of landing, false insertions.
- Location of LZ or alternate LZ.
- Marking of LZ (panel, smoke, SM, lights).
- Formation of landing.
- Codewords: LZ name, alternate LZ name.
- TAC air/artillery preparation, fire support coordination.
- Secure LZ or not.

(4) ***Service support:***
(a) Number of aircraft, times, number of lifts.
(b) Refuel/rearm during mission or not.
(c) Special equipment/aircraft configuration for weapons earned by unit personnel.
(d) Bump plan.

(5) ***Command and signal:***
(a) Frequency and call signs.
(b) Location of air mission commander.

h. **Vehicle Movement Coordination.** The sniper coordinates with the supporting unit through the S3.
(1) Identification of the unit.
(2) Supporting unit identification.
(3) Number and type of vehicles and tactical preparation.
(4) Entrucking point.
(5) Departure/loading time.
(6) Preparation of vehicles for movement.

- Driver responsibilities.
- Sniper team responsibilities.
- Special supplies/equipment required.

(7) Availability of vehicles for preparation/rehearsal/inspection (time and location).

(8) Route
- Primary.
- Alternate.
- Checkpoints.

(9) Detrucking points.
- Primary.
- Alternate.

(10) March interval/speed.

(11) Communications (frequencies, call signs, codes).

(12) Emergency procedures and signals.

5-5. COMPLETION OF PLAN

After the warning order has been issued and a thorough map reconnaissance made, most coordination should be completed. The sniper makes an intelligence update while the observer prepares himself and the equipment for the mission. The sniper completes his plan based on his map reconnaissance and or any changes in the enemy situation. He may or may not alter the tentative plan, but he can add detail. The sniper uses the OPORD format as a guide to refine his concept. He places the main focus on actions in the objective area and carefully assigns the observer specific tasks for all phases of the operation. He ensures all actions work smoothly and efficiently.

5-6. OPERATION ORDER

The operation order is issued in the standard OPORD format. Extensive use of terrain models, sketches, and chalkboards should be made to highlight important details such as routes, planned rally points, and actions at known danger areas. All aspects of the OPORD should be thoroughly understood by the sniper team to include memorizing the following

- Intelligence acquisition tasks.
- Situation—both friendly and enemy.
- Mission.
- Execution plan.
- Administrative plans.
- Communications and electronics, including frequencies, call signs, and antennas to be used.
- SALUTE format.

5-7. BRIEFBACK

The sniper team rehearses the briefback until it is near-perfect before presenting it to the S3, sniper employment officer, or commander. A good briefback indicates the team's readiness for the mission. (Figure 5-1 is an example of a sniper team briefback outline.)

SNIPER BRIEFBACK OUTLINE

1. Team Leader.
   a. Introduction of patrol.
   b. Mission statement.
2. Team Leader—Execution Plan.
   a. Insertion.
      (1) Method.
      (2) Location.
      (3) Time.
   b. Reconnaissance.
      (1) Position.
      (2) Method.
   c. Extraction.
      (1) Method.
      (2) Location.
      (3) Time.
      (4) Evasion route/plan.
3. Observer—Administrative Plan (S1)/Logistical Plan (S4).
   a. PW.
   b. Wounded in action/killed in action.
   c. Camera information/film/log.
   d. Mission logbook.
   e. Allergies/aid kit.
   f. Customs and cultural problems.
   g. MEDEVAC plan (tactical).
   h. Rations.
   i. Water.
   j. Resupply.

Figure 5-1. Briefback outline.

4. Observer—Intelligence (S2).
   a. Enemy situation.
   b. Friendly situation.
   c. Weather.
   d. Terrain restrictions (including MEDEVAC restrictions).
   e. Light data, percentage of illumination.
   f. Order of battle.
   g. Intelligence acquisition task.
5. Observer—Communications and Electronics.
   a. Window times and format.
   b. Frequencies and call signs.
   c. Antenna-type and terrain considerations.
   d. Special equipment.
   e. Code words.
6. Team Leader—Conclusion.
   a. Restate mission.
   b. Combat medical plan.
   c. Special Instructions Relating to Training Mission.
      (1) Contact with civilians.
      (2) Local authority POC.
      (3) Maneuver restrictions.

Figure 5-1. Briefback outline (continued).

5-8. EQUIPMENT CHECK

The sniper team ensures needed equipment is operational before signing it out. Weapons are clean, functional, and test-fired to confirm zero. The team checks radios by making a communications check with the NCS of the net they will be using and night vision devices by turning them on (adding an extra battery). The team then double-checks all equipment. If they encounter problems, the sniper notifies the PSG or SEO.

5-9. FINAL INSPECTION

Inspections reveal the team's physical and mental state of readiness. The sniper ensures that all required equipment is present and functional, and that the observer knows and understands the mission. The following items should be inspected:

- Completeness and correctness of uniform and equipment.
- Items such as pictures, papers, marked maps, and sniper data book that contain confidential material.
- Hats and pockets.
- Shine, rattles, and tie-downs.
- Weapons (loaded or unloaded).
- Fullness of canteens.

If unauthorized items are found, the sniper immediately corrects any deficiencies. Then, he questions the observer to make sure he knows the team plan, what his job is, and when he is to do it.

5-10. REHEARSALS

Rehearsals ensure team proficiency. During rehearsals, the sniper rechecks his plans and makes any needed changes. It is through well-directed and realistic rehearsals that the team becomes thoroughly familiar with their actions on the mission.

a. The sniper team uses terrain similar to that on which they will operate (if available), rehearsing all actions if time permits. A good way to rehearse is to talk the team through each phase, describing the actions of each sniper, and then perform the actions as a dry run. When actions are understood, the sniper team goes through all the phases, using the signals and commands to be used during the mission.

b. If there is no time for rehearsals, the sniper team conducts a briefback/talk-through. This method is used to supplement rehearsals or when security needs or a lack of time preclude dry runs and wet runs. In this method, the *team leader* talks the *observer* through his actions and then has him orally repeat those actions. The sniper team establishes the sequence of actions to be rehearsed and, if time permits, conducts rehearsals in the same sequence as in the mission.

5-11. FINAL PREPARATIONS

The sniper makes any last-minute changes and corrects any deficiencies found during initial inspections. Final inspection should be made by the SEO and an S3 representative. Again, pockets and rucksacks are emptied and inspected. The inspection team looks for personal papers, marked maps, and other unauthorized items. The sniper ensures all previous discrepancies are corrected; equipment is still operational; all needed items are present; and the observer is ready for the mission. The inspection team randomly asks questions about the mission.

5-12. PREPARATION FOR DEBRIEFING

After the mission, the SEO or S3 representative directs the sniper team to an area where they prepare for a debriefing. The team remains in the area until called to the operations center. The sniper will bring the sniper data book that contains a log sheet, a field sketch, a range card, and a road/area sketch for debriefing.

a. The sniper team—

(1) Lays out and accounts for all team and individual equipment.

(2) Consolidates all captured material and equipment.

(3) Reviews and discusses the events listed in the mission logbook from insertion to return, including details of each enemy sighting.

(4) Prepares an overlay of the team's route, area of operations, insertion point, extraction point, and significant sighting locations.

b. An S3 representative controls the debriefing. He directs the sniper—

(1) To discuss any enemy sightings since the last communications with the radio base station.

(2) To give a step-by-step amount of each event listed in the mission logbook from insertion until reentry of the FFL, including details of all enemy sightings.

c. When the debriefing is complete, the S3 representative releases the sniper team back to platoon control.

5-13. COUNTERSNIPER OPERATION

When an enemy sniper threat has been identified in the sniper team's area of operations, the team is employed to eliminate the enemy sniper.

a. A sniper team identifies an existing sniper threat by using the following indicators:

(1) Enemy soldiers in special camouflage uniforms.

(2) Enemy soldiers seen carrying weapons in cases or drag bags or weapons with long barrel lengths, mounted telescopes, and bolt-action receivers.

(3) Single-shot fire.

(4) Lack or reduction of enemy patrols during single-shot fire.

(5) Light reflecting from optical lenses.

(6) Reconnaissance patrols reporting small groups of (one to three) enemy soldiers.

(7) Discovery of single expended casings, such as 7.62-mm ammunition.

b. The sniper team then determines the best method to eliminate the enemy sniper. To accomplish this, the team gathers information and determines the pattern.

(1) *Gathers information.*

(a) Time of day precision fire occurs.

(b) Location of encountered enemy sniper fire.

(c) Location of enemy sniper sightings.

(d) Material evidence of enemy snipers, such as empty brass casings or equipment.

(2) ***Determines patterns.*** The sniper team evaluates the information to detect established patterns or routines. The team conducts a map reconnaissance, studies aerial photographs, or carries out ground reconnaissance to determine the movement patterns. The sniper must place himself in the position of the enemy and ask, "How would I accomplish this mission?"

c. Once a pattern or routine is identified, the sniper team determines the best location and time to engage the enemy sniper. The team can also request the following:

(1) Coordinating routes and fires.

(2) Additional preplotted targets (fire support).

(3) Infantry support to canalize or ambush the enemy sniper.

(4) Additional sniper teams for mutual supporting fire.

(5) Baiting of likely engagement areas to deceive the enemy sniper into commitment by firing.

(6) All elements in place 12 hours before the expected engagement time.

During a countersniper operation, the team must ignore battle activity and concentrate on the enemy sniper.

d. When an enemy sniper is operating in a unit's area, the sniper team ensures the unit employs passive countermeasures to defend against enemy sniper fire.

(1) Do not establish routines. For example, consistent meal times, ammunition resupply, assembly area procedures, or day-to-day activities that have developed into a routine.

(2) Conduct all meetings, briefings, or gatherings of personnel undercover or during limited visibility.

(3) Cover or conceal equipment.

(4) Remove rank from helmets and collars. Do not salute officers. Leaders should not use authoritative methods.

(5) Increase OPs and use other methods to increase the unit's observation abilities.

(6) Brief patrols on what to look for, such as single, expended rounds or different camouflage materials.

(7) Do not display awareness of the enemy's presence at any time.

5-14. REACTION TO ENEMY SNIPER FIRE

Although the sniper team's mission is to eliminate the enemy sniper, the team avoids engaging in a sustained battle with the enemy sniper. If the team is pinned down by enemy sniper fire and the sniper's position cannot be determined, the sniper team attempts to break contact to vacate the enemy sniper's kill zone.

a. The sniper team uses either hand-held or artillery generated smoke to obscure the enemy sniper's view. If the smoke provides sufficient obscuration, the sniper team breaks contact and calls for indirect fire on the enemy sniper position. If the smoke does not provide sufficient obscuration, the sniper team calls for an immediate suppression mission against the enemy sniper position. The team then breaks contact under the cover of indirect fire.

b. The sniper team should expect indirect fire and increased enemy patrolling activity shortly after contact with an enemy sniper.

Section II
MISSION PACKING LISTS

The sniper team requires arms and ammunition as determined by METT-T Some of the equipment mentioned in the example lists may not be available. A sniper team carries only mission-essential equipment normally not associated with a standard infantryman.

5-15. ARMS AND AMMUNITION

As a minimum, the sniper team requires arms and ammunition that should include the following

a. **Sniper:**

- M24 sniper weapon system with M3A scope.
- M9 bayonet.
- 100 rounds M118 special ball.
- M9 pistol.

- 45 rounds 9-mm ball ammunition.
- 4 M67 fragmentation grenades; 2 CS grenades; 2 concussion grenades (MOUT).
- M18A1 mine, complete.

b. **Observer:**

- M16A1/A2/M203 with quadrant sight and AN/PVS-4 mounted.
- M9 9-mm pistol.
- M9 bayonet.
- 210 (plus) rounds 5.56-mm ball ammunition.
- 45 rounds 9-mm ball ammunition.
- 6 rounds 40-mm high-explosive ammunition.
- 3 rounds 40-mm antipersonnel ammunition.
- 4 M67 fragmentation grenade, 2 CS grenades; 2 concussion grenades (MOUT).

5-16. SPECIAL EQUIPMENT

The sniper team requires special equipment that may include, but not be limited to the following

a. **Sniper:**

- M24 sniper weapon system deployment kit (tools and replacement parts).
- M9 pistol cleaning kit.
- Extra handset for radio.
- Extra batteries for radio (BA 4386 or lithium, dependent on mission length).
- SOI.
- M15 tripod.
- M49 observation telescope.
- AN/PVS-5/7 series, night vision goggles.
- Extra BA-1567/U or AA batteries for night vision goggles.
- Pace cord.
- E-tool with carrier.
- 50-foot 550 cord.
- 1 green and 1 red star cluster.
- 2 HC smoke grenades.
- Measuring tape (25-foot carpenter-type).
- 3 each 9-mm magazines

b. **Observer:**

- M16A1/A2 cleaning kit.
- M203 cleaning kit.
- AN/PRC-77/AN-PRC-119/AN/PRC-104A radios.
- Radio accessory bag, complete with long whip and base, tape antenna and base, handset, and battery (BA-4386 or lithium).
- 300-feet WD-1 field wire (for field-expedient antenna fabrication).
- Olive-drab duct tape ("100-mph" tape).
- Extra batteries for radio (if needed).
- Extra batteries (BA-1567/U) for AN/PVS-4.
- M19/M22 binoculars.
- Sniper's data book, mission logbook, range cards, wind tables, and "slope dope."
- 7 each 30-round capacity (5.56-mm) magazines.
- 3 each 9-mm magazines.
- Calculator with extra battery.
- Butt pack.
- 10 each sandwich-size waterproof bags.
- 2 HC smoke grenades.
- Lineman's tool.

  Range estimation (sniper data book).

5-17. UNIFORMS AND EQUIPMENT

A recommended listing of common uniforms and equipment follows; however, weather and terrain will dictate the uniform. As a minimum, the sniper team should have the following

- Footgear (jungle/desert/cold weather/combat boots).
- 2 sets BDUs (desert/woodland/camouflage).
- Black leather gloves.
- 2 brown T-shirts.
- 2 brown underwear (optional).
- 8 pair olive-drab wool socks.
- Black belt.
- Headgear (BDU/jungle/desert/cold weather).
- ID tags and ID card.
- Wristwatch (sweep second hand with luminous dial/waterproof).
- Pocket survival knife.

- Extra large ALICE pack, complete with frame and shoulder straps.
- 2 waterproof bags (for ALICE pack).
- 2 two-quart canteens with covers.
- 1 bottle water purification tablets.
- LBE complete.
- Red-lensed flashlight (angle-head type with extra batteries).
- MREs (number dependent on mission length).
- 9-mm pistol holster and magazine pouch (attached to LBE).
- 2 camouflage sticks (METT-T dependent).
- 2 black ink pens.
- 2 mechanical pencils with lead.
- 2 black grease pencils.
- Lensatic compass.
- Map(s) of operational area and protractor.
- Poncho.
- Poncho liner.
- 1 each ghillie suit complete.
- 1 each protective mask/MOPP suit.
- Foot powder.
- Toiletries.
- FM 23-10.

5-18. OPTIONAL EQUIPMENT

Certain situations may require equipment for specialized tasks and is METT-T dependent. The following equipment may prove useful in different climates/operational areas:

- M203 vest.
- Desert camouflage netting.
- Natural-colored burlap.
- Glitter tape.
- VS-17 panel.
- Strobe light with filters.
- Special patrol insertion/extraction system harness.
- 12-foot sling rope.
- 2 each snap links.
- 120-foot nylon rope.

- Lip/sunscreen.
- Signal mirror.
- Pen gun with flares.
- Chemical lights (to include infrared).
- Body armor/flak jacket.
- Sniper veil.
- Sewing kit.
- Insect repellant.
- Sleeping bag.
- Knee and elbow pads.
- Survival kit.
- Rifle drag bag.
- Pistol silencer/suppressor.
- 2.5-pounds C4 with caps, cord, fuze, and igniter.
- Rifle biped/tripod.
- Empty sandbags.
- Hearing protection (earmuffs).
- Thermometer.
- Laser range finder.
- Thermal imager.
- Pocket binoculars.
- 35-mm automatic loading camera with appropriate lenses and film.
- ½-inch camcorder with accessories.
- Satellite communications equipment.
- Short-range radio with earphone and whisper microphone.
- Field-expedient antennas.
- Information reporting formats.
- Encryption device for radio.

5-19. SPECIAL TOOLS AND EQUIPMENT (MOUT)

For operations in urban areas, the following tools and equipment are most useful; however, they are subject to availability

- Pry bar.
- Pliers.
- Screwdriver.
- Rubber-headed hammer.

- Glass cutter.
- Masonry drill and bits.
- Metal shears.
- Chisel.
- Auger.
- Lock pick, skeleton keys, cobra pick.
- Bolt cutters.
- Hacksaw or handsaw.
- Sledgehammer.
- Axe.
- Ram.
- Power saw.
- Cutting torch.
- Shotgun.
- Spray paint.
- Stethoscope.
- Maps/street plans.
- Photographs, aerial and panoramic.
- Whistle.
- Luminous tape.
- Flex cuffs.
- Padlocks.
- Intrusion detection system (booby traps).
- Portable spotlights.
- Money.
- Civilian attire.

5-20. ADDITIONAL EQUIPMENT TRANSPORT

The planned use of air and vehicle drops and caching techniques eliminates the need for the sniper team to carry extra equipment. Another method is to use the stay-behind technique when operating with a security patrol. (See Chapter 7.) Through coordination with the security patrol leader, the team's equipment may be distributed among the patrol members. On arrival at the ORP, the security patrol may leave behind all mission-essential equipment. After completing the mission, the team may cache the equipment for later pickup, or it may be returned the same way it was brought in.

CHAPTER 6
OPERATIONS

The SEO aids the sniper team in coordination of air support available for the three phases of operations: insertion, execution, and extraction and recovery. These techniques may be limited by the type of unit to which the sniper team is assigned, depending on the unit's resources. The team should adhere to the plan outlined in this chapter.

Section I
INSERTION

Insertion is the first critical phase of sniper operations. Regardless of the mission, the team must pass through terrain where the enemy may use sophisticated detection devices. The selected method of insertion depends on the mission, enemy situation, resources available, weather and terrain, depth of penetration, and mission priority.

6-1. PLANNING INSERTION

The preferred method of insertion is the one that best reduces the chance of detection. To provide the most current and specific details on the target area and infiltration routes from all sources, the headquarters and the sniper team adhere to the following:

a. **Intelligence.** Base operational plans on timely and accurate intelligence. Place special emphasis on efforts to obtain information on the enemy's ability to detect forces inserted by air, water, or land. The location and capabilities of air defense radar and weapons systems are critical.

b. **Deception.** Make plans to deny the enemy knowledge of the sniper team's insertion or to deceive him as to the location or intent of the operation. False insertions and other cover operations (such as air strikes, ground attacks, and air assault operations), as well as the use of multiple routes and means of insertion, ECM, and false transmissions, contribute to sniper deception plans. Select unexpected means of insertion, times, places, and routes, coupled with speed and mobility to help deceive the enemy. Also include in plans diversionary fires to direct the enemy's attention away from the team. Specific techniques may include the following

(1) Multiple airdrops, water landings, or both.

(2) Dispersion of insertion craft (air or water) if more than one, both in time and location.

(3) Landing a force in an area closer to other potential targets than to the actual targets.

(4) Leaks of false information.

(5) False landings or insertions.

(6) Diversionary actions, such as air strikes in other areas.

(7) Increased reconnaissance flights over false areas.

c. **Speed and Mobility.** Tailor individual loads to enhance speed and mobility, and balance these loads with the mission-related items necessary to achieve success. Speed is essential to limit the amount of time required to insert the team. If possible, carry only what is needed immediately and cache the rest to be retrieved.

d. **Stealth.** Stress stealth to avoid detection or interception by the enemy at the time of insertion during movement along routes and while traveling from the insertion area to the target area.

e. **Suppression.** Suppress enemy detection devices, weapons systems, and command and control facilities by electronic jamming or by suppressive fires. This detracts from the enemy's ability to discover the team during infiltration. Deception techniques contribute to suppression activities.

f. **Security.** Emphasize security measures to prevent compromise of the impending operation during preparation. This includes the security of rehearsal and training sites. Some measures that maybe used to assist in maintaining security areas follows:

(1) Restrict access to the isolation area during planning.

(2) Brief details of the operation to the team in the isolation area.

(3) Limit knowledge of planned operations on a need-to-know basis.

g. **Reconnaissance, Surveillance, and Target Acquisition.** Increase the use of RSTA equipment to detect and avoid enemy forces and their detection devices. Use passive night vision devices to achieve rapid assembly and reorganization. Also use these devices to help control and speed of movement and to traverse seemingly impassable terrain.

h. **Rehearsals.** Ensure rehearsals parallel, as close as possible, the actual conditions of insertion or extraction. Conduct rehearsals on terrain similar to that in the target area.

i. **Sand Tables.** Use sand tables in the planning phase since they are effective for orienting the team on unfamiliar DZs and surrounding terrain. The use of sand tables and terrain models enhance orderly and rapid assembly on the ground during the issuance of prejump orders and briefings.

6-2. AIR INSERTION

Air insertion is the fastest way to infiltrate. Sniper teams and equipment may be delivered by parachute (static-line or free-fall technique), fixed-wing (air landing), or helicopter (air landing, rappelling, or parachuting).

a. **Special Factors.** When planning an air insertion, headquarters considers several factors.

(1) A primary danger area is the perimeter (frontier area) where the enemy uses the most sophisticated weapons systems and air defenses.

(a) Suppression of enemy air defense maybe necessary along the infiltration corridor. This is done by a variety of sophisticated countermeasures applied against enemy equipment and by strikes against known enemy positions. Artillery, aircraft, or naval gunfire may provide assistance.

(b) Fire support, smoke screens, and suppressive measures may be critical since most of the enemy's detection devices and air defense weapons may be near the point of entry. Special equipment may be required to counter the enemy's RSTA effort whether moving by air, water, or land.

(2) If this area is within artillery or NGF range, fires should be planned on known and suspected enemy antiaircraft locations and on prominent landforms along the route.

(3) All flights over enemy territory should be routed over unoccupied areas, if possible. Flights should be planned to complement cover and deception phases and to avoid enemy air defenses.

(4) Since the sniper team depends on the transporting unit during this phase, snipers must coordinate all aspects of the air insertion with these units. To lower the chances of detection, the team makes the greatest use of reduced visibility, tactical cover, and deception. Drop zones and landing zones should be behind tree lines, in small forest clearings, or on other inconspicuous terrain.

(5) The sniper team considers the chance of in-flight emergencies. It must know the route and the checkpoints along the route. The team establishes simple ground assembly plans for contingencies before boarding. In an emergency, the SEO decides whether to continue or

abort the mission. In the absence of the SEO, the sniper makes the decision based on METT-T factors, contingency plans, and the distance to the target as compared to the distance back to forward friendly lines. Contingency provisions should be made for air and water rescue as well.

b. **Special Airborne Assault Techniques.** In airborne insertions during limited visibility, the headquarters emphasizes special delivery or navigational techniques.

(1) With the AWADS, personnel and equipment can be air-dropped during bad weather, even during zero-visibility conditions. Insertions may be made (day or night) without a pre-positioned USAF combat control team or an Army assault team. The supporting air unit requires both extensive DZ intelligence and significant lead time. All forces involved must thoroughly plan and coordinate the operation.

(2) HALO or HAHO jumps with high-performance parachutes allow parachutists to maneuver to a specific point on the ground. During these operations, they can use midair assembly procedures.

c. **Assembly.** The sniper team must be able to assemble and reorganize quickly and precisely because of its vulnerability to detection. The team develops assembly plans after careful consideration of METT-T factors, especially the location of the enemy, visibility, terrain, DZ information, dispersion pattern, and cross-loading. The number of assembly areas depends on the location, the size of available assembly areas, and the enemy's detection ability.

(1) Terrain association may be used as a backup method of designating assembly areas, but it has obvious disadvantages if the unit misses the DZ or if an in-flight change in mission dictates use of a new drop zone.

(2) A night vision plan is needed during landing, assembly, and movement in reduced visibility.

(3) Cold weather airborne insertion is difficult. Allocated times must be increased by at least 30 minutes for cold weather insertions.

(4) The team must be aware of the location of the assembly areas in relation to the direction of flight of the insertion aircraft. The direction of flight is 12 o'clock.

(5) During parachute insertion, team members must be ready for enemy engagement at all times, especially on the DZ. Immediate-action drills are required to counter enemy contact on the DZ.

d. **Planning.** The reverse planning process is of paramount importance for the ground tactical plan. The ground tactical plan, as developed from the mission assessment, is the first planning area to be considered. All other planning begins from this point.

(1) The selection of PZs or LZs requires adequate planning and coordination for effective use of air assets. Site selection must be coordinated face-to-face between the sniper team and the supporting aviation commander. The tactical situation is the key planning factor; others include the following:

- Size of landing points.
- Surface conditions.
- Ground slopes
- Approach and departure directions.
- Aircraft command and control.
- PZ and LZ identification.
- Rehearsals.

(2) The air movement plan coordinates movement of the team into the zone of action in a sequence that supports the landing plan. Key considerations are flight routes, air movement tables, flight formation, in-flight abort plan, altitude, and air speed.

(3) The landing plan introduces the team into the target area at the proper time and place. Rehearsals cannot be overemphasized. The team rapidly assembles, reorganizes, and leaves the insertion site. Fire support, if available, may be artillery, NGF, attack helicopters, or USAF tactical aircraft. The fire support plan must support all other plans. Supporting fires must be thoroughly coordinated with the air mission commander. Other planning considerations are evasion and escape, actions at the last LZ, assembly plan, downed aircraft procedures, control measures, weather delays, deception plans, and OPSEC.

6-3. AMPHIBIOUS INSERTION

Water insertion may be by surface swimming, small boat, submarine, surface craft, helocasting, or a combination thereof. The sniper team needs detailed information to plan and execute a small-boat landing, which is the most difficult phase of a waterborne insertion. Close coordination is required with naval support units.

a. **Planning.** While on the transporting craft, the team plans for all possible enemy actions and weather. Initial planning includes the following:

(1) ***Time schedule.*** The time schedule of all events from the beginning until the end of the operation is used as a planning guide. Accurate timing for each event is critical to the success of the operation.

(2) ***Embarkation point.*** The embarkation point is the point where the team enters the transporting craft.

(3) ***Drop site.*** The drop site is the site where the team leaves the primary craft and loads into a smaller boat.

(4) ***Landing site.*** The landing site is the site where the team beaches the boat or lands directly from amphibious craft.

(5) ***Loading.*** Loads and lashings, with emphasis on waterproofing, are followed IAW unit SOPs. Supervisors must perform inspections.

b. **Beach Landing Site Selection.** The beach landing site must allow undetected approach. When possible, the team avoids landing sites that cannot be approached from several different directions. The site chosen allows insertion without enemy detection. If sand beaches are used, tracks and other signs must be erased that may compromise the mission. Rural, isolated areas are preferred. The coastal area behind the landing site should provide a concealed avenue of exit. Other factors considered in each selection include enemy dispositions, distance to the target area, characteristics of landing and exit sites, and availability of cover and concealment.

c. **Tactical Deception.** Besides the water approach route plan, plans must deny the enemy knowledge of the insertion. This may include use of ECM or diversionary fire support to direct the enemy's attention away from the insertion site.

d. **Routes.** The route to the drop site should be planned to deceive the enemy. If possible, the route should be similar to that used in other types of naval operations (minelaying, sweeping, or patrolling). A major route change immediately after the team's debarkation may compromise the mission.

e. **Navigation.** Ship-to-shore navigation (to the landing site) maybe accomplished by dead reckoning to a shoreline silhouette or radar.

f. **Actions at the Drop Site.** Primary and alternate drop sites must be agreed upon. The drop site should be at least 1,500 meters offshore to prevent compromise by noise during loading and launching. (Some operations may permit landing directly from the transporting craft on shore.) If the enemy has surface radar capability, the drop site may need to be several miles offshore, or the use of ECM may be required.

g. **Actions at the Beach Landing Site.** To plan actions at the landing site, the team must consider the following:

- Actions during movement to the beach.
- Noise and light discipline.
- Navigational techniques and responsibilities.
- Actions on the beach.
- Plan for unloading boats (SOP).
- Plan for disposal or camouflage of boats.

h. **Actions on the Beach.** Once on the beach, sniper team members move to a covered and concealed security position to defend the landing site. The sniper team then conducts a brief listening halt and checks the beach landing area for signs of enemy activity. The team may deflate, bury, or camouflage the boat near the landing site or away from it, depending on the enemy situation, the terrain, and the time available. If the boat is to be disposed of or hidden near the landing site, a member must be designated to dig a hole or cut brush for camouflage. After the boat is disposed of, a designated team member sweeps the beach to erase tracks and drag marks.

i. **Insertion by Air From Ship.** Helicopters launched from a ship may extend the range of sniper teams. They may be vectored from ships to a predetermined LZ. Once in the air, other aspects of landing and assembling are the same as for air movement operations.

j. **Helocasting.** Helocasting combines a helicopter and small boat in the same operation. It is planned and conducted much the same as air movement operations, except that the LZ is in the water. While a helicopter moves at low levels (20 feet) and low speeds (20 knots), the sniper team launches a small boat and enters the water. Members then assemble, climb into the boat, and continue the mission.

k. **Contingency Planning.** The following contingencies must be covered in the planning stage:

- Enemy contact en route.
- Hot helocast site.
- Flares.
- Aerial attack.
- Small arms.
- Indirect fire.
- Downed aircraft procedure (if applicable).
- Evasion and escape.
- High surf.
- Adverse weather.
- Separation.

l. **Rehearsals.** The team must rehearse all aspects of the amphibious insertion to include boat launching, paddling, boat commands, capsize drills, beaching, and assembly.

6-4. LAND INSERTION

Land insertion from a departure point to the target area sometimes may be the best (or only) way to accomplish a mission. Normally, this is so when the enemy has total air superiority or has established effective air defenses. The sniper team can accomplish land insertions over any type of terrain, in any climate. However, thick forests, swamps, and broken or steep terrain probably offer the best chance of success.

a. **Planning.** Plans for overland movement enable the sniper team to move to the target area with the least risk of detection. Planning considerations include the following

(1) Selecting concealed primary and alternate routes based on detailed map reconnaissance and aerial photographs, ground reconnaissance, and data on the enemy situation from other sources.

(2) Avoiding obstacles, populated areas, silhouetting enemy positions, main avenues of approach, and movements along heavily populated routes and trails.

(3) Selecting the time of insertion to take advantage of reduced visibility and reduced alertness. The time is especially important during critical phases while passing through populated areas.

(4) Knowing routes, rendezvous points (and alternates), time schedules, danger areas, and the enemy situation are critical to speed and stealth.

(5) Providing centralized coordination to ensure that members act IAW cover and deception plans. Insertion by land is characterized by centralized planning and decentralized execution.

b. **Actions on Enemy Contact.** Once beyond the FFL, the sniper team must be alert to avoid detection while en route to the target area. If the sniper team becomes aware of the enemy, it must try to move away without an alert. The sniper team fights only when there is no alternative. Then, it breaks contact as quickly as possible. Following enemy contact, the sniper contacts the SEO for a decision to abort or continue the mission. If continuing the mission, the sniper team may have to establish a temporary position for resupply, extraction, or evacuation of wounded.

c. **Stay-Behind Technique.** The sniper team applies the stay-behind technique when the team moves with a security patrol. The team establishes an ORP, caches nonessential equipment, and changes into ghillie suits to prepare for movement to the TFFP. Once this is accomplished, the security patrol departs for a predetermined location to act as a quick-reaction force for the team or returns to its operational base. Use of this technique requires the following considerations:

- Noise and light discipline.
- Avoidance of enemy contact.

- Timing.
- Rough, inaccessible terrain.
- Medical evacuation.
- Communications.
- Method of extraction.
- Evasion and escape.

d. **Actions at the Insertion.** The sniper team develops a detailed assembly plan, basing it on the insertion method and the terrain at the insertion site.

(1) The sniper team selects an assembly area that can be identified at night and is near the insertion site. It uses this assembly area if team members become separated during the insertion. During parachute insertion, the sniper team uses the assembly area as an assembly point.

(2) The sniper team also designates an initial rally point that can be identified at night. The rally point is normally no closer than several hundred meters from the insertion site. The team uses the IRP for assembly if the insertion site is attacked either on insertion or shortly after departing the insertion site.

(3) When the insertion is complete, the sniper team accounts for equipment and supplies, and ensures any injuries are treated. If a disabling injury occurs during insertion, the sniper must decide, based on guidance, whether to continue the mission or to request extraction.

(4) The sniper team's most critical task is verifying the team's location. The sniper verifies his location at the insertion site or after moving away from the site.

(5) The sniper team sterilizes the site and caches or discards nonessential equipment. The preferred method is to bury discards away from the insertion site. The sniper team must camouflage the cache site.

(6) The sniper team departs the insertion site, then halts to listen for sounds of pursuit and to become familiar with local sounds. It establishes a primary azimuth and immediately begins information collection activities and map update.

6-5. VEHICLE INSERTION

Vehicle insertion uses wheeled or tracked vehicles to transport the sniper team to its insertion site. Wheeled or tracked vehicle insertion requires the same planning considerations used in other insertion techniques. The team risks compromise if it uses vehicle insertion beyond the FLOT due to noise. Enemy OPs and scout elements can easily detect and prevent infiltration of the sniper team. However, this technique can be effectively used in support of immediate battle operations by using deceptive measures.

Section II
EXECUTION

The execution phase consists of movement from the insertion site to the target area, mission execution, and movement to the extraction site.

6-6. MOVEMENT TO TARGET AREA

After leaving the insertion site, the sniper team transmits an initial entry report as required by unit SOP. This report ensures operable radio equipment and provides the team's status at the same time.

a. **Route Selection.** No matter which means of insertion, the selection of the route to the target area is critical.

(1) Enemy location, detection devices, and defensive capabilities; terrain; weather; and man-made obstacles are all to be considered when selecting the primary and alternate routes. En route checkpoints are selected to keep track of the team.

(2) The team uses NODs to operate during reduced visibility. The team's extensive training and land navigation skills allow it to rapidly traverse rugged terrain and to avoid detection.

b. **Movement Interval.** The interval between sniper team members may vary during movement into the target area. It is based on visibility, terrain, and enemy disposition. The team keys movement to the following rules, which should be discussed in detail in the sniper SOP.

(1) Maintain visual contact at a normal interval. (Intervals can expand and contract based on terrain and visibility.)

(2) Always maintain noise and light discipline.

(3) Observe the assigned sector of responsibility.

(4) React together (for example, when one gets down, they both get down.)

(5) Ensure the sniper team leader positions himself to the rear of the observer.

(6) Move on routes that best conceal movement from enemy observation and cover movement from direct enemy fire.

(7) Ensure the interval between members closes when moving through obstructions (darkness, smoke, heavy brush, narrow passes, and mine fields); ensure the interval opens when obstructions to movement and control lessen.

c. **Movement Security.** Each sniper team member must be security conscious, maintaining constant all-round security. During movement, each team member is responsible for an assigned security sector. The sniper team's route makes the best use of cover and concealment, and security or listening halts are made, as needed. Personal and equipment camouflage is enforced at all times.

d. **Arm-and-Hand Signals.** The sniper establishes standard arm-and-hand signals to reduce oral communications and to assist in control. These signals should conform to those listed in FM 21-75 and the sniper SOP.

6-7. OCCUPATION OF POSITION

The tentative final firing position, ORP, and route are selected during the mission planning phase by map and aerial photograph reconnaissance. The sniper team moves close to the TFFP and sets up an objective rally point. It then moves forward to search for a TFFP, ensuring the site is suitable and the target area can be observed at ground level. At this point, the TFFP becomes an FFP. Reconnaissance should be made during limited visibility. The team returns to the ORP, secures all mission-essential equipment, and moves to the FFP and occupies it. The sniper team watches and listens for the enemy before constructing the hide position (METT-T dependent).

6-8. SITE SELECTION

Selection of the firing position is METT-T dependent. As a minimum, the sniper team uses the following criteria when selecting an FFP:

a. Ensures that an unrestricted observation of the target area is possible. The team can then place the designated target area under constant, effective surveillance and within the range of RSTA devices and the sniper's weapon system.

b. Selects an area that provides a concealed entrance and exit routes.

c. Avoids man-made objects.

d. Avoids dominant or unusual terrain features.

e. Selects an area that is dry, or has good drainage and is not prone to flooding.

f. Selects an area that the enemy would not occupy.

g. Avoids the skyline or blending backgrounds.

h. Avoids roads or trails.

i. Avoids natural lines of movement (gullies, draws, or any terrain that affords easy foot movement).

j. Selects an area in which the team cannot be easily trapped.

k. Ensures it has a natural obstacle to vehicles between the FFP and the target area, if possible (roadside ditch, fence, wall, stream, or river).

1. Selects an area downwind of inhabited areas, if possible.

m. Selects an area in or near a suitable communications site.

n. Avoids the normal line of vision of the enemy in the target area.

o. Selects an area near a source of water.

6-9. REPORTS

The sniper team follows the communications procedures as outlined in the unit SOP. The team members must ensure that communications are maintained throughout the mission by the use of directional antennas, masking, and burst transmissions.

a. The sniper team does not analyze information it only collects and reports based on SIR. The team must format information reporting IAW the unit SOP and the type of communications equipment used.

b. Other reports that the sniper team may use, such as emergency resupply, communication checks, and emergency extraction, should also be formatted IAW the SOP.

6-10. MOVEMENT TO EXTRACTION SITE

Movement to a planned extraction site will be necessary in many operations. The sniper team must observe the principles of route selection and movement security.

a. **Priorities.** The time that a sniper team remains beyond the FFL depends on its mission and equipment. The extraction is critical from a standpoint of morale and mission accomplishment. Plans for extraction by air, ground, or water are made before the operation, with alternate plans for contingencies such as the evacuation of sick or injured personnel. During the mission, the sniper may be faced with an unforeseen situation that may demand the utmost in flexibility, discipline, and leadership.

b. **Code Words.** Each sniper team is given code words in the OPORD for use during extraction. For example, one code word may mean that the team is at its pickup zone. Another may mean that both the primary and alternate pickup zones are compromised and to abort the extraction.

c. **No Communication.** When a sniper team has missed a certain number of required transmissions, the operations section assumes that the team has a communications problem, is in trouble, or both. At that time, the no-communication extraction plan is used.

d. **Alternatives.** Extraction of the sniper team may be by means other than air. The OPORD may specify to extract the team by land or water, or to link up with friendly forces in an offensive operation. Any of these means may also be planned as alternates to avoid capture or if the sniper team cannot be extracted by air.

e. **Ground Extraction.** Despite the desirability of extracting the team by aircraft or linkup, use of these methods may be prevented by security of the sniper team, poor communications, or enemy air defense. The sniper team must be thoroughly trained in exfiltration techniques so they can walk out, either one at a time or together.

Section III
EXTRACTION AND RECOVERY

The sniper team performs an extraction as quickly as possible after the mission is accomplished. An extraction site is always planned and coordinated with supporting forces. However, the situation may dictate that the sniper decides whether to use the planned extraction site or to exfiltrate.

6-11. PLANNING

The sniper team must be prepared to exfiltrate over predetermined land routes to friendly lines as a team (or individually) or to exfiltrate to an area for extraction by air or water. Planning includes the following:

a. **Distance.** Distance may prevent an all-land exfiltration. The initial phase may be by land, ending in extraction by air or water.

b. **Terrain.** The terrain is important in selecting extraction means. The extraction site must offer favorable tactical considerations, tide data, PZ suitability, and cover from enemy direct-fire weapons. The sniper team uses the most unlikely terrain for extraction such as swamps, jungles, and mountain areas.

c. **Enemy.** Enemy pressure can develop during the extraction. Detailed plans must be made for contingency exfiltrations forced by the enemy.

d. **Evasion and Escape.** Preinsertion planning must include the development of a viable evasion and escape plan. The sniper team must do the following

(1) Checks all factors that deal with survival and evasion opportunities.

(2) Devises an evasion and escape plan that provides the best chance of survival and return to friendly lines in view of the hazards involved and mission objectives.

(3) Becomes familiar with the evasion and escape plans.

6-12. EVASION AND ESCAPE PLAN

Each mission has its specific problems associated with evasion and escape. The plan must conform to these unique problems while exploiting individual abilities, training of sniper team members, and supporting air or boat crews. The following general rules apply to evasion and escape plans for sniper operations:

a. The purpose of the plan is to attempt to save the individual who can no longer complete the assigned mission.

b. When sniper teams are behind enemy lines, the most successful escapes may involve air or water movement away from enemy-held territory.

c. Evasion and escape plans involve the following three phases:

(1) Phase one occurs during entry into the target area.

(2) Phase two occurs near the target area. It allows the sniper team to pursue its mission with a reasonable chance of success.

(3) Phase three occurs after the mission is accomplished. It is often the most difficult time to evade and escape.

d. The sniper team may be required to hide for several days to allow the enemy to become complacent before the team tries to move.

e. In selecting extraction sites, the sniper considers the danger of compromising other activities. He must prepare alternate plans for unforeseen developments.

6-13. AIR OR WATER EXTRACTION

Extraction by air or water is favored when resources are available and when it will not compromise the mission.

a. Other considerations that favor this method areas follows:

(1) Long distances must be covered.

(2) The time of return is essential.

(3) The enemy does not have air and naval superiority.

(4) Heavily populated hostile areas obstruct exfiltration.

(5) The team cannot be resupplied.

(6) Casualties must be extracted.

b. Several techniques maybe used to extract the team.

(1) Helicopter landing is the best method since the sniper team and its equipment can board the helicopter quickly.

(2) The troop ladder is the second best method. It lets sniper team members board the helicopter, but the helicopter can liftoff while snipers are still on the ladder.

(3) The STABO extraction system allows rapid pickup of one to four soldiers, who are suspended on lines beneath the helicopter. Soldiers are picked up and moved to an area where the helicopter can land. The sniper team then boards the helicopter.

(4) The jungle penetrator retrieves soldiers from areas where helicopters cannot land. It can pickup 1 to 3 persons at a time.

(5) The SPIES can extract soldiers from areas where helicopters cannot land. It can pickup 1 to 10 soldiers at a time.

6-14. LAND EXFILTRATION

This method is favored when snipers are not too far from friendly lines or no other means of extraction is available. It is also used when the terrain provides cover and concealment for foot movement and limits the employment of enemy mobile units against the exfiltrating team. Other considerations favoring this method are as follows:

a. Areas along exfiltration routes are uninhabited.

b. The enemy force is widely dispersed or is under such pressure that it is difficult for them to concentrate against the exfiltrating team.

c. The enemy force can stop an air or water extraction.

6-15. VEHICLE EXTRACTION

Vehicle extraction involves the exfiltration of the sniper team to an extraction site for extraction by a wheeled or tracked vehicle. Planning and coordination must be made during the preinsertion phase. Contingency plans must also be made to avoid compromise or any unforeseen situations.

6-16. RECOVERY

Recovery is the last phase of a sniper operation. It consists of the sniper team's return to the operations base, debriefing, equipment maintenance and turn-in, and stand-down. At the end of this phase, the sniper team prepares for future missions. (See Chapter 5.)

CHAPTER 7
COMMUNICATIONS

The basic requirement of combat communications is to provide rapid, reliable, and secure interchange of information.

Section I
FIELD-EXPEDIENT ANTENNAS

Communications are a vital aspect in successful mission accomplishment. The information in this section helps the sniper team maintain effective communications and correct any radio antenna problems.

7-1. REPAIR TECHNIQUES

Antennas are sometimes broken or damaged, causing either a communications failure or poor communications. If a spare antenna is available, the damaged antenna is replaced. When there is no spare, the sniper team may have to construct an emergency antenna. The following paragraphs contain suggestions for repairing antennas and antenna supports and the construction and adjustment of emergency antennas.

> DANGER
> SERIOUS INJURY OR DEATH CAN RESULT FROM CONTACT WITH THE RADIATING ANTENNA OF A MEDIUM-POWER OR HIGH-POWER TRANSMITTER. TURN THE TRANSMITTER OFF WHILE MAKING ADJUSTMENTS TO THE ANTENNA.

a. **Whip Antennas.** When a whip antenna is broken into two sections, the part of the antenna that is broken off can be connected to the part attached to the base by joining the sections. (Use the method shown in A, Figure 7-1, when both parts of the broken whip are available and usable.) (Use the method in B, Figure 7-1, when the part of the whip that was broken off is lost or when the whip is so badly damaged that it cannot be used.) To restore the antenna to its original length, a piece of wire is added that is nearly the same length as the missing part of the whip. The pole support is then lashed securely to both sections of the antenna. The two antenna sections are cleaned thoroughly to ensure good contact before connecting them to the pole support. If possible, the connections are soldered.

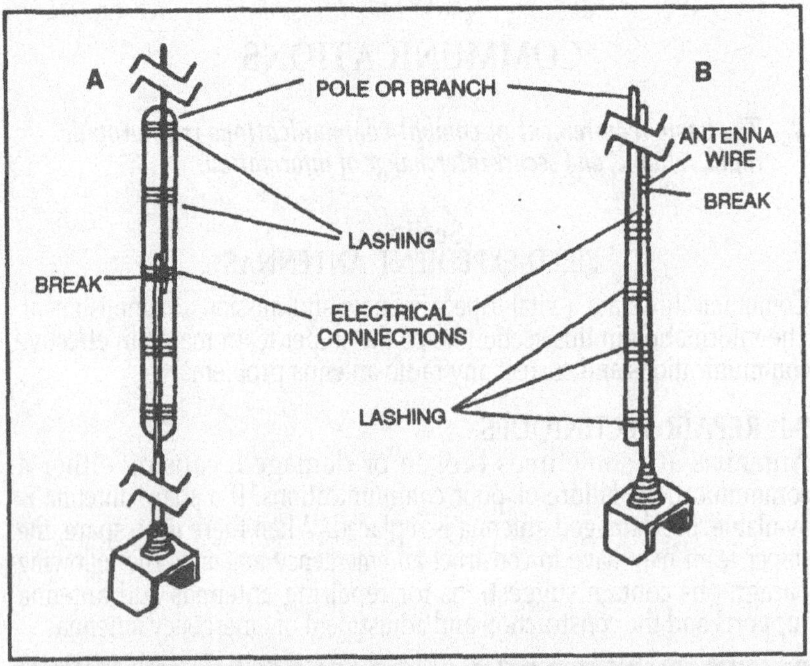

Figure 7-1. Emergency repair of broken whip antenna.

b. **Wire Antennas.** Emergency repair of a wire antenna may involve the repair or replacement of the wire used as the antenna or transmission line; or the repair or replacement of the assembly used to support the antenna.

(1) When one or more wires of an antenna are broken, the antenna can be repaired by reconnecting the broken wires. To do this, lower the antenna to the ground, clean the ends of the wires, and twist the wires together. Whenever possible, solder the connection.

(2) If the antenna is damaged beyond repair, construct a new one. Make sure that the length of the wires of the substitute antenna are the same length as those of the original.

(3) Antenna supports may also require repair or replacement. A substitute item may be used in place of a damaged support and, if properly insulated, can be of any material of adequate strength. If the radiating element is not properly insulated, field antennas may be shorted to ground and be ineffective. Many commonly found items can be used as field-expedient insulators. The best of these items are plastic or glass to include plastic spoons, buttons, bottle necks, and plastic bags. Though less effective than plastic or glass but still better than no insulator at all are wood and rope. The radiating element—the actual antenna wire-should touch only the antenna terminal and should be physically separated from all other objects, other than the supporting insulator. (See Figure 7-2 for various methods of making emergency insulators.)

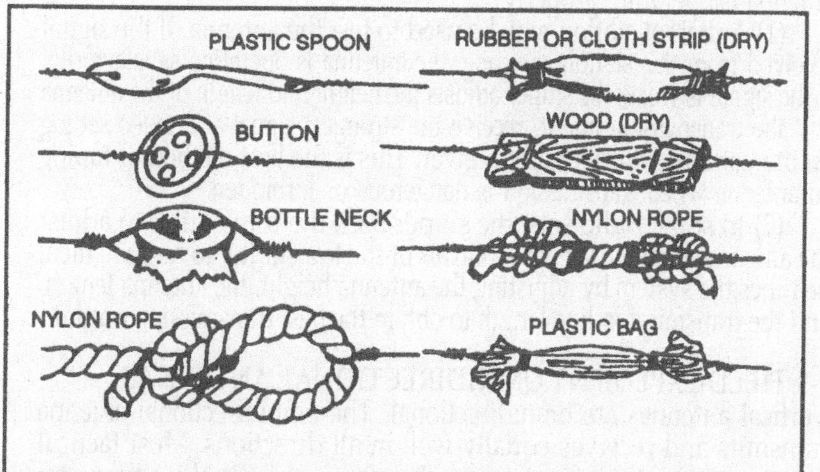

Figure 7-2. Improvised Insulators.

7-2. CONSTRUCTION AND ADJUSTMENT

Sniper teams may use the following methods to construct and adjust antennas.

a. **Construction.** The best kinds of wire for antennas are copper and aluminum. In an emergency, however, snipers use any type of wire that is available.

(1) The exact length of most antennas is critical. The emergency antenna should be the same length as the antenna it replaces.

(2) Antennas supported by trees can usually survive heavy wind storms if the trunk of a tree or a strong branch is used as a support. To keep the antenna taut and to prevent it from breaking or stretching as the trees sway, the sniper attaches a spring or old inner tube to one end of the antenna. Another technique is to pass a rope through a pulley or eyehook. The rope is attached to the end of the antenna and loaded with a heavyweight to keep the antenna tightly drawn.

(3) Guidelines used to hold antenna supports are made of rope or wire. To ensure the guidelines will not affect the operation of the antenna, the sniper cuts the wire into several short lengths and connects the pieces with insulators.

b. **Adjustment.** An improvised antenna may change the performance of a radio set. The following methods can be used to determine if the antenna is operating properly

(1) A distant station may be used to test the antenna. If the signal received from this station is strong, the antenna is operating satisfactorily. If the signal is weak, the sniper adjusts the height and length of the antenna and the transmission line to receive the strongest signal at a given setting on the volume control of the receiver. This is the best method of tuning an antenna when transmission is dangerous or forbidden.

(2) In some radio sets, the sniper uses the transmitter to adjust the antenna. First, he sets the controls of the transmitter to normal; then, he tunes the system by adjusting the antenna height, the antenna length, and the transmission line length to obtain the best transmission output.

7-3. FIELD-EXPEDIENT OMNIDIRECTIONAL ANTENNAS

Vertical antennas are omnidirectional. The omnidirectional antenna transmits and receives equally well in all directions. Most tactical antennas are vertical; for example, the man-pack portable radio uses a vertical whip and so do the vehicular radios in tactical vehicles. A vertical antenna can be made by using a metal pipe or rod of the correct length, held erect by means of guidelines. The lower end of the antenna should be insulated from the ground by placing it on a large block of wood or other insulating material. A vertical antenna may also be a wire supported by a tree or a wooden pole (Figure 7-3). For short vertical antennas, the pole may be used without guidelines (if properly supported at the base). If the length of the vertical mast is not long enough to support the wire upright, it may be necessary to modify the connection at the top of the antenna (Figure 7-4). (See FM 24-18.)

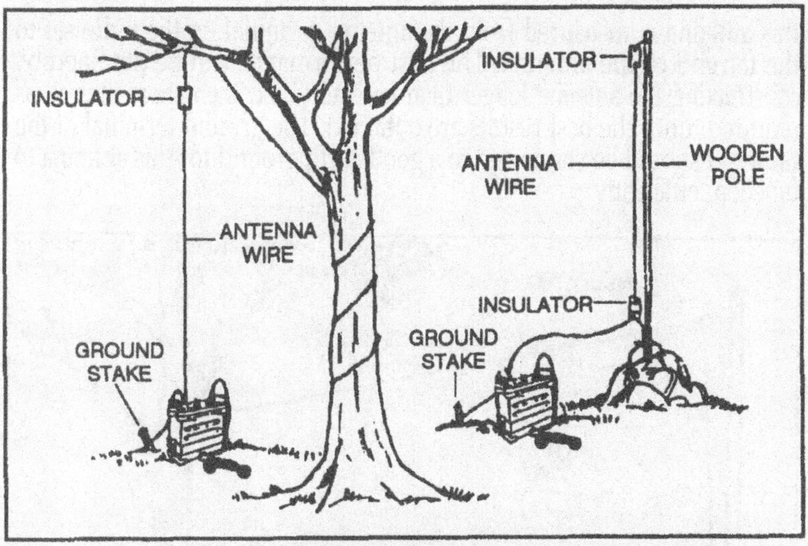

Figure 7-3. Field substitutes for support of vertical wire antennas.

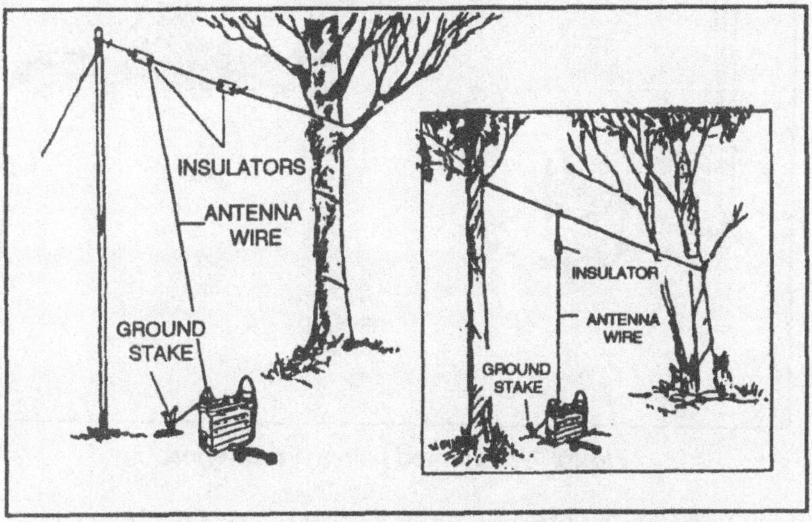

Figure 7-4. Additional means of supporting vertical wire antennas.

a. **End-Fed Half-Wave Antenna.** An emergency, end-fed half-wave antenna (Figure 7-5, page 7-6) can be constructed from available materials such as field wire, rope, and wooden insulators. The electrical length of this antenna is measured from the antenna terminal on the radio set to the far end of the antenna. The best performance can be obtained by constructing the antenna longer than necessary and then shortening it, as required, until the best results are obtained. The ground terminal of the radio set should be connected to a good earth ground for this antenna to function efficiently.

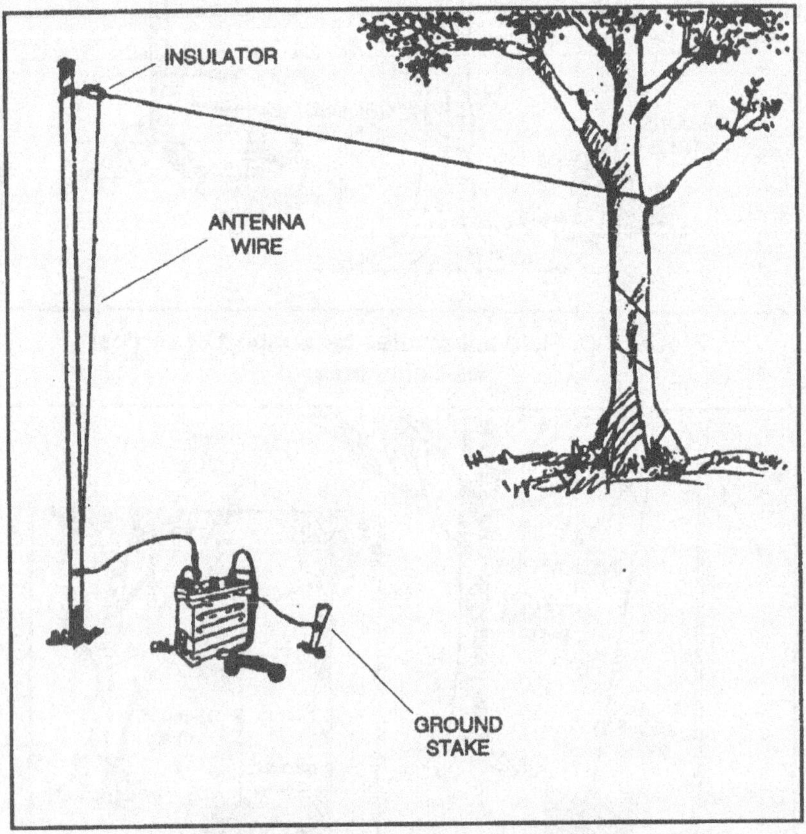

Figure 7-5. End-fed half-wave antenna.

b. **Center-Fed Doublet Antenna.** The center-fed doublet is a half-wave antenna consisting of two quarter wavelength sections on each side of the center (Figure 7-6). Doublet antennas are directional broadside to their length, which makes the vertical doublet antenna omnidirectional. This is because the radiation pattern is doughnut-shaped and bidirectional.

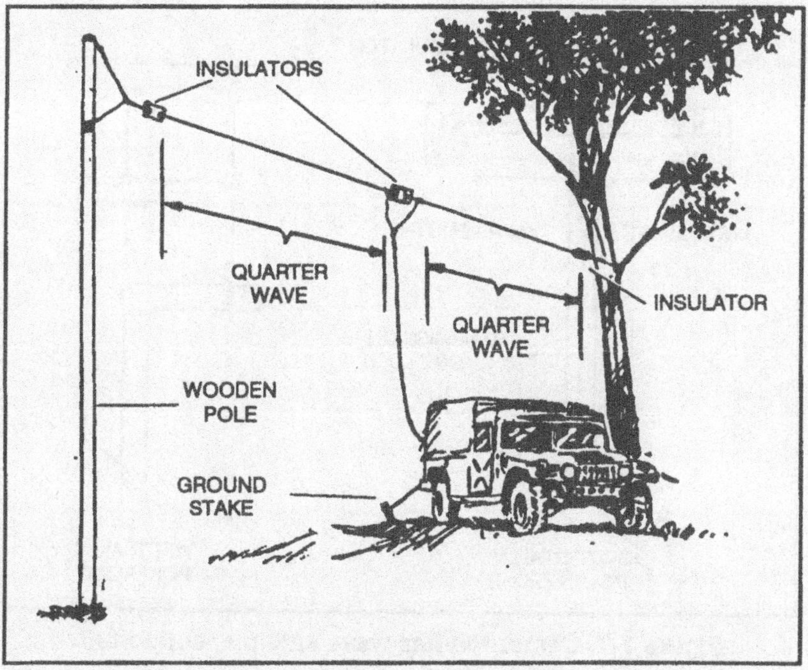

Figure 7-6. Center-fed half-wave doublet antenna.

(1) Compute the length of a half-wave antenna by using the formula in paragraph 7-5. Cut the wires as close as possible to the correct length; this is very important.

(2) Uses transmission line for conducting electrical energy from one point to another and for transferring the output of a transmitter to an antenna. Although it is possible to connect an antenna directly to a transmitter, the antenna is usually located some distance away.

(3) Support center-fed half-wave FM antennas entirely with pieces of wood. (A horizontal antenna of this type is shown in A, Figure 7-7, page 7-8, and a vertical antenna in B, Figure 7-7.) Rotate these antennas to any position to obtain the best performance.

(a) If the antenna is erected vertically, bring out the transmission line horizontally from the antenna for a distance equal to at least one-half of the antenna's length before it is dropped down to the radio set.

(b) The half-wave antenna is used with FM radios (Figure 7-8, page 7-8). It is effective in heavily wooded areas to increase the range of portable radios. Connect the top guidelines to a limb or pass it over the limb and connect it to the tree trunk or a stake.

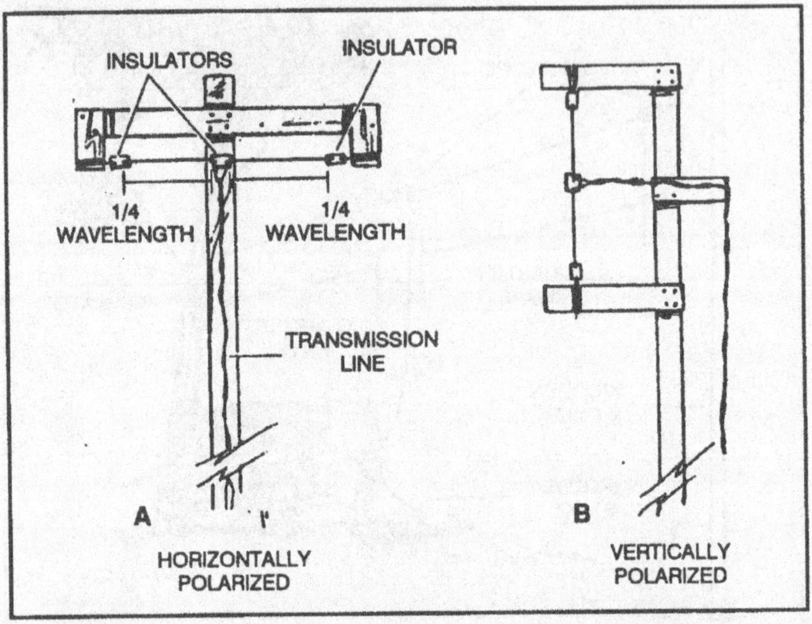

Figure 7-7. Center-fed half-wave antenna, supported.

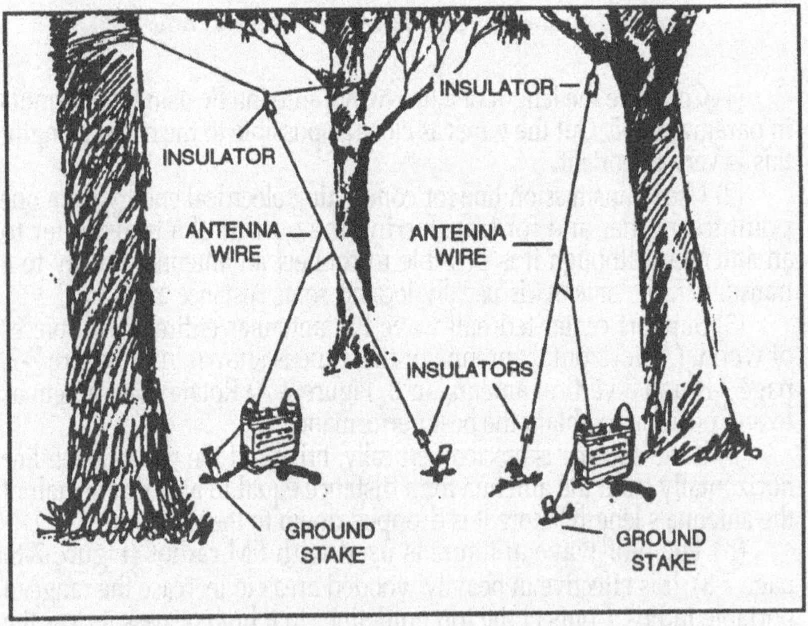

Figure 7-8. Improvised vertical half-wave antenna.

7-4. FIELD-EXPEDIENT DIRECTIONAL ANTENNAS

The vertical half-rhombic antenna (Figure 7-9) and the long-wire antenna (Figure 7-10) are two field-expedient directional antennas. These antennas consist of a single wire, preferably two or more wavelengths long, supported on poles at a height of 3 to 7 meters (10 to 20 feet) above the ground. The antennas will, however, operate satisfactorily as low as 1 meter (about 3 feet) above the ground—the radiation pattern is directional. The antennas are used mainly for either transmitting or receiving high-frequency signals.

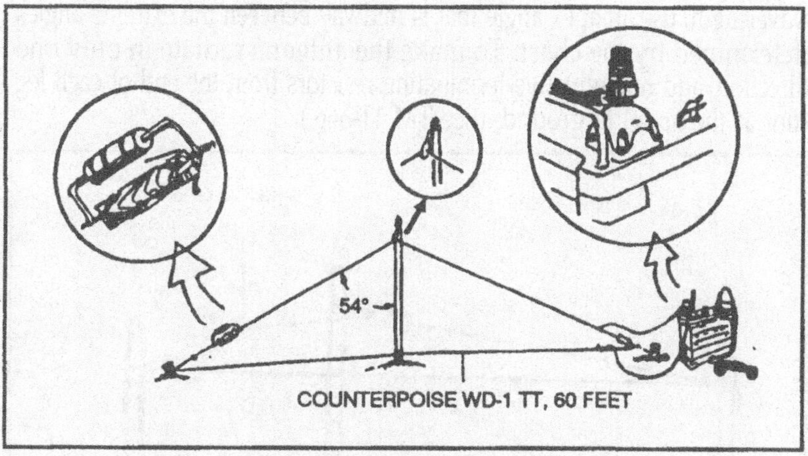

Figure 7-9. Verticle half-rhombic antenna.

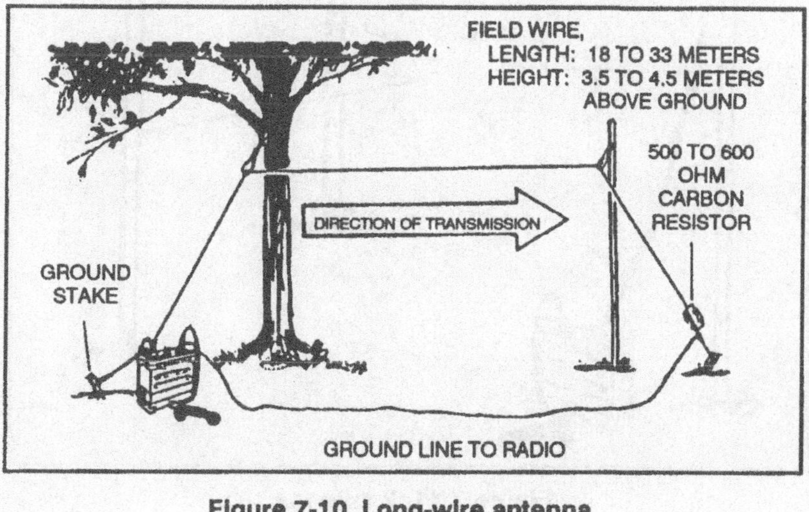

Figure 7-10. Long-wire antenna.

a. The V antenna (Figure 7-11) is another field-expedient directional antenna. It consists of two wires forming a V with the open area of the V pointing toward the desired direction of transmission or reception. To make construction easier, the legs should slope downward from the apex of the V; this is called a *sloping-V antenna* (Figure 7-12). The angle between the legs varies with the length of the legs to achieve maximum performance. (to determine the angle and the length of the legs, use the table in Table 7-l.)

b. When the antenna is used with more than one frequency or wavelength, use an apex angle that is midway between the extreme angles determined by the chart. To make the antenna radiate in only one direction, add noninductive terminating resistors from the end of each leg (not at the apex) to ground. (See TM 11-666.)

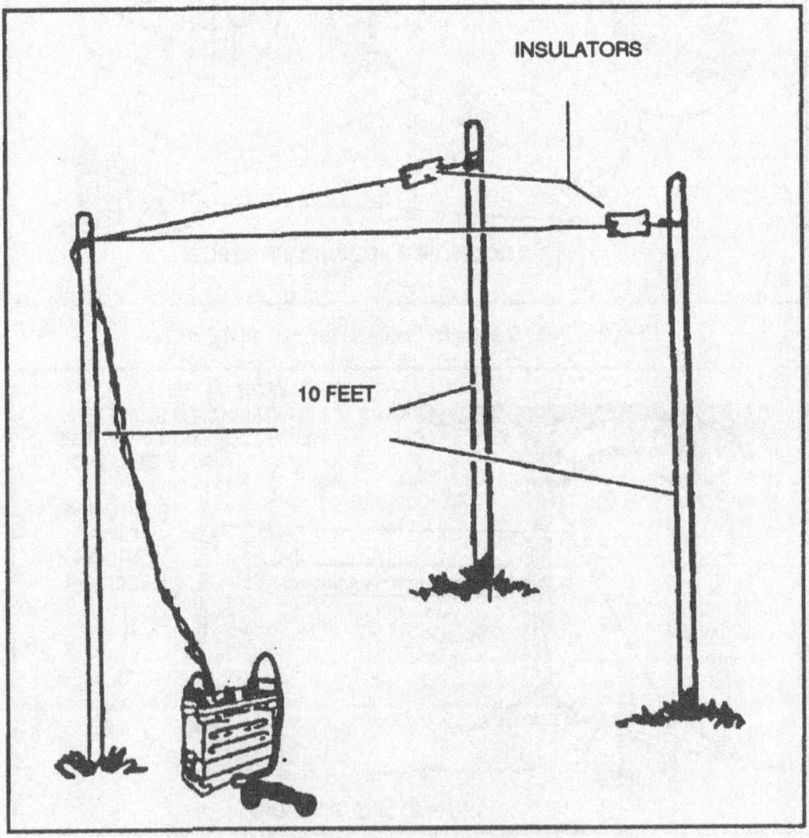

Figure 7-11. V antenna.

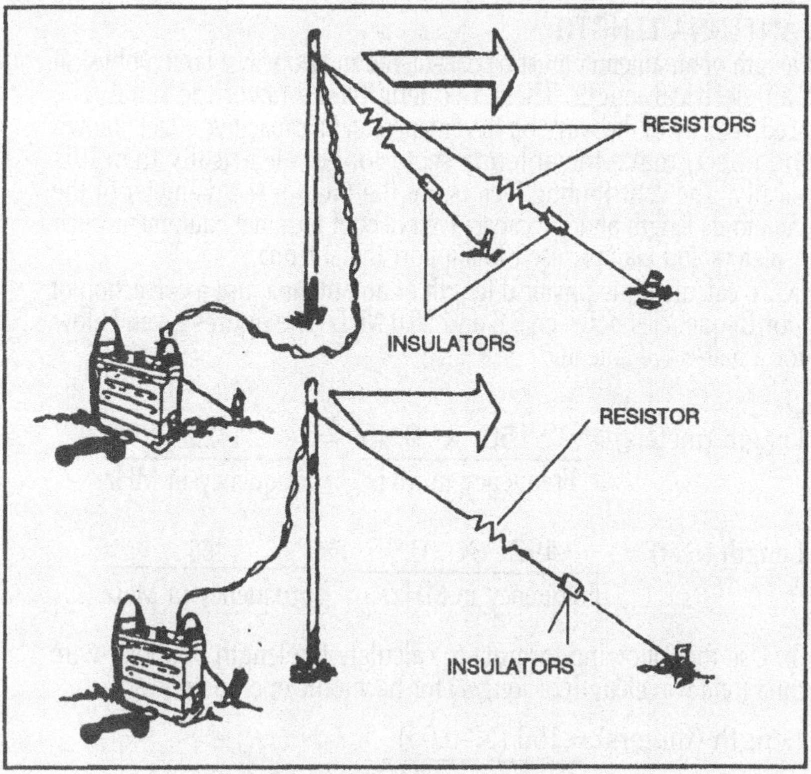

Figure 7-12. Sloping-V antenna.

ANTENNA LENGTH (wavelength)	OPTIMUM APEX ANGLE (degrees)
1	90
2	70
3	58
4	50
6	40
8	35
10	33

Table 7-1. Leg angle for V antennas.

7-5. ANTENNA LENGTH

The length of an antenna must be considered in two ways: both a physical and an electrical length. These two lengths are never the same. The reduced velocity of the wave on the antenna and a capacitive effect (known as end effect) make the antenna seem longer electrically than it is physically. The contributing factors are the ratio of the diameter of the antenna to its length and the capacitive effect of terminal equipment, such as insulators and clamps, used to support the antenna.

a. To calculate the physical length of an antenna, use a correction of 0.95 for frequencies between 3.0 and 50.0 MHz The figures given below are for a half-wave antenna.

$$\text{Length (meters)} = \frac{150 \times 0.95}{\text{Frequency in MHz}} = \frac{142.5}{\text{Frequency in MHz}}$$

$$\text{Length (feet)} = \frac{492 \times 0.95}{\text{Frequency in MHz}} = \frac{468}{\text{Frequency in MHz}}$$

b. Use the following formula to calculate the length of a long-wire antenna (one wavelength or longer) for harmonic operation:

$$\text{Length (meters)} = \frac{150\,(N-0.05)}{\text{Frequency in MHz}}$$

$$\text{Length (feet)} = \frac{492\,(N-0.05)}{\text{Frequency in MHz}}$$

N equals the number of half-wavelengths in the total length of the antenna. For example, if the number of half-wavelengths is 3 and the frequency in MHz is 7, then—

$$\text{Length (meters)} = \frac{150(N-0.05)}{\text{Frequency in MHz}} = \frac{150(3-.05)}{7} =$$

$$\frac{150 \times 2.95}{7} = \frac{442.50}{7} = 63.2 \text{ meters}$$

7-6. ANTENNA ORIENTATION

If the azimuth of the radio path is not provided, the azimuth should be determined by the best available means. The accuracy required in determining the azimuth of the path depends on the radiation pattern of the directional antenna. In transportable operation, the rhombic and V antennas may have such a narrow beam as to require great accuracy in azimuth determination. The antenna should be erected for the correct azimuth. Great accuracy is not required in erecting broad-beam antennas. Unless a line of known azimuth is available at the site, the direction of the path is best determined by a magnetic compass.

7-7. IMPROVEMENT OF MARGINAL COMMUNICATIONS

Under certain situations, it may not be feasible to orient directional antennas to the correct azimuth of the desired radio path. As a result, marginal communications may suffer. To improve marginal communications, the following procedure can be used:

a. Check, tighten, and tape cable couplings and connections.

b. Return all transmitters and receivers in the circuit.

c. Ensure antennas are adjusted for the proper operating frequency.

d. Change the heights of antennas.

e. Move the antenna a short distance away and in different locations from its original location.

Section II
RADIO OPERATIONS UNDER UNUSUAL CONDITIONS

The possibility of being deployed to different parts of the world presents many problems for the sniper team due to extremes in climate and terrain. This section informs the sniper team of these common problems and possible solutions to eliminate or reduce adverse effects.

7-8. ARCTIC AREAS

Single-channel radio equipment has certain capabilities and limitations that must be carefully considered when operating in cold areas. However, in spite of limitations, radio is the normal means of communications in such areas. One of the most important capabilities of the radio in Arctic-like areas is its versatility. Man-packed radios can be carried to any point accessible by foot or aircraft. A limitation on radio communications that radio operators must expect in extremely cold areas is interference by ionospheric disturbances. These disturbances, known as ionospheric storms, have a definite degrading effect on skywave propagation. Moreover, either the storms or the auroral (such as northern lights) activity can cause complete failure of radio communications. Some frequencies may be blocked completely by static for extended periods during storm activity. Fading, caused by changes in the density and height of the ionosphere, can also occur and may last from minutes to weeks. The occurrence of these disturbances is difficult to predict. When they occur, the use of alternate frequencies and a greater reliance on FM or other means of communications are required.

a. **Antenna Installation.** Antenna installation in Arctic-like areas presents no serious problems. However, installing some antennas may take longer because of adverse working conditions. Some suggestions for installing antennas in extremely cold areas areas follows:

(1) Antenna cables must be handled carefully since they become brittle in low temperatures.

(2) Whenever possible, antenna cables should be constructed overhead to prevent damage from heavy snow and frost. Nylon rope guidelines, if available, should be used in preference to cotton or hemp because nylon ropes do not readily absorb moisture and are less likely to freeze and break.

(3) An antenna should have extra guidelines, supports, and anchor stakes to strengthen it to withstand heavy ice and wind.

(4) Some radios (usually older generation radios) adjusted to a specific frequency in a relatively warm place may drift off frequency when exposed to extreme cold. Low battery voltage can also cause frequency drift. When possible, a radio should warmup several minutes before placing it into operation. Since extreme cold tends to lower output voltage of a dry battery, warming the battery with body heat before operating the radio set can reduce frequency drift.

(5) Flakes or pellets of highly electrically charged snow is sometimes experienced in northern regions. When these particles strike the antenna, the resulting electrical discharge causes a high-pitched static roar that can blanket all frequencies. To overcome this static, antenna elements can be covered with polystyrene tape and shellac.

b. **Maintenance Improvement in Arctic Areas.** The maintenance of radio equipment in extreme cold presents many problems. Radio sets must be protected from blowing snow since snow will freeze to dials and knobs and blow into the wiring to cause shorts and grounds. Cords must be handled carefully as they may lose their flexibility in extreme cold. All radio equipment must be properly winterized. The appropriate technical manual should be checked for winterization procedures. Some suggestions for maintenance in Arctic areas include:

(1) *Batteries.* The effect of cold weather conditions on wet and dry cell batteries depends on the following factors: the type and kind of

battery, the load on the battery, the specific use of the battery, and the degree of exposure to cold temperatures.

(2) ***Winterization.*** The radio set technical manual should rechecked for special precautions for operation in extremely cold climates. For example, normal lubricants may solidify and cause damage or malfunctions. They must be replaced with the recommended Arctic lubricants.

(3) ***Microphone.*** Moisture from the sniper's breath may freeze on the perforated cover plate of his microphone. Standard microphone covers can be used to prevent this. If standard covers are not available, a suitable cover can be improvised from rubber or cellophane membranes or from rayon or nylon cloth.

(4) ***Breathing and sweating.*** A radio set generates heat when it is operated. When turned off, the air inside the radio set cools and contracts, and draws cold air into the set from the outside. This is called *breathing*. When a radio breathes and the still-hot parts come in contact with subzero air, the glass, plastic, and ceramic parts of the set may cool too rapidly and break. When cold equipment is brought suddenly into contact with warm air, moisture condenses on the equipment parts. This is called *sweating*. Before cold equipment is brought into a heated area, it should be wrapped in a blanket or parka to ensure that it warms gradually to reduce sweating. Equipment must be thoroughly dry before it is taken into the cold air or the moisture will freeze.

7-9. JUNGLE AREAS

Radio communications in jungle areas must be carefully planned, because the dense jungle growth reduces the range of radio transmission. However, since single-channel radio can be deployed in many configurations, especially man-packed, it is a valuable communications asset. The capabilities and limitations of single-channel radio must be carefully considered when used by forces in a jungle environment. The mobility and various configurations in which a single-channel radio can be deployed are its main advantages in jungle areas. Limitations on radio communications in jungle areas are due to the climate and the density of jungle growth. The hot and humid climate increases maintenance problems of keeping the equipment operable. Thick jungle growth acts as a vertically polarized absorbing screen for radio frequency energy that, in effect, reduces transmission range. Therefore, increased emphasis on maintenance and antenna siting is a must when operating in jungle areas.

a. **Jungle Operational Techniques.** The main problem in establishing radio communications in jungle areas is the siting of the antenna.

The following techniques can be applied to improve communications in the jungle:

(1) Locate antennas in clearings on the edge farthest from the distant station and as high as possible.

(2) Keep antenna cables and connectors off the ground to lessen the effects of moisture, fungus, and insects. This also applies to all power and telephone cables.

(3) Use complete antenna systems, such as ground planes and dipoles, for more effect than fractional wavelength whip antennas.

(4) Clear vegetation from antenna sites. If an antenna touches any foliage, especially wet foliage, the signal will be grounded.

(5) When wet, vegetation acts like a vertically polarized screen and absorbs much of a vertically polarized signal. Use horizontally polarized antennas in preference to vertically polarized antennas.

b. **Maintenance Improvement in the Jungle.** Due to moisture and fungus, the maintenance of radio sets in tropical climates is more difficult than intemperate climates The high relative humidity causes condensation to form on the equipment and encourages the growth of fungus. Operators and maintenance personnel should check appropriate technical manuals for special maintenance requirements. Some techniques for improving maintenance in jungle areas follow:

(1) Keep the equipment as dry as possible and in lighted areas to retard fungus growth.

(2) Clear all air vents of obstructions so air can circulate to cool and dry the equipment.

(3) Keep connectors, cables, and bare metal parts as free of fungus growth as possible.

(4) Use moisture and fungus-proofing paint to protect equipment after repairs are made or when equipment is damaged or scratched.

c. **Expedient Antennas.** Sniper teams can improve their ability to communicate in the jungle by using expedient antennas. While moving, the team is usually restricted to using the short and long antennas that come with the radios. However, when not moving, snipers can use these expedient antennas to broadcast farther and to receive more clearly. However, an antenna that is not "tuned" or "cut" to the operating frequency is not as effective as the whips that are supplied with the radio. Circuits inside the radio "load" the whips properly so that they are "tuned" to give the greatest output. Whips are not as effective as a tuned doublet or tuned ground plane (namely RC 292-type), but the doublet or ground plane must be tuned to the operating frequency. This is especially critical with low-power radios such as the AN/PRC-77.

(1) ***Expedient 292-type antenna.*** The expedient 292-type antenna was developed for use in the jungle and, if used properly, can increase the team's ability to communicate. In its entirety, the antenna is bulky, heavy, and not acceptable for sniper team operations. The team can, however, carry only the mast head and antenna sections, mounting these on wood poles or hanging them from trees; or, the team can make a complete expedient 292-type antenna (Figure 7-13, page 7-18), using WD-1, wire, and other readily available material. The team can also use almost any plastic, glass, or rubber objects for insulators. Dry wood is acceptable when nothing else is available. (See Figure 7-2 for types of insulators that may be used.) The following describes how to make this antenna:

(a) Use the quick-reference table (Table 7-2, page 7-19) to determine the length of the elements (one radiating and three ground planes) for the frequency that will be used. Cut these elements (A, Figure 7-13, page 7-18) from WD-1 field wire (or similar wire). Cut spacing sticks (B, Figure 7-13) the same length. Place the ends of the sticks together to form a triangle and tie the ends with wire, tape, or rope. Attach an insulator to each corner. Attach a ground-plane wire to each insulator. Bring the other ends of the ground-plane wires together, attach them to an insulator (C, Figure 7-13, page 7-18), and tie securely. Strip about 3 inches of insulation from each wire and twist them together.

(b) Tie one end of the radiating element wire to the other side of insulator C and the other end to another insulator (D, Figure 7-13). Strip about 3 inches of insulation from the radiating element at insulator C.

(c) Cut enough WD-1 field wire to reach from the proposed location of the antenna to the radio set. Keep this line as short as possible, because excess length reduces the efficiency of the system. Tie a knot at each end to identify it as the "hot" lead. Remove insulation from the "hot" wire and tie it to the radiating element wire at insulator C. Remove insulation from the other wire and attach it to the bare ground-plane element wires at insulator C. Tape all connections and do not allow the radiating element wire to touch the ground-plane wires.

(d) Attach a rope to the insulator on the free end of the radiating element and toss the rope over the branches of a tree. Pull the antenna as high as possible, keeping the lead-in routed down through the triangle. Secure the rope to hold the antenna in place.

(e) At the radio set, remove about 1 inch of insulation from the "hot" lead and about 3 inches of insulation from the other wire. Attach the "hot" line to the antenna terminal (doublet connector, if so labeled). Attach the other wire to the metal case-the handle, for example. Be sure both connections are tight or secure.

(f) Set up correct frequency, turn on the set, and proceed with communications.

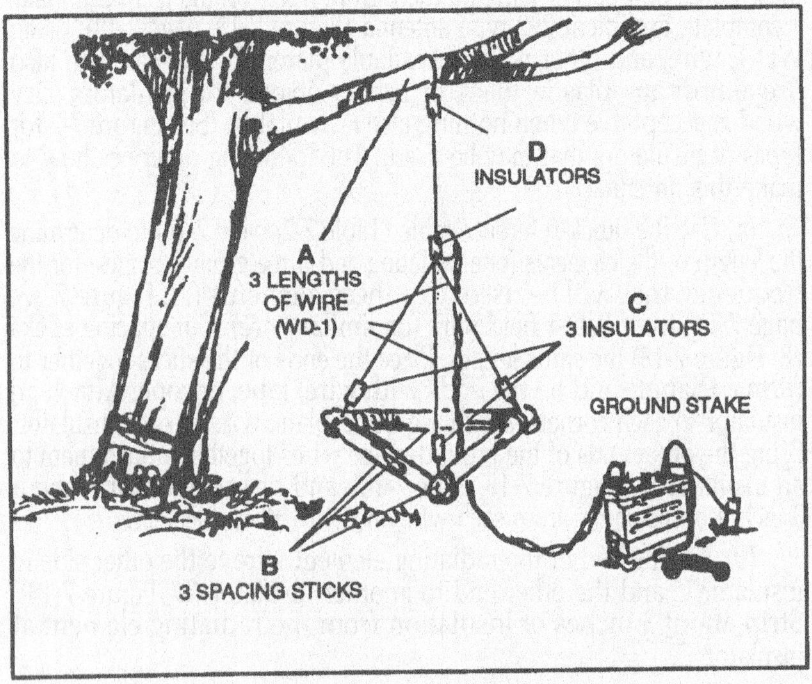

Figure 7-13. Expedient 292-type antenna.

(2) *Expedient patrol antenna. This* is another antenna that is easy to carry and quick to set up (Figure 7-14, page 7-20). The two radiating wires are cut to the length shown in Table 7-2 for the operating frequency. For the best results, the lead-in should extend at least 1.8 meters (6 feet) at right angles (plus or minus 30 degrees) to the antenna section before dropping to the radio set. The easiest way to set up this antenna is to measure the length of the radiating elements from one end of the lead-in (WD-1) and tie a knot at that point. The two wires are separated: one is lifted vertically by a rope and insulator;

the other is held down by a rock or other weight and a rope and insulator. The antenna should be as high as possible. The other end of the lead-in is attached to the radio set as described in paragraph 7-9c(l), expedient 292-type antenna.

OPERATING FREQUENCY IN MHz	ELEMENT LENGTH (radiating element and ground-plane elements)
30	2.38m (7 ft 10 in)
32	2.23m (7 ft 4 in)
34	2.1m (6 ft 11 in)
36	1.98m (6 ft 6 in)
38	1.87m (6 ft 2 in)
40	1.78m (5 ft 10 in)
43	1.66m (5 ft 5 in)
46	1.55m (5 ft 1 in)
49	1.46m (4 ft 9 in)
52	1.37m (4 ft 6 in)
55	1.3m (4 ft 3 in)
58	1.23m (4 ft 0 in)
61	1.17m (3 ft 10 in)
64	1.12m (3 ft 8 in)
68	1.05m (3 ft 5 in)
72	.99m (3 ft 3 in)
76	.94m (3 ft 1 in)

Table 7-2. Quick-reference chart.

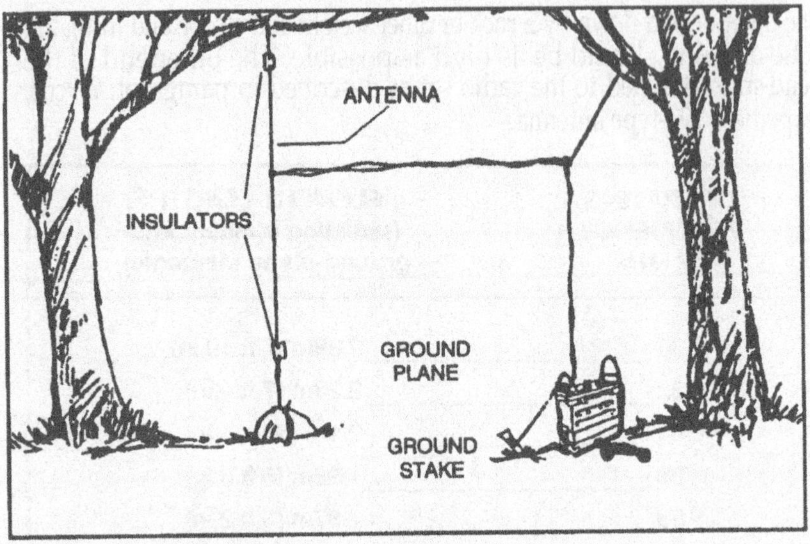

Figure 7-14. Expedient patrol antenna.

7-10. DESERT AREAS

Radio is usually the primary means of communications in the desert. It can be employed effectively in desert climate and terrain to provide a highly mobile means of communications demanded by widely dispersed forces.

a. **Techniques for Better Operations.** For the best operation in the desert, radio antennas should be located on the highest terrain available. In the desert, transmitters using whip antennas lose one-fifth to one-third of their normal range due to the poor electrical grounding common to desert terrain. For this reason, complete antenna systems must be used such as horizontal dipoles and vertical antennas with adequate counterpoises.

b. **Equipment Considerations.** Some radios automatically switch on their second blower fan if their internal temperature rises too high. Normally, this happens only in temperate climates when the radios are transmitting. This may disturb soldiers unaccustomed to radio operation in the desert environment. Operation of the second fan, however, is quite normal. Radio frequency power amplifiers used in AM and single sideband sets may overheat and burn out. Such equipment should be turned on only when necessary (signal reception is not affected). Since the RF power amplifiers take about 90 seconds to reach the operating mode, the SOP of units using the equipment allows for delays in replying. Dust affects communications equipment such as SSB/AMRF power amplifiers and radio teletypewriter sets. Radio teletypewriter sets are prone to damage due to the vulnerability of the oil lubrication system, which attracts and holds dust particles. Dust covers, therefore, should be used when possible. Some receiver-transmitter units have ventilating ports and channels that can get clogged with dust. These must be checked regularly and kept clean to prevent overheating.

c. **Batteries.** Dry battery supplies must be increased, since hot weather causes batteries to fail more rapidly.

d. **Electrical Insulation.** Wind-blown sand and grit damage electrical wire insulation over time. All cables that are likely to be damaged should be protected with tape before insulation becomes worn. Sand also finds its way into parts of items, such as "spaghetti cord" plugs, either preventing electrical contact or making it impossible to join the plugs together. A brush, such as an old toothbrush, should be carried and used to clean such items before they are joined.

e. **Condensation.** In deserts with relatively high dew levels and high humidity, overnight condensation can occur wherever surfaces are cooler than the air temperature, such as metals exposed to air. This condensation can affect electrical plugs, jacks, and connectors. All connectors likely to be affected by condensation should be taped to prevent moisture from contaminating the contacts. Plugs should be dried before inserting them into equipment jacks. Excessive moisture or dew should be dried from antenna connectors to prevent arcing.

f. **Static Electricity.** Static electricity is prevalent in the desert. It is caused by many factors, one of which is wind-blown dust particles. Extremely low humidity contributes to static discharges between charged particles. Poor grounding conditions aggravate the problem. All sharp edges (tips) of antennas should be taped to reduce wind-caused static discharges and the accompanying noise. If operating from a fixed position, teams ensure that equipment is properly grounded. Since static-caused noise lessens with an increase infrequency, the highest frequencies that are available and authorized should be used.

g. **Maintenance Improvement.** In desert areas, the maintenance of radio sets becomes more difficult due to the large amounts of sand, dust, or dirt that enter the equipment. Sets equipped with servomechanisms are especially affected. To reduce maintenance downtime, the team must keep sets in dustproof containers as much as possible. Air vent filters should also be kept clean to allow cool air to circulate to prevent overheating. Preventive maintenance checks should be made often. Also, the team should closely check the lubricated parts of the equipment. If dust and dirt mix with the lubricants, moving parts may be damaged.

7-11. MOUNTAINOUS AREAS

Operation of radios in mountainous areas have many of the same problems as in northern or cold weather areas. The mountainous terrain makes the selection of transmission sites a critical task In addition, terrain restrictions often require radio relay stations for good communications. Due to terrain obstacles, radio transmissions often have to be by line of sight. Also, the ground in mountainous areas is often a poor electrical conductor. Thus, a complete antenna system, such as a dipole or ground-plane antenna with a counterpoise, should be used. The maintenance procedures required in mountainous areas are the same as for northern or cold weather areas. The varied or seasonal temperature and climatic conditions in mountainous areas make flexible maintenance planning a necessity.

7-12. URBANIZED TERRAIN

Radio communications in urbanized terrain pose special problems. Some problems are similar to those encountered in mountainous areas. Some problems include obstacles blocking transmission paths, poor electrical conductivity due to pavement surfaces, and commercial power line interference.

a. Very high frequency radios are not as effective in urbanized terrain as they are in other areas. The power output and operating frequencies of these sets require a line of sight between antennas. Line of sight at street level is not always possible in built-up areas.

b. High frequency radios do not require or rely on line of sight as much as VHF radios. This is due to operating frequencies being lower and power output being greater. The problem is that HF radio sets are not organic to small units. To overcome this, the VHF signals must be retransmitted.

c. Retransmission stations in aerial platforms can provide the most effective means if available. Organic retransmission is more likely to be used. The antenna should be hidden or blended in with surroundings. This helps prevent the enemy from using it as a landmark to "home in" his artillery bombardment. Antennas can be concealed by water towers, existing civilian antennas, or steeples.

7-13. NUCLEAR BIOLOGICAL AND CHEMICAL ENVIRONMENT

One of the realities of fighting on today's battlefield is the presence of nuclear weapons. Most soldiers are aware of the effects of nuclear blast, heat, and radiation. The ionization of the atmosphere by a nuclear explosion will have degrading effects on communications due to static and the disruption of the ionosphere.

a. Electromagnetic pulse results from a nuclear explosion and presents a great danger to our radio communications. An EMP is a strong pulse of electromagnetic radiation, many times stronger than the static pulse generated by lightning. This pulse can enter the radio through the antenna system, power connections, and signal input connections. In the equipment, the pulse can break down circuit components such as transistors, diodes, and integrated circuits. It can melt capacitors, inductors, and transformers, destroying a radio.

b. Defensive measures against EMP call for proper maintenance, especially the shielding of equipment. When the equipment is not in use, all antennas and cables should be removed to decrease the effect of EMP on the equipment.

Section III
COMMUNICATIONS FORMATS

Timely, accurate information reporting reduces the unknown aspects of the enemy and the area of operations, contributing to the commander's risk assessment and successful application of combat power. This section provides the sniper team with a means of organized and rapid information delivery through reporting formats.

7-14. SPOT REPORT

This paragraph complies with STANAG 2022.

The sniper team uses the SPOTREP to report intelligence information. Each report normally describe a single observed event. When reporting groups of enemy vehicles, personnel report the location of the center of mass or indicate "from—to" coordinates. Higher headquarters sets the SPOTREP format, but the report usually follows the SALUTE format.

LINE 1 The size of the enemy force observed.

LINE 2 What the enemy was doing.

LINE 3 Where the enemy was located.

LINE 4 The unit to which the enemy belongs specified by markings on vehicles, distinctive features on uniforms, or special equipment that may identify the type enemy unit.

LINE 5 Time the enemy was observed.

LINE 6 Equipment the enemy carried, wore, or used.

Example: "C12, THIS IS STRIKER 1, SPOTREP, OVER."
"STRIKER 1, THIS IS Cl2 SEND MESSAGE, OVER."
"C12, THIS IS STRIKER 1. LINE 1: 3. LINE 2 MOVING IN A WESTERLY DIRECTION. LINE 3: GL024396. LINE 4: UNKNOWN. LINE 5: 2709911437. LINE 6: 1 SVD WITH PSO-1 TELESCOPE. CAMOUFLAGED OVERGARMENT AND RUCKSACK TWO INDIVIDUALS CARRYING AKM-74 RIFLES. 9-MM MAKAROV PISTOLS WITH SHOULDER HOLSTERS AND RUCKSACKS."

7-15. SITUATION REPORT

This paragraph complies with STANAG 2020.

The sniper team submits the SITREP to higher headquarters to report tactical situations and status. The team submits the report daily by 0600 hours after significant events or as otherwise required by the SEO or commander. The sender says, "SITREP," to alert the receiver of the type of report being sent. The following explains the reporting format according to line number:

LINE 1 Report as of date-time group.

LINE 2 Brief summary of enemy activity, casualties inflicted, prisoners captured.

LINE 3 Your location (encrypted—if not using secure communications).

LINE 4 Combat vehicles, operational.

a. Improved TOW vehicle.

b. M3 Bradley/Ml13Al.

c. Ml.

d. M60A3 tanks.

e. M106A1 mortar carriers.

f. Armored vehicle launched bridges (AVLB).

LINE 5 Defensive obstacles encoded.

a. Coordinates of mine fields.

b. Coordinates of demolitions executed.

c. Coordinates of reserve demolition targets.

LINE 6 Personnel strength.

a. Green (full strength, 90 percent or better on hand).

b. Amber (reduced strength, 80 to 89 percent on hand).

c. Red (reduced strength, 60 to 79 percent on hand, mission-capable).

d. Black (reduced strength, 59 percent or lesson hand).

LINE 7 Class III and V for combat vehicles.

a. Ammunition—green, amber, red, or black.

b. Fuel—green, amber, red, or black.

LINE 8 Summary of tactical intentions.

Example: "RED 1, THIS IS RED 5; BLUE 2. LINE 1: 062230. LINE 2: NEGATIVE CONTACT. LINE 3: I SET ES, STA NEL. LINE 4B: 1. LINE 5: ABATIS, 1 SET XB, RDJ ALT. LINE 6: GREEN. LINE 7A: GREEN. LINE 7B: AMBER. LINE 8: CONTINUING MISSION."

7-16. RECONNAISSANCE REPORT

This paragraph complies with STANAG 2096.

Due to the length and detail of a reconnaissance report, it should be sent by messenger rather than transmitted by radio. Graphic overlays and sketches are normally included with the report. The following explains the reporting format according to line number:

LINE 1 OR HEADING (collection data).

a. DTG information collected.

b. DTG information received.

c. Reporting unit.

LINE 2 OR 3 CAPITAL ROUTE CLASSIFICATION (data for a route classification).

a. Start point.

b. Checkpoint/release point.

c. Classification (code).

d. Trafficability (code).

e. Movement (code).

f. Location of critical points.

LINE 3 OR BRIDGE CLASSIFICATION (data for a bridge classification).

a. Location.

b. One-way class.

c. Two-way class.

d. Overhead clearance.

e. Bypass location.

f. Bypass (code).

g. Slope of entry bank.

h. Slope of exit bank.

LINE 4 OR FORDING/SWIM SITE (data for a ford or swim site).

a. Location.

b. Velocity (water speed).

c. Depth.

d. Type bottom (code).

e. Width.

f. Length.

g. Slope of entry bank.

h. Slope of exit bank.

LINE 5 OR TUNNEL CLASSIFICATION (data for a tunnel classification).

a. Location.

b. Usable width.

c. Overhead clearance.

d. Length.

e. Bypass location.

LINE 6 OR OBSTACLES (obstacle information).

a. Location.

b. Slope (code).

c. Type (code).

d. Length.

e. Bypass location.

f. Dimensions.

(1) From:

(2) To:

(3) To:

CODES: Classification

GREEN - all vehicles.
AMBER - no AVLBs.
RED - armed personnel carriers/BFVs.
BLACK -1 1/4-ton wheels or less.

Trafficability

X - all weather.
Y - limited weather.
Z - fair weather.

Movement

F - fast.
S - slow.

Bypass

E - easy.
D - difficult.

Type bottom

M - mud.
C - clay.
S - sand.
G - gravel.
R - rock.
P - paving.

Slope

A - less than 7 percent.
B - 7 or 10 percent.
C - 10 to 14 percent.
D - Over 14 percent.

Type obstacle

MF - mine field.

TD - tank ditch.

RF - rockfall or slide.

CH - chemical.

NBC - radiological.

RB - roadblock.

AB - abatis.

O - other.

NOTES: 1. During reconnaissance., report items as they occur, since they are time-sensitive.

2. If time permits, submit overlays to the S2 during briefing. The S2 routinely consolidates details of terrain features and passes them to higher headquarters at the end of the debriefing.

Example: "C12, THIS IS STRIKER 1, RECONREP OVER."
"STRIKER 1, THIS IS C12; SEND MESSAGE, OVER."
"C12, THIS IS STRIKER 1. LINE 1A: 2609910800. LINE 1C: ST 1. LINE 2A: I SET DL, JAR CMN. LINE 2B: SIL MNC. LINE 2C: GREEN. LINE 2D: X. LINE 2E: F."

7-17. MEACONING, INTRUSION, JAMMING, AND INTERFERENCE REPORT.

This paragraph complies with STANAG 6004.

When the sniper team knows or suspects that the enemy is jamming, or knows or suspects that the enemy is intruding on the net, the incident is reported immediately by secure means to higher headquarter. Such information is vital for the protection and defense of friendly radio communications. The sniper who is experiencing the MIJI incident forwards this report through the chain of command to the unit OP. He also submits a separate report for each MIJI incident. An example of a MIJI 1 report follows:

ITEM 1-022 (encrypted) or MIJI 1.

ITEM 2-3 (encrypted) or JAMMING.

ITEM 3 - 1 (encypted) or RADIO.

ITEM 4 - 46.45 (encyypted if being transmitted over a nonsecure communications means).

ITEM 5 - N6B85S.

ITEM 6 - FA86345964 (encrypted if being transmitted over a nonsecure communications means).

a. Item 1 - Type of Report. When transmitted over nonsecure communications means, the numerals 022 are encrypted as Item 1 of the MIJI report. When transmitted over secure communications means, the term MIJI 1 is used as Item 1 of the MIJI 1 report.

b. Item 2 - Type of MIJI Incident. When transmitted over nonsecure communications means, the appropriate numeral preceding one of the items below is encrypted as Item 2 of the MIJI report. When transmitted over secure communications means, the appropriate term below is used as Item 2 of the MIJI 1 report.

- Meaconing.
- Intrusion.
- Jamming.
- Interference.

c. Item 3 - Type of Equipment Affected. When transmitted over nonsecure communications means, the appropriate numeral preceding one of the terms below is encrypted as Item 3 of the MIJI 1 report. When transmitted over secure communications means, the appropriate term below is used as Item 3 of the MIJI report.

- Radio.
- Radar.
- Navigational aid.
- Satellite.
- Electro-optics.

d. Item 4 - Frequency or Channel Affected. When transmitted over nonsecure communications means, the frequency or channel affected by the MIJI incident is encrypted as Item 4 of the MIJI 1 report. When transmitted over secure communications means, the frequency or channel affected by the MIJI incident is Item 4 of the MIJI 1 report.

e. Item 5 - Victim Designation and Call Sign of Affected Station Operator. The complete call sign of the affected station operator is Item 5 of the MIJI 1 report over both secure and nonsecure communications means.

f. Item 6 - Coordinates of the Affected Station. When transmitted over nonsecure communications means, the complete grid coordinates of the affected station are encrypted as Item 6 of the MIJI 1 report. When transmitted over secure communications means, the complete grid coordinates of the affected station are Item 6 of the MIJI 1 report.

7-18. SHELLING REPORTS

This paragraph complies with STANAG 2934.

The sniper team prepares and submits a SHELREP when it receives incoming rockets, mortars, or artillery rounds (FM 6-121). The team also uses this format for bombing attacks and mortars. The SHELREP format is as follows:

- ALPHA: Unit call sign.
- BRAVO: Location of observer.
- CHARLIE: Azimuth to flash or sound.
- DELTA: Time shelling started.
- ECHO: Time shelling ended.
- FOXTROT: Location of shelled area.
- GOLF: Number, type, and caliber (fire support team personnel only).
- HOTEL: Nature of fire (barrage, harassment, or registration).
- INDIA: Number of rounds.
- JULIET: Time of flash to bang.
- KILO: Damage.

7-19. ENEMY PRISONER OF WAR/CAPTURED MATERIEL REPORT

This paragraph complies with STANAG 2084.

The sniper team immediately tags EPWs and captured materiel. This ensures that information of intelligence value (place, time, and circumstances of capture) is not lost during evacuation. Only EPWs or materiel of immediate tactical importance are reported to the troop or battalion TOG Snipers use the following formats to report EPWs and captured materiel:

a. Enemy Prisoners of War.

LINE 1 - Type of report.

LINE 2 - Item captured.

LINE 3 - Date/time of capture.

LINE 4 - Place of capture-grid coordinates.

LINE 5 - Capturing unit-all sign.

LINE 6 - Circumstances of capture (be brief).

b. Captured Materiel.

LINE 1 - Type of report.

LINE 2 - Item captured.

LINE 3 - Type document/equipment.

LINE 4 - Date/time captured.

LINE 5 - Place of capture-call sign.

LINE 6 - Capturing unit—call sign.

LINE 7 - Circumstances of capture (be brief).

After the report is given to the company/team/commander, disposition instructions will be provided if needed.

7-20. NBC 1 REPORT

This paragraph complies with STANAG 2103.

The sniper team uses the NBC 1 report to submit initial and subsequent information on an NBC attack, transmitting over the command or operation and intelligence net immediately after an NBC attack.

LINE 1 OR EVENT - Type of attack-nuclear, chemical, or biological.

LINE 2 OR BRAVO - Grid location of observer.

LINE 3 OR CHARLIE - Direction from observer to attack—mils or degree—true, grid, or magnetic.

LINE 4 OR DELTA - Date-time group of detonation or star of attack.

LINE 5 OR ECHO - Illumination time in seconds for nuclear attack.

LINE 6 OR ECHO BRAVO - End time for biological/chemical attack

LINE 7 OR FOXTROT - Actual or estimated (state which) grid coordinates for location of attack.

LINE 8 OR GOLF - Means of delivery.

LINE 9 OR HOTEL - Height of nuclear burst in feet or meters and or type of burst.

LINE 10 OR HOTEL BRAVO - Type of biological/chemical attack and height of burst.

LINE 11 OR INDIA BRAVO - Number of munitions or aircraft.

LINE 12 OR EFFECTS - Effects of burst/agent on personnel.

LINE 13 OR JULIETT - Flash-to-bang time in seconds for nuclear attack.

LINE 14 OR KILO - Crater (yes or no) and width in meters.

LINE 15 OR KILO BRAVO - Vegetation chemical/biological.

LINE 16 OR LIMA - Nuclear burst angular cloud width, measured at five minutes after detonation in mils or degrees.

LINE 17 OR MIKE - Stabilized cloud top height, in feet or meters, or angular cloud top angle, in degrees or mils, measured at H+10 minutes after detonation and stabilized cloud height, in feet or meters, or angular cloud bottom angle, in degrees or mils, measured at H+10.

LINE 18 OR PAPA ALPHA- Grid of predicted outline of external contours of hazardous cloud or area.

LINE 19 OR PAPA BRAVO - Downwind direction of nuclear cloud or duration of hazard in days.

LINE 20 OR SIERRA - Date-time group of reading for nuclear or detection time for biological/chemical.

LINE 21 OR YANKEE BRAVO - Effective downwind direction and wind speed.

LINE 22 OR ZULU ALPHA STABILITY - Air stability indicator.

LINE 23 OR ZULU ALPHA TEMPERATURE - Surface air temperature.

LINE 24 OR ZULU ALPHA HUMIDITY - Relative humidity range.

LINE 25 OR ZULU ALPHA WEATHER - Significant weather phenomena.

LINE 26 OR ZULU ALPHA COVER - Cloud cover.

LINE 27 OR NARRATIVE - Other significant observation.

LINE 28 - Not used.

LINE 29 OR AUTHENTICATION - Self-authentication, if required.

7-21. MEDICAL EVACUATION REQUEST

This paragraph complies with STANAG 3204.

The sniper team sends a MEDEVAC request to the medical team on the company command net.

a. When air assets are not available, the sniper team uses the ground evacuation format.

LINE 1 - Evacuation.

LINE 2 - Location for pickup (encode).

LINE 3 - Number of casualties.

LINE 4 - Category of patient(s).

A Urgent.
B Priority.
C Routine.

Use the letter of the appropriate subparagraph from Line 4 with the number of casualties in Line 3—for example, a2 means there are two urgent patients for evacuation.

b. When air assets are available, the sniper team uses the air evacuation format.

LINE 1 - Location.

LINE 2 - Radio frequency, call sign, and suffix.

LINE 3 - Precedence:

URGENT__ PRIORITY__ ROUTINE__ TACTICAL IMMEDIATE__

LINE 4 - Special equipment.

LINE 5 - Number of patients by type:

Little__ Ambulator__

LINE 6 - Security of pickup site.

LINE 7 - Method of marking pickup size.

LINE 8 - Patient's nationality and status.

LINE 9 - NBC contamination.

c. The definitions of the categories of precedence follow:

(1) ***Urgent.*** Used for emergency cases for evacuation as soon as possible and no more than two hours to save life, limb, and eyesight.

(2) ***Priority.*** Used when the patient should be evacuated within four hours or his medical condition will deteriorate to an URGENT precedence.

(3) ***Routine.*** Requires evacuation, but the patient's condition is not expected to deteriorate within the next 24 hours.

(4) ***Tactical immediate.*** Used when the patient's condition is not urgent or priority, but evacuation is required as soon as possible so as not to endanger the requesting unit's tactical mission.

CHAPTER 8

TRACKING/COUNTERTRACKING

When a sniper follows a trail, he builds a picture of the enemy in his mind by asking himself questions: How many persons am I following? What is their state of training? How are they equipped? Are they healthy? What is their state of morale? Do they know they are being followed? To answer these questions, the sniper uses available indicators to track the enemy. The sniper looks for signs that reveal an action occurred at a specific time and place. For example, a footprint in soft sand is an excellent indicator, since a sniper can determine the specific time the person passed By comparing indicators, the sniper obtains answers to his questions. For example, a footprint and a waist-high scuff on a tree may indicate that an armed individual passed this way.

Section I
TRACKING

Any indicator the sniper discovers can be defined by one of six tracking concepts: displacement, stains, weather, litter, camouflage, and immediate-use intelligence.

8-1. DISPLACEMENT

Displacement takes place when anything is moved from its original position. A well-defined footprint or shoe print in soft, moist ground is a good example of displacement. By studying the footprint or shoe print, the sniper determines several important facts. For example, a print left by worn footgear or by bare feet may indicate lack of proper equipment. Displacement can also result from clearing a trail by breaking or cutting through heavy vegetation with a machete. These trails are obvious to the most inexperienced sniper who is tracking. Individuals may unconsciously break more branches as they follow someone who is cutting the vegetation. Displacement indicators can also be made by persons carrying heavy loads who stop to rest; prints made by box edges can help to identify the load. When loads are set down at a rest halt or campsite, they usually crush grass and twigs. A reclining soldier also flattens the vegetation.

a. **Analyzing Footprints.** Footprints may indicate direction, rate of movement, number, sex, and whether the individual knows he is being tracked.

(1) If footprints are deep and the pace is long, rapid movement is apparent. Long strides and deep prints with toe prints deeper than heel prints indicate running (A, Figure 8-l).

(2) Prints that are deep, short, and widely spaced, with signs of scuffing or shuffling indicate the person is carrying a heavy load (B, Figure 8-l).

(3) If the party members realize they are being followed, they may try to hide their tracks. Persons walking backward (C, Figure 8-1) have a short, irregular stride. The prints have an unnaturally deep toe, and soil is displaced in the direction of movement.

(4) To determine the sex (D, Figure 8-l), the sniper should study the size and position of the footprints. Women tend to be pigeon-toed, while men walk with their feet straight ahead or pointed slightly to the outside. Prints left by women are usually smaller and the stride is usually shorter than prints left by men.

b. **Determining Key Prints.** The last individual in the file usually leaves the clearest footprints; these become the key prints. The sniper cuts a stick to match the length of the prints and notches it to indicate the width at the widest part of the sole. He can then study the angle of the key prints to the direction of march. The sniper looks for an identifying mark or feature, such as worn or frayed footwear, to help him identify the key prints. If the trail becomes vague, erased, or merges with another, the sniper can use his stick-measuring devices and, with close study, can identify the key prints. This method helps the sniper to stay on the trail. A technique used to count the total number of individuals being tracked is the box method. There are two methods the sniper can use to employ the box method.

(1) The most accurate is to use the stride as a unit of measure (Figure 8-2) when key prints can be determined. The sniper uses the set of key prints and the edges of the road or trail to box in an area to analyze. This method is accurate under the right conditions for counting up to 18 persons.

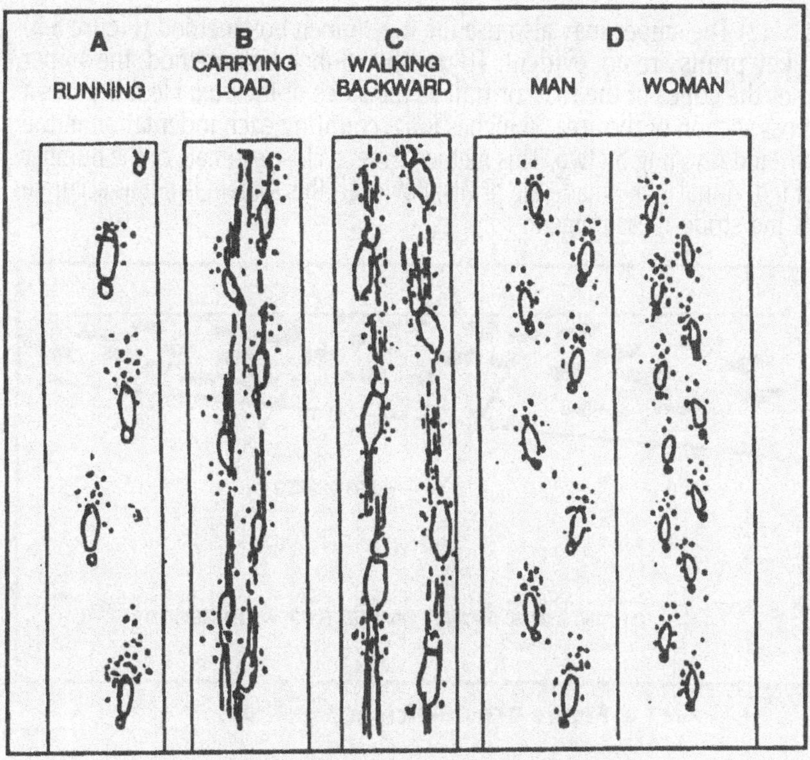

Figure 8-1. Different types of footprints.

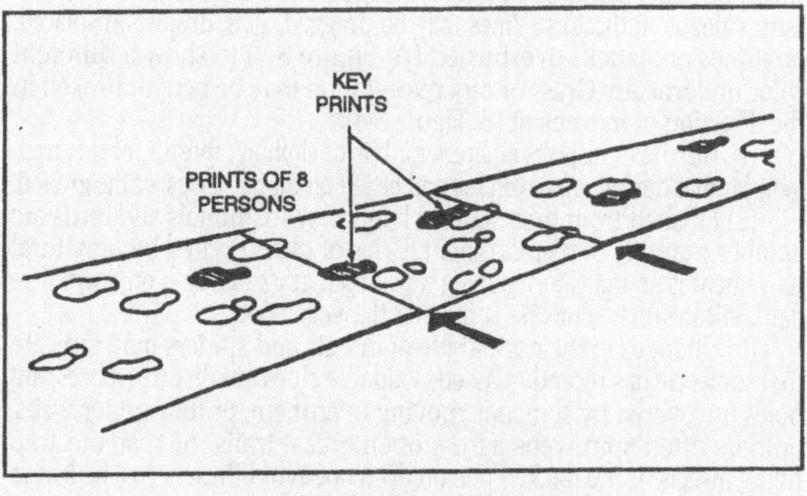

Figure 8-2. Stride measurement.

(2) The sniper may also use the the 36-inch box method (Figure 8-3) if key prints are not evident. To use the 36-inch box method, the sniper uses the edges of the road or trail as the sides of the box. He measures a cross section of the area 36 inches long, counting each indentation in the box and dividing by two. This method gives a close estimate of the number of individuals who made the prints; however, this system is not as accurate as the stride measurement.

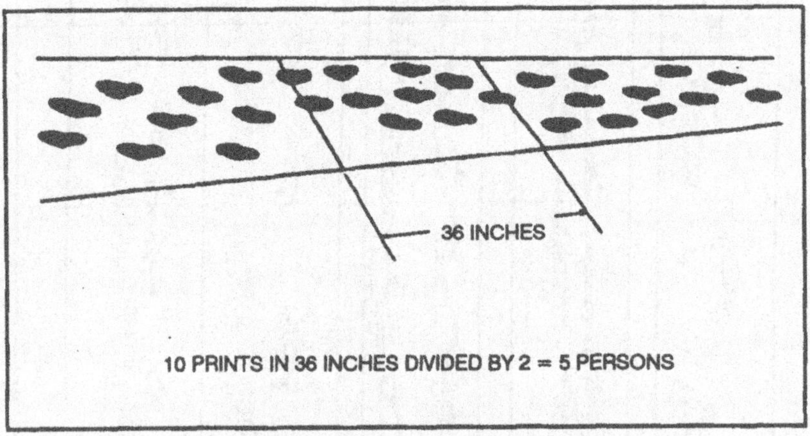

Figure 8-3. 36-Inch box method.

c. **Recognizing Other Signs of Displacement**Foliage, moss, vines, sticks, or rocks that are scuffed or snagged from their original position form valuable indicators. Vines may be dragged, dew droplets displaced, or stones and sticks overturned (A, Figure 8-4) to show a different color underneath. Grass or other vegetation may be bent or broken in the direction of movement (B, Figure 8-4).

(1) The sniper inspects all areas for bits of clothing, threads, or dirt from footgear that can be torn or can fall and be left on thorns, snags, or the ground.

(2) Flushed from their natural habitat, wild animals and birds are another example of displacement. Cries of birds excited by unnatural movement is an indicator; moving tops of tall grass or brush on a windless day indicates that someone is moving the vegetation.

(3) Changes in the normal life of insects and spiders may indicate that someone has recently passed. Valuable clues are disturbed bees, ant holes uncovered by someone moving over them, or tom spider webs. Spiders often spin webs across open areas, trails, or roads to trap flying insects. If the tracked person does not avoid these webs, he leaves an indicator to an observant sniper.

(4) If the person being followed tries to use a stream to cover his trail, the sniper can still follow successfully. Algae and other water plants can be displaced by lost footing or by careless walking. Rocks can be displaced from their original position or overturned to indicate a lighter or darker color on the opposite side. The person entering or exiting a stream creates slide marks or footprints, or scuffs the bark on roots or sticks (C, Figure 8-4). Normally, a person or animal seeks the path of least resistance; therefore, when searching the stream for an indication of departures, snipers will find signs in open areas along the banks.

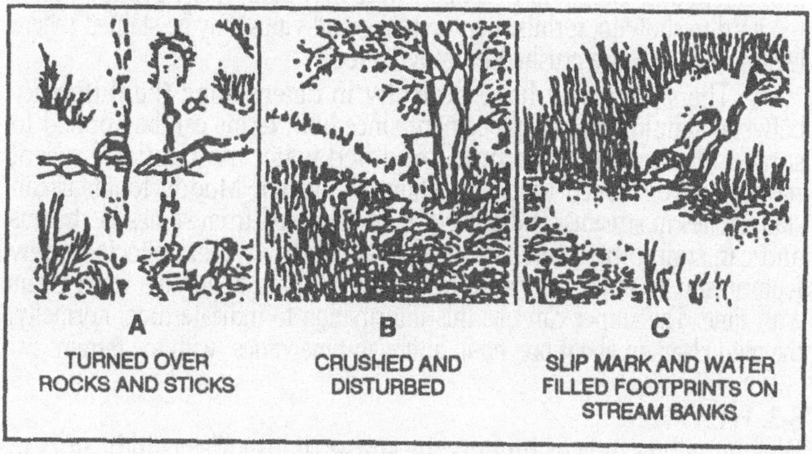

Figure 8-4. Other displacements.

8-2. STAINS

A stain occurs when any substance from one organism or article is smeared or deposited on something else. The best example of staining is blood from a profusely bleeding wound. Bloodstains often appear as spatters or drops and are not always on the ground; they also appear smeared on leaves or twigs of trees and bushes.

a. By studying bloodstains, the sniper can determine the wound's location.

(1) If the blood seems to be dripping steadily, it probably came from a wound on the trunk.

(2) If the blood appears to be slung toward the front, rear, or sides, the wound is probably in the extremity.

(3) Arterial wounds appear to pour blood at regular intervals as if poured from a pitcher. If the wound is veinous, the blood pours steadily.

(4) A lung wound deposits pink, bubbly, and frothy bloodstains.

(5) A bloodstain from a head wound appears heavy, wet, and slimy.

(6) Abdominal wounds often mix blood with digestive juices so the deposit has an odor and is light in color.

The sniper can also determine the seriousness of the wound and how far the wounded person can move unassisted. This proms may lead the sniper to enemy bodies or indicate where they have been carried.

b. Staining can also occur when muddy footgear is dragged over grass, stones, and shrubs. Thus, staining and displacement combine to indicate movement and direction. Crushed leaves may stain rocky ground that is too hard to show footprints. Roots, stones, and vines may be stained where leaves or berries are crushed by moving feet.

c. The sniper may have difficulty in determining the difference between staining and displacement since both terms can be applied to some indicators. For example, muddied water may indicate recent movement; displaced mud also stains the water. Muddy footgear can stain stones in streams, and algae can be displaced from stones in streams and can stain other stones or the bank. Muddy water collects in new footprints in swampy ground; however, the mud settles and the water clears with time. The sniper can use this information to indicate time; normally, the mud clears in about one hour, although time varies with the terrain.

8-3. WEATHER

Weather either aids or hinders the sniper. It also affects indicators in certain ways so that the sniper can determine their relative ages. However, wind, snow, rain, or sunlight can erase indicators entirely and hinder the sniper. The sniper should know how weather affects soil, vegetation, and other indicators in his area. He cannot determine the age of indicators until he understands the effects that weather has on trail signs.

a. By studying weather effects on indicators, the sniper can determine the age of the sign (for example, when bloodstains are fresh, they are bright red). Air and sunlight first change blood to a deep ruby-red color, then to a dark brown crust when the moisture evaporates. Scuff marks on trees or bushes darken with time; sap oozes, then hardens when it makes contact with the air.

b. Weather affects footprints (Figure 8-5). By carefully studying the weather process, the sniper can estimate the age of the print. If particles of soil are beginning to fall into the print, the sniper should become a stalker. If the edges of the print are dried and crusty, the prints are probably about one hour old. This varies with terrain and should be considered as a guide only.

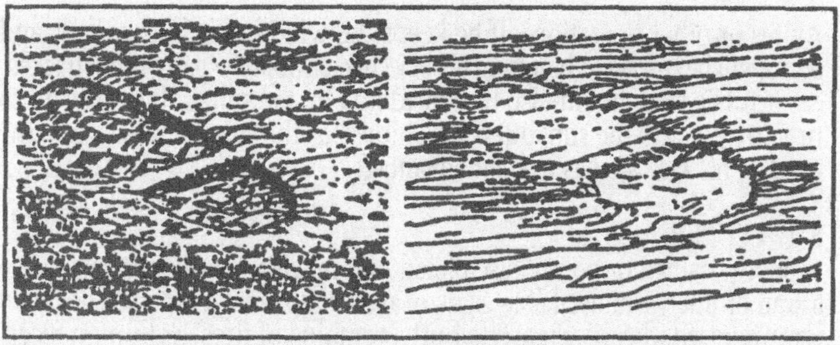

Figure 8-5. Weather effects on footprints.

c. A light rain may round the edges of the print. By remembering when the last rain occurred, the sniper can place the print into a time frame. A heavy rain may erase all signs.

d. Trails exiting streams may appear weathered by rain due to water running from clothing or equipment into the tracks. This is especially true if the party exits the stream single file. Then, each person deposits water into the tracks. The existence of a wet, weathered trail slowly fading into a dry trail indicates the trail is fresh.

e. Wind dries tracks and blows litter, sticks, or leaves into prints. By recalling wind activity, the sniper may estimate the age of the tracks. For example, the sniper may reason "the wind is calm at the present but blew hard about an hour ago. These tracks have litter in them, so they must be over an hour old." However, he must be sure that the litter was not crushed into them when the prints were made.

(1) Wind affects sounds and odors. If the wind is blowing toward the sniper, sounds and odors may be carried to him; conversely, if the wind is blowing *away* from the sniper, he must be extremely cautious since wind also carries sounds toward the enemy. The sniper can determine wind direction by dropping a handful of dust or dried grass from shoulder height. By pointing in the same direction the wind is blowing, the sniper can localize sounds by cupping his hands behind his ears and turning slowly. When sounds are loudest, the sniper is facing the origin.

(2) In calm weather (no wind), air currents that may be too light to detect can carry sounds to the sniper. Air cools in the evening and moves downhill toward the valleys. If the sniper is moving uphill late in the day or at night, air currents will probably be moving toward him if no other wind is blowing. As the morning sun warms the air in the valleys, it moves uphill. The sniper considers these factors when plotting patrol

routes or other operations. If he keeps the wind in his face, sounds and odors will be carried to him from his objective or from the party being tracked.

(3) The sun should also be considered by the sniper. It is difficult to fire directly into the sun, but if the sniper has the sun at his back and the wind in his face, he has a slight advantage.

8-4. LITTER

A poorly trained or poorly disciplined unit moving over terrain may leave a trail of litter. Unmistakable signs of recent movement are gum or candy wrappers, food cans, cigarette butts, remains of fires, or human feces. Rain flattens or washes litter away and turns paper into pulp. Exposure to weather can cause food cans to rust at the opened edge; then, the rust moves toward the center. The sniper must consider weather conditions when estimating the age of litter. He can use the last rain or strong wind as the basis for a time frame.

8-5. CAMOUFLAGE

Camouflage applies to tracking when the followed party employs techniques to baffle or slow the sniper. For example, walking backward to leave confusing prints, brushing out trails, and moving over rocky ground or through streams.

8-6. IMMEDIATE-USE INTELLIGENCE

The sniper combines all indicators and interprets what he has seen to form a composite picture for on-the-spot intelligence. For example, indicators may show contact is imminent and require extreme stealth.

a. The sniper avoids reporting his interpretations as facts. He reports what he has seen rather than stating these things exist. There are many ways a sniper can interpret the sex and size of the party, the load, and the type of equipment. Timeframes can be determined by weathering effects on indicators.

b. Immediate-use intelligence is information about the enemy that can be used to gain surprise, to keep him off balance, or to keep him from escaping the area entirely. The commander may have many sources of intelligence reports, documents, or prisoners of war. These sources can be combined to form indicators of the enemy's last location, future plans, and destination.

c. Tracking, however, gives the commander definite information on which to act immediately. For example, a unit may report there are no men of military age in a village. This information is of value only if it is combined with other information to make a composite enemy picture in the area. Therefore, a sniper who interprets trail signs and reports that he is 30 minutes behind a known enemy unit, moving north, and located at a specific location, gives the commander information on which he can act at once.

8-7. DOG/HANDLER TRACKING TEAMS

Dog/handler tracking teams are a threat to the sniper team. While small and lightly armed, they can increase the area that a rear area security unit can search. Due to the dog/handler tracking team's effectiveness and its lack of firepower, a sniper team may be tempted to destroy such an "easy" target. Whether a sniper should fight or run depends on the situation and the sniper. Eliminating or injuring the dog/handler tracking team only confirms that there is a hostile team operating in the area.

a. When looking for sniper teams, trackers use wood line sweeps and area searches. A wood line sweep consists of walking the dog upwind of a suspected wood line or brush line. If the wind is blowing through the woods and out of the wood line, trackers move 50 to 100 meters inside a wooded area to sweep the wood's edge. Since wood line sweeps tend to be less specific, trackers perform them faster. An area search is used when a team's location is specific such as a small wooded area or block of houses. The search area is cordoned off, if possible, and the dog/handler tracking teams are brought on line, about 25 to 150 meters apart, depending on terrain and visibility. The handler trackers then advance, each moving their dogs through a specific corridor. The handler tracker controls the dog entirely with voice commands and gestures. He remains undercover, directing the dog in a search pattern or to a likely target area. The search line moves forward with each dog dashing back and forth in assigned sectors.

b. While dog/handler tracking teams area potent threat, there are counters available to the sniper team. The beat defenses are basic infantry techniques: good camouflage and light, noise, and trash discipline. Dogs find a sniper team either by detecting a trail or by a point source such as human waste odors at the hide site. It is critical to try to obscure or limit trails around the hide, especially along the wood line or area closest to the team's target area. Surveillance targets are usually the major axis of advance. "Trolling the wood lines" along likely looking roads or intersections is a favorite tactic of dog/handler tracking teams. When moving into a target area, the sniper team should take the following countermeasures:

(1) Remain as faraway from the target area as the situation allows.

(2) Never establish a position at the edge of cover and concealment nearest the target area

(3) Reduce the track. Try to approach the position area on hard, dry ground or along a stream or river.

(4) Urinate in a hole and cover it up. Never urinate in the same spot.

(5) Bury fecal matter deep. If the duration of the mission permits, use MRE bags sealed with tape and take it with you.

(6) Never smoke.

(7) Carry all trash until it can be buried elsewhere.

(8) Surround the hide site with a 3-cm to 5-cm band of motor oil to mask odor; although less effective but easier to carry, garlic may be used. A dead animal can also be used to mask smell, although it may attract unwanted canine attention.

c. If a dog/handler tracking team moves into the area, the sniper team can employ several actions but should first check wind direction and speed. If the sniper team is downwind of the estimated search area, the chances are minimal that the team's point smells will probably be detected. If upwind of the search area, the sniper team should attempt to move downwind. Terrain and visibility dictate whether the sniper team can move without being detected visually by the handlers of the tracking team. Remember, sweeps are not always conducted just outside of a wood line. Wind direction determines whether the sweep will be parallel to the outside or 50 to 100 meters inside the wood line.

(1) The sniper team has options if caught inside the search area of a line search. The handlers rely on radio communications and often do not have visual contact with each other. If the sniper team has been generally localized through enemy radio detection-finding equipment, the search net will still be loose during the initial sweep. A sniper team has a small chance of hiding and escaping detection in deep brush or in woodpiles. Larger groups will almost certainly be found. Yet, the sniper team may have the opportunity to eliminate the handler and to escape the search net.

(2) The handler hides behind cover with the dog. He searches for movement and then sends the dog out in a straight line toward the front. Usually, when the dog has moved about 50 to 75 meters, the handler calls the dog back. The handier then moves slowly forward and always from covered position to covered position. Commands are by voice and gesture with a backup whistle to signal the dog to return. If a handler is eliminated or badly injured after he has released the dog, but before he has recalled it, the dog continues to randomly search out and away from the handler. The dog usually returns to another handler or to his former

handler's last position within several minutes. This creates a gap from 25 to 150 meters wide in the search pattern. Response times by the other searchers tend to be fast. Given the high degree of radio communication, the injured handler will probably be quickly missed from the radio net. Killing the dog before the handler will probably delay discovery only by moments. Dogs are so reliable that if the dog does not return immediately, the handler knows something is wrong.

(3) If the sniper does not have a firearm, one dog can be dealt with relatively easy if a knife or large club is available. The sniper must keep low and strike upward using the wrist, never overhand. Dogs are quick and will try to strike the groin or legs. Most attack dogs are trained to go for the groin or throat. If alone and faced with two or more dogs, the sniper should avoid the situation.

Section II
COUNTERTRACKING

If an enemy tracker finds the tracks of two men, this may indicate that a highly trained team may be operating in the area. However, a knowledge of countertracking enables the sniper team to survive by remaining undetected.

8-8. EVASION

Evasion of the tracker or pursuit team is a difficult task that requires the use of immediate-action drills to counter the threat. A sniper team skilled in tracking techniques can successfully employ deception drills to lessen signs that the enemy can use against them. However, it is very difficult for a person, especially a group, to move across any area without leaving signs noticeable to the trained eye.

8-9. CAMOUFLAGE

The sniper team may use the most used and the least used routes to cover its movement. It also loses travel time when trying to camouflage the trail.

a. **Most Used Routes.** Movement on lightly traveled sandy or soft trails is easily tracked. However, a sniper may try to confuse the tracker by moving on hard-surfaced, often-traveled roads or by merging with civilians. These routes should be carefully examined; if a well-defined approach leads to the enemy, it will probably be mined, ambushed, or covered by snipers.

b. **Least Used Routes.** Least used routes avoid all man-made trails or roads and confuse the tracker. These routes are normally magnetic azimuths between two points. However, the tracker can use the proper concepts to follow the sniper team if he is experienced and persistent.

c. **Reduction of Trail Signs.** A sniper who tries to hide his trail moves at reduced speed; therefore, the experienced tracker gains time. Common methods to reduce trail signs areas follows:

(1) Wrap footgear with rags or wear soft-soled sneakers, which make footprints rounded and leas distinctive.

(2) Brush out the trail. This is rarely done without leaving signs.

(3) Change into footgear with a different tread immediately following a deceptive maneuver.

(4) Walk on hard or rocky ground.

8-10. DECEPTION TECHNIQUES

Evading a skilled and persistent enemy tracker requires skillfully executed maneuvers to deceive the tracker and to cause him to lose the trail. An enemy tracker cannot be outrun by a sniper team that is carrying equipment, because he travels light and is escorted by enemy forces designed for pursuit. The size of the pursuing force dictates the sniper team's chances of success in employing ambush-type maneuvers. Sniper teams use some of the following techniques in immediate-action drills and deception drills.

a. **Backward Walking.** One of the basic techniques used is that of walking backward (Figure 8-6) in tracks already made, and then stepping off the trail onto terrain or objects that leave little sign. Skillful use of this maneuver causes the tracker to look in the wrong direction once he has lost the trail.

b. **Large Tree** A good deception tactic is to change directions at large trees (Figure 8-7). To do this, the sniper moves in any given direction and walks past a large tree (12 inches wide or larger) from 5 to 10 paces. He carefully walks backward to the forward side of the tree and makes a 90-degree change in the direction of travel, passing the tree on its forward side. This technique uses the tree as a screen to hide the new trail from the pursuing tracker.

NOTE: By studying signs, a tracker may determine if an attempt is being made to confuse him. If the sniper team loses the tracker by walking backward, footprints will be deepened at the toe and soil will be scuffed or dragged in the direction of movement. By following carefully the tracker can normally find a turnaround point.

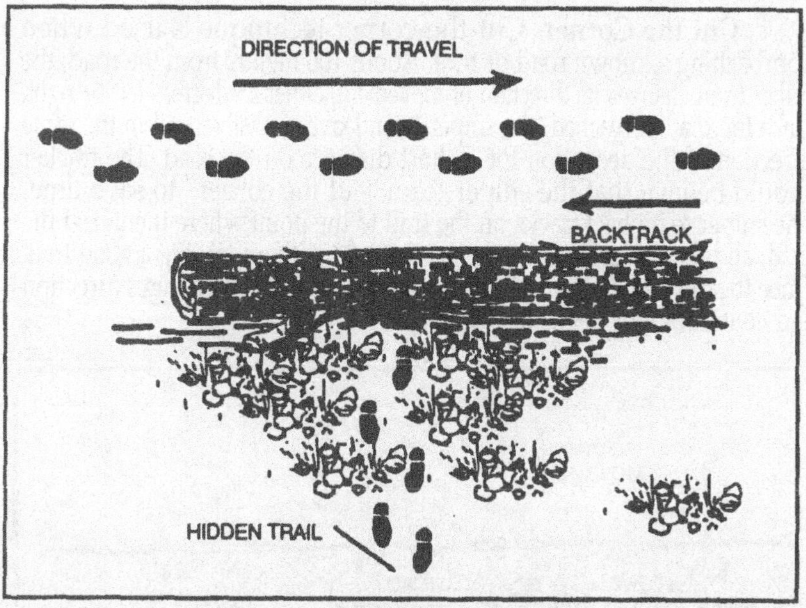

Figure 8-6. Walking backward.

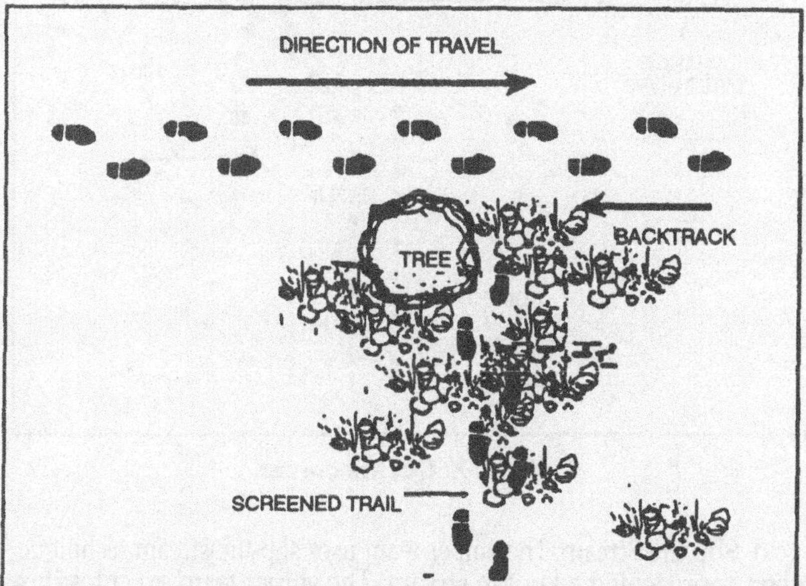

Figure 8-7. Large tree.

c. **Cut the Corner.** Cut-the-corner technique is used when approaching a known road or trail. About 100 meters from the road, the sniper team changes its direction of movement, either 45 degrees left or right. Once the road is reached, the sniper team leaves a visible trail in the same direction of the deception for a short distance on the road. The tracker should believe that the sniper team "cut the corner" to save time. The sniper team backtracks on the trail to the point where it entered the road, and then it carefully moves on the road without leaving a good trail. Once the desired distance is achieved, the sniper team changes direction and continues movement (Figure 8-8).

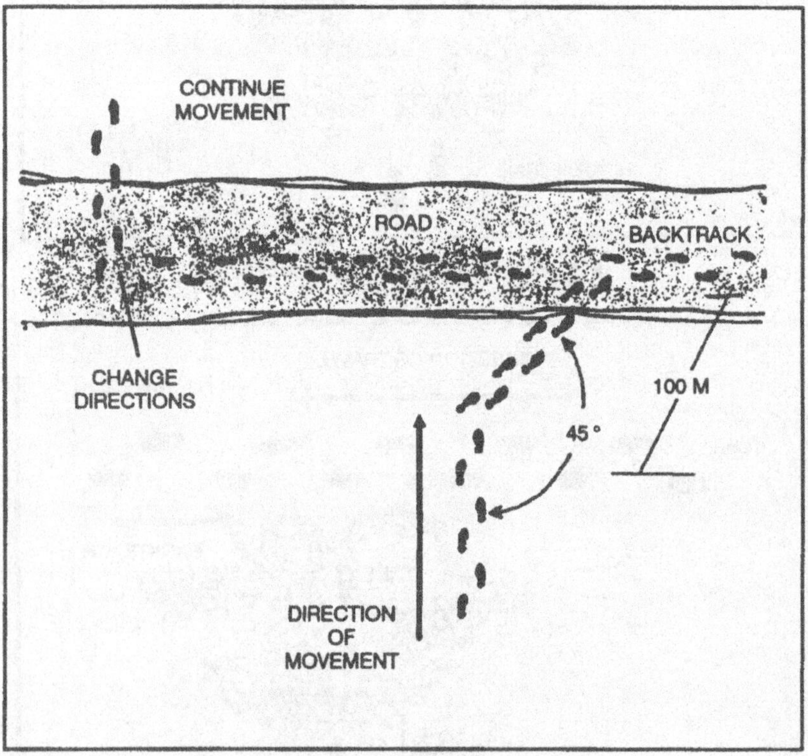

Figure 8-8. Cut the corner.

d. **Slip the Stream.** The sniper team uses slip-the-stream technique when approaching a known stream. The sniper team executes this method the same as the cut the comer technique. The sniper team establishes the 45-degree deception maneuver upstream, then enters the stream. The sniper team moves upstream to prevent floating debris and silt from compromising its direction of travel, and the sniper team establishes false trails upstream if time permits. Then, it moves downstream to escape since creeks and streams gain tributaries that offer more escape alternatives (Figure 8-9).

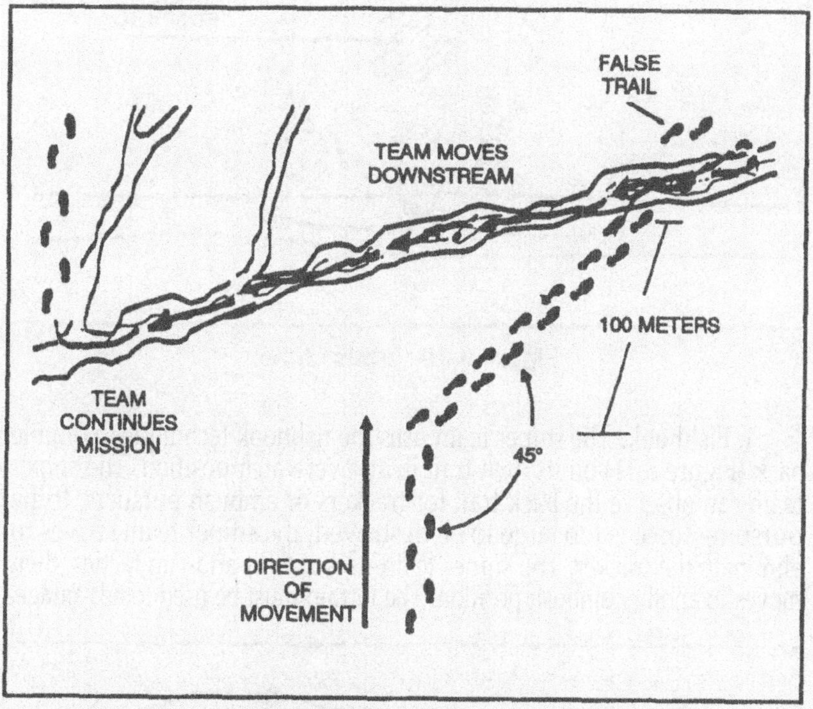

Figure 8-9. Slip the stream.

e. **Arctic Circle.** The sniper team uses the arctic circle technique in snow-covered terrain to escape pursuers or to hide a patrol base. It establishes a trail in a circle (Figure 8-10, page 8-16) as large as possible. The trail that starts on a road and returns to the same start point is effective. At some point along the circular trail, the sniper team removes snowshoes (if used) and carefully steps off the trail, leaving one set of tracks. The large tree maneuver can be used to screen the trail. From the hide position, the sniper team returns over the same steps and carefully fills them with snow one at a time. This technique is especially effective if it is snowing.

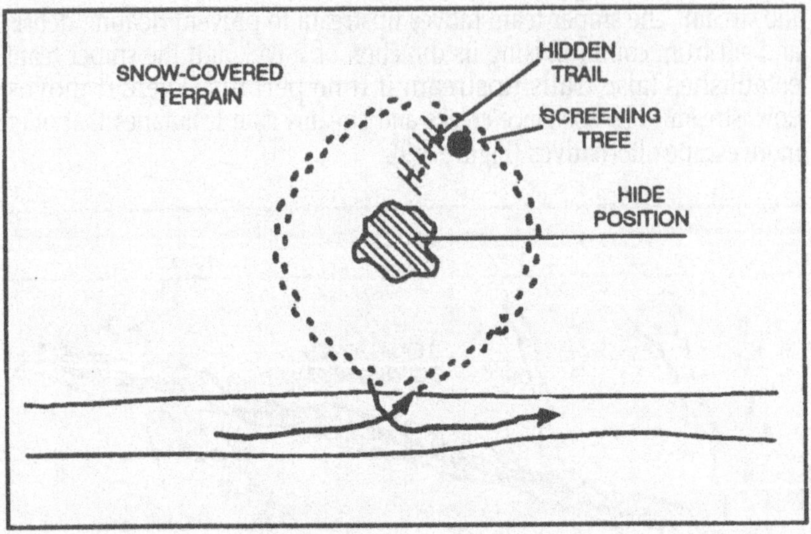

Figure 8-10. Arctic circle.

f. **Fishhook.** The sniper team uses the fishhook technique to double back (Figure 8-11) on its own trail in an overwatch position. The sniper team can observe the back trail for trackers or ambush pursuers. If the pursuing force is too large to be destroyed, the sniper team strives to eliminate the tracker. The sniper team uses the hit-and-run tactics, then moves to another ambush position. The terrain must be used to advantage.

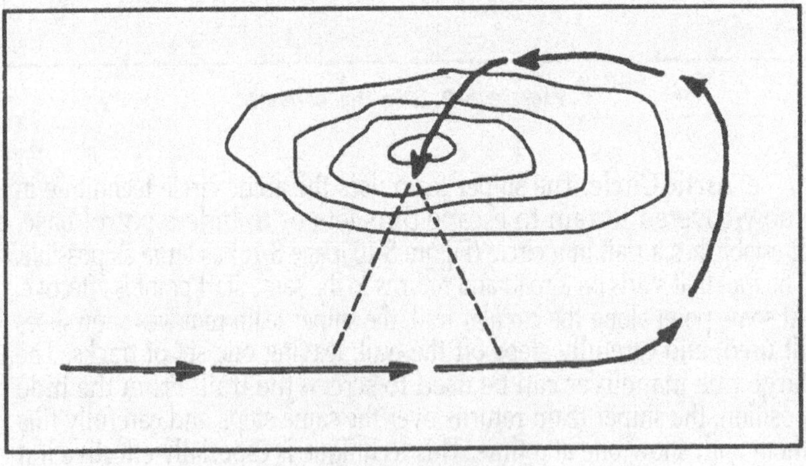

Figure 8-11. Fishhook.

CHAPTER 9

SNIPER SUSTAINMENT TRAINING

Repetitive training in long-range markmanship and field-craft skills ensures the best probability of effective engagement and the minimum risk of detection. Snipers must sustain basic soldier skills and master and sustain critical mission skills to accomplish their objectives. Both sniper and observer are trained snipers and should be highly skilled in the art of sniping. Sniping skills perish quickly; therefore, sniper teams must sustain and sharpen those skills regularly. To deny the importance and need to sustain sniper training deprives the commander of a valuable asset. This chapter also includes a 5-day sniper sustainment training program.

9-1. BASIC SKILLS SUSTAINMENT

Due to the primary and secondary missions of the sniper, minimum skill sustainment should include observation, range estimation, concealment, concealed movement, and rifle firing. Sustainment of these skills may best be accomplished through sniper training exercises and unit-level live-fire exercises. (DA Pamphlet 350-38 outlines the frequency and ammunition requirements needed to conduct sniper training.) Sniper training exercises provide snipers with practical experience in detecting and engaging realistic targets under field conditions on ranges comparable to a battlefield. This training also provides snipers with a means to practice the various sniper training fundamentals that has been taught previously, often collectively. These exercises mayor may not be graded; however, competition is a proven method to obtain the desired results. At the end of the exercises, the trainer critiques each sniper on his performance. These exercises include zeroing and practice fire, field fire (unknown distance), concealment, concealed movement target detection, range estimation, land navigation, memory enhancement exercise (KIM game), and communications. Each sniper will go through these training exercises.

a. **Zeroing and Pratice Fire.** To engage targets effectively during training exercises and in combat, the sniper must have his rifle accurately zeroed. For this reson the zeroing exercises are normally conducted on a measured known-distance range to ensure precise adjustment, recording, and practice under ideal conditions and to eliminate variables that may prevent achieving an effective zero. The sniper rifle is zeroed using both the telescopic andiron sights. A bull's-eye-type target should be used for zeroing. It is important to acquire a point-of-aim, point-of-impact zero at 100 meters using the M24. As the distance increases, the sniper must adjust his telescope to allow for elevation and wind to ensure the rounds stay in the center of the target.

b. **Field Fire.** Practical firing exercises are designed to develop sniper proficiency in the accurate and rapid engagement of various combat-type targets, as well as to provide practical work in other field techniques. Snipers should be given positions on the firing line and areas of the field fire course to observe and make range cards of the area.

(1) After the range cards have been completed, the snipers will be required to fire the course by having one member call the wind and adjust the other member's fire. The ability to call the wind is important as successful engagement of the targets. After one member fires the course, they switch positions and repeat the fire course.

(2) When firing the course, snipers should engage the targets in a sequence that starts with the 200-meter target, then engage each target out to 800 meters, then engage targets back to the 200-meter target. (Targets are engaged twice. Snipers will engage a target with no more than two rounds per target.) The course consists of engaging 20 targets with 30 rounds of ammunition within a 30-minute time iimit. The sniper should be scored as follows:

- 10 points for first-round hits.
- 5 points for second-round hits.
- 200 points maximum.
- 140 points needed to pass (70 percent).

(3) To enhance training, snipers should also fire the field fire course during limited visibility with overhead illumination such as parachute flares. This puts stress on the sniper to determine the range and to engage a target in a short amount of time.

(4) lb provide the most realistic training environment trainers do not use range commands to commence fire and cease fire in sniper exercises.

The only exception to this is when an unsafe condition exists. The command CEASE FIRE should be given immediately. Snipers must be given a thorough orientation on each exercise (to include safety requirements) before they are permitted to move into position. After the sniper has assumed his firing position in the designated location, he should be allowed to fire without further commands. Therefore, the range must be cleared for firing before the exercise begins. An NCO (assistant trainer) must be with each sniper to keep score and to maintain safety during the exercise. When the sniper completes firing, the NCO ensures the rifle is clear and signals the range officer.

NOTE: A blank copy of the forms that follow are located at the back of this manual for local reproduction.

c. **Concealment.** Concealment exercises develop and test the sniper's ability to conceal himself in an expedient firing position while observing and engaging an observer-instructor. Figure 9-1, page 9-4, is an example of completed DA Form 7325-R, Concealment Exercise Scorecard.

(1) In a cleared area with a wood line about 100 meters away, snipers conceal themselves within 10 minutes in the wood line. After the 10-minute preparation, an observer-instructor 100 meters away visually searches the area for 2 minutes without the aid of optics. After 2 minutes, the observer-instructor searches the wood line (from his position) for 18 minutes, using binoculars and the M49 observation telescope. If there are more than 10 snipers in the exercise, two observer-instructors and two assistant trainers may be needed. After the 20-minute period, an assistant trainer with a radio moves within 10 feet of a sniper, who is ready to fire at an observer-instructor.

(2) The sniper should be able to identify a white 5-inch number that is painted on an 8-inch by 8-inch international orange panel. This panel is held over a vital part of the observer-instructor, and two blanks are fired at him without the sniper being detected. If the target detects the sniper, he radios the assistant trainer and directs him to the sniper. The exercise should be scored on a 10-point system, with 7 points being a passing score. (See Paragraph 9-4, Day 3, to score the concealment exercise.)

CONCEALMENT EXERCISE SCORECARD

Exercise Number ___

For use of this form, see FM 23-10. The poponent agency is TRADOC.

DATA REQUIRED BY PRIVACY ACT OF 1974.

AUTHORITY: 10 USC 3012(g)/Executive Order 1974. PRINCIPAL PURPOSE(S): Evaluates individual training. ROUTINE USE(S): Evaluates individual proficiency. SSN is used for positive identification purposes only. MANDATORY OR VOLUNTARY DISCLOSURE AND EFFECT ON INDIVIDUAL NOT PROVIDING INFORMATION: Voluntary. Individuals not providing information cannot be rated/scored on a mass basis.

Last name	First	MI	Rank	SSN	Unit
SMITH	JOHN	D	SPC	457-28-1738	D Co 2-29 IN

Date	Weather\visibility	Score
17 FEB 93	CLEAR 67° UNLIMITED	9

If the sniper	Points	Deducted	Total
• Was detected without the aid of optics (first 2 minutes)	2	0	2
• Was detected with the aid of optics (18 minutes)	1	0	3
• Was detected when assistant trainer was within 10 feet of sniper	1	0	4
• Properly identied the number within 30 seconds	1	0	5
• Failed to properly identify the number	0	3	2
• Fired first shot, not detected	4	0	6
• Fired second shot, not detected	1	0	7
• Maintained stable firing position (support)	2	0	9
• Properly adjusted weapon's scope for range and windage	1	0	10

NOTES:

1. If the sniper was caught trying to identify the number, score 4 points.
2. If muzzle blast/flash is detected, deduct 1 point from total score.
3. Failing to comply with training standards and objectives (such as unnecessary movement, premature fire, outide prescribed boundries) will result in termination of the exercise and a score of zero.

(Check one of the target indicators.)

- ☐ Contrast to background
- ☐ Muzzle blast
- ☒ Muzzle flash
- ☐ Improper movement techniques
- ☐ Improper camouflage
- ☐ Shine
- ☐ Outline
- ☐ Sound

Roy B. [illegible]
Trainer's signature

John D. Smith
Sniper's signature

DA FORM 7325-R, JUL 94

Figure 9-1. Example of completed DA Form 7325-R, Concealment Exercise Scorecard.

d. **Concealed Movement.** Concealed movement exereise develops and tests the sniper's ability to move and occupy a firing position undetected. Trainers record scores on DA Form 7326-R, Concealed Movement Exercise Scorecard (Figure 9-2, page 9-6).

(1) This exercise requires the same amount of trainers and equipment as in the concealment exercises. Areas used should be observable for 1,000 meters and have easily recognizable left and right limits. Ideally, snipers should train in a different type of area each time they perform these exercises.

(2) The snipers move 800 to 600 meters toward two observer-instructors, occupy a firing position 100 to 200 meters away, identify in the same manner as the concealment exercise, and fire two blanks at the targets without being detected at any time. If one of the observer-instructors detects a sniper, he radios one of the assistant trainers and directs him to the sniper's position. The sniper is given three hours to complete the exercise. The exercise is scored on a lo-point system, with 7 points being a passing wore. (See Paragraph 9-4, Day 4, to score concealed movement exercise.)

CONCEALED MOVEMENT EXERCISE SCORECARD

Exercise Number____

For use of this form, see FM 23-10. The proponent agency is TRADOC.

DATA REQUIRED BY PRIVACY ACT OF 1974.

AUTHORITY: 10 USC 3012(g)/Executive Order 9397. PRINCIPAL PURPOSE(S): Evaluates individual training. ROUTINE USE(S): Evaluates individual proficiency. SSN is used for positive identification purposes only. MANDATORY OR VOLUNTARY DISCLOSURE AND EFFECT ON INDIVIDUAL NOT PROVIDING INFORMATION: Voluntary. Individuals not providing information cannot be rated/scored on a mass basis.

Last name	First	MI	Rank	SSN	Unit
SMITH	JOHN	D	SFC	457-38-1738	D Co 2-29 IN

Date	Weather\visibility	Score
17 FEB 93	CLEAR 50° UNLIMITED	

If the sniper	Points	Deducted	Total
• Was detected moving to FFL	0	0	0
• Was detected moving in FFL	6	0	6
• Fired first round shot, not detected	2	0	8
• Was not detected when assistant trainer is within 10 feet of sniper	2	0	10
• Properly identified number (within 30 seconds)	2	0	12
• Failed to properly identify number	2	0	14
• Was not detected when assistant trainer is within 5 feet of sniper	2	0	16
• Fired second shot, not detected	2	0	18
• Maintained stable firing position (support)	1	0	19
• Properly adjusted weapon's scope for range and windage	1	0	20

NOTES:

1. If muzzle blast/flash is detected, deduct 1 point from total score.
2. Failing to comply with training standards and objectives (such as unnecessary movement, premature fire, outside prescribed boundries) will result in termination of the exercise and a score of zero.

(Check one of the target indicators.)

- ☐ Contrast to background
- ☒ Muzzle blast
- ☐ Muzzle flash
- ☐ Improper movement techniques
- ☐ Improper camouflage
- ☐ Shine
- ☐ Outline
- ☐ Sound

REMARKS: Explain in detail on back the reason for sniper's detection.

Ron B. Wunla — Trainer's signature

John D. Smith — Sniper's signature

DA FORM 7326-R, JUL 94

Figure 9-2. Example of completed DA Form 7326-R, Concealed Movement Exercise Scorecard.

e. **Target Detection.** Target detection exercises sharpen the sniper's eyes by requiring him to detect, describe, and plot objects that cannot be easily seen or described without the skillful use of optics. Scores are recorded on DA Form 7327-R, Target Detection Exercise Scorecard (Figure 9-3, page 9-8).

(1) Areas used for target detection should be partly cleared at least 200 meters in depth and 100 meters in width with easily definable left and right limits. The area should have at least three TRPs that are easily recognized and positioned in different locations throughout the area. Ten military items are placed in the area. These items can be radio antennas, small-scale mock vehicles, batteries, map protractors, or weapons. Items should be placed so that they are undetectable with the naked eye, detectable but indescribable with the binoculars, and describable only by using the M49 observation telescope.

(2) Snipers are given an M49 observations telescope, M22 binoculars, pencil, clip board, and scorecard. Snipers are given 40 minutes to detect, describe, and plot each item in the area. Snipers remain in the prone position throughout the exercise. After 15 minutes, they will move to a different position, left or right of the centerline of observation and remain there for the next 15 minutes. For the last 10 minutes, they can choose a position anywhere along the line. When an object is detected, the sniper gives his location on the line of observation (A or B). Next, the sniper must describe the object using the categories of size, shape, color, condition, and appearance. Snipers receive 1/2 point for correctly plotting a target and 1/2 point for correctly describing it. They must achieve 7 points to receive a GO in this area.

NOTE: The trainer should sanitize the site before the exercise. If the sniper finds additional items to describe he may use the eleventh and twelfth lines of the scorecard. If the trainer allows the sniper can obtain credit for observation and detection skills.

ROSTER #: (11) **TARGET DETECTION EXERCISE SCORECARD** EX #: (PE 2)

For use of this form, see FM 23-10; the proponent agency is TRADOC

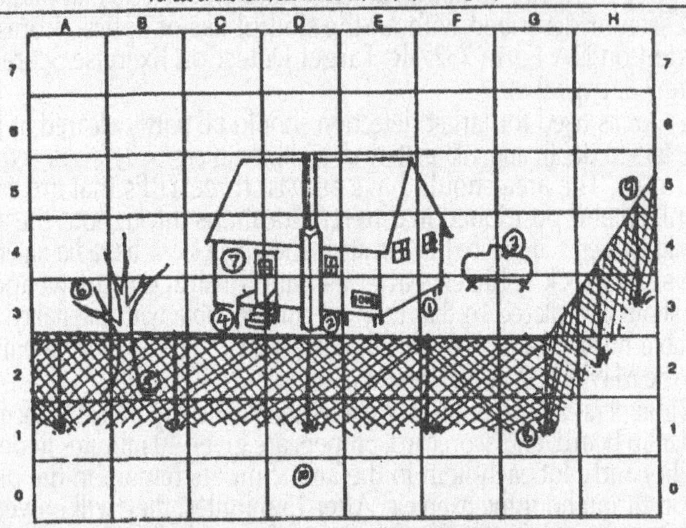

SKETCH NAME: BLDG 1002

GRID COORDINATE: PL217003

WEATHER: CLEAR 72°

75 Magnetic Azimuth

Sketch # 1 of 1

Block Scale: 1" = 10M

NAME: EDISON

RANK: SGT

DATE: 1 APR 93 TIME: 1050

#	SIZE	SHAPE	COLOR	CONDITION	APPEARS TO BE	GRID BOX LOC.
1	36"x4"x2"		BLACK	SERV	M16A2	F-3 (A)
2	10"x1"x4"		GREEN	SERV	RADIO	D-3 (A)
3	3"x3"x3"		BLUE	UNSERV	GRENADE	G-4 (A)
4	5"x2"x6"		BLACK	SERV	TANK	C-3 (B)
5	4"x4"x1/64"		CLEAR	SERV	PROTRACTOR	B-2 (B)
6	1"x1/8"x36"		YELLOW	SERV	STRAP	A-3 (B)
7	1/32"x1/2"x1/4"		BLACK	SERV	RANK INSIGNIA	C-3 (A)
8	6"x4"x12"		GREEN	SERV	AMMO BOX	G-1 (B)
9	1"x1"x1"		BLACK	SERV	CANTEEN CAP	H-4 (A)
10	1/2"x1/2"x1"		SILVER	SERV	PAPER CLIP	D-0 (B)

DA FORM 7327-R, JUL 94

Figure 9-3. Example of completed DA Form 7327-R, Target Detection Exercise Scorecard.

f. **Range Estimation.** Snipers must correctly estimate distance to effectively fire weapons, complete accurate range cards, and give reliable intelligence reports. Range estimation exercises should be conducted in an area that allows unobstructed observation of a human-size target up to 1,000 meters away. Scores are recorded on DA Form 7328-R, Range Estimation Exercise Scorecard (Figure 9-4, page 9-10). Personnel should be placed at various ranges and stages of concealment to give the sniper a challenging and realistic exercise. Snipers should be graded on their ability to estimate range by using the naked eye, M19/M22 binoculars, and the M3A scope. Snipers must correctly estimate the distance to 7 of 10 objects using their eyes (± 15 percent), 7 to 10 objects using the binoculars (± 10 percent), and 7 to 10 objects using the M3A telescope (± 5 percent). They must sketch their assigned sector on the back of the form, page 9-11.

RANGE ESTIMATION EXERCISE SCORECARD
Exercise Number____

For use of this form, see FM 23-10. The proponent agency is TRADOC.
DATA REQUIRED BY PRIVACY ACT OF 1974.
AUTHORITY: 10 USC 3012(g)/Executive Order 9397. PRINCIPAL PURPOSE(S): Evaluates individual training. ROUTINE USE(S): Evaluates individual proficiency. SSN is used for positive identification purposes only. MANDATORY OR VOLUNTARY DISCLOSURE AND EFFECT ON INDIVIDUAL NOT PROVIDING INFORMATION: Voluntary. Individuals not providing information cannot be rated/scored on a mass basis.

Last name	First	MI	Rank	SSN	Unit
SMITH	JOHN	D	SFC	457-28-1138	D Co 2-29 IN

Date	Weather\visibility	Score
17 FEB 93	CLEAR 60° UNLIMITED	27

	EYE ESTIMATION +- 15%	BINOCULAR ESTIMATION +- 10%	M3A TELESCOPE ESTIMATION +- 5%
1	108 M	105 M	108 M
2	130	X 142	127
3	200	217	215
4	225	230	X 215
5	275	250	258
6	300	315	315
7	350	330	326
8	400	410	405
9	X 500	420	440
10	475	405	488

1. Within three minutes, the range to the target is estimated at each point, using the naked eye, binoculars, and the M3A telescope. Estimations must be performed in the order listed.

2. Once an estimate is recorded, it cannot be changed; it will be counted as incorrect. However, the M3A telescope estimate may be changed before the next set of estimates are recorded.

3. The use of calculators is encouraged.

4. This is an individual exercise. Any sniper that talks or tries to look at another sniper's scorecard is terminated from the exercise.

5. If there are any questions, the trainer will assist you.

Roy B. Wueler
Trainer's signature

John D. Smith
Sniper's signature

DA FORM 7328-R, JUL 94

Figure 9-4. Example of completed DA Form 7328-R, Range Estimation Exercise Scorecard (front).

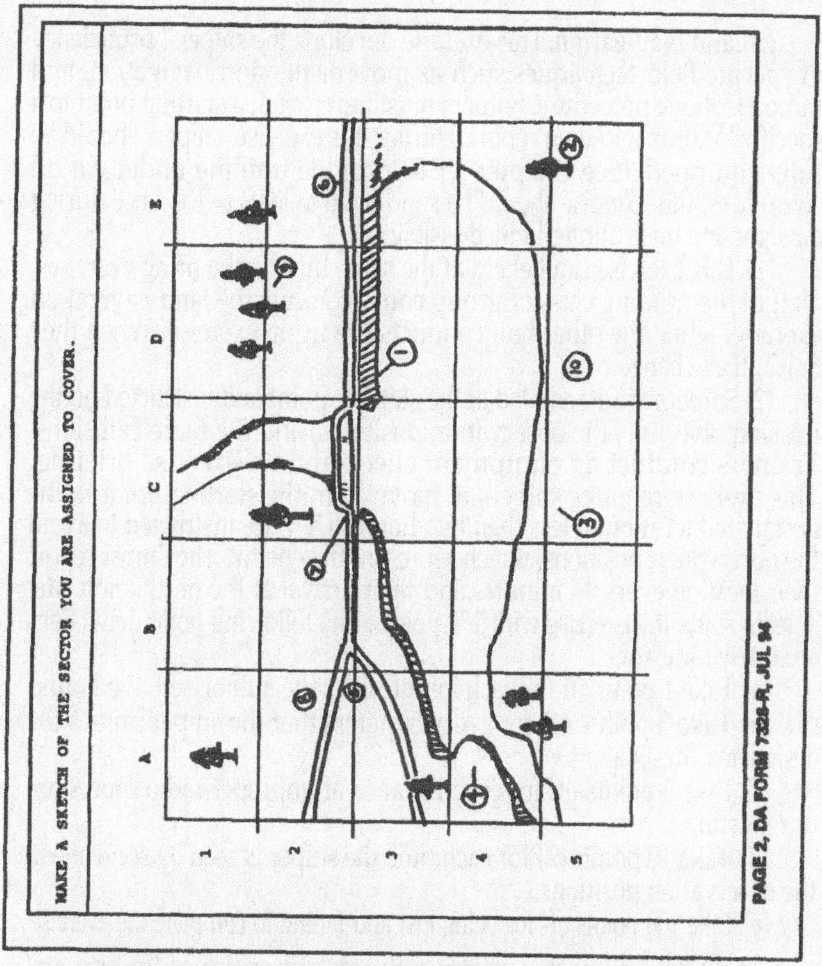

Figure 9-4. Example of completed DA Form 7328-R, Range Estimation Exercise Scorecard (back) (continued).

g. **Land Navigation.** This exercise develops the snipers' proficiency in specific field techniques such as movement, land navigation, and radiotelephone procedure. Snipers must move from a starting point to a specific location and then report. During this exercise, snipers should be fully equipped. (See Chapter 2.) To provide training under varied conditions, this exercise should be conducted at least twice, once during daylight and once during limited visibility.

(1) This exercise can beheld at the same time as the firing exercises. Half of the training class or group could conduct the land navigation exercise, while the other half conducts the firing exercise. When they finish, they change over.

(2) Snipers are assembled at the starting point and instructed on the mission objective, the observation positions, and the radio call signs. Trainers conduct an equipment check and an exercise briefing. This exercise requires snipers to move from the starting point to the designated location in less than two hours. They are instructed to avoid the observation positions, which represent the enemy. They must report their location every 15 minutes and their arrival at the destination site. A team starts the exercise with 100 points. The following point deductions are made for errors:

(a) Take 1 point off for each minute over the authorized two hours.

(b) Take 3 points off for every 5 meters that the sniper misses the designated destination.

(c) Take 5 points off for each instance of improper radio procedure or reporting.

(d) Take 10 points off for each time the sniper is seen by someone in the observation positions.

(e) Take 100 points off for being lost and failing to complete the exercise.

(3) At the end of this exercise, the trainer critiques the snipers' performance.

h. **Memory Enhancement Exercise (KIM Game).** A KIM game exercise consists of 10 variable military items on a table, covered with a blanket poncho, or anything suitable. Snipers observe the objects when uncovered but cannot touch the items or talk during the exercise. (Figure 9-5 is an example of a locally fabricated KIM game exercise scoresheet format.)

(1) After a prescribed time, the items are covered, and the snipers write their observations on a score sheet. They write the details that accurately describe the object, omitting unnecessary words. There are many variations that can be incorporated into a KIM game, such as PT, an extended amount of time between observing and recording, distractions

while observing and recording, or the use of different methods to display items. For example instead of a blanket uses towel or slides. At the end of the time limit, snipers turn in the score sheets, and trainers identify each item. Snipers describe each object in the following categories:

(a) *Size:* The sniper describes the object by giving the rough dimensions in a known unit of measure or in relation to a known object.

(b) *Shape:* The sniper describes the object by giving the shape such as round, square, or oblong.

(c) *Color:* The sniper records the color of the object.

(d) *Condition:* The sniper describes the object by giving the general or unusual condition of the object such as new, worn, or dented.

(e) *Appears to be:* The sniper describes what the object appears to be such as an AK-47 round or radio handset.

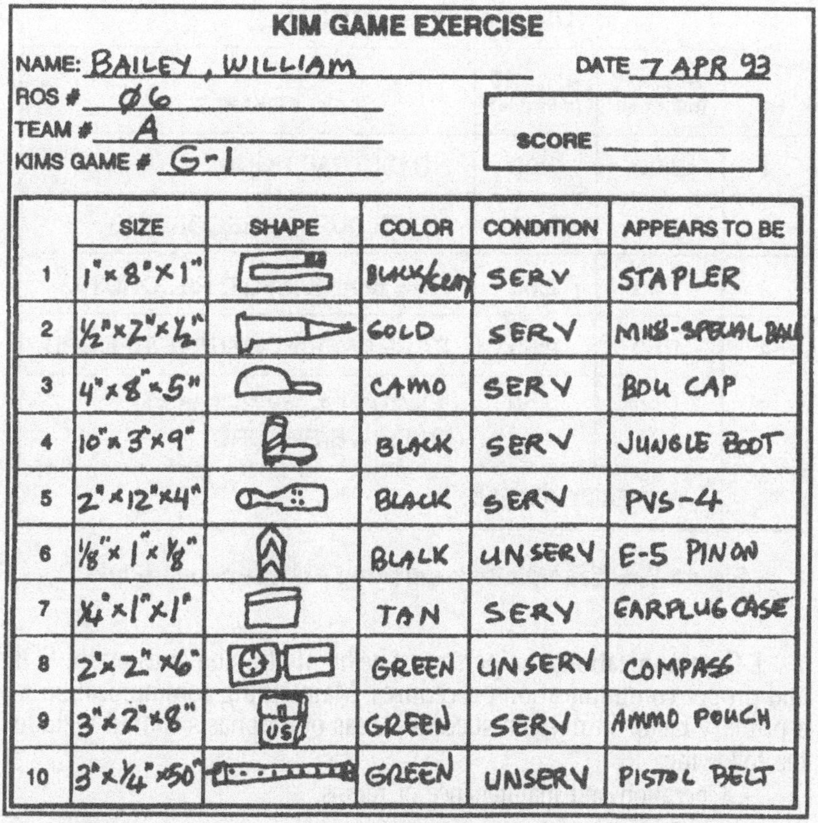

KIM GAME EXERCISE

NAME: BAILEY, WILLIAM DATE 7 APR 93

ROS # Ø6

TEAM # A

KIMS GAME # G-1

SCORE ______

	SIZE	SHAPE	COLOR	CONDITION	APPEARS TO BE
1	1"×8"×1"		BLACK/GRAY	SERV	STAPLER
2	1/2"×2"×1/2"		GOLD	SERV	M118-SPECIAL BALL
3	4"×8"×5"		CAMO	SERV	BDU CAP
4	10"×3"×9"		BLACK	SERV	JUNGLE BOOT
5	2"×12"×4"		BLACK	SERV	PVS-4
6	1/8"×1"×1/8"		BLACK	UNSERV	E-5 PIN ON
7	1/4"×1"×1"		TAN	SERV	EARPLUG CASE
8	2"×2"×6"		GREEN	UNSERV	COMPASS
9	3"×2"×8"	US	GREEN	SERV	AMMO POUCH
10	3"×1/4"×50"		GREEN	UNSERV	PISTOL BELT

Figure 9-5. Example of suggested format for KIM game exercise score sheet.

(2) Snipers receive 1/2 point for indicating that there was an item with some sort of description and the other 1/2 point for either exactly naming the item or giving a sufficiently detailed description using the categories listed above. The description must satisify the trainer to the extent that the sniper had never seen the object before. The total possible score is 10 points. Experience in the exercise, time restraints, and complexity of the exercise determines a passing score. This is the trainer's judgment based on his own experience in KIM games (Figure 9-6). The first few games should be strictly graded, emphasizing details. When the snipers are familiar with the game pattern, the trainer may make changes. The last game of the training should be identical to the first. In this way, the sniper can see if he improved.

KIM GAME SCHEDULE			
NO.	OBSERVE (minutes)	RECORD (minutes)	REMARKS
1	2:00	3:00	NO DISTRACTIONS
2	2:00	3:00	NOISE DURING RECORDING
3	1:50	2:50	FIRE BLANK WHILE RECORDING
4	1:50	2:50	PT BETWEEN OBSERVE/RECORD
5	1:30	2:30	2-HOUR DELAY BETWEEN OBSERVE/RECORD
6	REPEAT GAME NO. 1		

Figure 9-6. Example of suggested KIM game schedule.

i. **Communications.** Snipers must be highly trained in using the SOI and proper communication procedures. Maintaining communication is a primary factor in mission success. Areas of emphasis should include the following:

- Operation and maintenance of radios.
- Entering the net.
- Authentication.

- Encoding/decoding.
- Encrypting/decrypting.
- Antenna repair.
- Field-expedient antennas.

9-2. ADDITIONAL SKILLS SUSTAINMENT

Other than basic skills, the trainer must include additional skills into the sniper sustainment training program. Once mastered, these skills enhance the sniper's chance of surviving and accomplishing the mission.

a. **Call for Fire.** With advanced camouflage and movement techniques, snipers can move about the battlefield undetected. Snipers that have a working knowledge in the use and application of artillery, NGF, and CAS will bean asset to the commander. (See FM 6-30.)

(1) ***Artillery fire.*** Artillery fire is the secondary weapon of the sniper. Each sniper should master call-for-fire procedures (Figure 9-7, page 9-16), target location methods (Figure 9-8, page 9-17), and indirect-weapon system capabilities (Table 9-1, page 9-19). Separate radio stations may beset up with one being a simulated FDC. After the FDC receives the call for fire, it determines how the target will be attacked. That decision is announced to the FO as a message to the observer, which consists of three elements as follows:

- Unit to fire for effect.
- Any changes to requests in the call for fire.
- Method of fire (number of rounds to be fired).

Snipers can simulate calls for fire using the example format in Figure 9-7, page 9-16.

(2) ***Naval gunfire and close air support.*** In today's battlefield of "high-tech" munitions and delivery systems, a working knowledge of acquiring NGF and CAS (helicopter and fixed-wing) enables snipers to inflict heavy damage on enemy forces.

1. OBSERVER IDENTIFICATION

2. WARNING ORDER

   a. Type of Mission.
      (1) Adjust fire.
      (2) Fire for effect.
      (3) Suppress.

   b. Size of Element to Fire.

   c. Method of Target Location.
      (1) Grid.
      (2) Polar.
      (3) Shift from a known point.

3. TARGET LOCATION

   a. Grid. Six-digit grid coordinates.
   b. Polar. Direction and distance from the FO to the target.
   c. Shift. Direction to the target.

(1)	Lateral shift	L/R
(2)	Range shift	±
(3)	Vertical shift	U/D

4. TARGET DESCRIPTION

5. METHOD OF ENGAGEMENT

   a. Type of Adjustment.
      (1) Area fire.*
         - Bracketing.*
         - Creeping (DANGER CLOSE).
      (2) Precision fire.

   b. Trajectory.*

   c. Ammunition.
      (1) Projectile.*
      (2) Fuze.*
      (3) Volume of fire.*

   d. Distribution.*

Figure 9-7. Call-for-fire-format.

6. METHOD OF FIRE AND CONTROL

a. Method of Fire.

(1) Center platoon/center section.
(2) Battery/platoon right (left).
(3) Time interval.

b. Method of Control.

(1) Fire when ready.*
(2) At my command.
(3) Cannot observe.
(4) Time on target.

*Standard

Figure 9-7. Call-for-fire-format (continued).

1. GRID (a) Determine a six-digit grid coordinate to the designated target.
   (b) Determine the grid direction (observer-target) to the target, and ensure that the O-T direction is sent to the FDC after the call for fire is completed before the first correction.
2. POLAR (a) Determine the O-T direction to the target from the FO's position.
   (b) Determine the distance from the FO's position to the target.
3. SHIFT (a) Determine the O-T direction to the target.
   (b) Determine the lateral shift from the known point to the target.
   W = R x M (mil relation).
   W = Width of lateral shift in meters.

Figure 9-8. Target location methods.

R = Distance to the known point divided by 1,000. When shifting from a known point, the R is rounded to the nearest tenth.

M = Measured angle in mils between the known point and the target.

Example: $\frac{2,800}{1,000}$ (distance to known point) = 2/8 = R

M = 130 mils (measured angle from the known point to the target).

Therefore: W = R (2.8) × M (130)

W = 364 or LEFT 360 (nearest 10 mils)

(c) Determine the range shift from the known point to the target.

Example: 2,800 (distance to the known point)
-1,700 (distance to the target)
1,100 meters or DROP 1,100 meters (nearest 100 meters)

Figure 9-8. Target location methods (continued).

WEAPON	MAXIMUM RANGE (METERS)	MINIMUM RANGE (METERS)	MAXIMUM RATE (ROUNDS PER MINUTE FIRST MINUTE)	SUSTAINED RATE (ROUNDS PER MINUTE)
FIELD ARTILLERY				
105-mm HOWITZER M101A1, TOWED	11,000 14,500 (RAP)	0	10	3
105-mm HOWITZER M102, TOWED	11,500 14,500 (RAP)	0	10	3
155-mm HOWITZER M114A1, TOWED	14,600 14,600	0	4	1
155-mm HOWITZER M114A2, TOWED	14,600 19,400 (RAP)	0	4	1
155-mm HOWITZER M198, TOWED	24,000 30,000 (RAP)	0	4	TEMPERATURE DEPENDENT
155-mm HOWITZER M109, SP	14,600	0	4	1
155-mm HOWITZER M109A1/A2/A3, SP	18,100 23,500 (RAP)	0	4	1*
175-mm GUN M107, SP	32,800	0	1.5	0.5
203-mm HOWITZER M115, TOWED	16,800	0	1.5	0.5
203-mm HOWITZER M110, SP	16,800	0	1.5	0.5
203-mm HOWITZER M110A1, SP	20,600	0	1.5	0.5
203-mm HOWITZER M110A2, SP	22,900	0	1.5	0.5
MORTARS				
60-mm MORTAR	3,490 (HE) 1,472 (WP) 931 (ILLUM)	70 (HE) 33 (WP) 725 (ILLUM)	30	20
81-mm MORTAR	4,595 (HE) 4,850 (HE), track 4,737 (WP)	72 (HE) 70 (WP) 100 (ILLUM)	30 30	20 FOR 2 MINUTES, THEN 8 MINUTES
107-mm MORTAR	6,840 (HE) 5,650 (WP) 5,490 (ILLUM)	770 (HE) 920 (WP) 400 (ILLUM)	18	9 FOR 5 MINUTES, THEN 3 MINUTES

* CHG: ONE ROUND PER MINUTE FOR 60 MINUTES, THEN ONE ROUND EVERY 3 MINUTES THEREAFTER.

Table 9-1. Indirect-weapon systems capabilities.

WEAPON	MAXIMUM RANGE (METERS)	MINIMUM RANGE (METERS)	MAXIMUM RATE (ROUNDS PER MINUTE FIRST MINUTE)	SUSTAINED RATE (ROUNDS PER MINUTE)
NAVAL GUNFIRE				
5-inch/38	15,000	0	20	15
5-inch/54	22,500	0	30	20
16-inch/50	37,000	0	1	1

WEAPON	TRAVERSE LIMITS (mils)	SHELL/FUZE COMBINATIONS**
FIELD ARTILLERY		
105-mm HOWITZER	409R/400L	APICM, HE/PD, HE/DELAY, HE TRAINING INERT, HE/VT, HE/CP, RAP/PD, RAP/DELAY, WP/PD, WP/DELAY, WP/TRAINING INERT, SMOKE, ILLUM
105-mm HOWITZER M102, TOWED	6400	SAME AS ABOVE
155-mm HOWITZER M114A1, TOWED	448R/418L	CLGP, APICM, HE/PD, HE/DELAY, HE/TRAINING INERT, HE/VT, HE/CP, WP/PD, WP/DELAY, WP/TRAINING INERT, SMOKE, COLORED SMOKE, ILLUM
155-mm HOWITZER M114A2, TOWED	SAME AS ABOVE	CLGP, DPICM, RAAMS, ADAM, APICM, HE/PD, HE/DELAY, HE/TRAINING INERT, HE/VT, HE/CP, RAP/PD, RAP/DELAY, WP/PD, WP/DELAY, WP/TRAINING INERT, SMOKE, COLORED SMOKE, ILLUM
155-mm HOWITZER M198, TOWED	400R/400L 6400 SPEED	SAME AS ABOVE
155-mm HOWITZER M109, SP	6400	SAME AS ABOVE
155-mm HOWITZER M109A1/2/3, SP	6400	SAME AS ABOVE
175-mm GUN M107, SP	553R/533L	HE/PD, HE/DELAY, HE/TRAINING INERT, HE/VT
203-mm HOWITZER M115, TOWED	SAME AS ABOVE	APICM, HE/PD, HE/DELAY, HE/TRAINING INERT, HE/VT, HE/CP
203-mm HOWITZER M110, SP	SAME AS ABOVE	SAME AS ABOVE
203-mm HOWITZER M110A1, SP	SAME AS ABOVE	DPICM, APICM, HE/PD, HE/DELAY, HE/TRAINING INERT, HE/VT, RAP/PD, RAP/DELAY
203-mm HOWITZER M110A2, SP	SAME AS ABOVE	SAME AS ABOVE

** THESE REFLECT ONLY THOSE SHELL/FUZE COMBINATIONS THE OBSERVER MAY REQUEST—NOT ALL THOSE AVAILABLE.

Table 9-1. Indirect-weapon systems capabilities (continued).

WEAPON	TRAVERSE LIMITS (mils)	SHELL/FUZE COMBINATIONS**
MORTARS		
60-mm MORTAR	250R/250L	HE/PD, HE/DELAY, HE/VT, WP/PD, WP/DELAY, WP/TRAINING INERT, ILLUM
81-mm MORTAR	95R/95L 6400 TRACK	HE/PD, HE/DELAY, HE/VT, WP/PD, WP/DELAY, ILLUM
107-mm MORTAR	125R/125L 6400 TRACK	HE/PD, HE/DELAY, HE/TRAINING INERT, HE/VT, WP/PD, WP/DELAY, ILLUM
120-mm MORTAR		
NAVAL GUNFIRE		
5-inch/38	6400***	HE/PD, HE/TRAINING INERT, HE/VT, HE/CP, AP/DELAY, WP/PD, WP/TRAINING INERT, ILLUM
5-inch/54	6400***	SAME AS ABOVE
167-inch/50	6400***	HE/PD, HE/TRAINING INERT, HE/VT, HE/CP, AP/DELAY

* CHG: ONE ROUND PER MINUTE FOR 60 MINUTES, THEN ONE ROUND EVERY 3 MINUTES THEREAFTER.

** THESE REFLECT ONLY THOSE SHELL/FUZE COMBINATIONS THE OBSERVER MAY REQUEST—NOT ALL THOSE AVAILABLE.

*** WITH INCREASED MINIMUM RANGES WHEN FIRING OVER SHIP'S STRUCTURES.

Table 9-1. Indirect-weapon systems capabilities (continued).

b. **Insertion/Extraction Techniques.** Practical application of insertion/extraction techniques enables snipers to accomplish its mission and to exfiltrate with confidence. Leaders should tailor these techniques to unit assets; however, a working knowledge of all techniques listed in Chapter 7 is an invaluable tool to the team.

c. **Tracking/Counterattacking.** Footprints found by enemy trackers may indicate that snipers are in the area. A knowledge of countertracking techniques is a valuable tool to snipers not only to remain undetected but also to collect battlefield information. (See Chapter 8.)

d. **Survival Skills.** Survival training, incorporated with evasion and escape training, will better prepare the sniper in contingency planning during exfiltration and, possibly, infiltration. Judging enemy reaction is an impossible task therefore, the sniper may be forced to live off the land until linkup can be established with friendly forces.

e. **First Aid.** Adequate first-aid training can mean the difference between life and death until proper medical attention can be given.

f. **Communications Reporting Procedures.** A lack of timely, detailed reporting of battlefield information can hinder the overall success of maneuvering units. Properly formatted information (Chapter 6), precoordinated with communications personnel, ensures timely and accurate intelligence gathering. Snipers must train to use information reporting formats and procedures.

9-3. TRAINING NOTES

Snipers should be trained IAW DA Pamphlet 350-38. Training includes knowledge of equipment, ammunition, range and terrain requirements, and techniques of training and sustaining the skills of the sniper team.

a. **Equipment.** During all FIXs, each sniper should be equipped as indicated in Chapter 2. Team equipment should be available as needed.

b. **Known Distance Range Requirements.** A standard known-distance range, graduated in 100-meter increments from 100 to 1,000 meters, is required for zeroing and zero confirmation exercises. The target detection range facilities and procedures should permit observation and range determination to 800 meters.

c. **Field Firing Range Requirements.** The ideal field firing range should be on terrain that has been left in its natural state. The range should be a minimum of 800 meters in depth with provisions along the firing line for several sniper positions within each lane to provide a slightly different perspective of the target area (Table 9-2). Where time prevents construction of a separate range, it may be necessary to superimpose this facility over an existing field firing range.

(1) Iron maidens can be made out of 3/4-inch steel plate with a supporting frame. They should be cut out in the form of silhouettes 20 inches wide and 40 inches high. By painting these targets white, the sniper can easily detect where the bullet impacts on the target.

(2) Placing targets inside of window openings gives the sniper experience engaging targets that can be found in an urban environment. This is done by cutting a 15-inch by 15-inch hole in the center of a 36-inch by 48-inch plywood board. Then an E-type silhouette is emplaced on a hit-kill mechanism 2 to 4 meters behind the plywood.

(3) Targets placed inside a bunker-type position allows the sniper to gain experience firing into darkened openings. This position can be built with logs and sandbags with an E-type silhouette on a hit-kill mechanism placed inside.

METERS	TYPE TARGET
200	E-TYPE SILHOUETTE, HIT-KILL MECHANISM.
300	IRON MAIDEN SILHOUETTE; E-TYPE SILHOUETTE, HIT-KILL MECHANISM; MOVING TARGET MECHANISM.
325	E-TYPE SILHOUETTE, HIT-KILL MECHANISM.
375	E-TYPE SILHOUETTE, HIT-KILL MECHANISM, EMPLACED INSIDE A WINDOW.
400	E-TYPE SILHOUETTE, HIT-KILL MECHANISM, EMPLACED INSIDE A BUNKER.
500	IRON MAIDEN SILHOUETTE; MOVING TARGET MECHANISM, TRACKED VEHICLE WITH A HIT-KILL MECHANISM IN THE COMMANDER'S CUPOLA.
600 to 1,000	IRON MAIDEN SILHOUETTES.

Table 9-2. Field firing range requirements.

(4) Moving targets can be used at distances between 300 and 500 meters to give the sniper practical experience and to develop skill in engaging a moving target. Two targets, one moving laterally and one moving at an oblique, present a challenge to the sniper.

(5) Targets should be arranged to provide varying degrees of concealment to show enemy personnel or situations in logical locations (Figure 9-9, page 9-24). The grouping of two or more targets to indicate a crew-served weapon situation or a small unit is acceptable. Such arrangements, provided the targets can be marked, may require selective engagement by the sniper. The automatic target devices provide for efficient range operation and scoring.

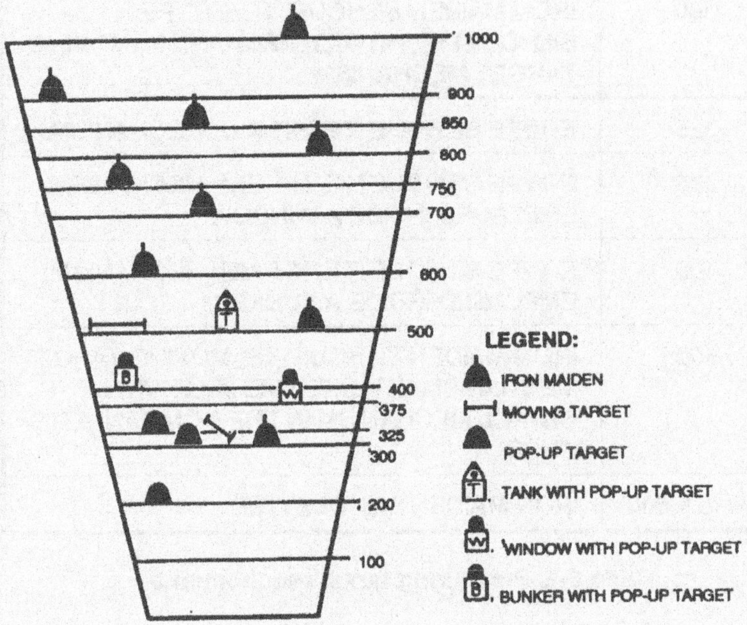

Figure 9-9. Lane layout.

9-4. EXAMPLE 5-DAY SNIPER SUSTAINMENT TRAINING PROGRAM

An example of a 5-day sniper sustainment training program is as follows:

DAY 1

TASK 1: Select sniper team routes and positions.

CONDITIONS: Given a review of selection of routes and positions, a situational sniper mission with a target area location that requires a minimum movement of 3,000 meters, a military map, a protractor, a felt-tip pen, an 8-inch-square clear plastic overlay, and one sheet of letter-size paper.

STANDARDS: Select and plot a primary and alternate route, objective rally point, and tentative final firing position that provides the best cover and concealment.

1. Prepare overlay with two grid reference marks; primary and alternate routes with arrows indicating direction of travel; minimum of three checkpoints, numbered in order; ORP; and a tentative final firing position.

2. Prepare a written log of movement. The sniper data book will contain the from-to grid coordinates, magnetic azimuths, distance, checkpoint number, objective rally point, and tentative final firing position.

3. Prepare overlay and written log of movement within 30 minutes.

TASK 2: Move while using individual sniper movement techniques.

CONDITIONS: Given a review of sniper movement techniques, a sniper weapon, a ghillie suit, and a flat, open area that allows trainers to observe movement techniques.

STANDARDS: Move correctly while using the designated movement technique.

1. Sniper low crawl.
2. Medium crawl.
3. High crawl.
4. Hands-and-knee crawl.
5. Walking.

NOTE: Trainers designate movement techniques and critique snipers on their movement.

TASK 3: React to enemy contact while moving as a member of a sniper team.

CONDITIONS: Given a review of sniper team movement techniques and reactions to enemy contact, sniper team's basic equipment and weapons, and an area of varying terrain with at least one danger area.

STANDARDS: React correctly to designated situations or danger areas.

1. Visual contact.
2. Ambush.
3. Indirect fire.
4. Air attack.
5. Danger area (linear and open area).

NOTE: Trainers designate situations and critique sniper teams on movement.

TASK 4: Describe target detection, selection, and observation techniques.

CONDITIONS: Given a review of target detection, selection, and observation techniques.

STANDARDS: Describe, orally or in writing, techniques used to observe, detect, and select targets.

TASK 5: Identify enemy uniforms, equipment, and vehicles.

CONDITIONS: Given a review of pictures or slides of enemy uniforms, equipment, and vehicles.

STANDARDS: Identify 7 of 10 enemy uniforms or rank insignia, 7 of 10 pieces of enemy equipment, and 7 of 10 enemy vehicles.

TASK 6: Describe range estimation techniques.

CONDITIONS: Given a review of range estimation techniques used by snipers.

STANDARDS: Describe, orally or in writing, range estimation techniques used by the sniper.

1. Eye methods.
2. Use of binoculars.
3. Use of M3A scope/M49 observation telescope.

TASK 7: Prepare a sniper range card.

CONDITIONS: Given a review of sniper range cards, a suitable target area, basic sniper equipment, and a sniper range card.

STANDARDS: Prepare a sniper range card complete with—

1. Grid coordinates of position.
2. Target reference point(s) (azimuth, distance, and description).
3. Left/right limits with azimuths.
4. Ranges throughout area.
5. Major terrain features.
6. Method of obtaining range/name.
7. Weather data.

TASK 8: Prepare a military sketch.

CONDITIONS: Given a review of sniper military sketching, a suitable area or object to sketch, and a blank military sketch sheet.

STANDARDS: Prepare a sketch complete with—

1. Grid coordinates of position.
2. Magnetic azimuth through center of sketch.
3. Sketch name and number.
4. Scale of sketch.
5. Remarks section.
6. Name/rank.
7. Date/time.
8. Weather data.

TASK 9: Maintain a sniper data book.

CONDITIONS: Given a review of the sniper data book and 20 blank sheets stapled together as a booklet.

STANDARDS: Maintain a sniper data book with a chronological listing of events that take place during the next three days and containing the following:

1. Grid coordinates of position.
2. Observer's name.

3. Date/time/visibility.
4. Sheet number/number of total sheets.
5. Series number/time and grid coordinate of each event.
6. Event.
7. Action taken.

NOTE: Trainers collect the sniper data books in three days.

DAY 2

TASK 1: Describe the fundamentals of sniper marksmanship.

CONDITIONS: Given a review of sniper marksmanship fundamentals.

STANDARDS: Describe, orally or in writing, the fundamentals of sniper marksmanship.

1. Position.
2. Breath control.
3. Aiming.
4. Trigger control.

TASK 2: Describe the effects of weather on ballistics.

CONDITIONS: Given a review of the effects of weather on ballistics.

STANDARDS: Describe, orally or in writing, the effects of weather on ballistics.

TASK 3: Describe the sniper team method of engaging targets.

CONDITIONS: Given a review of the sniper team method of engaging targets.

STANDARDS: Describe, orally or in writing, the sniper team method of engaging targets.

TASK 4: Describe methods used to engage moving targets.

CONDITIONS: Given a review of methods used to engage moving targets.

STANDARDS: Describe, orally or in writing, methods used to engage moving targets.

TASK 5: Describe methods used to engage targets at various ranges without adjusting the scope's elevation.

CONDITIONS: Given a review of methods used to engage targets at various ranges without adjusting the scope's elevation.

STANDARDS: Describe, orally or in writing, the methods used to engage targets at various ranges without adjusting the scope's elevation.

TASK 6: Zero rifle scope.

CONDITIONS: Given a sniper weapon, an M49 observation telescope, a suitable firing range, and 7 rounds of 7.62-mm special ball (Ml 18) ammunition.

STANDARDS: Zero rifle scope within 7 rounds.

DAY 3

TASK 1: Zero iron sights.

CONDITIONS: Given a sniper weapon, a suitable firing range, and 12 rounds of 7.62-mm special ball ammunition.

STANDARDS: Zero iron sights on a sniper weapon within 12 rounds.

TASK 2: Engage moving targets.

CONDITIONS: Given a sniper weapon, an M49 observation telescope, a suitable firing range, and 10 rounds of 7.62-mm special ball (M118) ammunition.

STANDARDS: Engage 10 moving targets, from 300 to 500 meters, achieving a minimum of 7 hits.

TASK 3: Estimate range.

CONDITIONS: Given a sniper weapon system (M24), M19 binoculars, and 10 targets out to 800 meters.

STANDARDS: Correctly estimate range to 7 of the 10 targets using eye estimation (± 15 percent), binoculars (± 10 percent), or the M24 sniper weapon (± 5 percent).

TASK 4: Detect targets.

CONDITIONS: Given a suitable area with 10 military objects, binoculars, M49 observation telescope, and a scorecard.

STANDARDS: Detect, plot, and describe 7 of 10 military objects within 40 minutes.

TASK 5: Participate in a concealment exercise.

CONDITIONS: Given a sniper weapon, ghillie suit, three 7.62-mm blank rounds of ammunition, an area to conceal a sniper position, and 10 minutes to prepare.

STANDARDS: Without being detected, occupy a position, identify, and fire three blank rounds at a target (located 100 to 200 meters away) who is equipped with binoculars and an M49 observation telescope. Must score 7 of 10 points (Figure 9-10).

IF THE SNIPER—	POINTS		
	GIVEN	DEDUCTED	TOTAL
WAS DETECTED WITHOUT THE AID OF OPTICS (FIRST 2 MINUTES)	2	0	2
WAS DETECTED WITH THE AID OF OPTICS (18 MINUTES)	1	0	3
WAS DETECTED WHEN ASSISTANT TRAINER IS WITHIN 10 FEET OF SNIPER	1	0	4
PROPERLY IDENTIFIED THE NUMBER WITHIN 30 SECONDS	1	0	5
FAILED TO PROPERLY IDENTIFY THE NUMBER	0	3	2
FIRED FIRST SHOT, NOT DETECTED	4	0	6
FIRED SECOND SHOT, NOT DETECTED	1	0	7
MAINTAINED STABLE FIRING POSITION (SUPPORT)	2	0	9
PROPERLY ADJUSTED WEAPON'S SCOPE FOR RANGE AND WINDAGE	1	0	10

NOTES:
1. IF THE SNIPER IS CAUGHT TRYING TO IDENTIFY THE NUMBER SCORE 4 POINTS.
2. IF MUZZLE BLAST/FLASH IS DETECTED, DEDUCT 1 POINT FROM TOTAL SCORE.
3. FAILING TO COMPLY WITH TRAINING STANDARDS AND OBJECTIVES (SUCH AS UNNECESSARY MOVEMENT, PREMATURE FIRE, OUTSIDE OF PRESCRIBED BOUNDARIES) WILL RESULT IN TERMINATION OF THE EXERCISE AND A SCORE OF ZERO.

DAY 4

TASK 1: Quality on Qualification Table No. 1.

CONDITIONS: Given a sniper weapon, M49 observation telescope, a suitable firing range, Qualification Table No. 1 scorecard, and 40 rounds of 7.62-mm special ball (Ml 18) ammunition.

STANDARDS: Engage targets from 200 to 700 meters, achieving a minimum of 140 points.

TASK 2: Engage targets in MOPP.

CONDITIONS: During daylight, given a sniper weapon, suitable firing range, MOPP suit, complete M25-series protective mask, M49 observation telescope, and 30 rounds of 7.62-mm special ball (M118) ammunition.

STANDARDS: While in MOPP, engage targets at 300 to 800 meters, achieving a minimum of 105 points.

TASK 3: Participate in a concealed movement exercise.

CONDITIONS: Given a sniper weapon, ghillie suit, two 7.62-mm blank rounds of ammunition, and a suitable area 1,000 meters long that is observable.

STANDARDS: Within 4 hours, move 600 to 800 meters; without being detected, occupy a position, identify, and fire two blank rounds at an enemy target who is equipped with binoculars and an M49 observation telescope. Must score 7 of 10 points (Figure 9-11).

IF THE SNIPER—	POINTS		
	GIVEN	DEDUCTED	TOTAL
FAILED TO CROSS THE FFL	0	0	0
CROSSED THE FFL	6	0	6
FIRED FIRST SHOT, NOT DETECTED	2	0	8
WAS NOT DETECTED WHEN ASSISTANT TRAINER IS WITHIN 10 FEET OF SNIPER	2	0	10
PROPERLY IDENTIFIED THE 1ST NUMBER	2	0	12
WAS NOT DETECTED WHEN ASSISTANT TRAINER IS WITHIN 5 FEET OF SNIPER	2	0	14
FIRED SECOND SHOT, NOT DETECTED	2	0	16
PROPERLY IDENTIFIED THE 2D NUMBER	2	0	18
MAINTAINED GOOD CAMOUFLAGE	1	0	19
MAINTAINED STABLE FIRING POSITION (SUPPORT)	1	0	20

NOTES: 1. IF MUZZLE BLAST/FLASH IS DETECTED, DEDUCT 1 POINT FROM THE TOTAL SCORE.

2. FAILING TO COMPLY WITH TRAINING STANDARDS AND OBJECTIVES (SUCH AS UNNECESSARY MOVEMENT, PREMATURE FIRE, OUTSIDE OF PRESCRIBED BOUNDARIES) WILL RESULT IN TERMINATION OF THE EXERCISE AND A SCORE OF ZERO.

Figure 9-11. Scoring for concealed movement exercise.

DAY 5

TASK 1: Qualify on Qualification Table No. 2.

CONDITIONS: Given a sniper weapon, M49 observation telescope, a suitable firing range, Qualification Table No. 2 scorecard, and 40 rounds of 7.62-mm special ball (Ml18) ammunition.

STANDARDS: Engage targets at 300 to 900 meters, achieving a minimum of 140 points.

TASK 2: Call for fire.

CONDITIONS: Given a review of call-for-fire procedures, two AN/PRC-77 radios, and a fire mission.

STANDARDS: Transmit the fire mission using proper radio procedures and the elements of the call-for-fire mission in sequence:

1. Observer identification.
2. Warning order.
3. Target location.
4. Target description.
5. Method of engagement (optional).
6. Method of fire and control (optional).

TASK 3: Locate target by grid coordinates.

CONDITIONS: Given a review of locating targets using the grid-coordinate method, a map of the target area, binoculars, compass, and a target.

STANDARDS: Determine and announce the six-digit coordinates of the target (within a 250-meter tolerance) within 30 seconds.

TASK 4: Locate a target by polar plot.

CONDITIONS: Given a review of target locating using the polar-plot method, a map of the target area, binoculars, a compass, and a target.

STANDARDS: Locate the target within 250 meters of the actual location. Announce the target location within 30 seconds after identification. Express direction to the nearest 10 roils and within 100 mils of actual direction. Express distance to the nearest 100 meters.

TASK 5: Locate target by shift from a known point.

CONDITIONS: Given a review of locating targets using the shift from a known-point method, a map of the target area, binoculars, a compass, a known point, and a target.

STANDARDS: Locate the target within 250 meters of the actual location and announce the target location within 30 seconds after identification. Express direction to the nearest 10 roils and within 100 roils of the actual direction. Express right or left corrections to the nearest 10 meters and range corrections to the nearest 100 meters.

TASK 6: Participate in a land navigation exercise during daylight.

CONDITIONS: Given a navigation course with at least four legs no less than 800 meters apart.

STANDARDS: Navigate the course without being detected by the observer-instructor. Preparing sketches, range cards, and or logs from the sniper data book can also be incorporated into the exercise.

NIGHT 5

TASK: Participate in a land navigation exercise during nightfall.

CONDITIONS: Given a navigation course (FM 21-26) with at least three legs no less than 500 meters apart. Observer-instructors can be placed on the course to detect any violations of noise and light discipline and deduct points from the sniper's score for violations.

STANDARDS: Navigate the course without being detected.

9-5. EMERGENCY DEPLOYMENT READINESS EXERCISE

Trainers use T&EOs from ARTEP 7-92-MTP: Move Tactically (7-5-1825); Select/Engage Targets (7-5-1869); Select/Occupy Firing Position (7-5-1871); Estimate Range (7-5-1872); and Debrief (7-5-1809) for additional sustainment training. An example of a battalion EDRE follows:

TIME	ACTION
0400	Battalion alerts sniper teams. 1. CQ relays uniform and packing list. 2. Sniper teams have two hours to report to battalion. 3. Sniper team leaders report to SEO when all of the team is accounted for. 4. Sniper team receives FRAGO from the SEO.
0600	Snipers depart battalion area by air, truck, or road march.
0800	Sniper teams arrive at range. 1. Sniper teams receive range/safety briefing. 2. Snipers receive issued ammunition. 3. Snipers zero weapons. 4. Sniper teams field/record fire on a range with targets positioned from 200 to 900 meters.
1100	Sniper teams depart range; move to concealed movement site by truck, road march, or tactical movement by teams.

TIME	ACTION
1200	Sniper teams arrive at conceded movement site. 1. Sniper teams receive briefing. 2. Site should be 800 to 1,000 meters long positioned with a observer- instructor as a target at one end with field table, M19 binoculars, M49 observation telescope, 8-inch by 8-inch international orange panels with white 5-inch number (1 to 9) painted on them, and two AN/PRC-77 radios for observer and assistant trainer. 3. Sniper will have four hours to move into his FFP, 50 to 200 meters from observer-instructor, and fire his first shot. 4. Sniper will have 30 seconds in which to identify number. 5. Sniper will fire second shot.

NOTE: All information is to be recorded in the sniper data book.

TIME	ACTION
	6. The entire exercise will be conducted without the sniper being detected by the observer-instructor.
1600	Sniper teams depart for day/night land navigation exercise. 1. Sniper teams start the exercise from a concealed movement site. 2. Sniper teams will be required to move to three different points. At each point they will perform one of the following • Draw a militaty sketch. • Draw a range card. • Do a target detection exercise. • Collect information/data. 3. All movement will be performed without being detected.
2000	Night navigation exercise. 1. Sniper teams start the exercise from the command post.

TIME	ACTION
	2. They will move undetected to three different points.
	3. They will perform a detection exercise with the use of NODS.
	4. They will record all information in the sniper data book.
	5. After collecting necessary data, they will move to an extraction point and construct a sniper hide position. They will prepare for target reduction.
0500-0600	Target reduction.
	1. Upon target reduction time, the sniper team will prepare for extraction.
	2. At extraction time, they will return to the battalion area.
	3. The SEO will debrief the sniper team.
	4. The SEO will conduct an after-action review.

NOTE: A written test could also be given as part of the EDRE.

9-6. RECORD FIRE TABLES

In accordance with DA Pamphlet 350-38, sniper qualification should occur quarterly. Sniper qualification involves the firing of two field fire tables. Qualification Table No. 1 grades target engagements primarily between 200 and 700 meters. Scores are recorded on DA Form 7329-R, Qualification Table No. 1 Scorecard (Figure 9-12, page 9-38). Qualification Table No. 2 grades on the longer ranges between 300 to 900 meters. Scores are recorded on DA Form 7330-R, Qualification Table No. 2 Scorecard (Figure 9-13, page 9-39). Although the sniper weapon system has an 800-meter maximum effective range, it can effectively hit targets at 1,000 meters. This is a challenge to the sniper and, with successful engagement, is a confidence builder in his ability. To qualify on firing tables No. 1 and No. 2, the sniper must adhere to the following standards:

NOTE: Completion of the DA Forms 7329-R and 7330-R is self-explanatory. Blank copies of these forms are located at the back of this manual for local reproduction.

- Achieve a 70 percent standard of 140 points out of a possible 200 points.

- Fire a first-round hit to equal 10 points.
- Fire a second round if the first round misses the target.
- Receive 5 points if the second round hits the target.
- Receive a score of O if the second round misses the target.
- Complete firing within 30 minutes. Total all first-round hits and multiply by 10; total second-round hits and multiply by 5.
- Add first-round and second-round hits for a total firing table score.
- Meet the 70 percent standard (140 points). Trainer checks satisfactory or unsatisfactory.

NOTE: Trainer and sniper sign the scorecard.

QUALIFICATION TABLE No. 1 SCORECARD

Exercise Number____

For use of this form, see FM 23-10. The proponent agency is TRADOC.

DATA REQUIRED BY PRIVACY ACT OF 1974.

AUTHORITY: 10 USC 3012(g)/Executive Order 9397. PRINCIPAL PURPOSE(S): Evaluates individual training. ROUTINE USE(S): Evaluates individual proficiency. SSN is used for positive identification purposes only. MANDATORY OR VOLUNTARY DISCLOSURE AND EFFECT ON INDIVIDUAL NOT PROVIDING INFORMATION: Voluntary. Individuals not providing information cannot be rated/scored on a mass basis.

(circle one)

Record or practice

Last name	First	MI	Rank	SSN	Unit
SMITH	JOHN	D	SFC	457-26-1738	Dco 2-29 IN

Date	Weather\visibility	Score
17 FEB93	CLOUDY 62° 3 MILES VISIBILITY	175

TARGET (meters)	1st Round	2d Round	Miss	TARGET (meters)	1st Round	2d Round	Miss
200	X			500	X		
300	X			400	X		
325	X			325	X		
375	X	X		400	X		
500	X			600	←	X	
600	X			500	X		
500	↗	X		700	X		
375	X			325	X		
600	↑	→	X	300	X		
700	X			200	X		

16 x10 3 x5= 175

Roy B. Wunks
Trainer's signature

John D. Smith
Sniper's signature

DA FORM 7329-R, JUL 94

Figure 9-12. Example of completed DA Form 7329-R, Qualification Table No. 1 Scorecard.

QUALIFICATION TABLE No. 2 SCORECARD

Exercise Number____

For use of this form, see FM 23-10. The proponent agency is TRADOC.

DATA REQUIRED BY PRIVACY ACT OF 1974.

AUTHORITY: 10 USC 3012(g)/Executive Order 9397. PRINCIPAL PURPOSE(S): Evaluates individual training. ROUTINE USE(S): Evaluates individual proficiency. SSN is used for positive identification purposes only. MANDATORY OR VOLUNTARY DISCLOSURE AND EFFECT ON INDIVIDUAL NOT PROVIDING INFORMATION: Voluntary. Individuals not providing information cannot be rated/scored on a mass basis.

(circle one)

Record or practice

Last name	First	MI	Rank	SSN	Unit
SMITH	JOHN	D	SPC	457-28-1738	D Co 2-29 IN

Date	Weather\visibility	Score
17 FEB 93	CLEAR 60° UNLIMITED	185

TARGET (meters)	1st Round	2d Round	Miss	TARGET (meters)	1st Round	2d Round	Miss
200	X			900		X	
300	X			850	↖	←	X
325	X			800	X		
375	X			750	X		
600	X			700	X		
500	X			900	↑	X	
600	X			500	X		
700	X			400	X		
750	↗	X		325	X		
800	X			300	X		
850	X						

17 x10 3 x5= 185

Roy B. Wnls
Trainer's signature

John D. Smith
Sniper's signature

DA FORM 7330-R, JUL 94

Figure 9-13. Example of completed DA Form 7330-R, Qualification Table No. 2 Scorecard.

9-7. M24 SNIPER MILES TRAINING

MILES training is an invaluable tool to simulate realistic combat training. Other than actual combat, the sniper's best means of displaying effectiveness as a force multiplier is through the use of the M24 sniper weapon system (MILES).

a. **Characteristics of the MILES Transmitter.** The M24 sniper weapon system MILES transmitter is a modified M16 transmitter. A special mounting bracket attaches the laser transmitter to the right side of the barrel (looking from the butt end) of the M24 and places it parallel with the line of bore. The laser beam output has been amplified and tightened to provide precision fire capability out to 1,000 meters. (For component information and instructions on mounting, zeroing, and operation, see TM 9-1265-211-10.)

b. **Training Value.** Using the M24 MILES, the trainer can enhance sustainment training in target engagement.

(1) ***Selection of firing positions.*** Due to transmitter modifications, the sniper must attain a firing position that affords clear fields of fire. Any obstruction (vegetation, terrain) can prevent a one-shot kill by deflecting or blocking the path of the laser beam. By attaining this type of position, the sniper will improve his observation and firing capabilities.

(2) ***Target detection/selection.*** Using MILES against multiple/cluster targets requires the sniper to select the target that will have the greatest effect on the enemy. The trainer provides instant feedback on the sniper's performance. Situations may be created such as bunkers, hostage situations, and MOUT firing. The hit-miss indicating aspects of MILES are invaluable in this type of training.

(3) ***Range estimation.*** The sniper must be highly skilled in range estimation (Chapter 3) to properly use the M24 sniper weapon system. The trainer's evaluation of this ability is as simple as the sniper pulling the trigger. When the range to the target is properly computed and elevation dialed on the M3A, one shot, either hit or miss, indicates a strength or weakness in the sniper's range estimation ability (if the fundamentals of marksmanship were properly applied).

(4) ***Markmanship.*** A target hit (kill) with MILES is the same as one with live ammunition. Applying marksmanship fundamentals results in a first-round kill; the training value is self-evident.

c. **MILES Training Limitations.** The concept of MILES is to provide realistic training however, MILES is limited in its capabilities as applied to the sniper's mission of long-range precision fire.

(1) ***Lack of external ballistics training.*** A laser is a concentrated beam of light emitted by the MILES transmitter. It travels from the sniper's

weapon undisturbed by outside forces such as temperature, humidity, and wind. Lack of these effects may lull the sniper into a false sense of confidence. The trainer should constantly reinforce the importance of these factors. The sniper should make a mental note of changes that should be applied to compensate for these effects.

(2) ***Engagement of moving targets.*** The engagement of moving targets (Chapter 3) requires the sniper to establish a target lead to compensate for flight time of his bullet. Traveling in excess of 186,000 miles per second (speed of light), the MILES laser nullifies the requirement for target lead. Again, the sniper may be lulled into a false sense of confidence. The trainer should enforce the principles of moving target engagement by having the sniper note appropriate target lead for the given situation.

APPENDIX A

PRIMARY SNIPER WEAPONS OF THE WORLD

Several countries have developed sniper weapon systems comparable to the United States systems. These weapon systems are sold to or copied by countries throughout the world. Within the everchanging world of politics, it is impossible to predict how the future enemy may be armed. The designs and capabilities of these weapon systems are sirnilar. However, the amount of training and experience separates the sniper the marksman. This appendix describes the characteristics and capabilities of prevalent sniper weapon systems.

A-1. AUSTRIA

The Austrian Scharfschutzengewehr 69 (SSG-69) is the current sniper weapon of the Austrian Army and several foreign military forces. It is available in either 7.62-mm x 51 or the .243 Winchester calibers. The SSG-69 is a manually bolt-operated, 5-round rotary or 10-round box, magazine-fed, single-shot repeating rifle. Recognizable features are synthetic stock hammer-forged, heavy barrel with a taper; two-stage trigger, adjustable for length and weight of pull; and a machined, longitudinal rib on top of the receiver that accepts all types of mounts. The sighting system consists of the Kahles ZF69 6-power telescope iron sights are permanently affixed to the rifle for emergency use. The telescope comes equipped with an internal bullet-drop compensator graduated to 800 meters, and a reticle that consists of an inverted V with broken cross hairs. The weapon, magazine, and telescope together weigh 10.14 pounds. This weapon has a barrel length of 25.59 inches and a total length of 44.88 inches with a muzzle velocity of 2,819 feet per second. It has an accuracy of 15.75 inches at 800 meters using RWS Match rounds.

A-2. BELGIUM

The Model 30-11 sniping FN rifle is the current sniper rifle of the Belgian and other armies. This weapon is a 7.62-mm x 51, 5-round internal or 10-round detachable box, magazine-fed, manually bolt-operated rifle with a Mauser-action heavy barrel and, through the use of butt-spacer plates, an adjustable stock. Its sighting system is the FN 4-power, 28-mm telescope and aperture sights with 1/6 MOA adjustment capability. The rifle weighs 10.69 pounds and, with its 19.76-inch barrel, is a total of 43.97 inches long. The Model 30-11 has a muzzle velocity of 2,819 fps. Accessories include the biped of the MAG machine gun, butt-spacer plates, sling, and carrying case.

A-3. THE FORMER CZECHOSLOVAKIA

The current sniper weapon system is the VZ54 sniper rifle. It is a manually bolt-operated, 10-round box, magazine-fed 7.62-mm x 54 rimmed weapon and built upon bolt-action with a free-floating barrel. This weapon is similar to the M1891/30 sniping rifle (former Russian weapon)-only shorter and lighter. The rifle is 45.19 inches long and weighs 9.02 pounds with the telescope. It has a muzzle velocity of 2,659 fps with a maximum effective range of 1,000 meters.

A-4. FINLAND

Finnish weapon technology introduces a 7.62-mm x 51 sniper rifle that is equipped with an integral barrel/silencer assembly. It is a bolt-action, 5-round box, magazine-fed weapon with a nonreflective plastic stock and a standard adjustable biped. Through the use of adaptors, any telescopic or electro-optical sight may be mounted. The weapon is not equipped with metallic sights. The 7.62-mm Vaime SSR-1 (silenced sniper rifle) weighs 9.03 pounds and is 46.45 inches long.

A-5. FRANCE

French sniper weapons consist of the FR-F1 and FR-F2.

a. **FR-F1.** The FR-F1 sniping rifle, known as the Tireur d'Elite, is a manually bolt-operated, 10-round detachable box, magazine-fed, 7.62-mm x 51 or 7.5-mm x 54 weapon. The length of the stock may be adjusted with the butt-spacer plates. This weapon's sighting system consists of the Model 53 bis 4-power telescopic sight and integral metallic sights with luminous spots for night firing. It weighs 11.9 pounds, has a barrel length of 21.7 inches, and a total length of 44.8 inches. This weapon has a muzzle velocity of 2,794 fps and a maximum effective range of 800 meters.

Standard equipment features a permanently affixed biped whose legs may be folded forward into recesses in the fore-end of the weapon.

b. **FR-F2.** The FR-F2 sniping rifle is an updated version of the F1. Dimensions and operating characteristics remain unchanged; however, functional improvements have been made. A heavy-duty biped has been mounted more toward the butt-end of the rifle, adding ease of adjustment for the firer. Also, the major change is the addition of a thick, plastic thermal sleeve around and along the length of the barrel. This addition eliminates or reduces barrel mirage and heat signature. It is also chambered for 7.62-mm x 51 NATO ammunition.

A-6. GERMANY

The FRG has three weapons designed mainly for sniping the Model SP66 Mauser, WA 2000 Walther, and Heckler and Koch PSG-1.

a. **Model SP66 Mauser.** The SP66 is not only used by the Germans but also by about 12 other countries. This weapon is a heavy-barreled, manually bolt-operated weapon built upon a Mauser short-action. Its 26.8-inch barrel, completely adjustable thumbhole-type stock, and optical telescopic sight provide a good-quality target rifle. The weapon has a 3-round internal magazine fitted for 7.62-mm x 51 ammunition and a Zeiss-Diavari ZA 1.5-6-variable power x 42-mm zoom telescopic sight. The muzzle of the weapon is equipped with a flash suppressor and muzzle brake.

b. **WA 2000 Walther.** The WA 2000 is built specifically for sniping. The entire weapon is built around the 25.6-inch barrel; it is 35.6 inches long. This uniquely designed weapon is chambered for .300 Winchester Magnum, but it can be equipped to accommodate 7.62-mm x 51 NATO or 7.5-mm x 55 Swiss calibers. It is a gas-operated, 6-round box, magazine-fed weapon, and it weighs 18.3 pounds. The weapon's trigger is a single- or two-staged type, and its optics consist of a 2.5-10-variable power x 56-mm telescope. It has range settings of 100 to 600 meters and can be dismounted and mounted without loss of zero.

c. **Heckler and Koch PSG-1.** The PSG-1 is a gas-operated, 5- or 20-round, magazine-fed, semiautomatic weapon and is 47.5-inches long with a 25.6-inch barrel and has a fully adjustable, pistol-grip-style stock. Optics consist of a 6-power x 42-mm telescopic sight with six settings for range from 100 to 600 meters. The 7.62-mm x 51 PSG-1 weighs 20.7 pounds with tripod and when fully loaded. The muzzle velocity is 2,558 to 2,624 fps.

A-7. ISRAEL

The Israelis copied the basic operational characteristics and configuration of the 7.62-mm Galil assault rifle and developed a weapon to meet the demands of sniping. The 7.62-mm x 51 Galil sniping rifle is a semiautomatic, gas-operated, 20-round bolt magazine-fed weapon. Like most service rifles modified for sniper use, the weapon is equipped with a heavier barrel fitted with a flash suppressor it can be equipped with a silencer that fires subsonic ammunition. The weapon features a pistol-grip-style stock, a fully adjustable cheekpiece, a rubber recoil pad, a two-stage trigger, and an adjustable biped mounted to the rear of the fore-end of the rifle. Its sighting system consists of a side-mounted 6-power x 40-mm telescope and fixed metallic sights. The weapon is 43.89-inches long with a 20-inch barrel without a flash suppressor and weighs 17.64 pounds with a biped, sling, telescope, and loaded magazine. When firing FN Match ammunition, the weapon has a muzzle velocity of 2,672 fps; when firing Ml18 special ball ammunition, it has a muzzle velocity of 2,557 fps.

A-8. ITALY

The Italian sniper rifle is the Berretta rifle. This rifle is a manually bolt-operated, 5-round box, magazine-fed weapon, and fires the 7.62-mm x 51 NATO rounds. Its 45.9-inch length consists of a 23-inch heavy, free-floated barrel, a wooden thumbhole-type stock with a rubber recoil pad, and an adjustable cheekpiece. Target-quality, metallic sights consist of a hooded front sight and a fully adjustable, V-notch rearsight. The optical sight consists of a Zeiss-Diavari-Z 1.5-power x 6-mm zoom telescope. The weapon weighs 15.8 pounds with biped and 13.75 pounds without the biped. The NATO telescope mount allows almost any electro-optical or optical sight to be mounted to the weapon.

A-9. SPAIN

The 7.62-mm C-75 special forces rifle is the current sniper rifle of Spain. This weapon uses a manually operated Mauser bolt-action. It is equipped with iron sights and has telescope mounts machined into the receiver to allow for the mounting of most electro-optic or optic sights. The weapon weighs 8.14 pounds. An experienced firer can deliver effective fire out to 1,500 meters using Match ammunition.

A-10. SWITZERLAND

The Swiss use the 7.62-mm x 51 NATO SG 510-4SIG rifle with telescopic sight. The SG 510-4 is a delayed, blow-back-operated, 20-round, magazine-fed, semiautomatic or fully automatic weapon. With biped, telescope, and empty 20-round magazine, the weapon weighs 1229 pounds. It is 39.9 inches long with a 19.8-inch barrel and a muzzle velocity of 2,591 fps.

A-11. UNITED KINGDOM

The United Kingdom has four weapons designed for use by military snipers: the L42A1, Models 82 and 85 Parker-Hale, and L96Al.

a. **L42A1.** The L42A1 is a 7.62-mm x 51 single-shot, manually bolt-operated 10-round box magazine-fed conversion of the Enfield .303, Mark 4. It is 46.49 inches long with a barrel length of 27.51 inches. It comes equipped with metallic sights and 6-power x 42-mm LIAl telescope, and has a muzzle velocity of 2,748 fps.

b. **Model 82.** The Model 82 sniper rifle is a 7.62-mm x 51 single-shot, manually bolt-operated, 4-round internal magazine-fed rifle built upon a Mauser 98-action. It is equipped with metallic target sights or the more popular V2S 4-variable power x 10-mm telescope. It can deliver precision fire at all ranges out to 400 meters with a 99 percent chance of first-round accuracy. The weapon weighs 10.5 pounds and is 45.7 inches long. It is made of select wood stock and has a 25.9-inch, freefloated heavy barrel. An optional, adjustable biped is also available.

c. **Model 85.** The Model 85 sniper rifle is a 7.62-mm x 51 single-shot, manually bolt-operated, 10-round box magazine-fed rifle designed for extended use under adverse conditions. Its loaded weight of 30.25 pounds consists of an adjustable-for-length walnut stock with a rubber recoil pad and cold-forged, free-floated 27.5-inch heavy barrel. The popular telescope is 6-power x 44-mm with a ballistic cam graduated from 200 to 900 meters. This weapon is guaranteed first-round hit capability on targets up to 600 meters. It also provides an 85 percent first-round capability at ranges of 600 to 900 meters. Features include:

(1) An adjustable trigger.

(2) A silent safety catch.

(3) A threaded muzzle for a flash suppressor.

(4) A biped with lateral and swivel capabilities.

(5) An integral dovetail mount that accepts a variety of telescopes and electro-optical sights.

d. The L96A1 sniper rifle is a 7.62-mm x 51 single-shot, manually bolt-operated, 10-round box magazine-fed rifle weighing 13.64 pounds. It features an aluminum frame with a high-impact plastic, thumbhole-type stock, a free-floated barrel; and a lightweight-alloy, fully adjustable biped. The rifle is equipped with metallic sights that can deliver accurate fire out to 700 meters and can use the LIA1 telescope. The reported accuracy of this weapon is 0.75 MOA at 1,000 meters. One interesting feature of the stock design is a spring-loaded monopod concealed in the butt. Fully adjustable for elevation, the monopod serves the same purpose as the sand sock that the US Army uses.

A-12. UNITED STATES

The US Army sniper weapons are the M21 and M24 SWS. As with other countries, earlier production sniper rifles are still being used abroad to include the Ml, MIA-EZ and the M21. Other sniper weapon systems used by US forces are the USMC M40A1 and special application sniper rifles such as the RAI Model 500 and the Barrett Model 82.

a. **M40A1.** The M40A1 sniping rifle is a manually bolt-operated, 5-round internal magazine-fed 7.62-mm x 51 NATO weapon. This weapon is equipped with a Unertyl lo-power fixed telescope with a roil-dot reticle pattern as found in the M24's M3A telescope. The M40A1 is 43.97 inches long with a 24-inch barrel and weighs 14.45 pounds. It fires Ml 18 special ball ammunition and has a muzzle velocity of 2,547 fps and a maximum effective range of 800 meters.

b. **RAI Model 500.** The RAl Model 500 long-range rifle is a manually bolt-operated, single-shot weapon, and it is chambered for the caliber .50 Browning cartridge. Its 33-inch heavy, fluted, free-floating barrel, biped, and fully adjustable stock and cheekpiece weigh a total of 29.92 pounds. The weapon is equipped with a harmonic balancer that dampens barrel vibrations, a telescope with a ranging scope base, and a muzzle brake with flash suppressor. The USMC and USN use this weapon, which has a muzzle velocity of 2,912 fps.

c. **Barrett Model 82.** The Barrett Model 82 sniping rifle is a recoil-operated, 1 l-round detachable box, magazine-fed, semiautomatic weapon chambered for the caliber .50 Browning cartridge. Its 36.9-inch fluted barrel is equipped with a six-port muzzle brake that reduces recoil by 30 percent. It has an adjustable biped and can also be mounted on the M82 tripod or any mounting compatible with the M60 machine gun. This weapon has a pistol-grip-style stock, is 65.9 inches long, and weighs 32.9 pounds. The sighting system consists of a telescope, but no metallic sights are provided. The telescope mount may accommodate any telescope with l-inch rings. Muzzle velocity of the Model 82 is 2,849 fps.

A-13. THE FORMER RUSSIA

The Russians have a well-designed sniper weapon called the 7.62-mm Dragunov sniper rifle (SVD). The SVD is a semiautomatic, gas-operated, 10-round box, magazine-fed, 7.62-mm x 54 (rimmed) weapon. It is equipped with metallic sights and the PSO-1 4-power telescopic sight with a battery-powered, illuminated reticle. The PSO-1 also incorporates a metascope that can detect an infrared source. Used by the former Warsaw Pact armies, this thumbhole/pistol-grip-style stocked weapon weighs 9.64 pounds with telescope and lo-round magazine. This weapon is 48.2 inches long with a 21.5-inch barrel, a muzzle velocity of 2,722 fps, and a maximum effective range of 600 to 800 meters.

A-14. THE FORMER YUGOSLAVIA

The former Yugoslav armed forces use the M76 semiautomatic sniping rifle. The M76 is a gas-operated, 10-round detachable box, magazine-fed, optically equipped 7.92-mm weapon. Variations of the weapon may be found in calibers 7.62-mm x 54 and 7.62-mm x 51 NATO. Believed to be based upon the FAZ family of automatic weapons, it features permanently affixed metallic sights, a pistol-grip-style wood stock, and a 4-power telescopic sight much the same as the Soviet PSO-1. It is graduated in NM-meter increments from 100 to 1,000 meters and has an optical sight mount that allows the mounting of passive nightsights. The M76 is 44.7 inches long with a 21.6-inch-long barrel. It weighs 11.2 pounds with the magazine and telescope, and it has a muzzle velocity of 2261 fps. A maximum effective range for the M76 is given as 800 meters with a maximum range of 1,000 meters.

APPENDIX B

M21 SNIPER WEAPON SYSTEM

The National Match M14 rifle (Figure B-1) and its scope makeup the M21 sniper weapon system. The M21 is accurized IAW United States Army Marksmanship Training Unit specifications and has the same basic design and operation as the standard M14 rifle (FM 23-8), except for specially selected and hand-fitted parts.

Section I
M21 SNIPER WEAPON SYSTEM

This section describes the general characteristics of the M21 SWS. The M21 has been replaced by the M24 (Chapter 2); however, the M21 is still in use throughout the US Army.

B-1. M21 DIFFERENCES

Significant differences exist between the M21 SWS and M24 SWS. These differences are as follows:

a. The barrel is gauged and selected to ensure correct specification tolerances. The bore is not chromium plated.

b. The stock is walnut and impregnated with an epoxy.

c. The receiver is individually custom fitted to the stock with a fiberglass compound.

d. The firing mechanism is reworked and polished to provide for a crisp hammer release. Trigger weight is between 4.5 to 4.75 pounds.

e. The suppressor is fitted and reamed to improve accuracy and eliminate any misalignment.

f. The gas cylinder and piston are reworked and polished to improve operation and to reduce carbon buildup.

g. The gas cylinder and lower band are permanently attached to each other.

h. Other parts are carefully selected, fitted, and assembled.

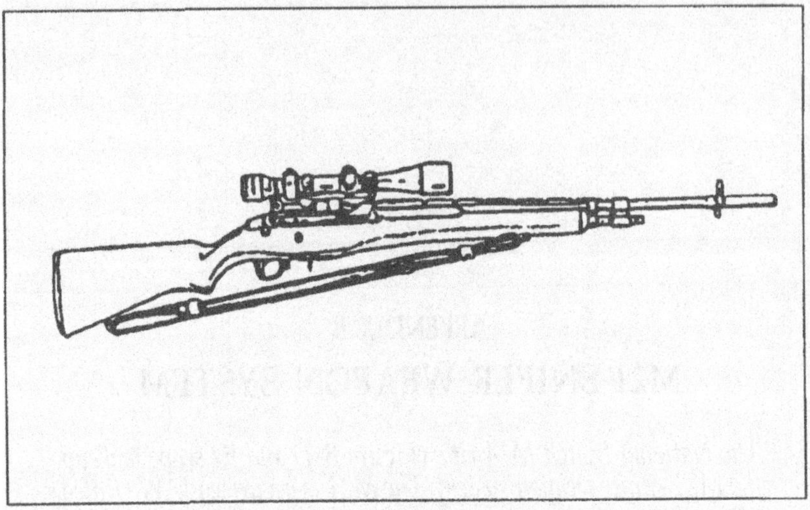

Figure B-1. National Match M14 rifle.

B-2. INSPECTION

If the sniper discovers a deficiency while inspecting the rifle, he reports it to the unit armorer. The following areas should be inspected:

a. Check the appearance and completeness of all parts. Shiny surfaces should be treated.

b. Check the flash suppressor for misalignment, burrs, or evidence of bullet tipping. The suppressor should be tight on the barrel.

c. Check the front sight to ensure that it is tight, that the blade is square, and that all edges and comers are sharp.

d. Check the gas cylinder to ensure it fits tightly on the barrel. The gas plug should be firmly tightened.

e. Check the forward band on the stock to ensure it does not bind against the gas cylinder front band.

f. Check the handguard. It should not bind against the receiver, the top of the stock, or the operating rod.

g. Check the firing mechanism to ensure the weapon does not fire with the safety "on," and that it has a smooth, crisp trigger pull when the safety is "off."

h. Check the rear sight tension by turning the aperture up to the "10" position. Then press down on top of the aperture with a thumb. If the aperture can be pushed down, the tension must be readjusted.

i. Check the stock for splits or cracks.

B-3. CARE AND MAINTANCE

Extreme care has been used in building the sniper rifle. A similar degree of attention must be devoted to its daily care and maintenance.

a. The rifle should not be disassembled by the sniper for normal cleaning and lubrication. Disassembly is performed only by the armorer during the scheduled inspections or repair. The armorer thoroughly cleans and lubricates the rifle at that time.

b. The following materials are required for cleaning the rifle

(1) Lubricating oil, general purpose (PL special).
(2) Lubricating oil, weapons (for below zero operation).
(3) Rifle bore cleaner.
(4) Rifle grease.
(5) Patches.
(6) Bore brush.
(7) Shaving brush.
(8) Toothbrush.
(9) Cleaning rags.

c. The recommended procedures for cleaning and lubricating the rifle are as follows:

(1) Wipe old oil, grease, and external dirt from the weapon.

(2) Clean the bore by placing the weapon upside down on a table or in a weapon cradle. hen, push a bore brush dipped in bore cleaner completely through the bore. Remove the bore brush and pull the rod out. Repeat this process four or five times.

(3) Clean the chamber (Figure B-2) and bolt face with bore cleaner and a chamber brush or toothbrush.

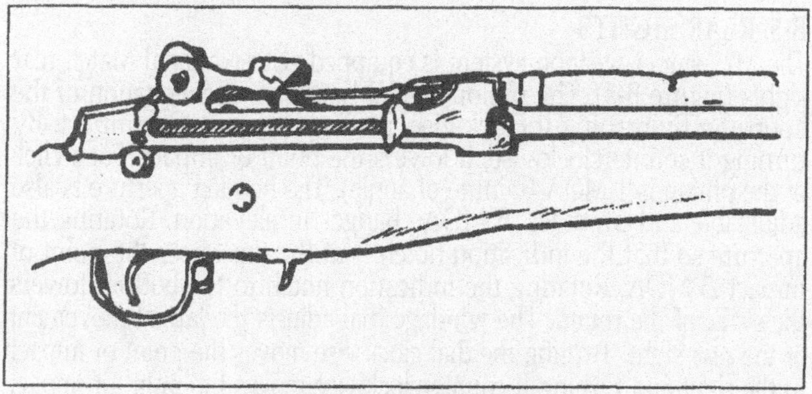

Figure B-2. M21 chamber.

(4) Clean the chamber, receiver, and other interior areas with patches dipped in RBC.

(5) Clean the bore by pulling clean patches through the bore until they come out of the bore clean.

(6) Wipe the chamber and interior surfaces with patches until clean.

(7) With the bolt and gas piston to the rear, place one drop of bore cleaner in between the rear band of the gas system and the lower side of the barrel. DO NOT PUT BORE CLEANER in the gas port. It will increase carbon buildup and restrict free movement of the gas piston.

(8) Lubricate the rifle by placing a light coat of grease on the operating rod handle track, caroming surfaces in the hump of the operating rod, the bolt's locking lug track, and in between the front band lip of the gas system and the metal band on the lower front of the stock.

(9) Place a light coat of PL special on all exterior metal parts.

B-4. LOADING AND UNLOADING

To load the M21, the sniper locks the bolt to the rear and places the weapon in the SAFE position. He inserts the magazine into the magazine well by pushing up, then pulling the bottom of the magazine to the rear until the magazine catch gives an audible click. To chamber a round, the sniper pulls the bolt slightly to the rear to release the bolt catch, then releases the bolt. To unload the M21, he locks the bolt to the rear and places the weapon in the SAFE position. Then he depresses the magazine release latch, and moves the magazine in a forward and downward motion at the same time.

B-5. REAR SIGHTS

The M21 sniper weapon system is equipped with National Match rear sights (Figure B-3). The pinion assembly adjusts the elevation of the aperture. By turning it clockwise, it raises the point of impact. By turning it counterclockwise, it lowers the point of impact. Each click of the pinion is 1 MOA (minute of angle). The hooded aperture is also adjustable and provides .5 MOA changes in elevation. Rotating the aperture so that the indication notch is at the top raises the point of impact .5 MOA. Rotating the indication notch to the bottom lowers the strike of the round. The windage dial adjusts the lateral movement of the rear sight. Turning the dial clockwise moves the point of impact to the right and turning it counterclockwise moves the point of impact to the left. Each click of windage is .5 MOA.

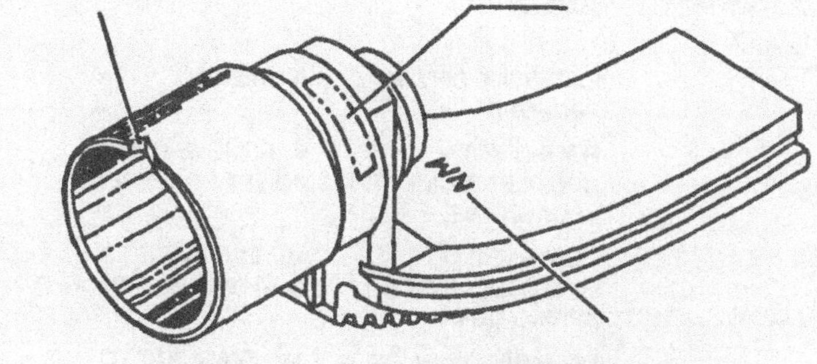

Figure B-3. National Match rear sight.

B-6. MALFUNCTIONS AND CORRECTIONS

Table B-1 contains pertinent information for the operator and serves as an aid to personnel who are responsible for restoring worn, damaged, or inoperative materiel to a satisfactory condition. If the weapon becomes unserviceable, it must be turned in for service by a school-trained National Match armorer.

MALFUNCTION	CAUSE	CORRECTION
FAILURE TO LOAD	DIRTY OR DEFORMED MAGAZINE	1. CLEAN OR REPLACE
	DAMAGED MAGAZINE TUBE	2. REPLACE MAGAZINE
	DIRTY MAGAZINE	3. CLEAN
	DAMAGED OR BROKEN MAGAZINE SPRING	4. REPLACE MAGAZINE
	DAMAGED OR BROKEN FOLLOWER	5. REPLACE MAGAZINE
	LOOSE OR DAMAGED FLOOR PLATE	6. REPLACE MAGAZINE

Table B-1. M21 malfunctions and corrections.

MALFUNCTION	CAUSE	CORRECTION
MAGAZINE INSERTS WITH DIFFICULTY	BENT OR DAMAGED MAGAZINE	7. REPLACE MAGAZINE
	EXCESSIVE DIRT IN RECEIVER	8. CLEAN
	ROUND NOT COMPLETELY SEATED IN MAGAZINE	9. REMOVE ROUND; INSERT PROPERLY
	DEFORMED OR OPERATING ROD SPRING GUIDE	10. EVACUATE TO AUTHORIZED ARMORER
	DAMAGED MAGAZINE LATCH	11. EVACUATE TO AUTHORIZED ARMORER
	MAGAZINE LATCH MOVEMENT RESTRICTED	12. CHECK MOVEMENT; CLEAN IF NECESSARY; IF BENT OR DISTORTED, EVACUATE TO AUTHORIZED ARMORER
MAGAZINE CANNOT BE RETAINED IN WEAPON	MAGAZINE LATCH DAMAGED	13. EVACUATE TO AUTHORIZED ARMORER.
	MAGAZINE LATCH SPRING DAMAGED	14. EVACUATE TO AUTHORIZED ARMORER
	MAGAZINE LATCH PLATE DAMAGED OR MISSING	15. REPLACE MAGAZINE
	DEFORMED OR DAMAGED OPERATING ROD SPRING GUIDE	16. EVACUATE TO AUTHORIZED ARMORER
	LOCKING RECESS AT TOP FRONT OF	17. REPLACE MAGAZINE
	MAGAZINE NOT FULLY INSTALLED	18. REMOVE; INSTALL CORRECTLY (MAKE SURE LATCH CLICKS)
FAILURE TO FEED	WEAK OR BROKEN SPRING	19. REPLACE MAGAZINE
	DAMAGED MAGAZINE	20. REPLACE MAGAZINE

Table B-1. M21 malfunctions and corrections (continued).

MALFUNCTION	CAUSE	CORRECTION
FAILURE TO FEED (CONTINUED)	DAMAGED OR DEFORMED STRIPPING LUG ON BOLT	21. EVACUATE TO AUTHORIZED ARMORER
	SHORT RECOIL	22. (SEE SHORT RECOIL)
	DIRTY AMMUNITION AND/OR MAGAZINE	23. CLEAN AMMUNITION AND/OR MAGAZINE
	WEAK OR BROKEN OPERATING ROD SPRING	24. EVACUATE TO AUTHORIZED ARMORER
	RESTRICTED MOVEMENT OF, OR DAMAGED OPERATING ROD25.	25. EVACUATE TO AUTHORIZED ARMORER
BOLT FAILS TO LOCK	CARTRIDGE CASE HOLDING BOLT OUT OF BATTERY	26. PULL BOLT CLOSE ASSEMBLY TO REAR AND REMOVE DEFORMED CARTRIDGE; CLEAN AMMUNITION AND/OR BARREL AND CHAMBER
	DIRTY CHAMBER	27. CLEAN CHAMBER
	EXTRACTOR DOES NOT SNAP OVER RIM OF CARTRIDGE CASE	28. EVACUATE TO AUTHORIZED ARMORER
	FROZEN OR BLOCKED EJECTOR SPRING AND PLUNGER	29. EVACUATE TO AUTHORIZED ARMORER
	RESTRICTED MOVEMENT OF, OR DAMAGED OPERATING ROD SPRING	30. EVACUATE TO AUTHORIZED ARMORER
	BOLT NOT FULLY ROTATED AND LOCKED IN RECEIVER	31. EVACUATE TO AUTHORIZED ARMORER
	WEAK OR BROKEN OPERATING ROD SPRING	32. EVACUATE TO AUTHORIZED ARMORER
	DAMAGED RECEIVER	33. EVACUATE TO AUTHORIZED ARMOR
FAILURE TO FIRE	BOLT NOT FULLY FORWARD AND LOCKED	34. (SEE BOLT FAILS TO LOCK)

Table B-1. M21 malfunctions and corrections (continued).

MALFUNCTION	CAUSE	CORRECTION
FAILURE TO FIRE (CONTINUED)	DEFECTIVE AMMUNITION	35. REMOVE AMMUNITION
	FIRING PIN WORN, DAMAGED, OR MOVEMENT RESTRICTED	36. EVACUATE TO AUTHORIZED ARMORER
	BROKEN HAMMER	37. EVACUATE TO AUTHORIZED ARMORER
	WEAK OR BROKEN HAMMER SPRING	38. EVACUATE TO AUTHORIZED ARMORER
	HAMMER AND TRIGGER LUGS OR SEAR WORN OR BROKEN, CAUSING HAMMER TO RIDE THE BOLT ASSEMBLY FORWARD	39. EVACUATE TO AUTHORIZED ARMORER
SHORT RECOIL	GAS PLUG LOOSE OR MISSING	40. TIGHTEN PLUG IF LOOSE; EVACUATE TO AUTHORIZED ARMORER IF MISSING
	RESTRICTED MOVEMENT OF OPERATING ROD ASSEMBLY	41. EVACUATE TO AUTHORIZED ARMORER
	BOLT BINDING	42. CLEAN
	GAS CYLINDER LOCK NOT FULLY INSTALLED	43. EVACUATE TO AUTHORIZED ARMORER
	GAS PISTON RESTRICTED	44. EVACUATE TO AUTHORIZED ARMORER
	DAMAGED CONNECTOR ASSEMBLY	45. EVACUATE TO AUTHORIZED ARMORER
	PARTIALLY CLOSED SPINDLE VALVE	46. EVACUATE TO AUTHORIZED ARMORER
	DEFECTIVE AMMUNITION	47. REPLACE AMMUNITION
FAILURE TO EXTRACT	SPINDLE VALVE CLOSED	48. EVACUATE TO AUTHORIZED ARMORER
	CARTRIDGE SEIZED IN CHAMBER	49. REMOVE

Table B-1. M21 malfunctions and corrections (continued).

MALFUNCTION	CAUSE	CORRECTION
FAILURE TO EXTRACT (CONTINUED)	SHORT RECOIL	50. (SEE SHORT RECOIL)
	DAMAGED OR DEFORMED EXTRACTOR	51. EVACUATE TO AUTHORIZED ARMORER
	WEAK, DEFORMED, OR FROZEN EXTRACTOR PLUNGER ASSEMBLY	52. EVACUATE TO AUTHORIZED ARMORER
	RUPTURED OR SEPARATED CARTRIDGE	53. EVACUATE TO AUTHORIZED ARMORER
FAILURE TO EJECT	SHORT RECOIL	54. (SEE SHORT RECOIL)
	WEAK, DEFORMED OR FROZEN EXTRACTOR PLUNGER ASSEMBLY	55. EVACUATE TO AUTHORIZED ARMORER
FAIL TO HOLD BOLT REARWARD	DAMAGED OR DEFORMED MAGAZINE FOLLOWER	56. REPLACE MAGAZINE
	DAMAGED BOLT LOCK	57. EVACUATE TO AUTHORIZED ARMORER
	BOLT LOCK MOVEMENT RESTICTED	58. EVACUATE TO AUTHORIZED ARMORER
	WEAK OR BROKEN MAGAZINE SPRING	59. REPLACE MAGAZINE

Table B-1. M21 malfunctions and corrections (continued).

Section II
M21 SIGHTING DEVICES

A scope mounted on the rifle allows the sniper to detect and engage targets more effectively. The target's image in the scope is in focus with the aiming point (reticle). This allows for a more focused picture of the target and aiming point at the same time. Another advantage of the scope is its ability to magnify the target. This increases the resolution of the target's image, making it clearer and more defined. Keep in mind, a scope does not make a soldier a better sniper, it only helps him to see better.

B-7. AUTO-RANGING TELESCOPE

Auto-ranging telescopes are part of the M21 system. The two types of ARTs on the M21 system are the ART I and ART II. The basic design and operating principle of both scopes are the same. Therefore, they will be described together, but their differences will be pointed out.

B-8. ART I AND ART II SCOPES

The ART has a commercially procured 3- to 9-variable-power telescopic scopesight, modified for use with the sniper rifle. This scope has a modified reticle with a ballistic earn mounted to the power adjustment ring on the ART I (Figure B-4). The ART II (Figure B-5) has a separate ballistic cam and power ring. The ART is mounted on a spring-loaded base mount that is adapted to fit the M14. It is transported in a hard carrying case when not mounted to the rifle. The scopes on the M21 sniper weapon system can also be used for rough range estimation. Once the sniper team is familiar with the M21 and is accustomed to ranging out on targets, it makes a mental note of where the power adjust ring is set at various distances.

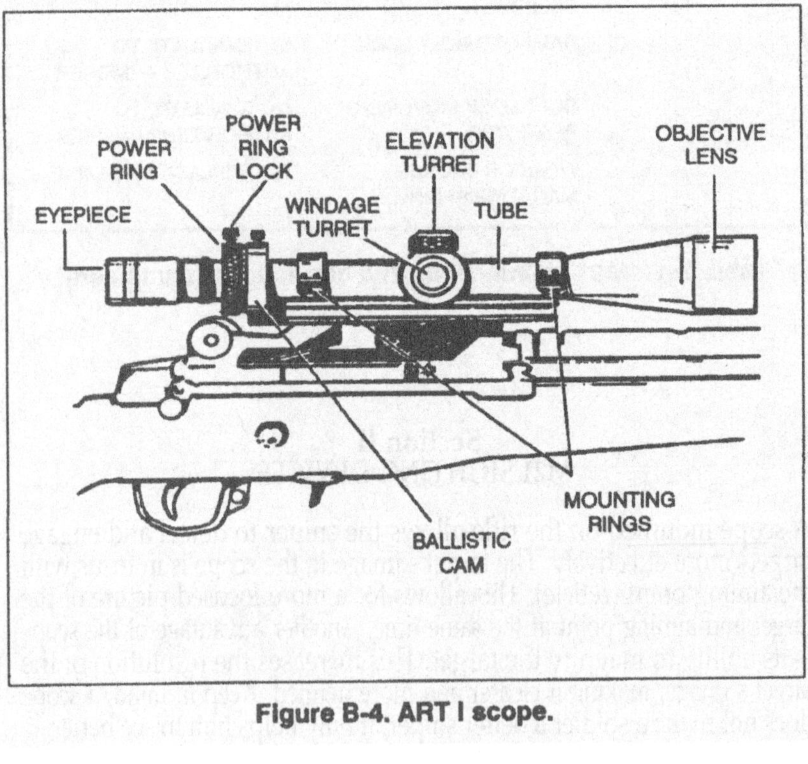

Figure B-4. ART I scope.

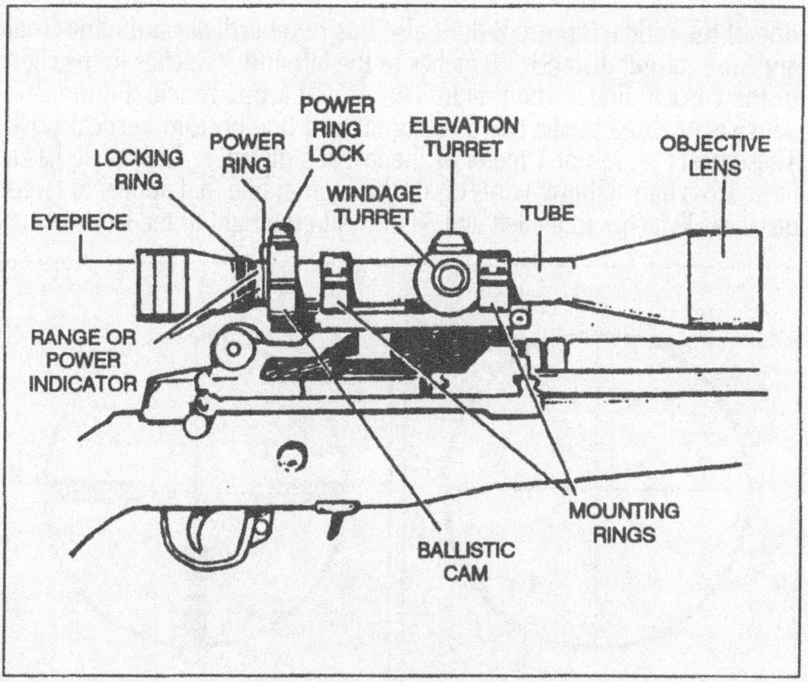

Figure B-5. ART II scope.

a. **Magnification.** The ART's increased magnification allows the sniper to seethe target clearly.

(1) The average, unaided human eye can distinguish detail of about 1 inch at 100 meters (1 MOA). Magnification, combined with well-designed optics, permit resolution of this 1 inch divided by the magnification. Thus, a 1/4 MOA of detail can be seen with a 4-power scope at 100 meters, or 1 inch of detail can be seen at 600 meters with a 6-power scope.

(2) The lens surfaces are coated with a hard film of magnesium fluoride for maximum light transmission.

(3) The elevation and windage turrets have dials on them that are located midway on the scope barrel and are used for zeroing adjustments. These dials are graduated in .5 MOA increments.

(4) These telescopes also have modified retitles The ART I scope has a basic cross hair design with two horizontal stadia lines that appear at target distances, 15 inches above and 15 inches below the horizontal

line of the reticle (Figure B-6). It also has two vertical stadia lines that appear at target distance, 30 inches to the left and 30 inches to the right of the vertical line of the reticle. The ART II scope reticle (Figure B-7) consists of three posts: two horizontal and one bottom vertical post. These posts represent 1 meter at the target's distance. The reticle has a basic cross hair with two dots on the horizontal line that appear at target distance, 30 inches to the left and 30 inches to the right of the vertical line.

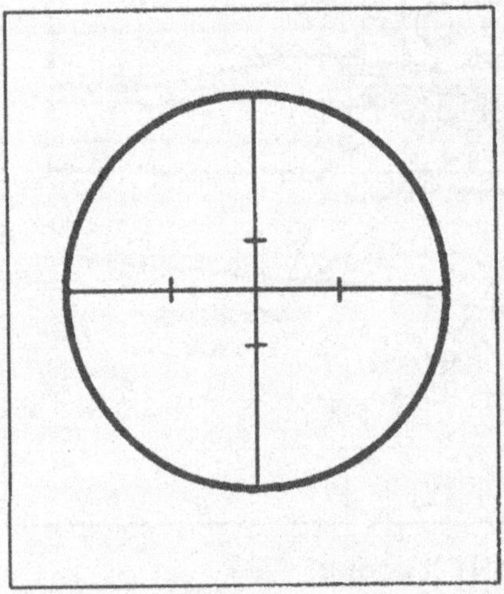

Figure B-6. ART I reticle.

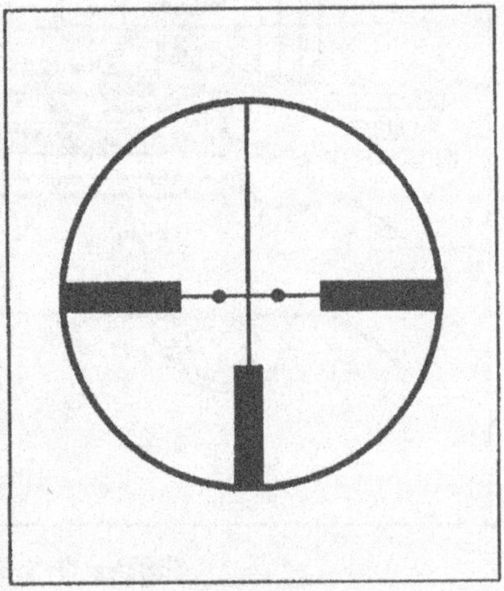

Figure B-7. ART II reticle.

(5) A ballistic cam is attached to the power adjustment ring on the ART I scope. The ART II scope has a separate power ring and ballistic cam.

(6) The power ring on both scopes increases and decreases the magnification of the scope, while the ballistic cam raises and lowers the scope to compensate for elevation.

(7) Focus adjustments are made by screwing the eyepiece into or away from the scope barrel until the reticle is clear.

b. **Scope Mounts.** The ART mounts are made of lightweight aluminum consisting of a side-mounting plate and a spring-loaded base with scope mounting rings. The scope mount is designed for low-profile mounting of the scope to the rifle, using the mounting guide grooves and threaded hole(s) on the left side of the receiver. The ART I has one thumbscrew that screws into the left side of the receiver (Figure B-8). The ART II mount has two thumbscrews; one is screwed into the left side of the receiver, and the other is screwed into the cartridge clip guide in front of the rear sight (Figure B-9).

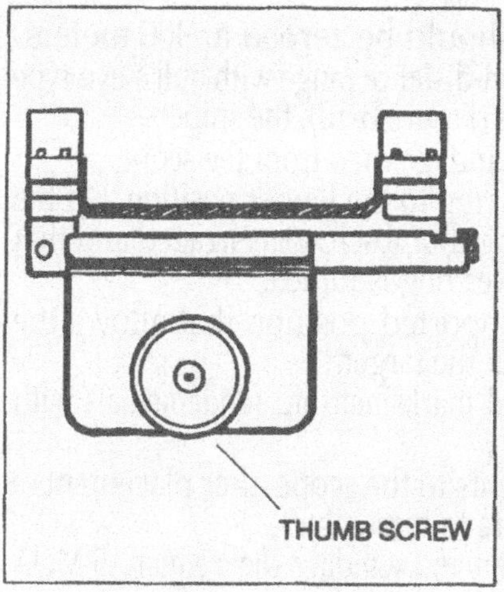

Figure B-8. ART I mount.

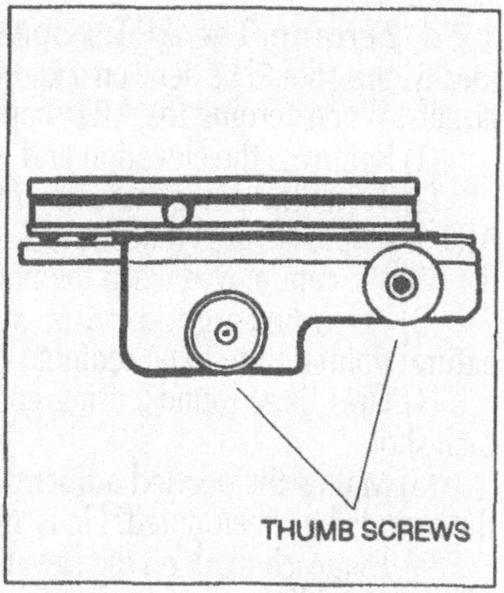

Figure B-9. ART II mount.

c. **Design and Operation.** The ART scopes are designed to automatically adjust for the needed elevation at ranges of 300 to 900 meters. This is done by increasing or decreasing the magnification of the scope until a portion of the target's image matches the represented measurement of the scope's reticle.

(1) For example, the power ring on the ART I scope can be adjusted until 30 inches of an object or a person's image (beltline to top of head) fits exactly in between the horizontal stadia lines (top stadia line touching top of the head and bottom stadia line on the beltline).

(2) Another example is to adjust the power ring on the ART II scope until 1 meter (about 40 inches) of a person's or an object's image appears equal to one of the posts in the reticle.

(3) When turning the power ring to adjust the target's image to the reticle, the sniper is also turning the ballistic cam. This raises or lowers the scope itself to compensate for elevation. Therefore, once the scope's magnification is properly adjusted in proportion to the target's image, the ballistic cam has at the same time adjusted the scope for the proper elevation needed to engage the target at that range.

(4) The ART II scope has a locking thumbscrew located on the power ring used for connecting and disconnecting the power ring from the ballistic cam. This allows the sniper to adjust the scope on target (auto-ranging mode), and then disengage the locking thumbscrew to increase magnification (manual mode) without affecting the elevation adjustment.

d. **Zeroing.** The ART scope should be zeroed at 300 meters. Ideally, this should be done on a known-distance range with bull's-eye-type targets. When zeroing the ART scope (Figure B-10), the sniper—

(1) Removes the elevation and windage caps from the scope.

(2) Turns the power adjustment ring to the lowest position. On the ART II scope, ensures that the locking thumbscrew is engaged and that the ballistic cam moves when the power ring is turned.

(3) Assumes a good prone-supported position that allows the natural point of aim to be centered on the target.

(4) Fires three rounds, using good marksmanship fundamentals with each shot.

(5) Makes the needed adjustments to the scope after placement of the rounds has been noted. He is sure to remember—

(a) That each mark on the elevation and windage dials equals .5 MOA (.5 MOA at 300 meters equals 1.5 inches.)

(b) That turning the elevation dial in the direction of the UP arrow will raise the point of impact; turning it the other direction will lower it.

(c) That turning the windage dial in the direction of the R arrow will move the point of impact to the right; turning it the other direction will move it to-the left.

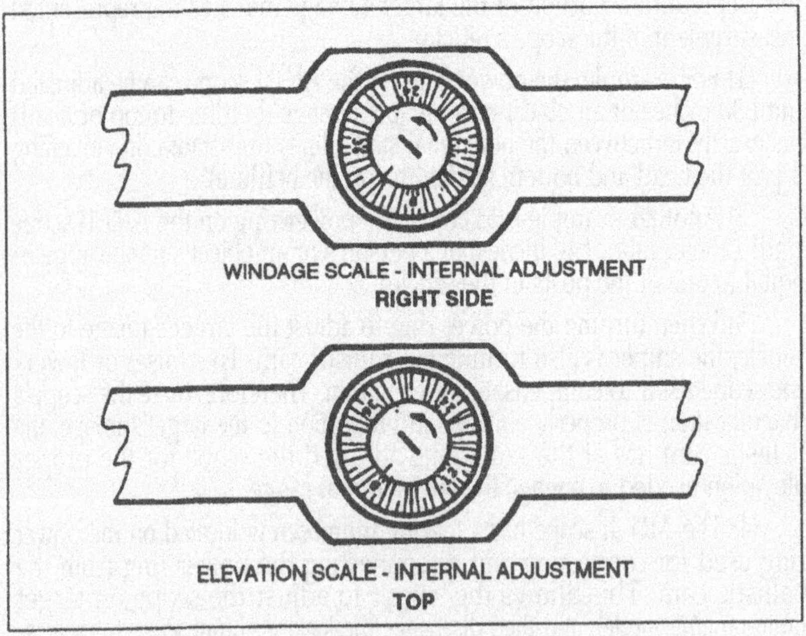

Figure B-10. Elevation and windage scales.

(6) Repeats the steps in paragraphs (4) and (5) above until two 3-round shot groups are centered on the target.

After the scope is properly zeroed, it will effectively range on targets out to 900 meters in the auto-ranging mode.

e. **Zeroing and Calibrating of the M21 Iron Sights.**If the telescope is damaged, the sniper must use his backup sighting system-iron sights. Due to time constraints, it may be impossible or impractical to search through the data book to determine the needed elevation setting to engage a target at a specific range. Once the elevation dial has been calibrated to the sniper's individual zero for that particular rifle, targets can be engaged anywhere between 0 and 1,080 meters by using index lines.

(1) The index lines on the elevation dial designate hundreds of yards to the target. Every other line is numbered with an even number, lines in between are the odd hundreds of yards-that is, the line marked with a number "2" is the 200-yard index line. The index line between the numbers 2 and 4 is the 300-yard index line. If the distance to the target is not in exact hundreds of yards, the elevation dial should be clicked between index lines to approximate the distance. If the target distance is less than 100 yards, the 100-yard setting should be used-the difference in impact is minimal.

(2) To calibrate the elevation dial, the sniper must first zero the rifle at a known distance that correlates to one of the index lines on the elevation dial. (The recommended distance is 300 yards.) Once zeroing is completed, calibration involves the following steps:

STEP 1: Turn the elevation dial forward (down, away from the sniper), and move the rear sight aperture assembly to its lowest setting (mechanical zero), counting the number of clicks. This number of clicks is elevation zero and must be remembered for use in the calibration process—for example, the number will be 10 clicks.

STEP 2: Loosen the screw in the center of the elevation dial using a dime or screwdriver (about one turn) until the dial can be rotated forward Be careful not to loosen the screw too much or it may fall and become lost. It is critical that once the screw is loosened to never rotate the elevation dial clockwise (up or toward the sniper) during calibration. This could result in improper calibration.

STEP 3: Turn the elevation dial forward (down, away from the sniper) until the index line on the receiver lines up with the index line on the dial that correlates to the distance at which the rifle was zeroed-for example, 300 yards. This is the index line between 2 and 4.

If the setting is passed (even by one click), rotate the elevation dial counterclockwise (down, away from the sniper) until the index lines match up. Never rotate the dial in the UP direction (clockwise, toward the sniper) with the screw in the elevation dial loose.

STEP 4: Remember the number of clicks (for example, 10) when zeroing the rifle and begin rotating the elevation dial counterclockwise (down, away from sniper). Count the clicks until the elevation dial has been rotated the same number of clicks that were on the rifle when zeroed. If too many clicks are used, start over at Step 3.

STEP 5: Now, hold the elevation dial, being careful not to allow it to rotate, then tighten the screw in the center of the elevation dial as tight as possible. Hold the elevation dial carefully with a pair of pliers to ensure the screw is tight.

STEP 6: To check the calibration, rotate the elevation dial to mechanical zero (all the way down), then count the number of clicks to zero. This should result in the index line on the receiver being lined up with the correct index line on the elevation dial (between 2 and 4). If this happens, the rear sight is now calibrated for elevation. If not, repeat Steps 1 through 5.

GLOSSARY

AARTY...... Army artillery
ADAM . . . . . . artillery delivered antipersonnel mine
aiming . . . . . . a marksmanship fundamental; refers to the precise alignment of the rifle sights with the target
ALICE . . . . . . all-purpose lightweight individual carrying equipment
AM . . . . amplitude modulation
antenna . . . . a device used to radiate or receive electromagnetic energy (usually RF)
antijamming . . a device, method, or system used to reduce or eliminate the effects of jamming
AP . . . . antipersonnel
APFT . . . . . . Army Physical Fitness Test
APICM . . . . antipersonnel improved conventional munition
armorer . . . . one who services and makes repairs on small arms and performs similar duties to keep small arms ready for use
ARNG . . . . . . .Army National Guard
ART . . . . . . .auto-ranging telescope
ARTEP . . . . . . .Army Training and Evaluation Program
AVLB . . . . . . armored vehicle launched bridge
AWADS . . . . adverse weather aerial delivery system

ball . . . . the projectile the bullet
ballistics . . . . a science that deals with the motion and flight characteristics of projectiles
BDU . . . . . . battle dress uniform
BMNT . . . . . . beginning morning nautical twilight
breath control a marksmanship fundamental refers to the control of breathing to help keep the rifle steady during firing
bullet drop . . how far the bullet drops from the line of departure to the point of impact

bull's-eye target . any target with a round black circle and scoring rings Normally used in competitive marksmanship training
butt plate . . . . . metal or rubber covering of the end of the stock on the rifle
CALFEX . . . . . combined arms live-fire exercise
cartridge . . . . . a complete round of ammunition
CAB . . . . close air support
CLGP . . . . cannon-launched guided projectile
CAP . . . . cleaner, lubricant, preservative
cm . . . . centimeter
CMF . . . . career management field
counterpoise . . . a conductor or system of conductors used as a substitute for a ground in an antenna system
CP . . . . concrete-piercing
CQ . . . . charge of quarters
crack and thump . a method to determine the general direction and distance to an enemy firer who is shooting at you
cradle . . . . a vise-like mechanism that holds a weapon in a secured position during test firing
CS . . . . a chemical agent (tear gas)
CW . . . . continuous wave

dia . . . . diameter
dipole . . . . a radio antenna consisting of two horizontal rods in line with each other with their ends slightly separated
DPICM . . . . . . dual-purpose improved conventional munition
DIG . . . . date-time group
DZ . . . . drop zone

E&E . . . . evasion and escape
ECM . . . . electronic countermeasures
EDGE . . . . emergency deployment readiness exercise
EEL . . . . essential elements of information
EVENT....... end of evening nautical twilight
effective wind . . the average of all the varying winds encountered

electromagnetic wave . . . a wave propagating **as** a periodic disturbance of the electromagnetic field and having a frequency in the electromagnetic spectrum

elevation adjustment . . . rotating the front sight post to cause the bullet to strike the higher or lower on the target

EMP . . . . electromagnetic pulse

EPW . . . . enemy prisoner of war

ERP . . . . end-route rally point

eye relief . . . . . the distance from the firing eye to the rear sight; eye relief is a function of stock weld

F . . . . Fahrenheit

FDC . . . . . . fire direction center

FFL . . . . . . final firing line

FFP . . . . . . final firing position

FLOT . . . . . . forward line of own troops

FM . . . . frequency modulated

FO . . . . forward observer

fps . . . . feet per second

FRAGO . . . . fragmentary order

freq . . . . frequency

FSK . . . . . . frequency-shift keying

ft . . . . feet

FTX . . . . . . field training exercise

ground . . . . . . a metallic connection with the earth to establish ground (or earth) potential

HAHO . . . . . . high altitude, high opening

half-wave antenna . . . . . an antenna whose electrical length is half the wavelength of the transmitted or received frequency

HALO . . . . . . high altitude, low opening

HC . . . . hydrogen chloride

HE . . . . high explosive

HF . . . . high frequency

hrs .... hours
Hz .... hertz

IAW .... in accordance with
ID .... identification
illum .... illumination
in .... inches
insulator ..... a device or material that has a high electrical resistance
interference ... any undesired signal that tends to interfere with the desired signal

IRP .... initial rally point

jamming .... deliberate interference intended to prevent reception of signals in a specific frequency band

KIM .... keep-in-memory (exercise game)

laser .... light amplification by simulated emission of radiation
LBE .... loading-bearing equipment
LAX .... live-fire exercise
line of departure . the line defined by the bore of the rifle or the path the bullet would take without gravity
line of sight .... a straight line from the eye through the aiming device to the point of aim
L/R.. .... left/right
LR .... laser range finder
LSA .... lubricating oil, weapons, semifluid
LZ .... landing zone

m .... meters
MEDEVAC .... medical evacuation
METT-T ..... mission, enemy, terrain, troops and time available
MHz .... megahertz

midrange trajectory/maximum ordinate. . the highest point the bullet reaches on its way to the target this point must be known to engage a target that requires firing underneath an overhead obstacle, such as a bridge or a tree; inattention to midrange trajectory may cause the sniper to hit the obstacle instead of the target

MIJI . . . . meaconing, intrusion, jamming, and interference

MILES . . . . . . multiple-integrated laser engagement system

min . . . . minute(s)

mm . . . . millimeter

MOA . . . . an angle that would cover 1 inch at a distance of 100 yards, 2 inches at 200 yards, and so on; each click of sight adjustment is equal to one minute of angle

MOPP . . . . . . mission-oriented protection posture

MOUT . . . . . . .military operations on urbanized terrain

mph . . . . miles per hour

MRE . . . . meal, ready-to-eat

MTP . . . . mission training plan

muzzle velocity . the speed of the bullet as it leaves the rifle barrel, measured in feet per second; it varies according to various factors, such as ammunition type and lot number, temperature, and humidity

NATO . . . . . . North Atlantic Treaty Organization

NBC . . . . nuclear, biological, chemical

NCO . . . . noncommissiotted officer

NGF . . . . naval gunfire

NOD . . . . night observation device

NSN . . . . national stock number

OIR . . . . other intelligence requirements

OP . . . . observation post

OPORD . . . . . operation order

OPSEC . . . . . operations security

optical sight. . sight with lenses, prisms, or mirrors used in lieu of iron sights

ORP . . . . .objective rally point

O-T . . . . observer-target

PD . . . . point-detonating
PFC . . . . private first class
P I R . . . . priority intelligence requirements
POC . . . . point of contact
point of aim . . . the exact spot on a target the rifle sights are aligned with
point of impact . . the point that a bullet strikes; usually considered in relation to point of aim
PSG . . . . platoon sergeant
PT . . . . physical training
PW . . . . prisoner of war
PZ . . . . pickup zone

QRF . . . . quick-reaction force
quarter-wave antenna . . an antenna with an electrical length that is equal to one-quarter wavelength of the signal being transmitted or received

RMMS . . . . . . remote antiarmor mine system
range card . . . . small chart on which ranges and directions to various targets and other important points in the area under fire are recorded
RAP . . . . rocket-assisted projectile
RBC . . . . . rifle bore cleaner
recoil . . . . the rearward motion or kick of a gun upon firing
RECONREP . . . reconnaissance report
retained velocity . the speed of the bullet when it reaches the target due to drag, the velocity will be reduced
RF . . . . radio frequency
RFA . . . . restrictive fire area
RFL . . . . restrictive fire line
round . . . . may refer to a complete cartridge or to the bullet
RSTA . . . . reconnaissance, surveillance, and target acquisition

S1 . . . . adjutant
S2 . . . . intelligence officer
S3 . . . . operations and training officer

S4 . . . . .supply officer
SALUTE . . . . . size, activity, location, unit, time, and equipment
SAW . . . . squad automatic weapon
SEO . . . . sniper employment officer
servomechanism . . . . . .an automatic device for controlling large amounts of power by using small amounts of power
SFC . . . . sergeant first class
SGT . . . . sergeant
SHELREP . . . . shelling report
shot group . . . . a number of shots fired using the same aiming point which accounts for rifle, ammunition, and firer varibility three shots are enough, but any number of rounds may be fired in a group
sight alignment . placing the center tip of the front sight post in the exact center of the rear aperature
silhouette target . a target that represents the outline of a man
single sideband . a system of radio communications in which the carrier and either the upper or lower sideband is removed from AM transmission to reduce the channel width and improve the signal-to-noise ratio
SIR . . . . specific information requirements
SITREP . . . . . situation report
SM . . . . smoke munitions
SOI . . . . signal operation instructions
SOP . . . . standing operating procedure
SP . . . . self-propelled
SPC . . . . specialist
SPIES . . . . . . special patrol insertion/extraction system
SPOTREP . . . . spot report
SSB . . . . single sideband
STAB . . . . . . a system for extracting personnel by helicopter
STANAG . . . . . Standardization Agreement
static . . . . sharp, short bursts of noise on a radio receiver caused by electrical disturbances in the atmosphere or by electrical machinery

steady position . . the first marksmanship fundamental, which refers to the establishment of a position that allows the weapon to be held still while it is being fire

stock weld . . . . the contact of the cheek with the stock of the weapon

STRAC . . . . . . standards in training commission

STX . . . . situational training exercise

supported position . . . . . . any position that uses something other than the body to steady the weapon (artificial support)

S W S . . . . sniper weapon system

TAB . . . . tactical air command

TFFP . . . . tentative final firing position

time of flight . . . the amount of time it takes for the bullet to reach the target from the time the round exits the rifle

TOC . . . . tactical operations center

TOW . . . . tube-launched, optically tracked wire-guided (missile)

trajectory . . . . . the path of the bullet as it travels to the target

TRC . . . . training readiness condition

TAP . . . . target reference point

unidirectional . . in one direction only

unsupported position . . . any position that requires the firer to hold the weapon steady using only his body (bone support)

USAF . . . . United States Air Force

USAR . . . . United States Army Reserve

U S C . . . .. United States Code

USMC . . . . . . United States Marine Corp

USN . . . . United States Navy

VHF . . . . very high frequency

VT . . . . variable time

wavelength . . . . the distance a wave travels during one complete cycle; it is equal to the velocity divided by the frequency

windage adjustment . . . moving the rear sight aperture to cause the bullet to strike left or right on the target

WP. . . . . white phosphorus

zeroing . . . . . . adjusting the rifle sights so bullets hit the aiming point at a given range

REFERENCES

SOURCES USED

These are the sources quoted or paraphrased in this publication.

*STANAG 2020. Operational Situation Reports. 13 February 1986

*STANAG 2022. Intelligence Reports. 29 September 1988.

*STANAG 2084. Handling and Reporting of Captured Enemy Equipment and Documents. 26 June 1986.

*STANAG 2096. Reporting Engineer Information in the Field. 13 Jul 1988.

*STANAG 2103. Reporting Nuclear Detonations, Radioactive Fallout, and Biological and Chemical Attacks, and Predicting Associated Hazards. 12 July 1988.

*STANAG 2934. Artillery Procedures—AARTY-1. 26 November 1990.

*STANAG 3204. Aeromedical Evacuation.

STANAG 6004. Meaconing, Intrusion, Jamming, and Interference Report. 20 March 1984.

DOCUMENTS NEEDED

These documents must be available to the intended users of this publication.

ARTEP 7-92-MTP. Infantry Scout Platoon/Squad and Sniper Team. 16 March 1989.

DA Form 5785-R. Sniper's Data Card. June 1989.

DA Form 5786-R. Sniper's Observation Log. June 1989.

DA Form 5787-R. Sniper's Range Card. June 1989.

DA Form 5788-R. Military Sketch. June 1989.

DA Form 7325-R. Concealment Exercise, July 1994.

DA Form 7326-R. Concealed Movement Exercise Scorecard, July 1994.

*This source was also used to develop this publication.

DA Form 7327-R. Target Detection Exercise Scorecard, July 1994.

DA Form 7328-R. Range Estimation Exercise Scorecard, July 1994.

DA Form 7329-R. Qualification Table No. 1 Scorecard, July 1994.

DA Form 7330-R. Qualification Table No. 2 Scorecard, July 1994.

*DA Pam 350-38. Training Standards in Weapon Training. 1990.

*TM 9-1005-306-10. Operator's Manual for 7.62 mm M24 Sniper Weapon System (SWS). 23 June 1989.

*TM 9-1265-211-10. Operator's Manual for Multiple Integrated Laser Engagement System (MILES)—Simulator System, Firing Laser, M89... Squad Automatic Weapon (SAW). 28 February 1989.

READINGS RECOMMENDED

These reading contain relevant supplemental information.

*FM 5-20. Camouflage. 20 May 1968.

*FM 5-36. Route Reconnaissance and Classification. 10 May 1985.

*FM 6-30. Observed Fire Procedures. 17 June 1985.

FM 6-121. Field Artillery Target Acquisition. 13 December 1984.

*FM 7-93. Long-Range Surveillance Unit Operations. 9 June 1987.

FM 8-10-4. Medical Platoon Leader's Handbook. 16 November 1990

FM 8-35. Evacuation of the Sick and Wounded. 22 December 1983.

*FM 17-98-1. Scout Leader's Handbook. 24 September 1990.

FM 21-26. Map Reading and Navigation. 5 July 1993.

FM 21-75. Combat Skills of the Soldier. 3 August 1984.

FM 23-8. M14 and M14A1 Rifles and Rifle Marksmanship. 15 April 1974.

FM 23-9. M16A1 and M16A2 Rifle Marksmanship. 3 July 1989.

FM 23-31. 40-mm Grenade Launchers M203 and M79. (To Be Published.)

*FM 24-1. Signal Support in the AirLand Battle. 15 October 1990.

*FM 24-18. Tactical Single-Channel Radio Communications Techniques. 30 September 1987.

*This source was also used to develop this publication.

FM 34-3. Intelligence Analysis. 15 March 1990.

FM 90-3(HTF). Desert Operations (How to Fight). 19 August 1977.

FM 90-4. Air Assault Operations. 16 March 1987.

FM 90-5(HTF). Jungle Operations (How to Fight). 16 August 1982.

FM 90-6. Mountain Operations. 30 June 1980.

FM 90-10(HTF). Military Operations on Urbanized Terrain (MOUT) (How to Fight). 15 August 1979.

TC 31-24. Special Forces Air Operations. 9 September 1988.

TC 31-25. Special Forces Waterborne Operations. 30 October 1988.

TM 11-666. Antennas for Radio Propagation. 9 February 1953.

TM 11-5855-262-10-1. Litton Model M972/M973. 15 June 1987.

TM 11-5860-201-10. Operator's Manual: Laser Infrared Observation Set, AN/GVS-5. 2 February 1982.

INDEX

SNIPER'S DATA CARD

For use of this form, see FM 23-10; the proponent agency is TRADOC

DISTANCE TO TARGET __________ METERS

RANGE	RIFLE AND SCOPE NO	DATE	ELEVATION		WINDAGE	
			USED	CORRECT	USED	CORRECT

AMMO	LIGHT	MIRAGE	TEMP	HOUR	HOLD

LIGHT

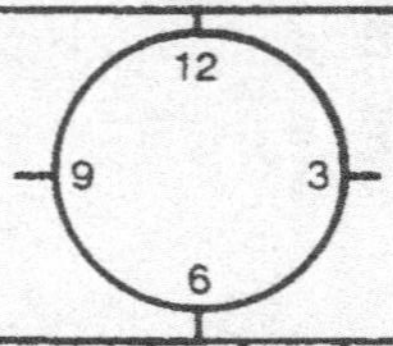

WIND

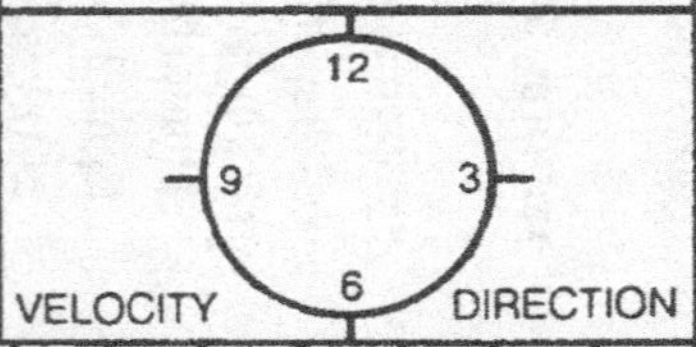

SHOT	1	2	3	4	5	6	7	8	9	10	REMARKS
ELEV											
WIND											
CALL											

NOTE: THE REQUIRED TARGETS WILL BE DRAWN IN BY HAND TO MEET THE NEEDS OF THE UNIT.

DA FORM 5788-R, JUN 89

SNIPER'S OBSERVATION LOG

For use of this form, see FM 23-10; the proponent agency is TRADOC

SHEET_____OF_____SHEETS

ORIGINATOR:	DATE/TIME:	LOCATION:

SERIAL	TIME	GRID COORDINATE	EVENT	ACTIONS OR REMARKS

DA FORM 5785-R, JUN 89

SNIPER'S RANGE CARD

For use of this fom,, see FM 23-10; the proponent agency is TRADOC

NAME:

METHOD OF OBTAINING RANGE

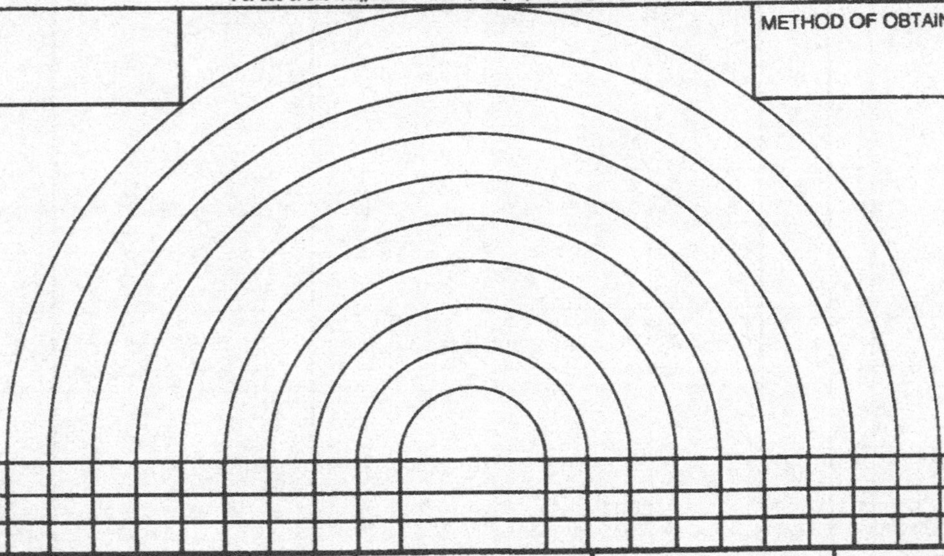

RANGE	RANGE
ELEVATION	ELEVATION
WINDAGE	WINDAGE

TEMPERATURE		WIND		TRP 1		TRP 2		TRP 3	
HIGH	LOW	VELOCITY	DIRECTION	AZIMUTH	DISTANCE	AZIMUTH	DISTANCE	AZIMUTH	DISTANCE
				DESCRIPTION		DESCRIPTION		DESCRIPTION	

DA FORM 5786-R, JUN 89

MILITARY SKETCH

For use of this form, see FM 23-10; the proponent agency is TRADOC

REMARKS:		REMARKS:

SKETCH NAME: ______ GRID COORDINATE: ______ WEATHER: ______	↑ = ______ MAGNETIC AZIMUTH	SKETCH # ______ OF ______ SCALE ___ : ___	NAME: RANK: DATE/TIME:

DA FORM 5787-R, JUN 89

Made in United States
Orlando, FL
04 November 2026

90206675R00351